ADVANCES IN GENETICS

VOLUME 24

Molecular Genetics of Development

ADVANCES IN GENETICS

Edited by

JOHN G. SCANDALIOS

Department of Genetics
North Carolina State University
Raleigh, North Carolina

E. W. CASPARI

Department of Biology
University of Rochester
Rochester, New York

VOLUME 24

Molecular Genetics of Development

Edited by

JOHN G. SCANDALIOS

Department of Genetics
North Carolina State University
Raleigh, North Carolina

ACADEMIC PRESS, INC.
Harcourt Brace Jovanovich, Publishers
San Diego New York Berkeley Boston
London Sydney Tokyo Toronto

ACADEMIC PRESS, INC.
1250 Sixth Avenue
San Diego, California 92101

United Kingdom Edition published by
ACADEMIC PRESS INC. (LONDON) LTD.
24-28 Oval Road, London NW1 7DX

LIBRARY OF CONGRESS CATALOG CARD NUMBER: 47-30313

ISBN 0-12-017624-6 (alk. paper)

PRINTED IN THE UNITED STATES OF AMERICA
87 88 89 90 9 8 7 6 5 4 3 2 1

CONTENTS

Heat-Shock Proteins and Development

URSULA BOND AND MILTON J. SCHLESINGER

Mechanisms of Heat-Shock Gene Activation in Higher Eukaryotes

MARIANN BIENZ AND HUGH R. B. PELHAM

Regulatory Gene Action during Eukaryotic Development

JOEL M. CHANDLEE AND JOHN G. SCANDALIOS

The Genetics of Biogenic Amine Metabolism, Sclerotization, and Melanization in *Drosophila melanogaster*

THEODORE R. F. WRIGHT

Developmental Control and Evolution in the Chorion Gene Families of Insects

F. C. KAFATOS, N. SPOEREL, S. A. MITSIALIS, H. T. NGUYEN, C. ROMANO, J. R. LINGAPPA, B. D. MARIANI, G. C. RODAKIS, R. LECANIDOU, AND S. G. TSITILOU

The Significance of Split Genes to Developmental Genetics

ANTOINE DANCHIN

Gene Transfer into Mice

GEORGE SCANGOS AND CHARLES BIEBERICH

The Molecular Basis of the Evolution of Sex

H. BERNSTEIN, F. A. HOPF, AND R. E. MICHOD

Gene Dosage Compensation in *Drosophila melanogaster*

JOHN C. LUCCHESI AND JERRY E. MANNING

Developmental Morphogenesis and Genetic Manipulation in Tissue and Cell Cultures of the Gramineae

FIONNUALA MORRISH, VIMLA VASIL, AND INDRA K. VASIL

PREFACE

This is the second, aperiodic "thematic" volume to be published in *Advances in Genetics*. The first such volume (Volume 22) appeared in 1984 under the title "Molecular Genetics of Plants." The present volume, "Molecular Genetics of Development," continues our efforts to summarize critically and highlight some of the exciting and significant developments which have been taking place in recent years in the science of genetics. The success of the first topical volume has been such that we feel the concept for these volumes and the timeliness of the topics covered are of use to the scientific community, and thus, serve the purpose for which they are intended. This format, to review periodically a central topic in a highly thematic volume, is necessitated because of the extremely rapid developments in genetics and molecular biology, which have led to an unparalleled information explosion in recent years. However, the long established tradition of *Advances in Genetics* to publish, in each regular volume, a series of outstanding, but largely unrelated articles, will be fostered so that the diversity of genetics as a science can be well represented.

The intent of this volume is not to present a comprehensive overview of the rapidly developing area of developmental genetics. Instead, each author was asked to summarize and highlight some of the more interesting and significant advances in his/her respective area of expertise. The ten articles constituting this volume include state-of-the-art discussions on heat-shock genes and their role in development, regulation of eukaryotic gene expression, the chorion gene families of insects, the significance of split genes, gene transfer in mammals, the molecular bases of sex, gene-dosage compensation, biogenic amines and insect development, and the genetic aspects of plant morphogenesis. Many of these are presented as model systems of the role of genes and their expression in developing organisms. Our intention is not merely to inform, but also to stimulate the reader, whether a beginning or an advanced researcher, to explore, question, and, whenever possible, test the various hypotheses presented herein. It is hoped that the material covered and presented is of lasting value in view of the very rapid developments in this very exciting field.

With the advent of the new and sophisticated knowledge and technology in the broad area of molecular biology, the question of development has come to the forefront of contemporary experimental biology. The resolution of the age-old question of how phenotypic diversity in form and function can be generated from the initial condition of genotypic constancy characteristic of the zygote is perhaps the ultimate frontier in modern biology. Answers to this basic question are incomplete, but the available data suggest that development is driven by the differential qualitative and quantitative expression of specific genes and their products at very specific times and places. We hope the articles in this volume will help to enhance our thinking and efforts in the direction of resolving this question.

And the end of all our exploring
Will be to arrive where we started
And know the place for the first time.

(From "Little Gidding" in *Four Quartets* by T. S. Eliot, copyright 1936 by Harcourt Brace Jovanovich, Inc.; copyright © 1963, 1964 by T. S. Eliot. Reprinted by permission of the publisher.)

I wish to acknowledge with gratitude the generous cooperation of the individual contributors. I thank several colleagues who aided me in the review process and the Academic Press staff for their gracious cooperation. Finally, I wish to thank Penelope, Artemis, Melissa, and Nikki for their help, love, patience, and understanding.

JOHN G. SCANDALIOS

ERNST W. CASPARI: GENETICIST, TEACHER, AND MENTOR

Eva M. Eicher

The Jackson Laboratory, Bar Harbor, Maine 04609

In this biographical sketch of Ernst Caspari, I hope to interweave his recollections about people and events that were instrumental in forming his career with some of the facts from his career as a research scientist and teacher. I have taken the liberty to include some of my impressions of Caspari as a teacher and as a major professor.

Ernst Wolfgang Caspari was born in Berlin on October 24, 1909 to Gertrud and Wilhelm Caspari. His memories of his mother are of a very loving, warm woman who cared for her children and her home. He remembers that she was very intelligent and had varied interests. She was always trying to procure some kind of work for which she was not trained or which, at that time, was not acceptable for a woman in her position. His father was head of the Cancer Department of the Institute for Experimental Therapy Frankfurt am Main from 1920 until 1936, when he was dismissed by the Nazi administration under the Nuremberg decrees. While Caspari was growing up, his father frequently talked about his research. One story he remembers involved experiments his father conducted with a friend, Emil Aschkinass, that were designed to study the effects of X rays on living systems. They found that X rays killed bacteria but not muscle or nerve cells. They also observed that after irradiating the pigmented bacteria *Serratia marcescens* unpigmented bacteria were found. These experiments, published in 1901, are the first reports of induced mutations. His father also conducted cancer research using mice and transplantable tumors, and he was the first to apply X rays to the treatment of human cancer. Caspari remembers frequent visits as a boy to his father's laboratory to see the mice in the "mouseoleum."

In addition to frequent discussions with his father about science and trips to his laboratory, Caspari also read books about biology and evolution. When Caspari was 14, he received for Christmas an introductory biology book *Ascaris, eine Einführung in die Wissenschaft vom Leben für Jedermann* (later translated into English as *Ascaris, the Biologist's Story of Life*), written in 1922 by Richard Goldschmidt.

This book directed his interest toward genetics. When Caspari told his father that he wanted to be a geneticist, his father was not enthusiastic because there was not much money to be earned in science, and he would have to support his son for a long time. However, his father finally agreed that Caspari could go to the university to pursue a career in zoology or botany. Caspari chose zoology.

To obtain his Ph.D. degree in zoology, Caspari went to Freiburg to study with Hans Spemann, an embryologist, but soon learned that Spemann knew little about genetics. So he decided to go to Berlin to get a degree with Goldschmidt only to find that Goldschmidt did not take graduate students. Then he decided to take a genetics course from Curt Stern, but Stern was in the United States at that time. By now, most students would have been frustrated and chosen to work with someone in another field. Undeterred, Caspari went to the University of Göttingen to work with Alfred Kühn.

Kühn gave Caspari a problem to investigate that involved wing pattern formation in the Mediterranean meal moth *Ephestia kühniella*. The year was 1931. One of Kühn's students had found that if *Ephestia* pupae were treated with heat at very specific times after pupation, aberrations developed in the wing pattern, with the type of aberration observed dependent on the time during pupation when heat was applied. Kühn suggested that Caspari try similar experiments using larvae. Caspari heat-treated larvae that were exactly timed as to their last molt, and after they hatched as adults, he mounted their wings onto microscope slides in order to analyze them. His total experiment consisted of about 5000 pairs of wings. After careful analysis of the results, Caspari told Kühn that he had found nothing unusual. Kühn wanted to see for himself, so Caspari organized all his material into a demonstration. When Kühn noted that some individual wings had abnormalities, Caspari pointed out that these abnormalities were variations among individual *Ephestia* and not related to the experimental heat treatment. Kühn agreed that this experiment was not worth pursuing and proposed another experiment.

Caspari began again. After some time passed, he decided this experiment was not going to work either, and without discussing his results with Kühn, he decided to try something else. Another graduate student in the laboratory had found that the dark-eyed *Ephestia* males had pigmented testes, while the red-eyed ones had nonpigmented testes. Caspari decided to investigate this observation further. First, he looked at the *Ephestia* that were in the "morgues" and confirmed the initial observations. He then determined that he could first observe pigmentation in the testes during the last larval instar. He proposed the idea of transplanting testes from one larva to another, and after discussing the possibilities with Karl Henke, who was knowledgeable about microsurgery, he began his transplantation experiments. As often happens in science, the unexpected occurred. Caspari found that the eyes and testes of the red-eyed hosts that received a pigmented testis transplant developed full pigmentation. He considered the possibility that he had mixed up the labels or that someone had played a joke on him. More experiments, however, revealed that there had been no labeling mix-up and no prank: the red-eyed *Ephestia* that developed pigmented eyes and testes had a testis transplant present and those that retained red eyes and nonpigmented testes were transplant failures.

Sometime after he was certain of his results, Caspari happened to take a walk on the city wall of Göttingen and met Kühn there. When Kühn asked him how his research was progressing, Caspari reported his exciting results. Kühn agreed with Caspari about the importance of his discovery and encouraged him to complete his experiments. Caspari completed his experiments, organized his data, and, after three days of virtual nonstop writing, finished his thesis. In 1933, his thesis research was published in the paper "Über die Wirkung eines pleiotropen Gens bei der Mehlmotte *Ephestia kühniella* Zeller," which appeared in *Wilhelm Roux' Archiv für Entwicklungsmechanik der Organismen*.

Caspari remained in Kühn's laboratory for a time after his doctoral degree was awarded. Efforts in Kühn's laboratory turned to trying to identify the cause of pigment development in the nonpigmented testes and eyes of the red-eyed *Ephestia* larvae when they received a pigmented testis transplant. Kühn considered this substance a hormone. Caspari thought of it as a pigment precursor. As a compromise, Caspari referred to it as "Wirkstoff" in his thesis paper, a word unfortunately not translatable into English. Later the missing substance in the red-eyed larvae was found to be an enzyme involved in the conversion of tryptophan to kynurenine. Kühn was able to obtain a research grant from the Rockefeller Foundation, based primarily on Caspari's findings, and wanted to offer Caspari the opportunity to continue his research. By then, however, the National Socialist German Worker's party was in power, and Kühn was not allowed legally to hire Caspari. Caspari needed a paying position and knew he had to leave Germany because of the political situation. He had heard that there were positions open in Istanbul and wrote a letter of inquiry to a zoologist at the University of Istanbul. While waiting for a reply from the zoologist, he was offered and accepted a position from Hugo Braun, a microbiologist also at the University of Istanbul.

Braun had already made a number of important contributions to bacteriology, including finding a method to grow pathogenic bacteria on chemically defined media and showing that typhoid bacillus required tryptophan for growth, which allowed it to be distinguished from the nontryptophan-requiring paratyphoid bacilli. Braun's long-range goal was to develop a protective vaccine against malaria, and he decided to begin using avian malaria. Caspari was hired as the zoologist on the project. The research plan included injecting canaries with a laboratory strain of avian malaria obtained from Germany and then transmitting the malaria from infected canaries to other canaries using mosquitoes that fed on infected birds. Unfortunately, no one in the laboratory realized that the males of laboratory strains of the avian malaria parasite often lose their ability to form gametes. Caspari's experiments were doomed to failure before they started because it is the gametocyte stage of malaria that is transferred by mosquitoes. Although Caspari's time in Istanbul was not as scientifically successful as it might have been, it was an important time in his life. In terms of his scientific career, he learned microbiological and immunological techniques. The most important thing that happened in Istanbul, however, was personal: he met and fell in love with Hermine (Hansi) Abraham. They were married on August 16, 1938.

On the day of his marriage, Caspari received a telegram from the United States. He assumed that the telegram was a congratulatory message and did not take time to read it. When he returned from his honeymoon, he opened the telegram to find that it was from President Lewis and included a fellowship offer from Lafayette College in Easton, Pennsylvania, with pay of $1000 per year. Caspari still laughs at the confusion he first had when he read this telegram that offered him the opportunity to come to the United States: at that time he did not know that heads of colleges are called "president" but he did know that Roosevelt was President of the United States. Caspari decided to accept the offer from Lafayette College. In 1938, he traveled alone to their new homeland, because Hansi had to remain in Istanbul until her contract expired. It was L. C. Dunn, a mouse geneticist, who had been instrumental in Caspari's obtaining a fellowship at Lafayette.

One of the saddest events in Caspari's life occurred three years after he came to the United States. In 1941, he successfully obtained a Cuban visa and funds to get his parents out of Germany, but the visa arrived at the American Consulate in Frankfurt two days after his parents had been deported to a concentration camp in Lodz, Poland.

His father served at Lodz as a physician treating the sick until he died from starvation. Years later, someone called Caspari's attention to a book written about the Lodz concentration camp. Caspari was able to obtain the book and found that it included a picture of his father. Caspari's mother was taken from the concentration camp in 1942, and he was never able to find out when and how she died.

During the first few years Caspari was at Lafayette College he worked for Paul David. David was very well trained in statistics and shared his knowledge with Caspari. Caspari's assignment in David's laboratory was to investigate the inheritance of a new mutation that David had obtained from a mouse breeder by the name of Holman in Sarasota, Florida. Caspari established that this mutation, named kinky (*Ki*), was inherited as an autosomal dominant and was homozygous lethal. He decided to determine the relationship of *Ki* to other known genes and found that *Ki* was linked to two other dominant mutations on chromosome 17 (then called linkage group IX), one being brachyury (*T*) and the other fused (*Fu*). This research led to a co-authorship with L. C. Dunn on two papers: "Close linkage between mutations with similar effects," published in 1942 in *Proceedings of the National Academy of Sciences of the United States of America*, and "A case of neighboring loci with similar effects," published in 1945 in *Genetics*.

The chairman of the Biology Department at Lafayette College had been promised a sabbatical leave. Caspari agreed to teach his comparative anatomy course and thus began his role as teacher. He did not receive additional pay for teaching, but they did give him money to hire an undergraduate student to help take care of the mice. In 1941, David's position in the department became available. Caspari was appointed at the level of Assistant Professor and was assigned to teach introductory biology and genetics. During this time, he also resumed his work on *Ephestia* and was able to show that red-eyed mutant *Ephestia* had higher levels of tryptophan than did normal pigmented *Ephestia*, which was in agreement with the hypothesis that the red-eyed mutation in *Ephestia* involved a block in the pathway leading from tryptophan to kynurenine.

Lafayette College was a men's school. When the United States entered World War II, Lafayette suddenly found itself with few students. In order to keep the college open, the decision was made to accept soldiers as students. Caspari's role was to teach them German. During this time, Caspari also taught biology courses, which now brought his teaching load to 18 contact hours per week. When the president announced that Lafayette College would no longer be able to pay its faculty, Caspari began to look for another position, which took him to Terre Haute, Indiana, to interview with Commercial Solvents for a job involving the manufacture of penicillin. During his interview, he met David Goddard, a plant physiologist from the University of Rochester, who was a consultant to Commercial Solvents. They happened to travel back on the same train, and Caspari told Goddard about his *Ephestia* research. Soon after that, an opening became available at the University of Rochester on the Manhattan Project, and Goddard suggested to Curt Stern that he hire Caspari. Stern made the offer, and Caspari accepted.

From 1944 to 1946, Caspari worked at the University of Rochester investigating the effects of irradiation on the rate of mutation in *Drosophila*. A publication in *Genetics*, "The influence of chronic irradiation with gamma rays at low dosages on the mutation rate in *Drosophila melanogaster*," resulted from these experiments, which Caspari co-authored with Stern in 1948.

Needless to say, Caspari found the mutation experiments he conducted with *Drosophila* very boring—looking all day into little creamers trying to determine whether any lacked wild-type males, which would be indicative of an induced lethal mutation on the X chromosome. To give himself something to do when he was tired of peering into creamers, Caspari read everything he could find about cytoplasmic inheritance. For a long time, Caspari had been interested in the question of maternal inheritance. While in Kühn's laboratory, he noticed that the ocelli of red-eyed *Ephestia* larvae were more darkly pigmented if they were produced from a mating of a dark-eyed female to a red-eyed male than from the reciprocal mating. He wondered if this difference was because the dark-eyed female parent provided something in the egg that was not provided by the red-eyed female, and in 1936, he had published these findings in an article entitled "Zur Analyse der Matroklinie der Vererbung in der a-Serie der Augenfarbenmutationen bei der Mehlmotte *Ephestia kühniella*" in *Zeitschrift für Induktive Abstammungs- und Vererbungslehre*. Caspari's discussions with Stern and his investigation of the literature on cytoplasmic and maternal inheritance resulted in a review article entitled "Cytoplasmic inheritance," which was published in 1948 in Volume 2 of *Advances in Genetics*. Caspari concluded in his review that pure cytoplasmic inheritance does not exist but rather that cytoplasmic elements interact with nuclear genes.

Caspari had become friends with Hubert Goodrich, a fish geneticist at Wesleyan University in Middletown, Connecticut, and Goodrich recommended to Butterfield, the president of Wesleyan, that he hire

Caspari. During Caspari's interview with Butterfield, they discussed Aristotle and Aristotelian teleology. Caspari's ability to discuss philosophical subjects must have impressed Butterfield, who had a Ph.D. in philosophy, because he offered Caspari a position at Wesleyan at the level of Associate Professor. In 1946, Caspari and Hansi moved to Middletown. During his first year at Wesleyan, Caspari had a heavy teaching load and was unable to carry out any experiments himself. But he did have two undergraduate students in his laboratory that year working on research projects, one by the name of David Prescott.

In 1947, Caspari took a sabbatical leave from Wesleyan University to work in the Department of Genetics at the Carnegie Institution of Washington in Cold Spring Harbor, New York. Theodosius Dobzhansky recommended to Mileslav Demerec, who was director of the laboratory, that he offer Caspari a position. Again Caspari found himself in the wonderful position of being able to pursue research full time. During the two years Caspari was at Carnegie, he tried to purify the enzyme from *Ephestia* that was responsible for transforming tryptophan to kynurenine. His efforts failed. With a postdoctoral fellow in his laboratory, H. Clark Dalton, he also investigated why the white axolotl mutant was unpigmented. They found that the defect lay in the failure of the precursor pigment cells to migrate from the neural crest. At this time, Caspari also began to investigate the possibility of finding immunological differences associated with the mutations *Ki* and *T* and their wild-type alleles in mice.

With his two-year sabbatical from Wesleyan at an end, Caspari had to decide whether he would return to Wesleyan or stay at Carnegie. He was very frustrated and discouraged by his failure to isolate the enzyme responsible for the conversion of tryptophan to kynurenine in *Ephestia*. He was also having trouble accepting the fact that he was getting paid to do research and that his research was not going very well. Wesleyan promised him a full professorship if he would agree to return. After much thought, Caspari agreed to return to Wesleyan if Hansi was admitted to graduate school as a master's degree candidate in psychology. Although Demerec thought that Wesleyan had offered Caspari more money, his decision to return to Wesleyan was based on the fact that Hansi was unchallenged by life at Cold Spring Harbor and wanted to study psychology. In addition, Caspari could teach and continue to do research at Wesleyan. He reasoned that, if his research did not go well, he would have his teaching and his students. So in 1949, he and Hansi packed up and returned to Connecticut. Hansi successfully completed her M.A. degree in psychology and had a productive career as a clinical psychologist.

While at Carnegie, Caspari had become very good friends with Demerec, B. P. Kaufmann, and Barbara McClintock. Saying goodbye to these friends was not as hard as it might have been because Caspari spent a number of summers at Cold Spring Harbor after returning to Wesleyan. In fact, his first summer back was one of the most memorable: he met an enthusiastic, young scientist by the name of Kenneth Paigen, who had come to the laboratory as a postdoctoral student to learn cytology from Kaufmann. Paigen had hypothesized that there were different kinds of mitochondria and thought he might be able to separate them by size. Caspari was also interested in mitochondria. One of their interactions that summer consisted of teaching each other different biological techniques: Paigen taught Caspari how to separate mitochondria, and Caspari taught Paigen immunological methods. Caspari's interactions with Paigen were later instrumental in a series of experiments Caspari conducted in mice showing that mitochondria of F_1 individuals were inherited from their female, not male, parent.

In 1949, Caspari was promoted to Professor of Biology at Wesleyan University, and in 1956, he was named the Daniel Ayres Professor of Biology. At this time in his career, he found himself more and more involved in committee assignments and administrative duties. He especially remembers two committee assignments. One involved an investigation into the education of American scientists. It was the 1950s, and Sputnik had already been launched. There was much concern in the United States at this time because it appeared that the United States was falling behind the Soviet Union in science. A well-funded faculty committee was formed by the Board of Trustees to determine whether Wesleyan was doing a good job training scientists. The committee decided to investigate this question in a larger context and included all American colleges in its study. They concluded from their investigation that the most successful colleges were generally poor midwestern colleges with enrollments consisting mainly of middle-income Protestant students. They further concluded that these colleges were successful because their faculties gave much time and energy to the education of their students.

Caspari was also involved in a three-member faculty committee appointed by the President to address the question of how Wesleyan could admit more students and still retain close contact between faculty and students. The recommendations of the committee members, who were known on campus as "The Three Wise Men," were never fully accepted by the rest of the faculty as a solution to Wesleyan's problem, although the committee's recommendations did lead to changes at some other American universities.

During his last years at Wesleyan, Caspari became interested in the biology of behavior. Caspari's first publication involving the genetics of behavior followed from an invitation from George Gaylord Simpson and Ann Roe to present a paper at a symposium on the evolution of behavior. Caspari's presentation, entitled "Genetic basis of behavior," appeared in *Behavior and Evolution* in 1958 and still is regarded as a classic paper in behavior genetics. I personally think that this area of behavior challenged Caspari intellectually more than any other area in biology. From 1956 to 1957, he was a Fellow at the Center for Advanced Studies in the Behavioral Sciences in Palo Alto, California. He returned to the center again from 1965 to 1966. In 1979, Caspari received the Th. Dobzhansky Award for Research in Behavior Genetics.

In 1960, Caspari was offered the chairmanship of the Department of Biology at the University of Rochester in Rochester, New York, and remained as chairman until 1966. Allen Campbell, Arnold Ravin, and Richard Lewontin, all well-known geneticists, were at Rochester at that time. Having such a group at Rochester influenced Caspari's decision to take the position. Caspari's appointment of Wolf Vishniac, a distinguished biochemist, added more eminence to the department. Later, Caspari strengthened the genetics faculty by appointing Uzi Nur.

One of the challenges facing Caspari as chairman was the need to upgrade the quality of the graduate students. One of the ways he accomplished this was to obtain a National Institute of Health (NIH) training grant and to use this money to attract good graduate students and to reward excellence.

Another problem facing Caspari as departmental chairman was the increasingly high ratio of undergraduate students interested in majoring in biology to departmental faculty members. In addition, there was an insufficient number of lecture rooms and teaching laboratories available to meet student needs. Caspari was unable to win administrative approval for hiring more faculty and for constructing a new building. In 1966, Caspari resigned as departmental chairman, thinking that someone else might be better able to solve the department's problems.

In 1965, Caspari was elected Vice President of the Genetics Society of America, and in 1966, he was elected President. Physical problems began to accumulate for Caspari during the late 1960s. In 1968, just after he had accepted the editorship of *Genetics*, he fell and broke his hip while on vacation in Canada. The next year he developed bladder cancer, and the year after that he had a heart attack. These years of sickness sapped Caspari's energy and interfered with his performance as editor. Although he was doing the best that he could under the circumstances, members of the Genetics Society complained bitterly about how long it took their papers to appear in print after they were submitted to *Genetics* for publication. Caspari hired an assistant and involved the editorial board in sending out papers for review and deciding whether to accept them for publication. Both of these changes helped shorten the time between submission of a paper and its appearance in print. In spite of the difficulties that Caspari faced as editor of *Genetics,* he found his duties both challenging and rewarding.

I would like now to share some of my memories of Caspari as teacher and as major professor.

In the fall of 1961, I entered graduate school at the University of Rochester. During my first semester I was employed as a teaching assistant. My first assignment was to help teach the laboratory given for introductory biology and grade exams. The teacher for the course was Ernst Caspari. His lectures were always well organized, and his rich voice filled the room while several hundred students sat quietly listening as they hurriedly wrote in their notebooks. Caspari paced back and forth, holding his notes in one hand while pumping the air with the other to accentuate the points he was trying to make. Caspari covered many topics, from cellular metabolism to genetic linkage. The students received a good introduction to zoology, and I was introduced to a number of areas of biology unfamiliar to me. Caspari "fired up" many of the freshman students and influenced their decision to major in biology. With time, some of the advanced biology courses also became overcrowded.

At the end of the second semester, I needed a job. Caspari hired me to clean and feed his mice and wash *Ephestia* cages. After my duties were finished, I was free to pursue other things. I especially remember the mouse room, a small room located in the basement. Steam pipes crisscrossed the ceiling, making it hot, especially in summer. The mice were housed in metal cages, the kind that rust with time, and the bedding fell out onto the floor through holes in the corners. One day, I was supposed to feed the mice lettuce and wash and fill small jars in their cages with water; the next day, I was supposed to wash the small jars again and this time fill them with diluted milk. Washing these jars was not a pleasant task because the mice used them as latrines. One strain of mice always tried to bite, and removing those dirty, smelly jars from their cages was a test of speed. There was no running water in the room, so everything had to be taken to another part of the building to be cleaned. Bedding for the mice consisted of peat moss. It often arrived wet and was spread out on the floor to dry. Washing *Ephestia* cages was a pleasure compared to tending the mice. By the end of the summer, I knew I never wanted to work with mice again.

The next summer, I again needed a job and Caspari gave me additional responsibilities in the mouse room, including keeping pedigrees up-to-date, setting up matings, and recording the litters born. I decided there had to be a better way to maintain mice. Caspari agreed that I could try to make the job easier and offered some money to help with my plan. I bought glass bottles to hold the water and hand-bored rubber stoppers to hold the sipper tubes I made from glass tubing. In a few weeks, there were no more little water jars for latrines. I also bought some commercial mouse bedding that eliminated dealing with the dirty, wet peat moss.

I have included my early mouse room experiences to convey how Caspari responded to efforts by his students to improve his laboratory. Caspari would listen to their ideas and encouraged them to try them. He always made sure that students knew he appreciated their efforts and always gave them credit for their successes.

Caspari believed that the best way to direct his graduate students' research was to allow them to make their own mistakes. He knew that the mistakes his students made themselves were the ones from which they learned the most. I first experienced Caspari's philosophy during the second summer I was caring for his mice. I found a litter containing two young with abnormal eyes. I asked Caspari if I could investigate whether this abnormality was inherited. I now know that this eye abnormality is a characteristic of this inbred mouse strain. However, at the time I did not know, and although Caspari knew, he did not tell me. Rather, he encouraged me to investigate whether the eye defect was caused by a mutation and, if so, how the mutation was inherited. Often, senior scientists discourage young scientists because they have already determined that the experiments will not work or the question is not worth answering. My research on the eye defect never came to anything publishable, but that was not important to Caspari. What was important to him was that I learned how to approach a scientific question and what are controls.

Caspari was always patient with his students, whether they were freshmen taking his introductory biology course or graduate students in his laboratory: in this respect, I remember the outcome of attending his developmental genetics course. No final exam was required for the course, but he expected each of us to write a term paper on a developmental genetics question. Our grade depended on the quality of our paper. One of Caspari's lectures dealt with the genetics and development of the *t* locus in the mouse. I decided I would write a review paper on this topic. I read every paper I could find on the *t* locus, but when it came time to write my paper, I just could not seem to get any thoughts down. The course ended and I still had not written a paper so I received an "incomplete." A year later, Caspari told me that I had to turn in a paper within a couple of weeks or receive a "failure" as a grade. I told him I just could not write a review on the subject I had originally chosen because I could think of nothing else to say about it beyond what had already been said. Caspari sat there quietly for a few minutes, obviously thinking. Then he told me he had noticed an interesting article about the mammalian X chromosome in *Nature* and maybe this subject would interest me. The article was Mary Lyon's classic paper dealing with her hypothesis of X-inactivation in female mammals. I read this paper. Then I read every available published paper on heterochromatin, Barr bodies, and late-replicating X chromosomes. I wrote my review and passed the course. It would have been easy for Caspari to have been indifferent to my problem, to have just "written me off" and flunked me. Instead, he listened and then thought of something that might stimulate my interests. Because I found Caspari intellectually challenging as a teacher and the type of person who would let students grow at their own pace, I asked him if he would accept me as a graduate student to work on the problem of X-inactivation. He agreed.

The way I decided on my thesis project reflects the way most of Caspari's graduate students chose their thesis projects. If a beginning student asked him for an idea, he would suggest a problem. If the student presented a potential research proposal to him, Caspari would encourage the student to try a few experiments to see if his/her idea was feasible. In either case, the student was responsible for figuring out how to approach the problem and for designing the experiments. Caspari was available to help when a student needed it, but he remained in the background and let the student struggle. When a student did ask for help, Caspari would usually suggest something in the form of a question: "What do you think would happen if you tried. . . ."

Many different kinds of experiments went on simultaneously in Caspari's laboratory. One year, a research scientist from Germany investigated pattern formation in *Galleria melonella,* a moth that feeds on bee honey. Another year, a Japanese investigator worked out methods to accomplish DNA transformation in *Ephestia.* In addition, Caspari had a postdoctoral fellow investigating a developmental problem involving behavior in *Habrobracon,* a wasp parasitic on *Ephestia,* and another visitor to the laboratory working on a behavioral genetic problem in *Ephestia.* There were two graduate students, an undergraduate honors major and a visiting scientist from Germany, who worked on the effects of 5-fluorouracil on the development of eye pigmentation in *Ephestia,* while at the same time, another graduate student was investigating the developmental changes of lymph proteins in *Ephestia* larvae. And, as if this was not enough diversity, one graduate student conducted λ phage genetic experiments, another investigated a behavioral genetics problem in mice, and I conducted cytogenetic experiments involving X-inactivation in mice. Caspari also conducted a number of experiments during this time, some involving *Ephestia* and others *Drosophila.*

Caspari ran his laboratory in the way he approached science: ask a question and proceed from there, utilizing as many organisms or as many biological techniques as are necessary. This approach did not result in hundreds of papers directed at answering a few specific questions. However, as a student, I received a broader education in genetics than I would have received had we used one organism for our experiments or shared just a few common techniques to answer our questions.

Caspari held weekly research meetings consisting of research presentations by members of his laboratory. During this time, his graduate students learned to critically appraise each other's research. More important, however, each of us learned to look at our own work as critically as we appraised each others.

My contacts with Caspari during my graduate student years permanently influenced me as a scientist. His stories about geneticists from the past made these people and their science "come alive," giving me "roots" as a geneticist. His willingness to let me explore and make mistakes only strengthened me as a scientist and gave me an advantage over those students who had the questions and answers handed to them. And the personal time and attention he showed me resulted in my wanting to give my best.

As with all research scientists, there were a number of people in Caspari's life who helped him advance his career. Two were his parents, who provided an intellectual atmosphere in which he could grow up and who supported his desire to be a geneticist. Another was Alfred Kühn, who allowed him freedom in the laboratory to explore and who taught him how to approach experimental problems. Another was Mileslav Demerec. It was Demerec who recommended Caspari for the chairmanship of the Program Committee for the Tenth International Congress of Genetics, held in Montreal in 1958. One of the things that Demerec wanted to do before he retired was to find a replacement for himself as editor of *Advances in Genetics*. One day, while Demerec and Caspari were walking together on the beach at Cold Spring Harbor, Demerec asked Caspari if he would assume the editorship. Caspari agreed. I think that Demerec realized that Caspari had the breadth of knowledge that was required to ensure the success of *Advances in Genetics*.

Caspari became Professor Emeritus of Biology at the University of Rochester in 1975. Acceptance of retirement is often difficult for someone who has found so much enjoyment in his work, and it was a difficult time for Caspari. He had been invited to work for a year in the Department of Genetics at Justus-Liebig-University in Giessen, and after much thought, he and Hansi returned to Germany. This was the first time either of them had been back to their native country since the 1930s.

After Hansi died in 1979, Caspari felt a great loss. He thought about what he could do with his life. Because he and Hansi had enjoyed their stay in Germany, he decided to return to Giessen. He did so in 1981 and returned to the United States in 1982. In 1983, the Justus-Liebig-University at Giessen awarded Caspari the honorary degree Dr. rer. nat., and in the same year, the University of Göttingen awarded him the Golden Ph.D. degree.

Often, when we look at successful people, we explain their success by suggesting that they were in the right place at the right time. However, many of Caspari's successes were accomplished at a time when he was in the wrong place at the wrong time. Certainly it was not the right time to be a young intellectual in Germany when the Nazi regime came to power, especially a young intellectual Jew. I also cannot imagine that a young, gifted geneticist would have chosen to pursue his career by going to Istanbul to work with a microbiologist, nor can I imagine he would have chosen to go to Lafayette College to do postdoctoral work. But these choices were open to Caspari, choices that allowed him to survive during these times, and he made the most of the situations in which he found himself. I think that the most important contributions Ernst W. Caspari has made to genetics are his experiments with *Ephestia* that contributed to the concept of one gene, one polypeptide; his critical review on the question of cytoplasmic inheritance; his involvement in bringing genetics to the study of behavior; his willingness to serve as editor of *Genetics* and *Advances in Genetics*; his service on the editorial board of a number of journals; his dedication and service to the Genetics Society of America; and most important, his dedication to teaching young people about science and the scientific method both in the classroom and in the laboratory.

ACKNOWLEDGMENTS

The writing of this biographical sketch of Ernst W. Caspari was partially funded by NIH research grant GM20919. I am indebted to Barbara Sanford, Muriel Davisson, Fay Lawson, Barbara Lee, and Linda Washburn of the Jackson Laboratory and to Anne Napier for helpful suggestions concerning this biography. I am especially grateful to Ernst Caspari for sharing with me many memories about events and people that helped shape his life. Some of the facts concerning the scientific career of Caspari's father were taken from a 1947 *Science* 45 article entitled "Wilhelm Caspari 1872–1944" written by E. Schwarz and R. Chambers.

Ernst W. Caspari

This volume is dedicated with deep appreciation to my friend and fellow Editor, Professor Ernst W. Caspari, who has maintained the tradition of excellence characteristic of Advances in Genetics

It is with a great deal of respect and affection that all contributors to this volume dedicate it to him for his many and significant contributions to developmental genetics

JOHN G. SCANDALIOS

HEAT-SHOCK PROTEINS AND DEVELOPMENT

Ursula Bond* and Milton J. Schlesinger

Department of Microbiology and Immunology, Washington University, School of Medicine,
St. Louis, Missouri 63110

At the simplest level there is little doubt that the heat shock response is homeostatic, to protect the cell against the ravages of the environmental insult and ensure that the cell can continue its normal life after the crisis has passed [M. Ashburner in Schlesinger *et al.* (1982)].

I. Introduction and Overview

Virtually every organism—from bacteria to mammals—has a set of genes that allows cells of that organism to tolerate damage imposed by various types of environmental stress which overwise will lead to irreversible injury to the organism. These genes were initially recognized in *Drosophila* embryos as puffs in polytene chromosomes arising very shortly after subjecting the embryos to a heat shock (Ritossa, 1962). Subsequently, the protein products of these genes were identified (Tissieres *et al.*, 1974) and they were, appropriately, called heat-shock proteins (HSPs). Now, we recognize these proteins and their genes as ones responding to more general kinds of stress, in addition to temperature shifts, and often refer to them as stress proteins. Recent studies show also that many of the stress protein genes are activated in the absence of stress, with some HSPs appearing at specific stages of

* Present address: Department of Molecular Biophysics and Biochemistry, Yale University, New Haven, Connecticut 06510.

development, in specific tissues, and even during the normal cell growth cycle.

Relatively little is known about how HSPs function but there is a substantial body of literature detailing the organization and sequence of HSP genes from a variety of organisms, including man, plants, worms, flies, and bacteria. This information has been reviewed in detail by several investigators active in research on the cellular stress response (Craig, 1985; Lindquist, 1986; Neidhardt *et al.*, 1985; Nover, 1984). A recent review by one of us (Schlesinger, 1986) summarizes current information about HSP structure and function. In this article we have focused on the appearance of the HSPs and activation of their genes during normal development of an organism and during the life cycle of various microorganisms that experience widely different temperatures in their environment. We also include a discussion of the role of HSPs during viral infection.

First, we briefly describe some general properties of HSPs and their genes. Prokaryotes—*Escherichia coli* is the prototype organism—have a small set of HSP genes [17 to date (see Neidhardt *et al.*, 1984)] which are controlled by the product of a gene [called *htpR* (Yura *et al.*, 1984)] encoding a "sigma" factor, a protein that modifies the bacterial DNA-dependent RNA polymerase (Grossman *et al.*, 1984). Functions of several bacterial HSPs are known: some appear to be involved in DNA replication and one is an ATP-dependent protease. The functions of HSPs in eukaryotic cells are less clear; however, a variety of studies has characterized the activation and regulation of the eukaryotic HSP genes. These genes have a highly conserved 14-bp promoter element, situated in the 5′-flanking sequences of the gene (Pelham, 1982). This element is a binding site for a conserved stress-activated factor (Parker and Topol, 1984; Wu, 1984, 1985). The prototype eukaryote for HSP studies is *Drosophila*. Some HSP genes in this organism occur in several copies and most of them are members of a "multigene family" with closely related protein-coding sequences but different promoter–enhancer elements. Those genes expressed by stress agents have no introns in the coding regions and the mRNA transcripts of some of these genes have regions in the 5′-noncoding portion that enable them to be translated selectively during stress (McGarry and Lindquist, 1985; Klementz *et al.*, 1985; Ballinger and Pardue, 1983). The major HSPs have subunit molecular weights of 80K–100K, 65K–75K, and 15K–30K. Most appear to perform a structural role and have been found in complexes with a variety of cellular proteins. The smaller proteins have a region in their sequence that is closely homologous to mammalian lens α-crystallin (Ingolia and Craig, 1982), a molecule capable of forming very high-molecular-weight (~10^6) polymers. Large aggregates of the small HSPs are seen in stressed cells of plants, animals, and flies. These proteins as well as the higher molecular-weight HSPs are found in various states of phosphorylation and members of the HSP70 family contain an ATP-binding site. HSPs from widely divergent species have similar structural properties, sharing an unusually high degree of homology. For example, the *Drosophila* 70-kDa HSP is 50% homologous to the *E. coli* and yeast HSP of similar molecular weight. Another very highly conserved eukaryote polypeptide, ubiquitin, has recently been shown to be a HSP (Bond and Schlesinger, 1985). This strong conservation of genetic regulatory elements and protein structure argues for an essential role of these proteins in cell survival.

What initiates the cell's heat-shock response is a mystery at this time. A number of physical conditions and chemical reagents (see Nover, 1984, p. 9) can activate the signal, which might be any one change or a combination of changes known to occur very shortly after the stress. A few of the latter are noted by Nover (1984), but the list continues to grow. Recent additions include an increase in inositol trisphosphate and intracellular calcium (Stevenson *et al.*, 1986) and lowering of cell pH immediately upon imposition of the stress agent (Drummond *et al.*, 1986). Once a cell senses a stress, much of the cell's biosynthetic activity shuts down. However, some parts of the cell's metabolic machinery respond to low levels of stress while others remain intact and functional until a more severe insult is encountered. Similarly, the threshold level for stress-signal sensitivity varies among the different cells of a complex organism as well as in cells that are in a different state of growth or differentiation. A comprehensive study of which cellular activities are most sensitive to a particular stress would be valuable; one has not been reported for any system although we know that DNA and RNA synthesis shuts down quickly, before protein synthesis. Certain reactions involving multicomponents of nascent polypeptides and oligonucleotides, i.e., ribosome assembly and mRNA splicing, stop under conditions in which polypeptide synthesis continues. The latter is clearly necessary for HSP production.

II. Heat-Shock Proteins and Development in Animal Systems

A. *Drosophila*

The first indications of the involvement of the small HSPs in normal embryonic development resulted from the molecular analysis of the small-molecular-weight HSP genes in *Drosophila melanogaster*. Recombinant clones for the small HSPs revealed that all four genes are clustered at cytological position *67B* on chromosome 4 (Craig and McCarthy, 1980; Corces *et al.*, 1980; Wadsworth *et al.*, 1980). In an unrelated study of developmentally regulated genes, Sirotkin and Davidson (1982) characterized a DNA segment containing a cluster of five genes whose transcripts were more abundantly expressed in early pupal stages than in embryos and showed that it mapped to *67B*. Further analyses revealed that two of the developmentally regulated transcripts corresponded to two of the previously identified HSP genes: *HSP26* and *HSP22*. More recent data show that the cytological locus *67B* contains a cluster of seven genes, four heat-shock genes, and three developmentally regulated genes (Ayme and Tissieres, 1985).

The four small HSP genes can be activated by the developmental hormone, ecdysterone, in the *D. melanogaster* embryonic cell line, S3, and in the imaginal discs isolated from pupariating larvae (Ireland and Berger, 1982; Ireland *et al.*, 1982; Vitex and Berger, 1984). In fact, the expression of the four small HSPs under non-heat-shocked conditions closely parallels the rise in ecdysterone titer during early development in *D. melanogaster*. mRNA transcripts corresponding to HSP26 and HSP28 are found in early embryos (see below) and the levels peak again in late third instar larvae and remain elevated through to midpupae when they decline again. Elevated levels of a transcript corresponding to HSP23 are also found in late third instar larvae through to midpupae but not in embryos. A less significant increase in HSP22 transcript occurs in early and midpupae (Table 1). Thus, the four small HSPs appear to be under dual regulation of heat shock and ecdysterone (Sirotkin and Davidson, 1982; Mason *et al.*, 1984; Cheney and Shearn, 1983; Lawson *et al.*, 1985; Thomas and Lengyel, 1986). A similar dual regulation has been found for the three other genes, noted *1*, *2*, and *3*, at *67B* that were originally described as being developmentally regulated (Ayme and Tissiere, 1985). All three genes, *1*, *2*, and *3*, are inducible by heat shock during specific development stages. The levels of gene *1* mRNA increase dramatically in heat-shocked larvae, prepupae, late pupae, and young and 3-day-old males and females. Gene *2*, on the other hand, is heat inducible in larvae and in adults. The basal level of gene *3* mRNA is higher than genes *1* and *2* in larvae and prepupae but the steady-state levels of this mRNA also increase after heat shock in these stages and in late pupae and adults. A comparison of the DNA sequence of gene *1* (Ayme and Tissieres, 1985) revealed a striking homology to the four heat-shock genes within a region that has been shown to be homologous to a

TABLE 1
Expression of HSP and HSC mRNAs during Embryonic Development in *Drosophila melanogaster*[a,b]

Gene	Embryo	Third instar larvae		Pre-pupae	Mid-pupae	Late pupae	Adults	
		Early	Late				Male	Female
HSP27/HSP28	++	−	+	++	++	−	−	+
HSP26	++	−	+	++	++	−	−	+
HSP22	−	−	−	+	+	−	−	−
Gene *1*	−	ND	+	++	++	−	−	−
Gene *2*	+	ND	−	−	−	−	−	−
Gene *3*	−	ND	−	+	+	−	−	−
HSP83	++	+	+	+	++	+	+	+
HSC1	−	−[c]		ND	ND	ND	++[d]	
HSC2	−	−[c]		ND	ND	ND	++[d]	
HSC4	++	++[c]		ND	ND	ND	++[d]	

[a] Adapted from data by Ayme and Tissieres (1985), Craig *et al.* (1983), Mason *et al.* (1984), and Thomas and Lengyel (1986).
[b] −, Undetectable amounts of mRNA; +, detectable amounts of mRNA; ++, increased amounts of mRNA; ND, not determined.
[c] Larvae were not classified as early or late.
[d] Adults were not classified by sex.

stretch of amino acids in the mammalian α-crystallin gene (Ingolia and Craig, 1982). The sequences of genes *2* and *3* have not yet been determined. Thus, five of the seven *67B* genes are interrelated by coding sequences and control elements; however, subtle differences in their regulation exist and may account for completely separate functions of these genes during development.

The correlation between the expression of the small HSPs during development and the rise in ecdysterone titers has been examined in the conditional mutant strain *ecd-1* (Thomas and Lengyel, 1986). This mutant exhibits a temperature-sensitive, nonpupariating phenotype which can be partially rescued by feeding high concentrations of 20-OH ecdysone. The rise in ecdysterone levels at pupariation can be inhibited in this mutant by a temperature shift to 29°C at 120 hours post egg laying. Under these conditions, the levels of HSP26 decrease over the next 60 hours following the temperature shift. Pupariation usually occurs at 166 hours in the wild type while the *ecd-1* larvae remain in the food and do not pupariate up to 212 hours post egg laying. When the remaining nonpupariating *ecd-1* larvae at 192 hours were placed on food containing 20-OH ecdysone, approximately 20% began to pupariate, coincident with a dramatic increase (40-fold) in the level of HSP26 mRNA, suggesting an *in vivo* role of ecdysteroids in HSP gene expression during normal development.

The *Drosophila* HSP28 and HSP26 are also expressed during oogenesis (Zimmermann *et al.*, 1983; Ambrosio and Schedl, 1984). *In situ* hybridizations indicate that mRNA transcripts homologous to these small HSPs are present in non-heat-shocked nurse cells and oocytes at all developmental stages. Since the latter are transcriptionally inactive, the presence of HSP mRNA transcripts in the oocyte reflects the transport of these transcripts from the nurse cells to the growing oocyte. At stages 8–9 the film-grain density, which indicates the extent of *in situ* hybridized mRNA, is substantially lower over the oocyte than over the nurse cells, while at stage 10 both show equal grain density. By stages 12–14 the oocyte shows markedly more grains than the nurse cells. In fact, there is approximately 10 times more hybridizable HSP mRNA in the oocyte at any given stage after stages 7–9 than in the average nurse cells. The amount of HSP28 and HSP26 mRNA in follicle cells throughout oogenesis is considerably less than in the nurse cells and oocytes.

The spatial distribution of HSP26 at various developmental stages in *Drosophila* has also been examined by *in situ* hybridization using single-stranded probes (Glaser *et al.*, 1986). These studies confirm the presence of HSP26 in nurse cells, oocytes, and the developing imaginal discs and also show that HSP26 is expressed in spermatocytes throughout all stages of development and in epithelium, proventriculus, and the central nervous system in late third instar larvae and white prepupae, the period corresponding to the major ecdysterone peak in *Drosophila* embryos.

To examine whether the developmental regulation of *HSP26* is a consequence of its position at the developmentally regulated locus *67B*, a fusion gene containing the upstream 2 kb of *HSP26* and the 5′ portion of *HSP26* fused to a β-galactosidase gene was constructed and reintroduced into *Drosophila* by P-factor germ-line transformation (Glaser *et al.*, 1986). The expression of the gene was assayed by β-galactosidase activity in the various tissues and was found to mimic the expression of the endogenous *HSP26* gene, showing correct temporal expression in nurse cells, oocytes, spermatocytes, epithelium, proventriculus, and central nervous system. Interestingly, no expression of the fusion gene was found in imaginal discs. This lack of expression in imaginal discs of the fusion protein in these transformants remains unexplained.

The molecular basis of the dual regulation of the small HSP gene expression has been the subject of a number of recent independent investigations. In particular, it is important to establish whether the induction of the HSP genes by ecdysterone administration and by heat shock is controlled by similar or distinct promoter sequences. This problem is being addressed by two methods of deletion analyses of the upstream sequences of the *HSP27* genes in *Drosophila*. Using P-element-mediated germ-line transformation of *D. melanogaster* embryos, Hoffman and Corces (1986) showed that sequences >1.1 kb upstream of *HSP27* are required for 80% of heat-shock inducibility. Deletion of sequences between −1.1. and −227 bp caused a further reduction in RNA accumulation, while deletion of sequences between −227 and −124 completely abolished heat-inducible transcription. The nature of the sequences responsible for ecdysterone induction of *HSP27* was also examined, and about 50% of ecdysterone-induced activity was lost in deletions between −111 and −227 bp and sequences downstream of −124 bp were responsible for the remaining activity. Thus ecdysterone induction is controlled by regulatory sequences distinct from those required for heat-shock induction.

The upstream sequences of *HSP27* responsible for heat inducibility and ecdysteroid responsiveness have also been analyzed by a transient-expression assay in *Drosophila* tissue-culture cells (Riddihough and Pelham, 1986). These studies established that heat-inducible promoters lie between −370 and −270 bases relative to the start of transcription. Sequences mediating ecdysterone induction were located in multiple elements between −579 and −455.

That the sequences responsible for the majority of heat inducibility in the *HSP27* gene reside >1.1. kb upstream of the start of transcription differs sharply from the position of similar sequences for most of the other HSPs. In the *HSP70* gene, the "heat-shock" promoter is located at −95 to −40 relative to the start of transcription (Pelham, 1982). The sequences in the −2.1-kb position upstream of *HSP27* were scanned for potential heat-shock promoter sequences (Hoffman and Corces, 1986; Riddihough and Pelham, 1986) and four sets of putative promoter sequences were located between positions −353 and −253, the region shown to be necessary for heat inducibility in the transient-expression assay experiments described above. Surprisingly, two sequence elements located at positions −1.9 and −1.84 kb also show a striking homology to sequences upstream of the *HSP70* gene. While the data obtained from these studies are contradictory in the location of regulatory elements, both experiments clearly indicate that sequences responsible for heat and ecdysterone responsiveness are separate and distinct. The discrepancies between the two sets of data may arise from the different experimental procedures: germ-line transformation versus transient expression.

The distinct nature of the heat-shock and ecdysterone promoters has been verified by analysis of the upstream sequences of the *HSP23*, *HSP22*, and *HSP26* genes. Here again, multiple distinct promoter elements appear to be required to elicit hormone and heat induction. In *HSP26*, promoter sequences responsible for heat-shock induction are located in sequences downstream of ~−350 bp while ovarian expression is governed by sequences distal to position −340 bp (Cohen and Meselson, 1985). At least four separate regions upstream of *HSP23* have been defined as containing sequences conferring heat inducibility on this gene (Mestril *et al.*, 1986). Three of these contain sequences closely related to the Pelham consensus sequence and appear to act in concert to control heat-shock induction as successive deletions of the upstream sequences cause further reductions in heat-shock inducibility (Pauli *et al.*, 1986). Two upstream regions, distinct from those governing heat-shock induction, have been shown to contain control elements responsible for hormone induction of the *HSP23* genes (Mestril *et al.*, 1986). Sequences related to the Pelham consensus sequence have been located in three separate positions upstream of *HSP22* (Klementz and Gehring, 1986). Sequences beyond the distal repeat are also required for full heat-shock inducibility and deletion of the distal repeat results in a fivefold to sixfold reduction in gene expression. Here again separate sequences appear to control heat-shock and hormone responsivenss of the gene.

In addition to the small HSP genes, the *HSP83* gene is developmentally regulated in *D. melanogaster*. *In situ* hybridization studies show extensive labeling in somatic follicle cells, in nurse cells, and in oocytes in non-heat-shocked ovarian chambers (Zimmermann *et al.*, 1983). In follicle cells the levels of HSP83 mRNA remain relatively constant throughout all stages of egg development. In contrast, the mRNA increases 10-fold in nurse cells between stages 3 and 10. As with the small HSPs, HSP83 mRNA synthesized in developing nurse cells is ultimately transported into the oocyte where it accumulates.

The levels of HSP83 mRNA in the fertilized egg remain relatively high, presumably reflecting accumulation of the transcript during oogenesis. As embryogenesis proceeds, the levels decrease and then remain constant, except for a slight rise (1.5-fold) at the midpupal ecdysteroid peak period (Thomas and Lengyel, 1986). The temporal expression of HSP83 differs from that of the small HSPs, which are predominantly expressed in third instar larvae, prepupae, and pupae.

HSP70 is the dominant stress protein expressed in *Drosophila* as well as other organisms. There are five stress-activated *HSP70* genes in *D. melanogaster* (Livak *et al.*, 1978; Schedl *et al.*, 1978; Craig *et al.*, 1979; Moran *et al.*, 1979; Spradling *et al.*, 1977; Henikoff and Meselson, 1977); thus, it is surprising that little if any HSP70 mRNA is expressed during oogenesis and embryogenesis. However, a number of non-heat-inducible genes closely related in coding sequences to *HSP70* (they are called heat-shock cognates) are developmentally expressed (Craig *et al.*, 1983). Three heat-shock cognate genes, *HSC1, HSC2,* and *HSC4,* have been characterized thus far in *D. melanogaster:* they map to the cytological loci, *70C, 87D,* and *88E* on the third chromosome, respectively. Primer extension experiments with DNA fragments unique to each gene were performed to determine the relative abundance of the HSC transcripts at different developmental stages (Craig *et al.*, 1983). In the adult, HSC1 and HSC2 transcripts were 20-fold more abundant than in larvae and embryos, whereas HSC4 mRNA was equally abundant in embryos, larvae, and adults. The expression of the *Drosophila* HSC4 has been further analyzed with monoclonal antibodies (Palter *et al.*, 1986). A protein, HSC70, which corresponds to the HSC4 transcripts was equally abundant in ovaries of adult females and in embryos but was fivefold less abundant in larvae, pupae, and adults. A protein, designated HSC72 was expressed throughout all developmental stages. Since neither HSC1 nor HSC2 mRNA transcripts were expressed in embryos and larvae, the possibility remains that HSC72 is encoded by a fourth, as yet unidentified, cognate gene. In fact, there are preliminary data indicating four additional *HSP70* cognate genes in *Drosophila* (E. A. Craig, personal communication). The ratio of HSC70/HSC72 has been estimated to be 10 : 1 in unfertilized eggs and early embryos, 1 : 1 in late larvae, and 3 : 1 in adults. Indirect immunofluorescence analysis using a monoclonal antibody directed against HSC70 indicates that this protein is found uniformly dispersed in the cytoplasm of the egg following fertilization. During blastoderm formation the protein concentrates in the egg cortex, and once cell formation is complete, HSC70 is heavily concentrated in filamentous structures of the cytoplasm surrounding the nucleus.

The relative abundance of HSC70 during normal development is best illustrated by comparing its levels to the major HSP70 following heat shock in *Drosophila*; surprisingly, following heat shock the steady-state level of HSP70 is only 30% of HSC70!

In addition to studying HSP gene expression during normal *Drosophila* development, investigators have examined the ability of the organism to respond to heat shock at different developmental stages.

A dramatic difference in response to heat shock is found in somatic and germ-line cells in the developing egg chamber of *Drosophila*. Follicle cells and nurse cells synthesize the normal complement of HSPs following heat shock, whereas no synthesis occurs in the developing oocyte and even during the earlier stages of embryogenesis. This period coincides with silencing of nuclear transcription as characterized by a decline in nurse cell nuclear volume and the passage of nurse cell cytoplasm into the oocyte, and the period when the nuclei of early embryos are engaged in a fast mitotic cycle. In fact, preblastoderm embryos are highly temperature sensitive (Dura, 1981; Bergh and Arking, 1984; Graziosi *et al.*, 1983). Heat treatment of eggs causes disintegration of nuclei and of cytoplasmic islands, displacement and swelling of the nucleus, and a blockage in mitosis (Graziosi *et al.*, 1983). Possibly, the fatality of heat-treated early embryos is a reflection of the inability of the embryos to mount a heat-shock response.

After blastoderm formation, embryos survive heat treatment but display delayed development (Bergh and Arking, 1984; Dura, 1981). The interruption of developmental regulation in *Drosophila* has been explored through the analysis of the phenomenon of phenocopies. The term phenocopies describes the generation of mutantlike morphological alterations of *Drosophila* that are induced by imposition of stress during development (Mitchell and Lipps, 1978). Application of stress, such as heat shock, during a narrow window of a specific stage in development can induce mutants specific to that developmental stage. Thus, heat shock at blastoderm stage induces dominant mutants of the Bithorax Complex (Santamaria, 1979). The most well-characterized heat-induced phenocopy is "multihair," which results from mutation of wing hairs in *Drosophila* pupae (Mitchell and Petersen, 1983). The wing of the pupal fly contains ~30,000 hair cells. About 90% of these cells responds to a heat shock at 38 hours in the pupal stage to yield the multihair phenotype. In fact, a temporal gradient of sensitivity to heat is found in hairs in each part of the fly: a strong multihair phenotype is observed in legs 1 and 2 44 hours after puparium formation. However, 2 hours later, the multihair phenocopy is most apparent in leg 3 while normal hairs are found on legs 1 and 2. A similar sensitivity window is found for hairs on the abdominal segments and on the head, thorax, and haltere. Presumably the appearance of specific mutants is a reflection of an interruption of normal gene expression by the heat shock which results in a repression of genes required for that specific stage or for progression to later stages of development.

B. Other Organisms

Relatively little is currently known about the developmental induction of HSP genes and their cognates in organisms other than *Drosophila*. But in accordance with the highly conserved nature of the heat-shock response and the stress proteins, we would expect strong similarities between the temporal expression of HSPs and cognates under non-heat-shock conditions in the developing embryos of other eukaryotes.

1. *Xenopus*

In *Xenopus laevis* oocytes, the mRNA for HSP70 is abundant in the absence of stress (Bienz, 1984a,b) and, as is the case in *Drosophila*, this mRNA appears to be maternally inherited. Synthesis of HSP70 in the oocyte, however, is regulated at the level of translation and occurs constitutively (Bienz and Gurdon, 1982). Following fertilization, no HSP70 synthesis is apparent and in fact, analogous to the situation in the early *Drosophila* embryo, HSP70 is not synthesized following a heat shock until the late blastula stage. Unlike *Drosophila* embryos, the corresponding small HSP30 in *X. laevis* in not constitutively expressed during any stage of embryo development (Bienz, 1984a), and even heat-treated embryos prior to the tadpole stage do not synthesize HSP30 (Bienz, 1984b).

Again, as observed in *Drosophila*, the fertilized egg before midblastula stage is extremely thermosensitive and these cleave-stage embryos rapidly become abnormal and degenerate after heat shock. Mid-cell blastulae can partially survive heat shock but exhibit exogastrulation of the vegetally derived endoderm while the animal hemisphere-derived ectoderm and mesoderm develop normally. Fine cell blastulae and early gastrulae are completely thermotolerant. The acquisition of thermotolerance is coincident with the ability to synthesize heat-shock proteins (Heikkila *et al.*, 1985; Nickells and Browder, 1985).

The early postfertilization heat-shock incompetence period has been detected in a variety of species, including sea urchin (Roccheri *et al.*, 1982) and mouse (Morange *et al.*, 1984; Muller *et al.*, 1985; Wittig *et al.*, 1983; see also Section II,B,2 below).

2. *Rodents*

HSP70 and HSP68 are among the first proteins detected at the early two-cell stage of postfertilization development of the egg during mouse embryogenesis (Bensaude *et al.*, 1983). High levels of HSP70 and HSP89 were found also in early mouse embryos (Wittig *et al.*, 1983; Morange *et al.*, 1984). Despite this, the early embryo (two to eight cells) is thermosensitive and does not induce HSPs after a heat shock. Blastocytes are able to induce HSPs after stress and develop thermotolerance (Muller *et al.*, 1985).

In the developing postnatal rat, *in vivo* hyperthermia results in the synthesis of HSP71 in heart and adrenals at all ages (1 day to 25 weeks old) (Currie and White, 1983). However, HSP71 was not synthesized in the brain of animals 2 weeks old or younger. In an earlier study in which organ cultures of postnatal rats were examined for their ability to induce HSP71, White (1981) showed that the mechanical trauma of slicing the tissue was sufficient to induce HSP71 and hyperthermia was unnecessary. HSP71 was synthesized in heart and lung slices at all stages of postnatal development but brain slices failed to induce HSP71 until 3 weeks postnatal. During this period overall protein synthesis decreased but the amount of HSP71 remained relatively constant in heart and lung slices and increased approximately 50-fold in brain slices. The synthesis of HSP71 in 3-week-old rats (and in older rats) coincided with the maturation of brain capillaries and the blood–brain barrier, and high levels of HSP71 synthesis were found in fractions enriched in cerebral microvessels.

There is other evidence that a protein related to HSP71 may be synthesized very early in postnatal rat brain. Brown (1983) detected a 74-kDa protein that is induced in 2-day-old rat pups subjected to an elevation of body temperature from 32 to 38°C. Eight-day-old rat pups did not synthesize the 74-kDa protein at 38°C and temperatures had to be raised to 43°C to elicit synthesis.

We have examined the levels of ubiquitin mRNA in tissues of young mice and gerbils subjected to a brief hyperthermia (U. Bond, T. Nowak, and M. J. Schlesinger, unpublished data). Twofold to threefold increases were measured in liver and kidney but brain tissue showed a much lower increase. In all cases, the changes in ubiquitin mRNA were accompanied by similar changes in HSP70 mRNA. Ubiquitin mRNA levels also change dramatically during late embryonic development and postnatal growth in the rat (U. Bond, C. Langner, J. Gordon, and M. J. Schlesinger, unpublished data). In kidney, liver, and gut, ubiquitin mRNA levels fall as the embryo develops, while heart levels rise after birth. Brain showed little change through the development period examined.

A variety of developmental abnormalities has been observed in heat-treated rat embryos (Walsh *et al.*, 1985). Presomite (9.5-day) rat embryos were explanted and cultured and exposed to various elevated temperatures. Heat exposure resulted in four phenotypes: microphthalmia, microcephaly, gross reduction in forebrain region, and open neural tubes. The severity of the deformity was dose dependent, with all the defects due to failure of normal ectoderm induction. Embryos treated with a mild, nonteratogenic exposure to heat were protected against a subsequent exposure which otherwise caused severe craniofacial defects. The acquired thermotolerance coincided with the synthesis of HSPs. German (1984) has discussed the hypothesis that exposure to stress, such as heat shock, during embryonal development could account for many intrauterine developmental abnormalities. The precise period during gestation when the stress is administered would determine the nature of the abnormalities.

3. *Dictyostelium*

Differentiation of *Dictyostelium discoideum* can be induced by conditions of high cell density. A variety of developmentally regulated genes is induced at about 8 hours, at the time of cell–cell aggregation (Chung *et al.*, 1981; Landfear *et al.*, 1982). There is also an accumulation of a number of cytoplasmic mRNAs during the first hour of differentiation. Lodish and co-workers have isolated a 2.5-kb genomic DNA clone (pB41-6) that hybridizes to a number of mRNAs that accumulate at the onset of the differentiation process: a mRNA, E1, is present only during the first 3 hours of development while two other hybridizing mRNAs, designated L1 and L2, accumulate later in the differentiation process. When growing cells which do not normally contain mRNAs hybridizing to clone pB41-6 are heat shocked, the patterns of mRNAs are similar to those observed in early stages of development; E1 mRNA is induced while L1 and L2 are not (Zuker *et al.*, 1983). DNA sequence analysis revealed that clone pB41-6 contained sequences homologous to the repetitive gene family represented by a 4.5-kb transposable element, DIRS-1, that is repeated ~40 times in the haploid genome; pB41-6 also contained a heat-shock promoter element that mapped to the right terminal repeat of the transposable element (Zucker *et al.*, 1984). In fact, three heat-shock promoter elements were found in DIRS-1, one each within the left and right terminal repeats and a third within the repeated element. The heat-shock promoter elements in the terminal repeats are orientated in a manner which would enable them to direct transcription outward into the flanking DNA sequences and may also direct heat inducibility of the internal sequences of DIRS-1. Recent studies by Bienz and Pelham (1986) have shown that the heat-shock promoter sequences can in fact act as heat-shock enhancers, allowing for heat-shock gene synthesis regardless of position or orientation relative to the gene.

Since it is not yet known if DIRS-related sequences code for proteins, the role of these heat-inducible transcripts in the process of differentiation remains unclear.

4. *Volvox*

Perhaps one of the most extraordinary circumstances in which HSPs are alleged to play a role in development is in the switch from asexual to sexual reproduction in *Volvox carteri*. During a study of light-induced cytodifferentiation, Kirk and Kirk (1986) discovered that the progeny of illuminated, heat-shocked asexual females exhibited a low but significant frequency of sexual development. Under normal circumstances, the asexual life cycle is controlled by a light–dark cycle (approximately 32 hours light to 16 hours dark) (Kirk and Kirk, 1985). Toward the end of the light period gonidia (asexual reproductive cells) initiate the sequence of rapid cleavage divisions that constitute the first phase of embryogenesis. The switch from asexual to sexual reproduction is initiated by the action of a glycoprotein pheromone "inducer" that is produced by the sexual male *Volvox*. The protein accumulates in sperm packets and is released along with the sperm into the medium. Under the influence of this inducer, the gonidia of the female strain form egg-bearing spheroids by unequal cleavages in the anterior two-thirds of the embryo at the division of the 64-cell stage while the androgenodia (sperm-producing cells) of the male embryo form by unequal cleavages of all the cells at the last divisions in the embryo. The final stages of embryogenesis occur early in the dark period when the fully cleaved embryo turns inside out, a process referred to as inversion. The two cell types, somatic cells and gonidia, remain relatively undifferentiated until the start of the next light–dark cycle when cytodifferentiation begins.

Previously, it was thought that only sexual males could produce the inducer. However, after heat shock, asexual females generated egg-bearing sexual daughters. Asexual males and also a "sterile" strain incapable of responding to inducer produced by wild-type males could also release inducer in response to a heat shock. Unlike most heat-shock responses, the induction of the sexual inducer is confined to heat exposure and cannot be activated by other forms of stress such as treatment with amino acid analogs or heavy metals. The latter result suggests that induction of the sexual inducer occurs through a mechanism distinct from HSP initiation, possibly by thermolabile repressors. The authors suggest that heat-shock autoinduction of sexuality may reflect a physiological adaptation to temperature differentials that *Volvox* is exposed to in the wild, where the algae reproduce asexually in temporary ponds in spring but later become sexual and produce dormant desiccation-resistant zygospores during the dry summer months. In this case thermotolerance may be manifested by the production of the tough drought-resistant zygospores.

III. Heat-Shock Proteins and Development in Microorganisms

A. Protozoan Parasites

Many microorganisms are exposed to extreme temperature differentials during the course of their normal life cycle and HSPs might well be used by the organism during the initial stress period. In particular, parasites often exist in multiple hosts that vary in body temperature. The protozoan parasite, *Leishmania*, exhibits two morphologically distinct stages during its life cycle. The promastigotes are flagellar forms living freely in the alimentary canals of the vector, the poikilothermic sand fly, at 25°C. Upon entering the homeothermic mammalian host (37°C), the promastigote must successfully differentiate into the amastigote form in order to survive. The differentiation from promastigote to amastigote can be accomplished in culture solely by a temperature shift from 24 to 34°C (Hunter *et al.*, 1984). Remarkably, the differentiation process is accompanied by the synthesis of a set of polypeptides corresponding precisely in molecular weights to HSPs (Hunter *et al.*, 1984; Lawrence and Robert-Gero, 1985).

A similar correlation exists between the induction of HSPs and the differentiated state of the parasitic protozoa, *Trypanosoma brucei*. mRNA transcripts homologous to HSP70 and HSP83 are 25–100 times more abundant in trypomastigotes (mammalian bloodstream forms) than in the procyclic stages (insect form) (Van der Ploeg *et al.*, 1985). The similarities in heat-shock expression in these two parasitic protozoans suggest that HSPs may have an adaptive function in intracellular survival of the parasites within the mammalian host.

B. Yeast Sporulation

The finding that a subset of HSPs is expressed during oogenesis in *D. melanogaster* prompted an analysis of the corresponding developmental period in yeast, namely, spore formation. Meiosis and spore formation in yeast are generally triggered by starvation conditions, such as nitrogen deprivation in the presence of a nonfermentable carbon source. These growth conditions lead to the synthesis of a number of proteins which include HSP26 and HSP84 (Kurtz and Lindquist, 1984). Two proteins related to HSP70 also increase during this time period (Kurtz *et al.*, 1986). While the synthesis of a specific subset of HSPs in growth conditions which lead to sporulation may indicate a role for these proteins at this stage in development, the induction of HSPs may not be directly related to spore formation. In fact, the nonsporulating isogenic strains, a/a and α/α, when exposed to similar growth conditions, also show increased synthesis of HSPs (Kurtz and Lindquist, 1984). Surprisingly, deletion or disruption of the single *HSP26* gene in yeast had no effect on sporulation in the mutant strains, nor did this gene appear to be required for growth and survival at high temperatures (Petko and Lindquist, 1986).

Commitment to sporulation affects the ability of yeast to mount a "classical" heat-shock response, and sporulating cells are thermosensitive. This is reminiscent of the thermosensitive period in oogenesis and in postfertilized *Drosophila* embryos. This thermosensitivity, and the similarities in the pattern of temporal expression of a subset of HSPs during premeiotic sporulation–induction conditions and oogenesis in *Drosophila,* have led to the suggestion that the involvement of HSPs during development is highly conserved and may be one of the earliest developmental pathways to evolve in eukaryotic cells.

C. Bacteria and Bacteriophage

A substantial body of knowledge now exists for the *E. coli* heat-shock genes and for the molecular mechanism which regulates their expression (reviewed in Neidhardt *et al.,* 1984). Table 2 lists the *E. coli* HSPs whose functions have been identified. Transcription of the genes for these proteins requires the product of the *E. coli htpR* gene. The latter encodes the sigma 32, which replaces the normal sigma 70 on the *E. coli* RNA polymerase during the heat shock. What regulates expression of the sigma 32 is not clear but it is made constitutively by the normal sigma 70–polymerase complex. Sigma 70, in turn, can be regulated by the *htpR* system. Thus, *E. coli* HSP induction might occur by inhibiting a "normal" rapid turnover of constitutively made sigma 32. One mechanism that might function in the putative turnover of sigma 32 is the ATP-dependent proteolytic system, and a transient deficiency in this system (i.e., a decrease in ATP levels or an increase in substrates for the protease system) could lead to higher steady-state levels of the sigma 32 and subsequent induction of the HSPs. The

TABLE 2
Functions of Some *Escherichia coli* HSPs[a]

Name	Possible functions
groEL	Morphology of coliphage (weak ATPase activity); some role in RNA and DNA synthesis
dnaK, dnaJ, grpE[b]	Phage DNA replication (weak ATPase activity); modulation of heat-shock response; necessary for RNA and DNA synthesis
Sigma	Promoter recognition; subunit of RNA polymerase
groES	Morphology of coliphage; some role in RNA and/or DNA synthesis
lon, La	ATP-dependent protease
Lysyl-tRNA synthetase, form II	Charging of tRNA; synthesis of diadenosine tetraphosphate

[a] Based on material published by Neidhardt *et al.* (1984).
[b] Ang *et al.* (1986).

extent of HSP gene activation would be determined by the levels of sigma 32. It is worth noting that an analogous mechanism involving a stabilization of a "heat-shock trans-activating element" has been proposed to account for induction of eukaryotic HSP genes (Munro and Pelham, 1986a).

It is not at all clear why the particular enzymatic activities known to be encoded by *htpR*-regulated genes are needed for the heat-shock response. But based on the role of some of these HSPs in bacteriophage replication (Friedman *et al.*, 1984; and see also below), we can hypothesize that they are important in the recovery of bacterial growth and cell division that would be temporarily halted by the initial effects of the sharp temperature rise.

The discovery that several of the HSP genes are also activated by virus infection of *E. coli*, in particular λ-phage, has provided clues to some functions for HSPs. The bacterial genes activated by phage were initially detected by noting that there were bacterial mutations which blocked λ-phage replication. Some of these mutants were found to code for proteins that also were induced by a heat shock. Quite a bit is now known about a few of these proteins including their sequence and how they participate in λ-replication. The protein with strong homology to eukaryotic cell HSP70 is the product of the *dnaK* gene (Bardwell and Craig, 1984). This protein has a subunit molecular weight of 70K and has several regions in its sequence that are 80% homologous to yeast and *Drosophila* HSP70, with an overall homology of 50% to the *Drosophila* HSP70. It also shares the property of having an ATPase activity and is autoregulated (Tilly *et al.*, 1983). In addition, the *E. coli* HSP70 has nonspecific 5′-nucleotidyl phosphatase and autophosphorylation activities (Zylicz *et al.*, 1983). The dnaK protein is found in a complex with the *E. coli* dnaJ protein and λ-phage proteins O and P (Lebowitz *et al.*, 1985). The latter are part of the λ-phage DNA replication complex, thus the bacterial HSP70 would appear to be involved in DNA replication. In fact, the defective *dnaK* mutation can be suppressed by alterations in the λ-phage P-protein. Both dnaK and dnaJ products act at a step following that involving the λ O- and P-proteins, the *E. coli* dnaB protein, and transcriptional activation, but before the action of the dnaG—the primase (Dodson *et al.*, 1986). Another *E. coli* HSP, the *grpE* gene product with a molecular weight of 24K, also interacts with the λ P-replication protein (Ang *et al.*, 1986). Two other *E. coli* HSPs have been shown to be involved in phage morphogenesis. The *groES* and *groEL* genes code for proteins acting during the assembly of λ-phage heads. The five proteins (groEL, groES, grpE, dnaK, and dnaJ) are also essential for bacterial colony formation at high temperatures (Georgopoulos and Eisen, 1974), and both RNA and DNA syntheses are affected in mutants grown at the higher temperatures.

Recent data from J. C. A. Bardwell and E. A. Craig (personal communication) indicate that an *E. coli* HSP of 68 kDa has a sequence with 40% homology to the HSP83–HSP90 of eukaryotes. What role this protein has in the bacteria is not yet known. There is as yet no indication which of the *E. coli* proteins corresponds to the small eukaryotic HSPs. However, both *E. coli* and the eukaryotes have an ATP-dependent protease that is under heat-shock control. For *E. coli*, it is the *lon* gene encoding the La protease (Chung and Goldberg, 1981), and for the eukaryotic cell it is the ubiquitin gene that responds to a heat shock (Bond and Schlesinger, 1985).

D. Animal Viruses and Heat Shock

Bacteriophages are not unique as viral activators of the host cell HSPs. Several kinds of animal DNA viruses induce the mammalian cell HSPs. Infection of monkey and mouse cells by the papovaviruses, polyoma and SV40, induce the formation of 92- and 72-kDa polypeptides, and these same proteins plus one of 70 kDa are synthesized in cells stressed at 43°C for 30–60 minutes (Khandjian and Turler, 1983). Viral activation of these cellular genes appears to result from an interaction of the virus T antigen binding to a regulatory site of the HSP gene (Kingston *et al.*, 1986). This site is at −110 to −170 from the start of HSP70 mRNA transcription and is distinct from the consensus HSP regulatory element. The T-antigen-responsive element consists of two GPuGGC pentanucleotide sequences. When this element was joined to the bacterial chloramphenicol acetyltransferase (CAT) gene, CAT could be activated in cells expressing polyoma virus early proteins. The element may also function as a "constitutive" promoter to form basal levels of HSP70 in cells not carrying virus proteins, for it has been found in both the human and chicken *HSP70* gene in 5′-flanking sequence (Morimoto *et al.*, 1986) (Fig. 1). Mutants of herpes simplex virus also induce HSPs during infection of chicken embryo fibroblasts (Notarianni and Preston, 1982). Induction required the formation of a viral product which appears to be an "immediate early" viral protein. Other *ts* herpes virus mutants but not the wild type also induce stress protein synthesis.

The earliest virus protein formed during adenovirus infection, noted E1A, is also an inducer of the cellular *HSP70* gene (Nevins, 1982; Kao and Nevins, 1983; Wu *et al.*, 1986). Induction is at the level of transcription and could be shown dependent on an E1A product in nonstressed cells. The *E1A* gene product can also transform cells and there is speculation that HSP70 synthesis is associated with cell growth. By analogy to the prokaryotic system, some of the HSPs might be involved in viral DNA replication and assembly. Since the host cell HSP genes can be activated by a DNA virus gene product, in the absence of virus replication, it is unlikely that cell HSP formation is an indirect effect of a stress resulting from virus infection. Some RNA

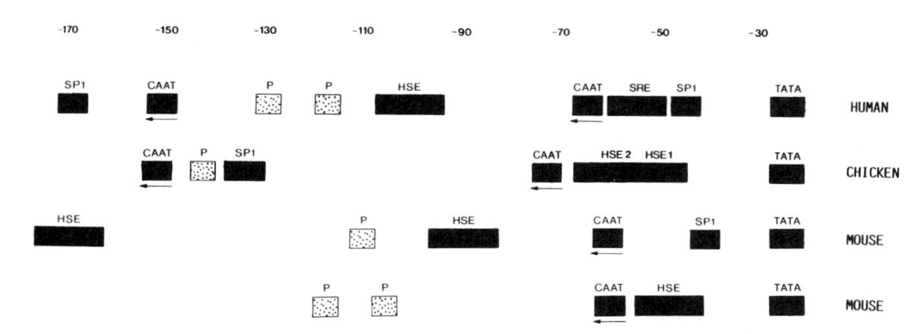

FIG. 1. Comparison of promoter elements of vertebrate *HSP70* genes. Numbers indicate base pairs in the 5′-flanking sequences upstream from the site of initiation of transcription. HSE is the 14-bp "Pelham" consensus sequence (Pelham, 1982). P is the sequence GPuGGC which reacts with SV40 large T antigen (DeLucia *et al.*, 1983). SRE is an element responding to serum (Wu and Morimoto, 1985). SP1 is the site described by Dynan and Tjian (1983). This figure was kindly supplied by R. Morimoto.

viruses also induce cell stress proteins during the infectious cycle and in this case, induction could well be the result of a stress response. Collins and Hightower (1982) reported that Newcastle disease virus infection of chicken embryo cells stimulated the formation of the major stress proteins. Avirulent virus strains were stronger inducers and they also stimulated the synthesis of the glucose-regulated proteins. One of the latter, the gp78, has recently been shown to be an isoform of the HSP70 (Munro and Pelham, 1986b), and the other, gp94, appears to be related to HSP90 (Pelham, 1986; M. Green, personal communication). Infection by other paramyxoviruses also stimulates the glucose-regulated proteins (Peluso *et al.*, 1978).

IV. Tissue-Culture Systems

A number of tissue-culture systems which allow for differentiation *in vitro* have been tested for heat-shock response, and some striking differences are found with regard to the stage of differentiation and the sensitivity to HSP induction. In a few cases, changes in constitutive levels of HSPs have been found as the various cell lines undergo differentiation.

A. Myogenesis

Embryonic quail myoblasts will spontaneously differentiate into myotubes in culture (Atkinson, 1981). This differentiation is characterized by a shift from mononucleated to multinucleated cells. During this process of differentiation, the response of the cells to heat shock changes. Myoblasts constitutively express HSP94, HSP88, HSP82, and HSP64. Upon heat shock, the synthesis of these HSPs increases and three new HSPs of molecular weight 94K, 64K, and 25K are synthesized. The induced HSP 94K and 64K proteins differ in their isoelectric points from the HSPs of similar molecular weights.

Cultures containing mostly myotubes required a more severe stress (i.e., a longer period of the higher temperature) to elicit the same response as that observed in myoblast cultures. However, the small HSP25 was not synthesized under any form of stress in myotubes, whereas HSP82 appeared in higher amounts in heat-treated myotubes than in myoblasts.

Since myogenesis involves the fusion of cells to form multinucleated cells, the possibility existed that the differential heat-shock response was due to multinucleation as opposed to differentiation. In fact, these two processes can be distinguished by treating myoblast cells with EGTA, which inhibits cell fusion but not differentiation. Heat-treated, EGTA-treated myogenic cells synthesize HSP25, while in the absence of EGTA no HSP25 is made. Thus multinucleation, which in itself is an integral step in myogenesis, accounts for some of the alterations in the heat-shock response.

Primary cell cultures from a *Drosophila* gastrulation stage embryo also can undergo differentiation *in vitro*. The differentiation of two cell types has been well characterized: myoblasts divide to yield myocytes, which fuse after elongation to form myotubes, and neuroblasts differentiate to form clusters of neurons, which send out axons and form miniature ganglia. Treatment of these embryonic cells with a variety of teratogenic drugs, including coumarin and diphenylhydantoin, inhibits both muscle and neuron differentiation. Concomitant with the inhibition of differentiation is the synthesis of a subset of the small HSPs, HSP23 and HSP22 (Buzin and Bournias-Vardiabasis, 1984). Since the administration of these teratogenic agents does not result in the synthesis of HSP70, HSP83, and the other small HSPs, the results cannot simply be attributed to a general "stress" response. Instead, these data indicate a relationship between the inhibition of differentiation and the synthesis of HSP22 and HSP23.

B. Teratocarcinomas

Mouse embryonic carcinoma cell lines can be induced to differentiate by the administration of retinoic acid, followed by the addition of dibutyryl cyclic AMP. In several of these cell lines, a subset of HSPs is constitutively expressed in the undifferentiated state, for example, the cell lines PCC4-Aza-RI and PCC7-S-1009 have high constitutive levels of HSP89, HSP70, and HSP59 proteins and fail to increase HSP synthesis after heat shock (Morange *et al.*, 1984). In this regard, they mimic early embryonal stages in *Drosophila* in their lack of responsiveness to heat shock. After retinoic acid-mediated differentiation, both cell lines respond to heat shock by induction of HSPs. Another cell line, F9, also shows high levels of HSP89, HSP70, and HSP59 in the absence of stress but a marked reduction in HSP89 occurs after differentiation (Bensaude and Morange, 1983). The decrease in HSP89 expression following *in vitro* differentiation has also been reported in the murine PSA-G teratocarcinoma cell line (Levine *et al.*, 1984).

Two other murine teratocarcinoma stem cell lines (OTT6050-2125 and OTT6050-2127) also illustrate a "differentiation state"-dependent heat-shock response (Wittig *et al.*, 1983). Undifferentiated cells are unable to mount a heat-shock response. When the teratocarcinoma cells are transplanted subcutaneously into mice, they grow as solid tumors that typically contain differentiated tissue (glandular structures and neural tissue). In this state, the tumors synthesize HSPs in response to heat shock.

C. Erythroid Differentiation

A model system used to study erythroid differentiation is the cell line K562, which can be induced to differentiate by hemin. A 70-kDa protein accumulates during hemin-induced erythroid maturation, and this protein is structurally related to the heat-inducible HSP70 in K562 cells (Singh and Yu, 1984). Late erythroblasts from human bone marrow have a protein antigenically related to HSP70.

V. Conclusion

Since the initial discovery of the induction of HSP synthesis by heat shock and other forms of stress, it has become apparent that HSPs also play a role in the normal life-span of the cell. In this article we have emphasized the role of HSPs in cellular differentiation and reviewed data showing that HSPs are expressed at specific stages of embryonic development and during tissue differentiation. In addition, the recent finding that some HSPs, in particular HSP70, are transiently expressed during the normal cell cycle (Kao *et al.*, 1985; Milarski and Morimoto, 1986) implicates these proteins in a basic metabolic activity of the cell.

The evidence presented in this review suggests that heat-shock gene expression is regulated by multiple control elements responding to varying events in the cell, the organism, and the surrounding environment. Microorganisms apparently utilize the HSP genes during their normal life cycle as well as during a stress response. Bacteriophages have co-opted the cell HSPs for their own development, employing the proteins in DNA replication and phage assembly. It is presumed that the HSPs perform functions in the cell which are analogous to those described for the phage.

In the eukaryote, HSP genes have evolved to form families consisting of isoproteins with distinct but related functions. In one of the simplest nucleated cells, the yeast, members of the same protein family are regulated by promoter elements which respond to different stimuli—nutrient depletion, temperature, etc. In some higher eukaryotes, there appear to be fewer members of the same family, but single members can have several different promoter elements in the gene. The result is that a specific HSP function, i.e., DNA replication, can be affected by multiple stimuli, ranging from virus infection to hormones to cell-cycle activity.

The fact that proteins of identical structure are used by the organism for normal growth and development as well as for survival from environmental insults suggests two different explanations for the function of these proteins. The obvious one is that the organism needs to "protect" itself at specific stages of its life from potential damage by stress agents, and HSPs perform this role. There is a vast body of literature supporting the concept that HSPs are responsible for thermotolerance; however, a number of systems have been described in which thermotolerance is found in the absence of HSP formation. The alternative hypothesis is that the HSPs are used for normal growth and metabolic activity and their role in stress is to enable the cell to "return" to a normal pattern of growth that has been temporarily blocked as the result of the stress. That is, they are essential elements in cell homeostasis. Since "tolerance" is a measure of the recovery of cell growth, this postulated role in homeostasis can account for thermotolerance. In other words, "protection" from stress-induced alterations in the cell takes the form of providing a cell with resources to regain its former state. Many of the biochemical properties that are now known for HSPs and the variety of biological activities associated with HSP induction strongly support this hypothesis. But the ultimate resolution of the role of HSPs in biology will come when we understand the true function of these proteins.

REFERENCES

Ambrosio, L., and Schedl, P. (1984). Gene expression during *Drosophila melanogaster* oogenesis: Analysis by *in situ* hybridization to tissue sections. *Dev. Biol.* **105**, 80–92.

Ang, D., Chandrasekhar, G. N., Zylicz, M., and Georgopoulos, C. (1986). The grpE gene of *Escherichia coli* codes for heat shock protein B25.3, essential for both lambda DNA replication at all temperatures and host viability at high temperatures. *J. Bacteriol.* **167**, 25–29.

Atkinson, B. G. (1981). Synthesis of heat shock proteins by cells undergoing myogenesis. *J. Cell Biol.* **89**, 666–673.

Atkinson, B. G., Cunningham, T., Dean, R. L., and Somerville, M. (1983). Comparison of the effects of heat shock and metal-ion stress on gene expression in cells undergoing myogenesis. *Can. J. Biochem. Cell Biol.* **61**, 404–413.

Ayme, A., and Tissieres, A. (1985). Locus 67B of *Drosophila melanogaster* contains seven not four closely related heat shock genes. *EMBO J.* **4**, 2949–2954.

Ballinger, D. G., and Pardue, M. L. (1983). The control of protein synthesis during heat shock in *Drosophila* cells involves altered polypeptide elongation rates. *Cell* **33**, 103–114.

Bardwell, J. C. A., and Craig, E. A. (1984). Major heat shock gene of *Drosophila* and the heat inducible dnaK gene are homologous. *Proc. Natl. Acad. Sci. U.S.A.* **81**, 848–852.

Bensaude, O., and Morange, M. (1983). Spontaneous high expression of heat-shock proteins in mouse embryonal carcinoma cells and ectoderm from day 8 mouse embryo. *EMBO J.* **2**, 173–177.

Bensaude, O., Babinet, C., Morange, M., and Jacob, F. (1983). Heatshock proteins, first major products of zygotic gene activity in mouse embryo. *Nature (London)* **305**, 331–333.

Bergh, S., and Arking, R. (1984). Developmental profile of the heat shock response in early embryos of *Drosophila. J. Exp. Zool.* **231**, 379–391.

Bienz, M. (1984a). Developmental control of the heat shock response in *Xenopus. Proc. Natl. Acad. Sci. U.S.A.* **81**, 3938–3942.

Bienz, M. (1984b). *Xenopus* hsp 70 genes are constitutively expressed in injected oocytes. *EMBO J.* **3**, 2477–2483.

Bienz, M., and Gurdon, J. B. (1982). The heat shock response in *Xenopus* oocytes is controlled at the translational level. *Cell* **29**, 811–819.

Bienz, M., and Pelham, H. R. B. (1986). Heat shock regulatory elements function as an inducible enhancer in the *Xenopus* hsp70 gene and when linked to a heterologous promoter. *Cell* **45**, 753–760.

Bond, U., and Schlesinger, M. J. (1985). Ubiquitin is a heat shock protein in chicken embryo fibroblasts. *Mol. Cell. Biol.* **5**, 949–956.

Brown, I. R. (1983). Hyperthermia induces the synthesis of a heat shock protein by polysomes isolated from the fetal and neonatal mammalian brain. *J. Neurochem.* **40**, 1490–1493.

Buzin, C. H., and Bournias-Vardiabasis, H. (1984). Teratogens induce a subset of small heat shock proteins in *Drosophila* primary embryonic cell cultures. *Proc. Natl. Acad. Sci. U.S.A.* **81**, 4075–4079.

Cheney, C. M., and Shearn, A. (1983). Developmental regulation of *Drosophila* imaginal disc proteins: Synthesis of a heat shock protein under non-heat-shock conditions. *Dev. Biol.* **95**, 325–330.

Chung, C. H., and Goldberg, A. L. (1981). The product of the lon (capR) gene in *E. coli* is an ATP-dependent protease, protease La. *Proc. Natl. Acad. Sci. U.S.A.* **78**, 4931–4935.

Chung, S., Handfear, S. M., Blumberg, D. D., Cohen, N. S., and Lodish, H. F. (1981). Synthesis and stability of developmentally regulated *Dictyostelium* mRNAs are affected by cell–cell contact and cAMP. *Cell* **24**, 785–797.

Cohen, R. S., and Meselson, M. (1985). Separate regulatory elements for the heat-inducible and ovarian expression of the *Drosophila* hsp 26 gene. *Cell* **43**, 737–746.

Collins, P. L., and Hightower, L. (1982). Newcastle Disease Virus stimulates the cellular accumulation of stress (heat shock) proteins. *J. Virol.* **44**, 703–707.

Corces, V., Holmgren, R., Freund, R., Morimoto, R., and Meselson, M. (1980). Four heat shock proteins of *Drosophila melanogaster* coded within a 12 kilobase region in chromosome sub-division 67B. *Proc. Natl. Acad. Sci. U.S.A.* **77**, 5390–5393.

Craig, E. A. (1985). The heat shock response. *Crit. Rev. Biochem.* **18**, 239–280.

Craig, E. A., and McCarthy, B. J. (1980). Four *Drosophila* heat shock genes at 67B: Characterization of a recombinant plasmid. *Nucleic Acids Res.* **8**, 4441–4458.

Craig, E. A., McCarthy, B. J., and Wadsworth, S. C. (1979). Sequence organization of two recombinant plasmids containing genes for the major heat shock induced protein of *D. melanogaster*. *Cell* **16**, 575–588.

Craig, E. A., Ingolia, T. D., and Manseau, L. (1983). Expression of *Drosophila* heat-shock cognate genes during heat shock and development. *Dev. Biol.* **99**, 418–426.

Currie, R. W., and White, F. P. (1983). Characterization of the synthesis and accumulation of a 71-kilodalton protein induced in rat tissues after hyperthermia. *Can. J. Biochem. Cell Biol.* **61**, 438–446.

DeLucia, A. L., Lewton, B. A., Tjian, T., and Tegtmeyer, R. (1983). Topography of Simian Virus 40 A protein–DNA complexes: Arrangement of pentanucleotide. *J. Virol.* **46**, 143–150.

Dodson, M., Echols, H., Wickner, S., Alfano, C., Mensa-Wilmot, K., Gomes, B., Lebowitz, J., Roberts, J. D., and McMacken, R. (1986). Specialized nucleoprotein structures at the origin of replication of bacteriophage λ: Localization unwinding of duplex DNA by a six-protein reaction. *Proc. Natl. Acad. Sci. U.S.A.* **83**, 7638–7642.

Drummond, I. A. S., McClure, S. A. D., Poenie, M., Tsien, R. Y., and Steinhardt, R. A. (1986). Large changes in intracellular pH and calcium observed during heat shock are not responsible for the induction of heat shock proteins in *Drosophila melanogaster*. *Mol. Cell. Biol.* **6**, 1767–1775.

Dura, J. (1981). Stage dependent synthesis of heat shock induced proteins in early embryos of *Drosophila melanogaster*. *Mol. Gen. Genet.* **184**, 381–385.

Dynan, W. S., and Tjian, R. (1983). The promoter-specific transcription factor Sp1 binds to upstream sequences in the SV40 early promoter. *Cell* **35**, 79–87.

Friedman, D. I., Olson, E. R., Georgopoulos, C., Tilly, K., Herskowitz, I., and Banuett, F. (1984). Interactions of bacteriophage and host macromolecules in the growth of bacteriophage (lambda). *Microbiol. Rev.* **48**, 299–325.

Georgopoulos, C., and Eisen, H. (1974). Bacterial mutants which block phage assembly. *J. Supramol. Struct.* **2**, 349–359.

German, J. (1984). Embryonic stress hypothesis of teratogenesis. *Am. J. Med.* **76**, 293–299.

Glaser, R. L., Wolfner, M. F., and Lis, J. T. (1986). Spatial and temporal pattern of hsp 26 expression during normal development. *EMBO J.* **5**, 747–754.

Graziosi, G., Di Cristini, F., Di Marcotullio, A., Marzari, R., Micali, F., and Savoini, A. (1983). Morphological and molecular modifications induced by heat shock in *Drosophila melanogaster* embryos. *J. Embryol. Exp. Morphol.* **77**, 167–182.

Grossman, A. D., Erikson, J. W., and Gross, C. (1984). The hptR gene product in *E. coli* is a sigma factor for heat shock promoters. *Cell* **38**, 383–390.

Heikkila, J. J., Kloc, M., Bury, J., Schultz, G. A., and Browder, L. W. (1985). Acquisition of the heat shock response and thermotolerance during early development of *Xenopus laevis*. *Dev. Biol.* **107**, 483–489.

Henikoff, S., and Meselson, M. (1977). Transcription of two heat shock loci in *Drosophila*. *Cell* **12**, 441–451.

Hoffman, E., and Corces, V. (1986). Sequences involved in temperature and ecdysterone-induced transcription are located in separate regions of a *Drosophila melanogaster* heat shock gene. *Mol. Cell. Biol.* **6**, 663–673.

Hunter, K. W., Cook, C. L., and Hajunga, E. G. (1984). Leishmanial differentiation *in vitro:* Induction of heat shock proteins. *Biochem. Biophys. Res. Commun.* **125**, 755–760.

Ingolia, T. D., and Craig, E. A. (1982). Four small heat shock genes are related to each other and to mammalian α-crystallin. *Proc. Natl. Acad. Sci. U.S.A.* **72**, 2360–2364.

Ireland, R. C., and Berger, E. M. (1982). Synthesis of low molecular weight heat shock

peptides stimulated by ecdysterone in a cultured *Drosophila* cell line. *Proc. Natl. Acad. Sci. U.S.A.* **79,** 855–859.
Ireland, R. C., Berger, E., Sirotkin, K., Yund, M. A., Osterbur, D., and Fristrom, J. (1982). Ecdysterone induces the transcription of four heat-shock genes in *Drosophila* S3 cells and imaginal discs. *Dev. Biol.* **93,** 498–507.
Kao, H., and Nevins, J. R. (1983). Transcriptional activation and subsequent control of the human heat shock gene during adenovirus infection. *Mol. Cell. Biol.* **3,** 2058–2065.
Kao, H., Capasso, O., Heintz, N., and Nevins, J. R. (1985). Cell cycle control of the human hsp 70 gene: Implications for the role of a cellular E1A-like function. *Mol. Cell. Biol.* **5,** 628–633.
Khandjian, E. W., and Turler, H. (1983). Simian virus 40 and polyoma virus induce synthesis of heat shock proteins in permissive cells. *Mol. Cell. Biol.* **3,** 1–8.
Kingston, R. E., Cowie, A., Morimoto, R., and Gwinn, K. A. (1986). Binding of polyoma virus large T antigen to the human hsp 70 promoter is not required for transactivation. *Mol. Cell. Biol.* **6,** 3180–3190.
Kirk, M. M., and Kirk, D. L. (1985). Translational regulation of protein synthesis in response to light, at a critical stage of *Volvox* development. *Cell* **41,** 419–428.
Kirk, D. L., and Kirk, M. M. (1986). Heat shock elicits production of sexual inducer in *Volvox. Science* **231,** 51–54.
Klementz, R., and Gehring, W. J. (1986). Sequence requirement for expression of the *Drosophila melanogaster* heat shock protein hsp 22 gene during heat shock and normal development. *Mol. Cell. Biol.* **6,** 2011–2019.
Klementz, R., Hultmark, D., and Gehring, W. J. (1985). Selective translation of heat shock mRNA in *Drosophila melanogaster* depends on sequence information in the leader. *EMBO J.* **4,** 2053–2060.
Kurtz, S., and Lindquist, S. (1984). Changing patterns of gene expression during sporulation in yeast. *Proc. Natl. Acad. Sci. U.S.A.* **81,** 7323–7327.
Kurtz, S., Rossi, J., Petko, L., and Lindquist, S. (1986). An ancient development induction: Heat-shock proteins induced in sporulation and oogenesis. *Science* **231,** 1154–1157.
Landfear, S. M., Lefebvre, P., Chung, S., and Lodish, H. F. (1982). Transcriptional control of gene expression during development of *Dictyostelium discoideum. Mol. Cell. Biol.* **2,** 1417–1426.
Lawrence, F., and Robert-Gero, M. (1985). Induction of heat shock and stress proteins in promastigotes of three leishmania species. *Proc. Natl. Acad. Sci. U.S.A.* **82,** 4414–4417.
Lawson, R., Mestril, R., Luo, J., and Voellmy, R. (1985). Ecdysterone selectivity stimulates the expression of a 23000-Da heat shock protein-β-galactosidase hybrid gene in cultured *Drosophila* cells. *Dev. Biol.* **110,** 321–330.
Lebowitz, J. H., Zylicz, M., Georgopoulos, C., and McMacken, R. (1985). Initiation of DNA replication on single-stranded DNA templates catalyzed by purified replication proteins of bacteriophage lambda and *Escherichia coli. Proc. Natl. Acad. Sci. U.S.A.* **82,** 3988–3992.
Levine, R. A., LaRosa, G. J., and Guda, L. J. (1984). Isolation of cDNA clones for genes exhibiting reduced expression after differentiation of murine teratocarcinoma stem cells. *Mol. Cell. Biol.* **4,** 2142–2150.
Lindquist, S. (1986). The heat shock response. *Annu. Rev. Biochem.* **55,** 1151–1191.
Livak, K. J., Freund, R., Schweber, M., Wensink, P. C., and Meselson, M. (1978).

Sequence organization and transcription at two heat shock loci in *Drosophila*. *Proc. Natl. Acad. Sci. U.S.A.* **78**, 5613–5617.
McGarry, T. J., and Lindquist, S. (1985). The preferential translation of *Drosophila* hsp 70 mRNA requires sequences in the untranslated leader. *Cell* **42**, 903–911.
Mason, P. J., Hall, L. M. C., and Gausz, J. (1984). The expression of heat shock genes during development in *Drosophila melanogaster*. *Mol. Gen. Genet.* **194**, 73–78.
Mestril, R., Schiller, P., Amin, J., Klapper, H., Ananthan, J., and Voellmy, R. (1986). Heat shock and ecdysterone activation of the *Drosophila melanogaster* hsp 23 gene: A sequence implied in developmental regulation. *EMBO J.* **5**, 1667–1673.
Milarski, K. L., and Morimoto, R. I. (1986). Expression of human hsp70 during the synthetic phase of the cell cycle. *Proc. Natl. Acad. Sci. U.S.A.* **83**, 9517–9521.
Mitchell, H. K., and Lipps, L. S. (1978). Heat shock and phenocopy induction in *Drosophila*. *Cell* **15**, 907–918.
Mitchell, H. K., and Petersen, N. S. (1983). Gradients of differentiation in wild-type and bithorax mutants of *Drosophila*. *Dev. Biol.* **95**, 459–467.
Moran, L., Mirault, M. E., Tissieres, A., Los, J., Schedl, P., Artavains-Tsakonas, S., and Gehring, W. J. (1979). Physical map of two *D. melanogaster* DNA segments containing sequences coding for the 70,000 dalton heat shock protein. *Cell* **17**, 1–8.
Morange, M., Dui, A., Bensaude, O., and Babinet, C. (1984). Altered expression of heat shock proteins in embryonal carinoma and mouse early embryonic cells. *Mol. Cell. Biol.* **4**, 730–735.
Morimoto, E. I., Hunt, C., Huang, S.-Y., Berg, K. L., and Banerji, S. S. (1986). Organization, nucleotide sequence and transcription of the chicken HSP gene. *J. Biol. Chem.* **261**, 12692–12699.
Muller, W. U., Li, G. C., and Goldstein, L. S. (1985). Heat does not induce synthesis of heat shock proteins or thermotolerance in the earliest stage of embryo development. *Int. J. Hyperthermia* **1**, 97–102.
Munro, S., and Pelham, H. R. B. (1986a). What turns on heat shock genes? *Nature (London)* **317**, 447–448.
Munro, S., and Pelham, H. R. B. (1986b). An HSP-70-like protein in the ER: Identity with the 78 kd glucose-regulated protein and immunoglobulin heavy chain binding protein. *Cell* **46**, 291–300.
Neidhardt, F. C., VanBogelen, R. A., and Vaughn, V. (1984). The genetics and regulation of heat shock proteins. *Annu. Rev. Genet.* **18**, 295–329.
Nevins, J. R. (1982). Induction of the synthesis of a 70,000 dalton mammalian heat shock protein by the adenovirus E1A gene product. *Cell* **29**, 913–919.
Nickells, R. W., and Browder, L. W. (1985). Region-specific heat shock protein synthesis correlates with a biphasic acquisition of thermotolerance in *Xenopus laevis* embryos. *Dev. Biol.* **112**, 391–395.
Notarianni, E. L., and Preston, C. M. (1982). Activation of cellular stress protein genes by herpes simplex virus temperature-sensitive mutants which overproduce immediate early polypeptides. *Virology* **123**, 113–122.
Nover, L. (1984). "Heat Shock Response of Eukaryotic Cells." Springer-Verlag, Berlin and New York.
Palter, K. B., Watanabe, M., Stinson, L., Mahowald, A. P., and Craig, E. A. (1986). Expression and localization of *Drosophila melanogaster* hsp 70 cognate proteins. *Mol. Cell. Biol.* **6**, 1187–1203.
Parker, C. S., and Topol, J. (1984). A *Drosophila* RNA polymerase II transcription factor binds to the regulatory site of a HSP 70 gene. *Cell* **37**, 273–283.

Pauli, D., Spierer, A., and Tissieres, A. (1986). Several hundred base pairs upstream of *Drosophila* hsp 23 and 26 genes are required for their heat induction in transformed flies. *EMBO J.* **5,** 755–761.

Pelham, H. R. B. (1982). A regulatory upstream promoter element in the *Drosophila* HSP 70 heat shock gene. *Cell* **30,** 517–528.

Pelham, H. R. B. (1986). Speculations on the functions of the major heat shock and glucose regulated proteins. *Cell* **46,** 959–961.

Peluso, R. W., Lamb, R. A., and Choppin, P. W. (1978). Infection with paramyxoviruses stimulates synthesis of cellular polypeptides that are stimulated in cells transformed by Rous sarcoma virus or deprived of glucose. *Proc. Natl. Acad. Sci. U.S.A.* **75,** 6120–6124.

Petko, L., and Lindquist, L. (1986). HSP 26 is not required for growth at high temperatures, nor for thermotolerance spore development or germination. *Cell* **45,** 885–894.

Riddihough, G., and Pelham, H. R. B. (1986). Activation of the *Drosophila* hsp 27 promoter by heat shock and by ecdysone involves independent and remote regulatory sequences. *EMBO J.* **5,** 1653–1658.

Ritossa, F. M. (1962). A new puffing pattern induced by heat shock and DNP in *Drosophila. Experientia* **18,** 571–573.

Roccheri, M. C., Sconzo, G., Di Carlo, M., Di Bernardo, M. G., Pirrone, A., Gambino, R., and Giudice, G. (1982). Heat-shock proteins in sea urchin embryos. Transcriptional and posttranscriptional regulation. *Differentiation* **22,** 175–178.

Santamaria, P. (1979). Heat shock induced phenocopies of dominant mutants of the bithorax complex in *Drosophila melanogaster. Mol. Gen. Genet.* **172,** 161–163.

Schedl, P., Artavanis-Tsakonas, S., Steward, R., and Gehring, W. J. (1978). Two hybrid plasmids with *D. melanogaster* DNA sequences complementary to mRNA coding for the major heat shock protein. *Cell* **14,** 921–929.

Schlesinger, M. J. (1986). Heat shock proteins: The search for functions. *J. Cell Biol.* **103,** 321–325.

Schlesinger, M. J., Ashburner, M., and Tissières, A., eds. (1982). "Heat Shock. From Bacteria to Man," p. 6. Cold Spring Harbor Laboratory, Cold Spring Harbor, New York.

Singh, M. K., and Yu, J. (1984). Accumulation of a heat shock-like protein during differentiation of human erythroid cell line K562. *Nature (London)* **309,** 631–633.

Sirotkin, K., and Davidson, N. (1982). Developmentally regulated transcription from *Drosophila melanogaster* chromosomal site 67B. *Dev. Biol.* **89,** 196–210.

Spradling, A., Pardue, M. L., and Penman, S. (1977). mRNA in heat shocked *Drosophila* cells. *J. Mol. Biol.* **109,** 559–587.

Stevenson, M. A., Calderwood, S. K., and Hahn, G. M. (1986). Rapid increases in inositol trisphosphate and intracellular Ca^{++} after heat shock. *Biochem. Biophys. Res. Commun.* **137,** 826–833.

Thomas, S. R., and Lengyel, J. A. (1986). Ecdysteroid-regulated heat-shock gene expression during *Drosophila melanogaster* development. *Dev. Biol.* **115,** 434–438.

Tilly, K., McKittrick, N., Zylicz, M., and Georgopoulos, C. (1983). The dnaK protein modulates the heat-shock response of *Escherichia coli. Cell* **34,** 641–646.

Tissieres, A., Mitchell, H. K., and Tracy, U. (1974). Protein synthesis in salivary glands of *Drosophila melanogaster:* Relation to chromosome puffs. *J. Mol. Biol.* **84,** 389–398.

Van der Ploeg, L., Giannini, S. H., and Cantor, C. R. (1985). Heat shock genes: Regulatory role for differentiation in parasitic protozoa. *Science* **228,** 1443–1446.

Vitek, M. P., and Berger, E. M. (1984). Steroid and high-temperature induction of the small heat-shock protein genes in *Drosophila*. *J. Mol. Biol.* **178**, 173–189.

Wadsworth, S., Craig, E. A., and McCarthy, B. J. (1980). Genes for three *Drosophila* heat shocked induced proteins at a single locus. *Proc. Natl. Acad. Sci. U.S.A.* **77**, 2134–2137.

Walsh, D. A., Hightower, L. E., Klein, N. W., and Edwards, M. J. (1985). The induction of heat shock proteins during early mammalian development. *In* "Heat Shock. From Bacteria to Man" (M. J. Schlesinger, M. Ashburner, and A. Tissières, eds.). Cold Spring Harbor Laboratory, Cold Spring Harbor, New York.

White, F. (1981). The induction of 'stress' proteins in organ slices from brain, heart and lung as a function of postnatal development. *J. Neurosci.* **1**, 1312–1319.

Wittig, S., Hensse, S., Keitel, C., Elsner, C., and Wittig, B. (1983). Heat shock gene expression is regulated during teratocarcinoma cell differentiation and early embryonic development. *Dev. Biol.* **96**, 507–514.

Wu, C. (1984). Two protein-binding sites in chromatin implicated in the activation of heat-shock genes. *Nature (London)* **309**, 229–234.

Wu, C. (1985). An exonuclease protection assay reveals heat shock element and TATA box DNA binding proteins in crude nuclear extracts. *Nature (London)* **317**, 84–87.

Wu, B. J., Hurst, H. C., Jones, N. C., and Morimoto, R. I. (1986). The E1A 13S product of adenovirus-5 activates transcription of the cellular human hsp 70 gene. *Mol. Cell. Biol.* **6**, 2994–2999.

Yura, T., Tobe, T., Ito, K., and Osawa, T. (1984). Heat shock regulatory gene (htpR) of *Escherichia coli* is required for growth at high temperature but is dispensable at low temperature. *Proc. Natl. Acad. Sci. U.S.A.* **81**, 6803–6807.

Zimmerman, J. L., Petri, W., and Meselson, M. (1983). Accumulation of a specific subset of *D. melanogaster* heat shock mRNAs in normal development without heat shock. *Cell* **32**, 1161–1170.

Zucker, C., Cappello, J., Chisholm, R. L., and Lodish, H. F. (1983). A repetitive dictyostelium gene family that is induced during differentiation and by heat shock. *Cell* **34**, 997–1005.

Zucker, C., Cappello, J., Lodish, H. F., George, P., and Chung, S. (1984). Dictyostelium transposable element DIRS-1 has 350-base-pair inverted terminal repeats that contain a heat shock promoter. *Proc. Natl. Acad. Sci. U.S.A.* **81**, 2660–2664.

Zylicz, M., Lebowitz, J. H., McMacken, R., and Georgopoulos, C. (1983). The dnaK protein of *Escherichia coli* possesses an ATPase and autophosphorylating activity and is essential in an *in vitro* DNA replication system. *Proc. Natl. Acad. Sci. U.S.A.* **80**, 6431–6435.

MECHANISMS OF HEAT-SHOCK GENE ACTIVATION IN HIGHER EUKARYOTES

Mariann Bienz* and Hugh R. B. Pelham†

*Zoological Institute, University of Zürich, CH-8057 Zürich, Switzerland

†Medical Research Council (MRC) Laboratory of Molecular Biology, Cambridge CB2 2QH, England

I. Introduction

Most cells respond to heat shock and similar environmental stresses by switching on a set of genes, the heat-shock genes. This gene activation is rapid and reversible, resulting in a transition from hardly detectable levels of transcription under normal conditions to extremely high transcription rates during heat shock. Consequently, heat-shock proteins (HSPs) accumulate to high levels; according to the current view, these serve some repair function with respect to the damage that has been caused by the heat shock. The structures of the major HSPs are strongly conserved among animals, plants, yeast, and bacteria, whereas the activation mechanism, including its components, is almost identical among higher eukaryotes, similar among higher and lower eukaryotes, but probably quite different among prokaryotes.

Three major types of HSPs are found in most organisms: the most highly conserved HSP with a molecular weight of 70K (HSP70), a larger HSP with a molecular weight of 80K–90K (HSP90; called HSP83 in *Drosophila*), and a group of small HSPs with molecular weights between 15K and 30K that are related among themselves. Growing cells normally express low levels of HSP90, and also one or more HSP70-like proteins. These proteins have been termed HSP70 cognates and they are encoded by separate genes, distinct from the strictly heat-inducible HSP70 genes. Small HSPs are found in certain cell types and stages of normal development. These observations have led to the conclusion that heat-shock proteins serve a specialized function during heat shock, related to a more general function which they, or closely related proteins, serve during the normal life of the cell.

Heat-shock gene induction represents a model case for gene activation in two respects. During transient heat induction, the heat-shock genes undergo reprogramming from the OFF- to the ON-state: heat-shocked and unshocked cells differ in their state of gene activation in a way analogous to two different cell types or cell lineages. Furthermore, some heat-shock genes are not only transiently heat inducible, but are also activated during normal cell growth or development, which leads to the question of how multiple control mechanisms act on the same gene. Heat-shock gene activation has been studied in the hope that understanding the regulation of these genes will provide a conceptual framework for the investigation of more complex genes, for example, those involved in developmental switches.

In this review, we will first consider the components of the heat-shock gene activation systems, i.e., the cis-acting elements and the trans-acting factors. We shall then summarize data on how these components act together to result in transcription activation and how multiple controls are achieved. We will finally address the question of how the cell detects the environmental stimulus and translates it into gene activation and how the functions of the gene products relate to this process. We will focus on heat-shock gene activation in higher eukaryotes and relate this to the field of higher eukaryotic transcription. Only those aspects of heat-shock genes and proteins which are relevant to the question of gene activation will be included. Shorter reviews and reviews covering different aspects of the heat-shock field have been published previously (Ashburner and Bonner, 1979; Schlesinger *et al.*, 1982; Nover, 1984; Bienz, 1985; Pelham, 1985, 1986; Voellmy, 1985; Craig, 1985; Lindquist, 1986).

II. The Heat-Shock Regulatory Element (HSE)

Early transfection experiments demonstrated that a cloned *Drosophila HSP70* gene comes under heat-shock control when it is stably integrated into the genome of mouse fibroblasts (Corces *et al.*, 1981).

A detailed analysis of the promoter of this gene identified a short DNA sequence upstream of the TATA box that is required for heat inducibility when the gene is assayed by transient transfection of monkey cells (Pelham, 1982; Mirault *et al.*, 1982). Similar sequence motifs were found in other heat-shock promoters (Pelham, 1982). A palindromic consensus sequence CT-GAA--TTC-AG was derived, synthesized, and shown to be sufficient for conferring heat inducibility on a heterologous gene (Pelham and Bienz, 1982). This sequence was called the heat-shock regulatory element (HSE) (Bienz, 1985; Pelham, 1985) and can be found within the first 400 bp upstream of every heat-shock gene from every higher eukaryote which has been sequenced so far (Holmgren *et al.*, 1981; Karch *et al.*, 1981; Hackett and Lis, 1983; Southgate *et al.*, 1983; Bienz, 1984a; Schoeffl *et al.*, 1984; Czarnecka *et al.*, 1985; Hunt and Morimoto, 1985; Nagao *et al.*, 1985; Russnak and Candido, 1985; Voellmy *et al.*, 1985; Hickey *et al.*, 1986; Morimoto *et al.*, 1986; Pauli *et al.*, 1986; Riddihough and Pelham, 1986; Rochester *et al.*, 1986). Various expression experiments have subsequently confirmed the importance of HSE in conferring heat inducibility (see below). Comparison of HSEs found in 20 heat-shock gene promoter regions establishes rules as to the structure of a HSE, its copy number, and its position within a promoter (Fig. 1).

There are eight nucleotides within a HSE which are highly conserved (Fig. 1b). Expression assays established that, in general, in order to constitute an individual functional HSE, seven nucleotides have to match the consensus sequence C--GAA--TTC--G (reviewed in Bienz, 1985). In the following sections, we shall refer to "strong" HSEs in the case of a seven or eight out of eight match (i.e., 7/8 match or 8/8 match), and to "weak" HSEs in case of a six out of eight match (i.e., 6/8 match), to this minimal consensus sequence. Furthermore, the HSE consensus sequence is symmetrical, suggesting that the putative binding protein consists of two identical subunits or monomers which recognize the same half-binding site on either strand of the DNA helix. Many HSEs occur in doublets and some occur in triplets, i.e., two or three HSEs overlap each other in a fixed spatial arrangement with 10 nucleotides between them (e.g., C--GAA--TTC--GAA--TTC--G). Thus, the centers of symmetry are located on the same side of the DNA helix. A double or triple HSE can probably accommodate more than one dimer of a putative binding protein (see Section III).

Most heat-shock promoters contain several HSEs. The most proximal one is usually found immediately 5′ to the TATA box in a preferred location 15–18 bp (exceptionally 14 or 28 bp) from it (Fig. 1a). This suggests that the proteins binding to these two promoter ele-

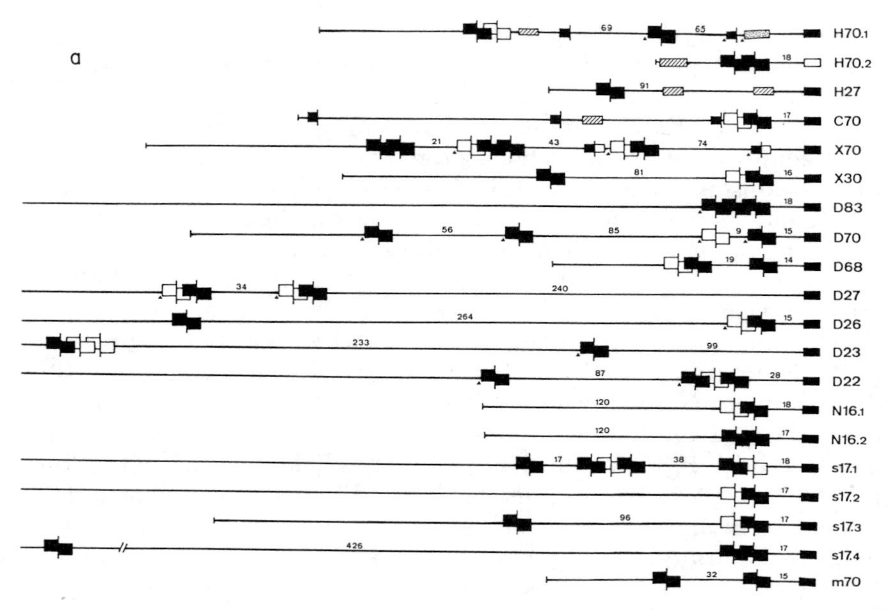

FIG. 1. Heat-shock gene promoters. Promoter regions of 20 different heat-shock genes from higher eukaryotes have been sequenced: three human (*H70.1*, Hunt and Morimoto, 1985; *H70.2*, Voellmy *et al.*, 1985; *H27*, Hickey *et al.*, 1986), one chicken (*C70*, Morimoto *et al.*, 1986), two *Xenopus* (*X70* and *X30*, Bienz, 1984a), seven *Drosophila* (*D83*, Hackett and Lis, 1983; *D70*, Karch *et al.*, 1981; *D68*, Holmgren *et al.*, 1981; *D27*, Riddihough and Pelham, 1986; *D26* and *D22*, Southgate *et al.*, 1983; *D23*, Pauli *et al.*, 1986), two nematode (*N16.1* and *N16.2*, Russnak and Candido, 1985), four soybean (*s17.1*, Schoeffl *et al.*, 1984; *s17.2*, Czarnecka *et al.*, 1985; *s17.3* and *s17.4*, Nagao *et al.*, 1985), and one maize gene (*m70*, Rochester *et al.*, 1986). The number in each short name refers to the molecular weight of the encoded protein; in each case, the publication containing the most 5′-flanking sequences is given as a reference. (a) Promoter structure. The promoters of the 20 genes are aligned at the TATA boxes (small solid boxes on the right-hand side; the open box in the *H70.2* gene represents a poor match to a TATA sequence). Large solid boxes with symmetry lines represent strong HSEs (7/8 or 8/8 match to the HSE consensus sequence C--GAA--TTC--G), the corresponding open boxes represent weak HSEs (6/8 match); weak HSEs are depicted only if they overlap a strong HSE (except with *D70* HSE2, which is known to be functional in conjunction with HSE1; see text). Small numbers indicate base pair distances between HSEs or between the proximal HSE and the TATA box. Small solid boxes with a vertical line represent CCAAT sequences (in each case ATTGG); the two CCAAT boxes in the *X70* gene are symmetrical, but the downstream part contains a poor match to the CCAAT sequence (open box), whereas all the other CCAAT boxes consist of a single ATTGG sequence. Hatched boxes represent good matches (at least 8/10) to the SP1-binding site consensus sequence (G/T GGGCGG G/A G/A C/T) (Dynan and Tjian, 1985); an additional SP1-binding site, not depicted in the figure, is found in the *H70.2*

b

```
Single HSEs                                                    Gene          Matches

                 *  ***  ***  *
+ggcgaaaccc CtgGAAtaTTCccG acctggcagc                          H70.1 HSE1     8
 ttccttaacg agaGAAggTTCcaG atgagggctg                          H27   HSE1     7
 ccagtgaatg CcaGAAgtTgCtaG cacagcctcc                          X30   HSE2     7
+gagagcgcgc CtcGAAtgTTCgcG AAaagagcgc                          D70   HSE1     8
+gcacactaTT CtcGttgcTTCgaG agagcgcgcc                          D70   HSE2     6
+atatataaat aaaGAAtaTTCtaG AAtcccaaaa                          D70   HSE4     7
 gcagggaaat CtcGAAttTTCccc tcccggcgac                          D68   HSE1     7
 tgcgctctTT CtaGAAacTTCggc tctctcactc                          D26   HSE2     7
+attttcagcc CgaGAAgtTTCgtG tcccttctcg                          D23   HSE1     8
+ccagagagcc CcaGAAacTTCcac ggagttcgct                          D22   HSE2     7
 atggtccaat CccGAAacTTCtaG ttgcggttcg                          s17.1 HSE3     8
 atatatatat CtaGAAggTTgtaG AAgactagct                          s17.3 HSE2     7
 atttttcat  CttcAAacTTCaaG ttgttgtagt                          s17.4 HSE2     7
 tcgccccgtg CccGAAtcTTCtgG acgcgccatc                          m70   HSE1     8
 ctccactcct CcaGAgccTTCcaG AAccccaatc                          m70   HSE2     7
```

```
Double HSEs
                          *  ***  ***  *
                 *  ***  ***  *
 ccagtgaatc CcaGAAgacTCtgGAgagTTCtga gcagggggcg            H70.1 HSE2   7/6
 ctcgactggg CggGAAggTgCggGAAggTTCgcG gcggcggggt            H70.2 HSE1   7/8
 ctggattggt CcttAgcgTTCtgGcAggTTCcaG AAgaaggcta            C70   HSE1   6/7
+ggctaacgaa atgGAAgccTCggGAAacTTCggG tcggttgcta            X70   HSE1   6/8
 acgactctct CgaGAAagcTCgcGAAtcTTCcgc gattgtgact            X70   HSE3   7/7
 cagagagcac atgGAAgtcTCggGAAcgTcCcaG AAcactaact            X30   HSE1   6/7
+atatataaat aaaGAAaacTCgaGAAatTTCtct ggccgttatt            D70   HSE3   6/7
 ctcgcacaca CacGAActgaCtgGAAtgTTCtga ccctttctcg            D68   HSE2   6/7
+agccgctgtg CcaGAAagagCcaGAAgaTgCgaG agaaaactgt            D27   HSE1   6/7
+aactcccaga aaaGAAatgTCaaGAAgtTTCtgG ttctttctcc            D27   HSE2   6/8
+tttcctttTT CtgtcActTTCcgGActcTTCtaG AAaagctcca            D26   HSE1   6/7
 cctcctttg  CaaGAAgcagCtcGAAtgTTCtaG AAaaaggtgg            N16.1 HSE1   6/8
 atgaatgcat CtaGgAccTTCtaGAAcaTTCtaa acggctgcag            N16.2 HSE1   7/7
 tgatgcataa CaaGgActTTCtcGAAagTaCtat attgctcctc            s17.1 HSE1   7/6
 ttgcaaaaag tagGAtttTTCtgGAAcaTaCaaG attatccttt            s17.2 HSE1   6/7
 atattgtaaa CaatAtttTTCtgGAAcaTaCaaG agtatccttt            s17.3 HSE1   6/7
 aaattgcaaa CacGAtttTTCtgGAAcgTaCacG attatccttt            s17.4 HSE1   7/7
```

```
Triple HSEs
                          *  ***  ***  *
                *  ***  ***  *       *  ***  ***  *
+actcagcaac CgtGAcacTgCcgGAAaccTCgcGAAagTTCttc gggtgatctc  X70   HSE2  6/7/7
+atccctgcat CcaGAAgccTCtaGAAgtTTCtaGAgacTTCcaG ttcgggtgcg  D83   HSE1  7/8/7
 gttctgctgt CtcGAAgtTTCgcGAAttTaCtccAtccTTCgtG gaatatactc  D23   HSE2  8/6/6
+agcaaaggyc gaaGAAaaTTCgaGAgagTgCcgGtAttTTCtaG attatatgga  D22   HSE1  7/6/7
 tcgaagaagt CcaGAAtgTTctGAAagTTtcaGAAaaTTCtaG ttttgagatt  s17.1 HSE2  7/6/7
```

FIG. 1b. See legend on pp. 34 and 36.

ments have to interact (or avoid interacting) with each other in a particular way. Preliminary expression assays suggest that increasing the separation to 21 bp (or generally to integral numbers of helical turns?) may result in increased transcriptional activity under non-heat-shock conditions (T. McGarry and S. Lindquist, personal communiation; M. Bienz, unpublished observations). Thus, a precise separation between the binding sites on the DNA helix may help to keep the promoters inactive in unshocked cells.

In several genes, the most proximal HSE is found far upstream from the TATA box (Fig. 1a). These genes are not just transiently heat inducible, but are also subject to developmental control, and other control elements, e.g., CCAAT boxes or SP1-binding sites, are frequently found in their promoters. Evidently HSEs can activate a heat-shock gene over large distances as well as from nearby (see Section IV).

III. The Heat-Shock Transcription Factor (HSTF)

A. *In Vitro* TRANSCRIPTION

Evidence that HSEs are the binding sites for a specific transcription factor came from *in vitro* transcription experiments with extracts of *Drosophila* cells. Nuclear extracts from heat-shocked cells were found

---

gene overlapping the HSE at its 5′ side. The stippled box in the *H70.1* gene indicates the SRE element (Wu *et al.*, 1986c). A small solid triangle at the left of a promoter element indicates that this element has been shown to be functional either by expression or by footprinting assays (see text). A short vertical line at the left-hand side indicates the end of the known sequence, which in some cases extends beyond the 400-bp flanking sequences contained in the figure (interruption in the *s17.4* gene means that 100 bp have been deleted in order to include the distal HSE in the figure). All boxes and sequence lines are drawn to scale. Note the nonrandom position of the proximal HSE (or promoter element) with respect to the TATA box. Also, many distal promoter elements (HSEs or others) appear to be located between six and eight helical turns or between twelve and fourteen helical turns from the TATA-proximal one. (b) HSE structures. The sequences of all HSEs depicted in (a) and additional 10 bp of flanking sequences on either side are given. Capital letters indicate matches to the consensus sequence (also highlighted by asterisks above the sequences). The short name of each gene and the number of the HSE (one equals proximal) and its match to the consensus sequence (6/8, 7/8, or 8/8) are given at the right-hand side. A plus at the left-hand side indicates functionality of this HSE [see (a) and text]. Note that the 5′-flanking two-T residues and the 3′-flanking two-A residues are sometimes conserved, extending the homology to the basic consensus sequence and possibly the binding site for HSTF (see Section III,A).

to transcribe the *Drosophila HSP70* gene efficiently. By fractionating the extract and using *in vitro* transcription as an assay, Parker and Topol (1984a) partially purified a heat-shock transcription factor (HSTF) and showed by DNase footprinting that it binds to a region of the HSP70 promoter that includes the two proximal HSEs (Fig. 1a). Subsequently, more detailed studies showed that it is indeed the HSE sequence which is recognized and that each of the four Gs (two on either strand) within the HSE consensus sequence is a close contact point for HSTF, as defined by methylation interference and protection experiments (Shuey and Parker, 1986). Binding to the two sites in the HSP70 promoter is cooperative: the upstream weak HSE (6/8 match) has a 12.5-fold lower intrinsic affinity for HSTF than the downstream strong HSE, but once the downstream site is filled, HSTF binds efficiently to the upstream site (Topol *et al.*, 1985). Binding of the second HSTF molecule appears to distort the interactions of the first one with the DNA, and Shuey and Parker (1986) suggested that this apparent change in the position of the factor is crucial to the activation of the promoter. Such interactions, however, cannot be of general importance, because the organization of different heat-shock promoters varies considerably, for example, the *Drosophila virilis* HSP70 promoter retains the proximal HSE, but lacks the weak adjacent one (R. Blackman, personal communication); furthermore, artificial constructs in which the weak HSP70 HSE is replaced by another HSE at some distance are also efficiently heat inducible (Cohen and Meselson, 1984; Mestril *et al.*, 1986).

It is not clear how HSTF interacts with a double HSE. In principle, two dimers could bind independently, each monomer recognizing one strand of the DNA:

```
       - - - 1 -->     - - - 2 -->
C--GAA--TTC--GAA--TTC--G
G--CTT--AAG--CTT--AAG--C
<-- 1 - - -     <-- 2 - - -
```

However, the finding that all the G residues in a single HSE are contacted by HSTF seems inconsistent with this, because the central two Gs in a double HSE would be protected by both HSTF dimers. Thus, maybe a HSTF dimer flanked by a HSTF monomer binds to a double HSE. It is possible that more than one HSTF dimer binds even to a single HSE, i.e., that individual monomers can be stacked on one or either side of a single HSTF dimer. This idea is suggested by various footprints of individual *Drosophila* HSEs which appear to extend over additional "half" HSEs (C. Parker, personal communication), and by the finding that, quite frequently, half a HSE consensus sequence can be found adjacent to functional HSEs (mostly resulting in double HSEs; Fig. 1b). Triple HSEs can accommodate two HSTF dimers in principle; however, studies of HSTF binding to the *Drosophila* HSP83 triple HSE suggest that the middle consensus sequence (8/8 match) is occupied first at low HSTF concentration, presumably by a HSTF dimer, and the two flanking HSEs are occupied later, presumably each by a HSTF monomer (C. Parker, personal communication).

The frequent occurrence of double and triple HSEs may indicate that they are preferred and especially strong binding sites for HSTF. Recent *in vitro* binding experiments with HeLa cell extracts show that a synthetic double HSE does in fact compete more efficiently for HSTF than a single HSE (M. Lewis and H. R. B. Pelham, unpublished observations).

Fractionation of the *Drosophila* cell extract revealed a second factor that binds to a region of the HSP70 promoter (and others) that includes the TATA box, protecting the DNA sequences between −40 and +30 from DNase (Parker and Topol, 1984b; Wu, 1984a, 1985). This factor (in partially pure form) stimulates HSTF-dependent transcription from the HSP70 promoter about threefold, but it is not essential for correct initiation (Parker and Topol, 1984b). The fact that HSTF can support initiation in the absence of any other detectable sequence-specific DNA-binding protein may indicate that HSTF is somehow able to direct RNA polymerase to the appropriate position on the template. It is easy to imagine how HSTF can do this when the HSE is adjacent to the presumed polymerase interaction site (the TATA box), but the mechanism is less obvious when the HSE is at a remote location, e.g., in the *Drosophila HSP27* gene (Fig. 1a). Indeed, this gene is not efficiently transcribed in these *in vitro* extracts (C. Parker, personal communication). This problem is discussed in detail in Section IV).

B. Chromatin Studies

An independent approach to the problem of identifying factors bound to heat-shock promoters was taken by Wu, who carried out a series of experiments on the *in vivo* chromatin structure of the *Drosophila HSP70* and *HSP83* genes. He originally showed, by an indirect end-labeling approach, that the 5′ ends of these genes are hypersensitive to DNase I in the nuclei of unshocked cells (Wu, 1980). Subsequently, by cutting with a restriction enzyme within the hypersensitive region and trimming the ends of the DNA with exonuclease III, he was able to map the boundaries of the exposed DNA with considerable accuracy. These boundaries presumably correspond to regions of the promoter that are protected by tightly bound proteins. Using this approach, Wu showed that the TATA-box region (from about −40 to −12) is protected in both heat-shocked and control cells, whereas the HSE-containing region is only protected during heat shock (Wu, 1984a). Moreover, when extracts from heat-shocked nuclei were added to unshocked nuclei, the HSEs became protected (Wu, 1984b). These results suggest that the presence of the TATA factor is not sufficient to activate transcription of the heat-shock genes, and that induction involves a change in the availability of a HSE-binding protein or, more likely, in its affinity for the HSE. This protein was termed heat-shock-activating protein (HAP) (Wu, 1984a); it is presumably identical to HSTF. For simplicity, we shall use the term HSTF throughout this review.

Similar conclusions were reached by Cartwright and Elgin (1986), who used a chemical cleavage method to map exposed regions in the small *Drosophila* heat-shock promoters in tissue-culture cells. They found that several regions which are hypersensitive in unshocked cells (see also Costlow and Lis, 1984) became protected during heat shock; these regions generally contain HSE sequences and, thus, binding of HSTF after heat shock is indicated.

All these results illustrate an unresolved paradox: if HSTF is not bound to DNA in unshocked cells, what prevents nucleosomes from covering the HSE regions? It is possible that all cells get intermittently stressed and thus have a history of HSTF binding that is retained in their chromatin structure. Alternatively, HSTF may bind weakly, but not detectably by the methods currently used, in unshocked cells. Another possibility is that some feature of the adjacent DNA sequence is incompatible with nucleosome formation and thus ensures accessibility to the HSEs; however, there are no good candidates for such sequences. It seems most likely that HSTF itself is responsible for the DNase I hypersensitivity of its binding sites and that the nucleosome organization is a consequence of HSTF binding in one way or another, but this remains to be proved.

C. Activation of HSTF

HSTF must exist in some form in unshocked cells, because heat-shock genes can be activated in the presence of protein synthesis inhibitors (Ashburner and Bonner, 1979). The chromatin studies sug-

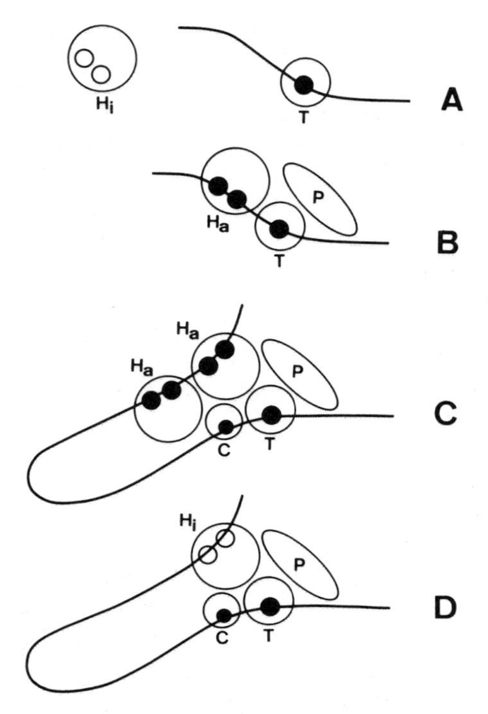

FIG. 2. Hypothetical transcription complexes. Curved or looped-out lines represent the DNA sequences of four heat-shock promoters in different states. Transcription factors are depicted by circles: H_i, inactive HSTF (low affinity to the HSE is indicated by small open circles); H_a, activated HSTF (high affinity to the HSE is indicated by small solid dots); C, CTF [or possibly other transcription factors (e.g., SP1)]; T, TATA-binding protein. Association of polymerase (P) with the promoter indicates transcriptional activity. (A) Before heat shock. Only the TATA factor is bound to the promoter and HSTF is inactive. (B) Heat-shock activation. HSTF has been activated and binds the HSE adjacent to the TATA box. A stable transcription complex is formed by the direct contact of a HSTF dimer and the TATA-binding protein (and possibly additional secondary transcription factors). In *Drosophila* and maybe other invertebrate cells, two HSTF dimers are required for the formation of this transcription complex. (C) Heat-shock activation at a distance. Two activated HSTF dimers binding HSEs at a distance from the TATA box and an additional transcription factor with anchoring function (CTF) are required for the formation of a stable transcription complex. The requirements for more transcription factor molecules in this case may arise since the intermediate DNA has to be looped out. (D) Model for developmental activation without heat shock, based on HSTF. Due to an auxiliary protein (CTF), inactive HSTF (or low levels of activated HSTF) can be attracted into the promoter to bind to the HSE, leading to the formation of a stable transcription complex without heat shock.

gest that it is unable to bind tightly to DNA (Fig. 2A). Thus, a current working hypothesis is that HSTF becomes physically modified in heat-shocked cells such that its affinity for the HSE is greatly increased (a model that predicts a specific modification of HSTF is discussed in Section VII). Alternatively, HSTF could be restricted to some subcellular compartment under normal conditions and it could be released from that compartment during heat shock and thus become available for HSE binding.

The current biochemical evidence cannot distinguish between these alternative models of heat-induced HSTF activation. There is a controversy whether HSTF capable of binding to HSEs or capable of stimulating heat-shock gene transcription can be extracted from unshocked cells. Recent data obtained with HeLa cell extracts show that these cells, after heat shock, contain an HSE-binding activity detectable by footprinting, reduction in the electrophoretic mobility of DNA fragments on gels, or nitrocellulose filter-binding assays. This activity, which presumably corresponds to HSTF, is virtually undetectable in nuclear extracts from unshocked cells (M. Lewis and H. R. B. Pelham, unpublished observations; R. Kingston, personal communication), agreeing with the results obtained by Wu (1984b, 1985) with *Drosophila* extracts (see above). On the other hand, Parker and Topol (1984a) and Morgan *et al.* (1987) reported that HSTF activity, capable of binding to HSEs, can be isolated from unshocked *Drosophila* or HeLa cells, but that this HSTF-containing extract is less active in stimulating heat-shock gene transcription *in vitro*. In order to explain the discrepancy between the different observations one can either assume that the cells have been stressed inadvertently during growth (e.g., by anoxia) or that HSTF has been activated inadvertently during preparation (e.g., by proteolysis); alternatively, if HSTF were bound to some subcellular compartment in unshocked cells, it might not be extractable by the procedures used in some of the studies.

Recently, *Drosophila* HSTF has been purified to apparent homogeneity by the use of affinity chromatography based on HSE oligomers. Its molecular weight may be 110K (C. Wu, personal communication) or 70K (C. Parker, personal communication). In the latter case it was demonstrated that the 70-kDa band eluted from a protein gel is capable of HSE binding; however, smaller protein fragments appear to be capable of HSE binding (C. Parker, personal communication), suggesting that HSTF is an unstable protein and that the molecular weight of 70K may still correspond to a metastable fragment of HSTF. The same purification procedure extracts a HSE-binding protein from yeast with an identical molecular weight of 70K (C. Parker, personal communication). However, when stringent precautions are taken against proteolysis, a 150-kDa protein is obtained (P. K. Sorger and H. R. B. Pelham, unpublished). Treatment of this protein with papain yields a prominent 70-kDa protein. Thus, some of the above-mentioned discrepancies may be accounted for by proteolysis of HSTF during its preparation. There are several examples in prokaryotes of genes that are activated or inactivated by the stabilization or destabilization of regulatory proteins (reviewed in Ptashne, 1986a). The availability of antibodies against purified HSTF will greatly aid studies of the mechanism of HSTF activation.

IV. Heat Induction of Heat-Shock Genes

A. HSE Requirements

Initial studies, in which the *Drosophila HSP70* gene was assayed on a replicating plasmid in monkey COS cells, showed that although this promoter contains four recognizable HSEs, a single TATA-proximal one was sufficient for maximal induction (Pelham, 1982; Mirault *et al.*, 1982). Similar results were obtained when the gene was assayed in HeLa cells without replication (M. Bienz and H. R. B. Pelham, unpublished observations). The conclusion was reinforced by the findings that one synthetic HSE could confer heat inducibility on the herpes virus thymidine kinase gene (Pelham and Bienz, 1982), and that the proximal HSE (double or triple) appeared to be sufficient for induction of the *Drosophila* HSP22 and HSP26 promoters in COS cells (Ayme *et al.*, 1985).

The first clue that more than one HSE might contribute to promoter function came from experiments in which the *Drosophila HSP70* gene was microinjected into *Xenopus* oocytes; it was observed that two tandem HSEs gave full heat induction more reproducibly than just one (Bienz and Pelham, 1982). A much greater difference was observed when the gene was assayed in flies using P-element-mediated germ-line transformation. In this case, a deletion mutant that retained only the TATA-proximal HSE was nearly two orders of magnitudes less active than one that retained two HSEs (Dudler and Travers, 1984; Simon *et al.*, 1985). A somewhat different result was seemingly obtained by Cohen and Meselson (1984), using essentially the same assay, who found only a small effect of deleting the upstream weak HSE; however, this can be explained by the fortuitous presence of a strong HSE (7/8 match) in the plasmid sequences that were placed adjacent to the deletion end point, 35 bp upstream of the TATA-proximal HSP70 HSE.

The difference between COS cells and *Drosophila* appears to reflect a genuine species difference, rather than a copy number effect, because transient-expression assays in *Drosophila* tissue-culture cells give the same result as assays in transformed flies (Amin *et al.*, 1985; Morganelli *et al.*, 1985). A possible explanation might be that the *Drosophila* HSTF binds more weakly than the mammalian one, and thus simultaneous binding of two HSTF molecules may be required for stabilization of the HSTF–DNA complex. This is unlikely to be the whole answer, however, because *in vitro* studies show that the weak upstream HSE contributes so little binding energy that it increases the affinity of HSTF for the strong adjacent HSE only by a factor of two (Topol *et al.*, 1985). It is thus more likely that it is the interaction of HSTF with some other entity (RNA polymerase or other transcription factors) that requires more than one molecule of HSTF in *Drosophila* but not in mammalian cells.

Subsequent studies have shown that the proximal HSE is never sufficient for strong induction in *Drosophila* cells, and even in mammalian cells multiple HSEs are sometimes required (see Section V,D). Plant cells seem similar to *Drosophila* cells in that they require multiple HSEs for maximal function (Gurley *et al.*, 1986).

B. Long-Range Effects of HSEs in *Drosophila* Cells

In the case of the *Drosophila HSP70* gene, the proximal HSE is adjacent to the TATA box, and it is reasonable to assume that binding of HSTF to this site allows polymerase to interact simultaneously with this factor and with the TATA factor and/or the initiation site on the DNA. In other genes, the proximal HSE is at a considerable distance from the TATA box: 99 bp in the case of the *HSP23* and 240 bp in the *HSP27* gene. In both cases, the proximal HSE is not sufficient for heat induction (Mestril *et al.*, 1986; Pauli *et al.*, 1986; Riddihough and Pelham, 1986). The results are particularly clear in the case of the *HSP27* gene, which has two double HSEs spread over about 100 bp (Fig. 1a). Deletion analysis showed that both are required for maximal induction. The distal one can be replaced by the two proximal HSEs from the *HSP70* gene, even if these are placed at the 3′ side of the gene (about 2 kb from the *HSP27*-proximal HSE), but efficient activation from a distance requires a duplication of the two *HSP70* HSEs (Riddihough and Pelham, 1986). Similar studies of the *HSP26* and *HSP22* genes whose proximal HSEs are adjacent to the TATA box

(Fig. 1a) show that, although these are important, the distal HSEs located 264 and 87 bp from the respective proximal HSE are essential for efficient induction (Cohen and Meselson, 1985; Klemenz and Gehring, 1986; Pauli *et al.*, 1986). Two general rules emerge from these experiments. First, multiple HSEs are invariably required in *Drosophila* cells, and more appear to be required when they are far from the TATA box (or from the nearest HSE) than when they are close to it. Second, HSEs act in a synergistic and distance-independent way with each other and with the TATA region of the promoter.

C. Long-Range Effects of HSEs in Vertebrate Cells

Early experiments in COS cells suggested quite different conclusions from those above. Not only was a single HSE sufficient, but there also appeared to be stringent restrictions on its position. In general, synthetic constructs and natural (*Drosophila*) promoters with HSEs 28 bp and more away from the TATA box showed less inducible activity than those with a shorter spacing (13–18 bp); the *Drosophila* HSP27 and HSP23 promoters were essentially inactive or constitutively active at a low level in mammalian cells (Pelham and Lewis, 1983; Ayme *et al.*, 1985).

An exception is the *Xenopus* HSP70 promoter, which contains three HSEs (two double and one triple HSE); the proximal one is located some seven helical turns upstream of the TATA box. In addition there are two CCAAT boxes, one of which is located in the position normally occupied by the TATA-adjacent HSE (Fig. 1a). Despite the large HSE–TATA distance, the gene is as active as the *Drosophila HSP70* gene in COS and HeLa cells; however, a single HSE is not functional and the whole proximal HSE (double HSE) is not sufficient—maximal activation requires the proximal and the first distal HSE. In addition, the proximal CCAAT box is essential for full heat inducibility; if this CCAAT box is present, the HSEs can activate the gene in a distance-independent way (Bienz and Pelham, 1986). A CCAAT box is found in many vertebrate promoters and is known to be the binding site for a transcription factor [CCAAT-binding transcription factor (CTF) (Jones *et al.*, 1985)]. The presence of this additional promoter element evidently allows multiple distance HSEs to act as an enhancer (for a review of enhancer effects, see Serfling *et al.*, 1985). Confirmation of this was provided by experiments in which two copies of a region encompassing the two proximal *Drosophila* HSP70 HSEs were placed 800 bp upstream of the human β-globin gene—the latter has its own upstream promoter elements, but is inactive in HeLa cells in the absence of an enhancer sequence (see in Dierks *et al.*,

1983). The globin gene then became very strongly heat inducible; this effect was abolished if the upstream elements of the globin promoter were deleted (Bienz and Pelham, 1986). These experiments suggest an explanation for the failure of HSP27 and HSP23 to work in mammalian cells: their promoters evidently lack additional TATA-proximal promoter elements that can be recognized in these cells.

The conclusion from these experiments is that HSEs and HSTF can activate vertebrate promoters in two ways: either from a position adjacent to the TATA box (Fig. 2B), in which case a single HSE can suffice, or from a distance (Fig. 2C). In the latter case, multiple HSEs are required for efficient activation, together with an additional TATA-adjacent promoter element, such as a CCAAT box, which is itself incapable of supporting high levels of transcription. A recently reported sequence of a human HSP27 promoter fits this general picture (Hickey *et al.*, 1986): the distance between the proximal HSE and the upstream TATA box (there are two TATA boxes and two RNA start sites in this gene) is 91 bp, but two SP1-binding sites are located within these sequences, one of them in the position where the proximal HSE is normally found (Fig. 1a). Even though SP1 is involved in the transcription of other genes in unshocked cells (reviewed in Dynan and Tjian, 1985), this promoter shows very little activity in the absence of heat shock, suggesting that the role of SP1 is to assist HSTF.

Xenopus oocytes seem to be somewhat intermediate between *Drosophila* and mammalian cells. The *HSP23* gene has been reported to be heat inducible in injected oocytes (Mestril *et al.*, 1985). This suggests that a single HSE can activate this promoter at a distance, at least to some extent, apparently without the assistance of additional promoter elements. In both the *Drosophila* and the *Xenopus HSP70* gene, a single HSE can be sufficient for full heat inducibility in injected oocytes (Bienz and Pelham, 1982; Bienz, 1986). These less stringent requirements, with respect to number of HSEs, additional promoter elements, and distance between HSEs and the TATA box, may point toward an activity in *Xenopus* oocytes which facilitates formation of stable transcription complexes or stabilizes these (see Section V,D).

D. Possible Mechanisms of HSTF Action

How does HSTF activate heat-shock genes? In most cases, activated HSTF binds to a HSE immediately adjacent to the TATA-binding protein (Fig. 2B). This probably leads to the formation of a stable transcription complex which itself may be recognized by RNA polymerase and other transcription factors that do not bind DNA directly (for a review, see Brown, 1984). As mentioned in Section III, HSTF itself may have the required ability to attract RNA polymerase to the template. In promoters where the proximal HSE is located at a distance from the TATA box, we can assume that the mechanism of transcriptional activation is based on the same principle: on the formation of a stable transcription complex. In order for this to happen, the intermediate DNA has to be looped out (Fig. 2C). Indirect and direct evidence for looping out of DNA has been obtained in prokaryotic promoters (Dunn *et al.*, 1984; Hochschild and Ptashne, 1986; Griffith *et al.*, 1986), and indirect evidence suggests that this may also occur in the SV40 early promoter (Takahashi *et al.*, 1985; for a review, see Ptashne, 1986b). Generation of a stable transcription complex involving proteins bound to distant sites on the DNA requires more energy. This in turn may explain why a single HSTF dimer is not sufficient for long-distance activation; the HSE-containing fragments with enhancer activity (Bienz and Pelham, 1986) can both accommodate at least three dimers of HSTF. It may also explain why a CCAAT box is required in mammalian cells for HSE-mediated long-distance activation: the role of CTF could be to anchor distantly bound HSTF dimers next to the TATA factor, thus facilitating formation of a stable transcription complex (Bienz and Pelham, 1986). It is not known whether the interaction between HSTF and CTF is specific, or whether other upstream element binding factors such as SP1 can also provide this anchoring function (by interacting with HSTF). It is also unclear whether or not the sequence of the DNA that has to be looped out is of any importance: some sequences may require less energy for bending than others.

In *Drosophila* cells, on the other hand, there does not seem to be any requirement for a factor analogous to CTF, because sequences immediately upstream of the TATA box can be removed without affecting the heat inducibility of the promoter (Riddihough and Pelham, 1986). It is conceivable that an "upstream element" analogous to the CCAAT box in the *Xenopus HSP70* gene may be located immediately downstream of the TATA box in the *Drosophila* genes: sequences in the most 5′ part of the mRNA leader are important for transcriptional activation of the *HSP22* gene (Klemenz *et al.*, 1985), although this may simply reflect contacts between the TATA factor (or polymerase) and the DNA in this region. An alternative explanation is that the TATA factor in *Drosophila* is itself capable of a sufficiently strong and stable interaction with HSTF to allow a transcription complex to form in the absence of additional DNA-binding proteins.

In both *Drosophila* and mammalian cells, cooperative interactions between multiple HSTFs in the transcription complex are likely, and indeed cooperativity between HSTF molecules binding to adjacent sites has been shown (Topol *et al.*, 1985). Combined with the fact that multiple HSTF dimers are usually required for full heat induction, this results in a switch system which can react to subtle environmental changes within a sharply defined window (as discussed in Ptashne, 1986a).

The heat-shock response is essentially a transient one, and thus the cell must be able to dismantle stable transcription complexes that have been formed on the heat-shock promoters. This may present a problem because the stability of the HSTF–DNA binding is very high—the time taken for HSTF to dissociate from a synthetic double HSE *in vitro* is measured in hours (M. Lewis and H. R. B. Pelham, unpublished observations). Possibly the factor can be modified while on the DNA, and thereby be induced to dissociate from it. Alternatively, there may be a separate mechanism which disrupts the transcription complex. As discussed in Section VI, HSP70 itself apparently has the ability to disrupt protein–protein interactions; perhaps HSP70 can help to turn off its own transcription by breaking cooperative interactions between HSTF molecules.

V. Developmental Activation of Heat-Shock Genes

With the term "developmental activation" we will refer to all those types of heat-shock gene activation which are distinct from heat induction and which occur in a cell during the normal cell cycle or during normal development. The known cases can be divided into those in which the induction is transient (analogous to heat induction) and those in which long-term activation takes place. Examples of the first group are the ecdysone inducibility of small *Drosophila* heat-shock genes and the temporary induction of a human *HSP70* gene during the normal cell cycle. Examples of the second group may be the low level of expression of the *HSP83* gene in unshocked *Drosophila* cells and the activation of the *Xenopus HSP70* gene in oocytes. It is conceivable that different activation mechanisms are used for transient as opposed to long-term activation. These examples will be discussed in the following paragraphs in more detail.

A. The Small *Drosophila* Heat-Shock Genes

It was observed that the small *Drosophila* heat-shock genes are expressed during the late larval and early pupal stages (Sirotkin and Davidson, 1982; Mason *et al.*, 1984). Furthermore, HSP27, HSP26, and HSP83 mRNA accumulates in the ovarian nurse cells (Zimmerman *et al.*, 1983; Mason *et al.*, 1984). It is likely that the late larval–prepupal induction is a consequence of high ecdysone titers during these stages, since it was shown that the small heat-shock genes are inducible in tissue-culture cells on addition of ecdysone to the medium (Ireland *et al.*, 1982). Although ovaries contain low levels of ecdysone (Handler, 1982), the possibility remains that the ovarian expression of HSP27 and HSP26 is connected with, and maybe even reflects, their ecdysone inducibility. Stabilization of mRNA during ecdysone treatment has also been shown (Vitek and Berger, 1984) and may play an important role in the developmental induction of the small heat-shock genes.

The main question is whether the different types of regulation of these genes act via distinct sites at the promoter level. Several groups have tried to identify sequence elements required for ecdysone inducibility and for ovarian expression, although, unfortunately, in none of these genes has both types of developmental expression been analyzed. It appears though that developmental control elements are generally separable from HSEs.

Cohen and Meselson (1985), analyzing flies transformed with *HSP26* gene, found that sequences required for the ovarian expression are located far upstream from the gene (between −341 and − 728 bp); sequences closer to the gene are dispensable for ovarian expression, but contain two strong HSEs (a single and a double HSE) far apart from each other which are required for heat inducibility. Glaser *et al.* (1986) transformed flies with fusions between the HSP26 promoter and the β-galactosidase gene and used histochemical staining to examine tissue-specific expression. They found that a deletion retaining 278 bp of upstream flanking sequence was no longer expressed in ovaries, as expected. However, this mutant was still expressed in spermatocytes, implying that there is a third, independent mode of expression of the *HSP26* gene.

A similar separation of developmental and heat-shock control elements was found in the *HSP27* gene: maximal heat inducibility is dependent on the presence of two double HSEs located between −270 and −370 bp upstream of the gene, whereas the regulatory region for ecdysone inducibility is located further upstream. In this case, it was shown that sequences between −579 and −455 are sufficient for conferring ecdysone inducibility on a heterologous gene (Riddihough and Pelham, 1986). This short region appears to contain multiple elements involved in ecdysone regulation; however, sequence inspection did not reveal any obvious candidate for such an element (Riddihough and Pelham, 1986). These results were obtained using transient-expression assays in tissue-culture cells, and thus the question of the relationship between ecdysone inducibility and ovarian expression could not be answered. Attempts to map regulatory elements in this promoter using P-element transformation have so far given inconsistent and variable results (Hoffmann and Corces, 1986). This seems to be due to chromosomal position effects which have been a problem in several studies of the small HSP promoters using P-element transformation. The activity of any given deletion can vary dramatically among different fly lines with different insertion sites. The reasons for the variability are not fully understood, but problems may arise because transcription from adjacent promoters through the test gene inhibits expression, or because enhancer elements several kilobases away can interact with the TATA box of the test gene, as has been demonstrated for both HSEs and other regulatory elements (Riddihough and Pelham, 1986; Garabedian *et al.*, 1986).

In these two well-documented cases (HSP26 and HSP27), the developmental control elements differ from the HSE; it seems that the sites for heat-shock control are generally located closer to the gene than the sites for developmental control. The regulatory proteins which confer developmental control are obviously distinct from HSTF. One candidate for such a protein is the ecdysone receptor. The latter probably belongs to a group of related proteins, the steroid receptors, some of which have been cloned (Miesfeld *et al.*, 1986, and references therein). Recently, a sequence homology has been found between the DNA-binding domain of the glucocorticoid receptor (Rusconi *et al.*, 1986) and a gene which is induced by the first ecdysone peak in early pupae (W. Segraves and D. Hogness, personal communication). It remains to be seen whether the latter represents the ecdysone receptor or a related protein.

It is easy to imagine a model for ecdysone induction that is analogous to that for heat induction. Binding of ecdysone presumably increases the affinity of the latter for DNA; it then binds to the appropriate functional element (which in the case of the *HSP27* gene is in a DNase-hypersensitive region of the chromatin), and then interacts directly with the TATA factor, the intervening DNA being looped out. The resultant transcription complex may be stabilized by the additional binding of factors, but such a mechanism would be independent of HSTF.

The HSP23 promoter appears to be organized somewhat differently, in that sequences required for ecdysone inducibility in transient *Drosophila* cell assays (between −379 and −143) (Mestril *et al.*, 1986) are flanked by the two HSEs required for full heat inducibility (Pauli *et al.*, 1986). Evidently, the regulatory elements for the two types of control are intermingled with each other. Furthermore, deletion of two short-sequence stretches around −200 strongly reduces the ecdysone response, but also affects the heat inducibility, although this region does not contain HSE sequences. It may well be that some transcription factors are involved in both responses, perhaps acting as general stabilizers of transcription complexes. There is no evidence that the ecdysone receptor itself is important for the heat-shock response, because a cell line lacking the receptor shows normal heat induction of HSP23 (Ireland and Berger, 1982).

An even closer connection between developmental and heat-shock control is suggested by the work of Klemenz and Gehring (1986) on the *HSP22* gene. By transforming flies with a naturally occurring variant of the *HSP22* gene (whose protein product can be distinguished from the endogenous HSP22 on two-dimensional gels), they found that expression of the protein in prepupae (presumably in response to ecdysone) was unaffected in mutants retaining at least 194 bp of promoter sequences, but was severely reduced in a mutant that had an additional 18 bp deleted. This critical region contains the distal HSE, suggesting that HSTF may be involved in this developmental regulation. It is possible, of course, that the binding site for another transcription factor overlaps this region. However, another possibility is that a rare molecule of active HSTF, or even inactive HSTF, is induced to bind to this HSE by a favorable interaction with another factor. A similar mechanism, involving an auxiliary factor which enhances HSTF affinity to the HSE, is discussed below in the section on *Xenopus HSP70* gene (Section V,D).

B. Mammalian *HSP70* Genes

Higher eukaryotes contain three different types of HSP70-like proteins (discussed further in Section VI): the heat-inducible HSP70, a cytoplasmic HSP70 cognate which is expressed in normally dividing cells, and a protein called grp78 whose synthesis is induced by glucose starvation and other inhibitors of glycosylation. Thus, different types of control are separated in different genes, maybe reflecting an evolutionary trend of specialization. Humans have an additional type of *HSP70* gene which codes for a protein that is very similar to the heat-inducible one, but is expressed at a significant basal level. This gene has been studied in considerable detail. It is inducible by heat shock and cadmium, but it is also expressed at elevated levels in serum-stimulated, dividing tissue-culture cells (Wu and Morimoto, 1985) and in the presence of the adenovirus protein E1A (Wu *et al.*, 1986a).

Deletion analysis of this human gene (Wu *et al.*, 1986b) established that heat and cadmium inducibility is dependent on a short-sequence stretch around −100; this region contains a HSE and, at its 3′ side, a sequence resembling the metal regulatory element found in metallothionein genes. Functional importance of the distal HSE further upstream has not been shown. Elements required for the basal level of expression, the response to serum stimulation and presumably to E1A, are all located in the 65 bp between the proximal HSE and the TATA box (Wu *et al.*, 1986b). These sequences include a CCAAT box and a purine-rich stretch (called SRE; Fig. 1a) which resembles the interferon regulatory element that has been implied in negative control (Goodbourn *et al.*, 1986). Footprinting analysis revealed HSTF and CTF binding to this promoter (Morgan *et al.*, 1987). An exonuclease analysis, using extracts from unshocked HeLa cell nuclei incubated with several DNA fragments, resolved three proteins bound in the promoter area (Wu *et al.*, 1987). The most upstream one protects the CCAAT box and thus probably corresponds to CTF. The adjacent protein at the 3′ side is probably identical to the TATA factor. With linker-scanning mutants, a third protein was found to bind to the purine-rich stretch and to overlap the footprints of CTF and possibly of the TATA factor. This protein may act as a repressor preventing other transcription factors from binding to the promoter (Wu *et al.*, 1987). According to this idea, it would have to be present in an altered form or at low levels in serum-stimulated cells and it thus should not be able to bind to the promoter in these cells. However, there is hardly any difference in the capacity for promoter binding, if extracts from quiescent and rapidly dividing cells are compared with respect to binding capacity of this protein (B. Wu, personal communication).

The response to serum stimulation most likely reflects a transient activation of this gene during the S phase of the cell cycle (Milarski and Morimoto, 1986). Replication is required for this activation (Wu and Morimoto, 1985). It is conceivable that the process of replication removes a repressor (see above) from the DNA. Alternatively, since replication temporarily results in histone-free templates, this could facilitate access of a transcription factor which is present in the cell at low concentrations or which has a low affinity to the promoter. The transcriptional stimulation by E1A may be the result of an increased level of this transcription factor, since E1A might generally act by increasing the effective concentration of limiting transcription factors (Yoshinaga *et al.*, 1986).

As in the small *Drosophila* heat-shock genes, heat-shock and developmental control of this human gene apparently involves different factors bound to distinct sites in the promoter. It is clear that the activation during the cell cycle does not require HSTF; on the other hand, it is possible that HSTF-mediated heat-shock induction may require CTF (by analogy to the *Xenopus HSP70* gene). It seems an economical strategy of evolution to use the same transcription factor for different types of control (see also Section V,D).

The finding that adenovirus protein E1A stimulates transcription of this human *HSP70* gene raises the possibility that other proteins involved in growth control might have similar effects. A sequence homologous to polyoma large T-antigen-binding site can be found in this promoter and stimulation of this gene by polyoma large T antigen has been reported; however, deletion of the above sequences does not have any effect on the expression of this gene, and their stimulatory effect, when these sequences are brought up close to a TATA box, may just be due to SP1 binding (Kingston *et al.*, 1986). An earlier finding was that the human c-myc protein stimulated expression from a heterologous *Drosophila* HSP70 promoter (Kingston *et al.*, 1984). This appears to be a fortuitous rather than a general stimulatory effect of *HSP70* genes, because it is dependent on sequences far upstream from this particular promoter which are not conserved between different *Drosophila HSP70* genes and which are presumably not present in the human HSP70 promoters (R. Kingston, personal communication).

Other members of the mammalian HSP70 family have recently been cloned. The partial sequence of a different human *HSP70* gene has been reported (Voellmy *et al.*, 1985): the promoter contains a double HSE and on its 3′ side, at a typical distance of 18 bp, a weak homology to the TATA-box sequence; it appears to be exclusively under heat-shock control. The sequence of the rat grp78 promoter, as expected since it is not heat inducible, does not contain any HSEs (Lin *et al.*, 1986). A rat *HSP70* cognate gene with introns has been isolated and characterized (P. Sorger and H. R. B. Pelham, unpublished). In growing cells, this gene is expressed at high levels and the corresponding mRNA is not significantly heat inducible nor is it heat inducible in quiescent cells, which express much lower levels of this mRNA. Its promoter contains several SP1-binding sites, two CCAAT boxes, and two double HSEs, all within 250 bp of the TATA box. The role of HSEs in a promoter that does not respond strongly to heat shock has yet to be determined. Finally, there is a report of a germ-cell-specific *HSP70* gene in mammals which is exclusively expressed during spermatogenesis, but this gene has not been cloned yet (Krawczyk *et al.*, 1987; Zakeri and Wolgemuth, 1987).

The mammalian HSP70 family represents a good example of a group of related genes, each of which exhibits a distinct pattern of regulation. Their promoters make use of common and ubiquitous transcription factors, such as CTF and SP1, as well as more specialized factors, such as HSTF. These factors have been biochemically well characterized. Further study of their action in regulating these genes may thus provide detailed insight into the ways in which combinatorial control mechanisms work.

C. The *Drosophila HSP83* Gene

Most cells synthesize detectable levels of a HSP90-like protein under normal conditions, but produce more of it during heat shock. In *Drosophila,* there is a single heat-inducible gene coding for HSP83 that is expressed at low levels in tissue-culture cells grown under normal conditions. Its promoter contains a triple HSE with three good matches to the heat-shock consensus sequence, adjacent to the TATA box (Fig. 1). Chromatin studies demonstrate that this HSE is inaccessible to restriction enzymes or exonuclease in tissue-culture cells, whether the cells are heat shocked or not (Wu, 1984a). These results suggest that HSTF can bind and thus activate transcription even under non-heat-shock conditions. The strong resistance of the HSP83 promoter–HSTF complex to exonuclease digestion indicates that HSTF binds more tightly to this HSE than to the one in the HSP70 promoter. The simplest explanation, therefore, is that there are low levels of activated HSTF in unshocked tissue-culture cells and that this is sufficient to activate a promoter with a very high-affinity binding site, but not sufficient to activate other heat-shock genes (Wu, 1984b).

Such a mechanism is quite distinct from the previously discussed examples of developmental heat-shock gene expression which mostly involve factors distinct from HSTF. With the *HSP83* gene, developmental control appears to involve HSTF, although the presumed HSE requirement for basal expression has not yet been demonstrated in any functional assay. It is possible that the occupation of the HSP83 HSE by HSTF in tissue-culture cells is simply due to the sensitivity of these cells to stress and to the high affinity of HSTF for this HSE. Alternatively, an auxiliary factor may facilitate binding of low levels of active HSTF (or even "inactive" HSTF) to the promoter. The efficiency of HSTF binding would then be due partly to protein–protein interactions and the basal level of expression could be regulated by the availability of the auxiliary factor (see Fig. 2D and Section V,D). There is some circumstantial evidence for the additional promoter element in the *HSP83* gene: a DNase-hypersensitive site has been observed in chromatin about 450 bp upstream of the transcription start site, in a region that does not contain any obvious HSE sequences (Wu, 1980; Costlow and Lis, 1984; Han *et al.*, 1985); however, it is not known whether this region is required for expression of the gene.

D. The *Xenopus HSP70* Gene

Heat shock induces the synthesis of HSP70 protein in *Xenopus* oocytes. Due to their especially unfavorable ratio of number of genes to cytoplasmic volume, this heat-shock response cannot be based on the transcriptional activation of the oocyte *HSP70* genes. It was therefore proposed that these genes are constitutively activated for at least a few days, if not during the whole period of oogenesis, resulting in the accumulation of HSP70 mRNA, which is stored and translationally activated only during heat shock (Bienz and Gurdon, 1982). Recently, it has been claimed that the observed synthesis of HSP70 protein in heat-shocked oocytes actually occurs only in the adhering follicle cells and not in the oocyte itself (King and Davis, 1987). This result questions the existence of the postulated translational control mechanism. Repetition of the experiments under well-defined conditions led to confirmation of the original results (M. Bienz and J. B. Gurdon, unpublished data), but the point remains somewhat uncertain (A. Colman, personal communication). In any case, ovulated eggs which are completely free from follicle cells synthesize HSP70 during heat shock (Browder *et al.*, 1987), suggesting that, at least in eggs, the stored oocyte HSP70 mRNA is translated during heat shock.

Stored HSP70 mRNA can be detected in unshocked oocytes and possibly eggs and early embryos, but in no other somatic cell which has not been heat shocked (Bienz, 1984b; Browder *et al.*, 1987); the 5′ end of this stored HSP70 mRNA is identical with the one found in heat-induced somatic HSP70 mRNA (M. Bienz, unpublished). Heat-shocked somatic *Xenopus* cells induce the synthesis of HSP30 proteins, but stored HSP30 mRNA or heat-inducible HSP30 protein has not been found in oocytes. Genes for both types of HSPs have been isolated (Bienz, 1984a). The promoter sequences of three different *HSP70* genes (from distinct chromosomal locations) are almost identical for the first 256 bp upstream of the RNA start site and diverge completely from there; this border of sequence conservation, just upstream of the most distal HSE, apparently delineates the whole promoter (see Fig. 1a and Section IV,C), which contains three HSEs and two CCAAT boxes. These *HSP70* genes are constitutively expressed in injected oocytes, although they are strictly heat inducible in transfected somatic cells; a functional *HSP30* gene is heat inducible in both cell types (Bienz, 1984a). Thus, a gene- and cell-type-specific control mechanism must exist which leads to constitutive activation of the *HSP70* genes in oocytes.

This activation requires both the proximal HSE (or any other substituting HSE) and the proximal CCAAT box (Bienz, 1986). Genes without a CCAAT box, e.g., the *Drosophila HSP70* or the *Xenopus HSP30* gene, cannot be activated efficiently in injected oocytes without heat shock. Thus, as in the *Drosophila HSP83* gene, HSTF is apparently involved in transcriptional activation of a gene without heat shock. However, in this case, neither the HSE by itself (or its specific sequence) nor the CCAAT box by itself, but only the combination of the two elements, can confer efficient constitutive activation. It may be that CTF acts as an auxiliary factor, interacting with "inactive" or low levels of activated HSTF and thus facilitating binding of HSTF by increasing its affinity for the promoter (Fig. 2D).

The same two elements are required for heat inducibility of the *Xenopus HSP70* gene in transfected somatic cells (Bienz and Pelham, 1986). This raises the question of how cell-type specificity is achieved. Oocytes may contain high enough levels of activated HSTF to result in promoter binding in conjunction with CTF, whereas the levels in somatic cells may be too low for this. Alternatively, a slightly different form of CTF and/or HSTF might exist in oocytes; the oocyte-specific forms could have an increased affinity for each other compared to the somatic forms, so that a transcription complex can be built more efficiently. A variation of the same idea is the hypothesis that a third, oocyte-specific factor exists that stabilizes the HSTF–CTF complex (see Section IV,C).

It is striking that in both the human and the *Xenopus HSP70* genes, a CCAAT box is required for developmental control. In both cases, activation occurs during or after replication: *Xenopus* oocytes are arrested in the first meiotic prophase; the activation of the human *HSP70* gene during S phase of the cell cycle requires replication. As discussed above, the "template-clearing effect" of replication may be important for CTF-dependent gene activation. On the other hand, the fact that replication generates two adjacent copies of each promoter element may be crucial. Duplication of HSEs within a promoter can have dramatic effects on head inducibility of a gene (see Section IV), presumably since transcription complexes involving two instead of one HSTF dimer are inherently more stable. It is conceivable that a transcription complex can be formed that includes factors bound to sequence elements present on both homologous chromatids.

If developmental activation of genes can be achieved with ubiquitous factors such as CTF, then variations of intracellular abundance or conformation of these factors must be sufficient to confer cell-type specificity (Bienz, 1986). The developmental regulation of the *Xenopus* 5 S RNA genes represents a precedent for this: the intracellular abundance of transcription factor IIIA determines the activation of the oocyte-specific 5 S genes in a crucial way (Brown and Schlissel, 1985). Promoter analysis of other cell-type-specific genes has identified so far mainly ubiquitous promoter elements, e.g., a CCAAT box in the globin genes or the ATGCAAAT octamer in the immunoglobulin genes (Dierks *et al.*, 1983; Mason *et al.*, 1985; Mattaj *et al.*, 1985). It thus seems possible that other developmentally regulated genes may use the same type of mechanism as the *Xenopus HSP70* gene, relying on subtle variations of factor availability or affinity to generate their observed pattern of cell-type-specific expression.

VI. Functions of Heat-Shock Proteins

It is reasonable to suppose that there is a logical connection between the regulatory pattern of a particular heat-shock gene and the function of the encoded heat-shock protein. In the final two sections we will address the question of how a cell senses the need for heat-shock proteins and how it translates this need into a gene activation mechanism. We will first summarize some aspects of HSP function and then present a working hypothesis connecting function with regulation.

A. The HSP70 Family

Although HSP70 was originally thought to be synthesized (and thus required) only in stressed cells, this view has changed with the discovery of a set of related genes (the so-called "cognates") which are expressed during normal growth. Two such proteins are known in *Drosophila* cells, and at least two other cognate genes have been isolated, although these are expressed at a much lower level (Ingolia and Craig, 1982a; Craig *et al.*, 1983; Palter *et al.*, 1986). In the yeast *Saccharomyces cerevisiae*, there are at least eight different genes encoding a group of related HSP70-like proteins; the individual genes are regulated somewhat differently, but their products can partially substitute for each other (Craig *et al.*, 1985). Because of this, most of them are not essential genes, but at least some HSP70-like protein seems to be required for growth at any temperature (Craig and Jacobson, 1984, 1985). Mammals have at least three major HSP70-like proteins: one that is strictly heat inducible, an HSP70 cognate (HSP70) that is abundant in all growing cells, and a protein called grp78, which is localized in the endoplasmic reticulum and is particularly abundant in secretory cells (Welch *et al.*, 1983; Lowe and Moran, 1984; Welch and Feramisco, 1985; Munro and Pelham, 1986). This complement of proteins is probably typical not only of vertebrates, but also of *Drosophila* and yeast cells. Indeed, one of the *Drosophila* cognates cross-reacts with antibodies specific for grp78 (M. Lewis, S. Munro, and H. R. B. Pelham, unpublished observations), and a yeast gene encoding a very similar protein has been identified (L. Moran, personal communication). On the other hand, *Escherichia coli* appears to have only a single HSP70-like protein, the product of the *dnaK* gene (Bardwell and Craig, 1984). However, this protein is abundant in cells grown at all temperatures and is induced only moderately by heat shock.

HSP70 is of central importance to the heat-shock response because it appears to be responsible for regulating the level of HSPs in the cell, at least in yeast and *E. coli*. For example, overproduction of dnaK protein from a multicopy plasmid attenuates the heat-shock response in *E. coli*, affecting not only its own synthesis, but also that of other HSPs (Tilly *et al.*, 1983). Similarly, disruption of the genes which encode the most abundant HSP70-like protein in yeast leads to over-expression of other, normally minor, members of the family, so that the total HSP70 complement of the cell is virtually unchanged (Craig and Jacobsen, 1984; Chappell *et al.*, 1986). At the same time, HSP90 is expressed at a higher level than normal. Lindquist and co-workers have presented indirect evidence for a similar kind of regulation in heat-shocked *Drosophila* cells: manipulation of the amount of heat-shock transcription and functional HSP synthesis with actinomycin D, canavanine, and cycloheximide led to the conclusion that the response is feedback regulated, and that the protein whose abundance most closely correlates with inhibition of HSP synthesis is HSP70 (DiDomenico *et al.*, 1982). It was originally suggested that HSP70 might directly interact with heat-shock promoters and reduce their activity. A simpler interpretation is that HSP70 is the protein most directly responsible for reducing the intracellular signal that activates HSTF.

What then does HSP70 do? It has been known for a long time that treatments that induce HSPs protect cells from the effects of a subsequent heat shock (Ashburner and Bonner, 1979). A general hypothesis predicted that macromolecular complexes, which are particularly prone to heat-shock damage, e.g., biosynthetic ribonucleoprotein complexes, can be prevented from unfolding and denaturation by association with HSPs (Schlesinger *et al.,* 1982). The properties of HSP70 are consistent with this. Heat shock causes precipitation of numerous nuclear proteins, damages the structure of partially assembled ribosomes, and completely blocks nucleolar function (Simard and Bernard, 1967; Nover, 1984; Pelham, 1984; Evan and Hancock, 1985). HSP70 binds first to the nuclear matrix (which contains the precipitated proteins) and then concentrates in nucleoli, where it remains associated with the preribosomes for several hours after the heat shock (Sinibaldi and Morris, 1981; Pelham, 1984; Welch and Feramisco, 1984; Welch and Suhan, 1986). The recovery of nucleoli from heat shock is at least partially aided by HSP70, because in COS cells transfected with plasmids that overproduce this protein, ribosome assembly and export resume much earlier than in untransfected cells. This recovery requires neither RNA nor protein synthesis and appears to involve the repair of preëxisting structures (Pelham, 1984).

A clue to the mechanism of HSP70 action comes from the observation that all HSP70-like proteins bind ATP and at least some of them can in certain circumstances hydrolyze it (Zylicz *et al.,* 1983; Welch and Feramisco, 1985; Chappell *et al.,* 1986). The tight binding of HSP70 to heat-shocked nuclei and nucleoli can be reversed *in vitro* by the addition of ATP but not nonhydrolyzable analogs (Lewis and Pelham, 1985). This suggested a model in which HSP70 first binds to heat-damaged proteins and then uses the energy of ATP hydrolysis to release itself, perhaps undergoing a conformational change in the process. By covering exposed hydrophobic surfaces, HSP70 could reduce the formation of aggregates and promote their dissociation. Perhaps it also uses ATP-derived energy to distort the proteins to which it binds, and thus to break their interactions with others. Upon release of HSP70, the denatured protein would have a chance to refold or reassemble into its usual state or, failing that, be degraded.

Other reactions involving members of the HSP70 family reveal a common underlying theme. It has been shown that mammalian HSC70 (and possibly HSP70) has the property, *in vitro,* of uncoating endocytic clathrin-coated vesicles (Ungewickell, 1985; Chappell *et al.,* 1986; Rothman and Schmid, 1986). This reaction involves the disruption of interactions between the clathrin trimers which make up the cagelike structure of the coat and is dependent on ATP hydrolysis. The *E. coli* dnaK protein, which has both ATPase and autokinase activities, has the ability to disrupt tight interactions between the λ P-protein and the host enzyme helicase (Dodson *et al.*, 1986). Finally, mammalian grp78 binds to newly synthesized immunoglobulin heavy chains prior to binding of light chains and can be released from them *in vitro* by the addition of ATP (Bole *et al.*, 1986; Munro and Pelham, 1986). It is likely that grp78 helps to prevent aggregation of heavy chains and thus aids immunoglobulin assembly.

Although the above examples are specific ones, our model would predict that HSP70-like proteins show a general preference for denatured or abnormal proteins, and there is some evidence for this. For example, grp78 binds to certain mutants of the influenza hemagglutinin molecule that fail to assemble correctly into trimers, and to a mutant of SV40 T antigen that has been fused to a leader sequence and is thus delivered to the endoplasmic reticulum (ER), where it is glycosylated and presumably fails to fold correctly (Sharma *et al.*, 1985; Gething *et al.*, 1986). In cells transformed with a gene encoding the p53 tumor antigen, the overproduced p53 binds to HSC70 (and HSP70 in heat-shocked cells) (Pinhasi-Kimhi *et al.*, 1986). In most cells, p53 is a good substrate for the ubiquitin-linked protein degradation system, having a half-life of less than an hour, and by this criterion it is an "abnormal" protein (Rotter and Wolf, 1985).

In unstressed cells, the proteins most likely to be unfolded are newly synthesized monomers or nascent polypeptides that have not yet assembled into their final form. Such proteins are found both in the cytoplasm and in the ER, and it is possible that HSC70 and grp78, respectively, are responsible for preventing their aggregation and for catalyzing their assembly. This is consistent with the much higher levels of HSC70 in rapidly growing cells than in resting cells (M. Lewis and P. Sorger, unpublished observations). HSC70 might also be involved in the rapid assembly of cellular structures from stored material that occurs during *Drosophila* embryogenesis, thus accounting for the very high abundance of this protein in embryos (Palter *et al.*, 1986). It appears that early embryos are generally not capable of inducing heat-shock gene transcription during heat shock, maybe due to insufficient levels of HSTF (reviewed in Heikkila *et al.*, 1985); thus, one of the functions of HSC70 in these cells may be to substitute for HSP70 during a potential heat shock.

Various other putative roles have been suggested for HSP70. In particular, there have been claims that it is a RNA-binding protein (Kloetzel and Bautz, 1983), and it was suggested that the protein that binds to the poly(A) tails of mRNA is a member of the HSP70 family (Schoenfelder *et al.,* 1985). However, the sequence of the yeast poly(A)-binding protein shows no homology to HSP70 (Sachs *et al.,* 1986), and the evidence that HSP70 associates with RNA remains largely circumstantial. Purified HSP70 shows little affinity for poly(A), rRNA, or heparin (M. Lewis and H. R. B. Pelham, unpublished observations).

B. Other Heat-Shock Proteins

Less is known about the properties of the other heat-shock proteins. The mammalian equivalent of *Drosophila* HSP83 is HSP90, and like HSP83 it is abundant in normal growing cells. This protein binds to steroid hormone receptors that are not occupied by hormone, and to several tyrosine protein kinases between the time of their synthesis and their association with the plasma membrane (Catelli *et al.,* 1985; Sanchez *et al.,* 1985; Schuh *et al.,* 1985, and references therein). It has been suggested that the function of these interactions is to keep the target proteins in an inactive state, or simply to keep them soluble. The latter function would be related to that proposed for HSP70, and it is interesting that, like HSC70, HSP83/HSP90 is abundant in mouse and *Drosophila* embryos and has a relative in the endoplasmic reticulum (grp94) (Bensaude *et al.,* 1983; Palter *et al.,* 1986; Sorger and Pelham, 1987).

The small HSPs are less well conserved than HSP70 or HSP90. They show structural homology to the α-crystallins (Ingolia and Craig, 1982b) and exist in particles in the cytoplasm similar or identical to prosomes (Arrigo *et al.,* 1985). Human spermatocytes have been found to contain high levels of mRNA coding for a HSP27-like protein (D. Wolgemuth, personal communication). The single small yeast HSP (HSP26) is an abundant spore protein (Kurtz *et al.,* 1986); however, deletion of the HSP26 gene does not result in thermosensitivity or any other recognizable phenotype of either the spores or cells (Petko and Lindquist, 1986).

It has been proposed that prosomes are involved in tRNA processing, transport, or storage (Castano *et al.,* 1986). However, an alternative role is suggested by the observation that a latent protease activity purified from rat liver cytosol has a structure and protein composition similar to that of prosomes (Kanaka *et al.,* 1986a,b). This would support the hypothesis of Hightower (1980) that the heat-shock response is induced by the presence of aberrantly folded proteins and that the heat-shock proteins are responsible for degrading such proteins. Even more striking is the finding that ubiquitin, which is known to be involved in such degradations, is a heat-shock protein in yeast and chick cells, although apparently not in *Drosophila* (Bond and Schlesinger, 1985; Finley and Varshavsky, 1985). The possible importance of ubiquitin in the heat-shock response is discussed in Section VII.

VII. The Connection between Gene Activation and Function: A Hypothesis

A fundamental problem in cell and developmental biology is how a cell interprets environmental stimuli or information and how it translates this into gene activation. The heat-shock response can be viewed as a model system. Although the effects of heat shock on cells on the one hand and heat-shock gene activation on the other hand have been studied extensively, the mechanistic connection between the two processes is not yet understood. The specific question is how HSTF is activated as a consequence of heat shock. A hypothetical answer was proposed previously (Munro and Pelham, 1985) and we will outline it here.

Several lines of evidence suggest that the production of abnormal proteins, either by heat shock and other stressors (e.g., amino acid analogs), results in overloading of the protein degradation system and switches on the heat-shock genes. First, truncated proteins that are normally very rapidly degraded in COS cells are stabilized when the cells are heat shocked, are treated with the amino acid analog 2-azetidine carboxylic acid, or are exposed to sodium arsenite. All these treatments induce the heat-shock response in these cells, and there is a good correlation between the degree of activation of heat-shock genes and the efficiency of stabilization (Munro and Pelham, 1984; H. R. B. Pelham, unpublished observations). Second, the ts85 mouse cell line that has a defective degradation system synthesizes HSPs at the nonpermissive temperature (39.5°C), even though this is below the normal heat-shock temperature (Ciechanover *et al.*, 1984; Finley *et al.*, 1984). Third, injection of denatured but not native bovine serum albumin into *Xenopus* oocytes induces the transcription of a coinjected heat-shock gene (Ananthan *et al.*, 1986). There are also examples of individual mutant or foreign proteins which induce the heat-shock response in the cells in which they are expressed (Karlik *et al.*, 1984; Goff and Goldberg, 1985; Hiromi and Hotta, 1985).

In eukaryotes, degradation of abnormal proteins involves ubiquitination of their N-termini as a first step, probably in order to tag them for the degradation enzymes (reviewed in Finley and Varshavsky, 1985). It appears to be this step that becomes limiting after heat shock. This can be deduced from the fact that the abnormal proteins that accumulate in COS cells do not have ubiquitin attached to them (S. Munro, unpublished observations). Histones, H2A and H2B, can be ubiquitinated at lysine residues; this modification is reversible and is probably unconnected with protein degradation, but it draws on the same intracellular pool of free ubiquitin. Following heat shock, the level of histone ubiquitination is markedly reduced, indicating that the pool of free ubiquitin is severely depleted (Glover, 1982). This has recently been shown directly by the introduction of ^{125}I-labeled ubiquitin into HeLa cells; upon heat shock, the label is chased from the histones and free ubiquitin pool into high-molecular-weight conjugates (Carlson and Rechsteiner, 1985; Parag *et al.*, 1987).

The model proposes that cells contain a sensor protein that, in an analogous fashion to H2A, is normally maintained in a ubiquitinated state, but becomes deubiquitinated when the degradation system is overloaded with substrates (Fig. 3). This sensor protein would be active when not ubiquitinated; it could either be responsible for activating HSTF or, more simply, it could be HSTF itself. This would provide a logical mechanistic link between the effects of heat shock on the cell and the subsequent activation of the heat-shock genes. The prediction is that HSTF itself, or possibly its activating enzyme, will

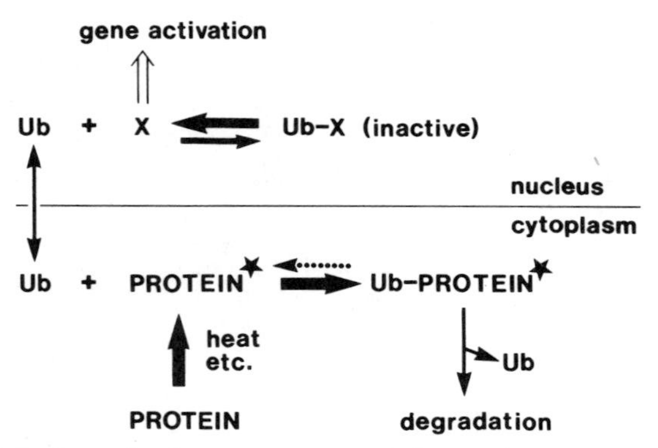

FIG. 3. Model for heat-shock gene activation. Abnormal proteins (PROTEIN*) are produced by stress and are substrates for the ubiquitin-linked degradation system (some ubiquitin–protein conjugation may be reversed without degradation). Ubiquitin (Ub) is also used to inactivate a protein, X, that is responsible for turning on the heat-shock genes. When abnormal protein is abundant, free ubiquitin levels fall, deubiquitination of X by isopeptidases outstrips the rate of its ubiquitination, and active X accumulates. In the simplest version of the model, X is HSTF.

be found in a ubiquitinated state in unstressed cells; at present, it is not known whether this is so.

A final feature of the ubiquitin model is that it can explain feedback regulation (DiDomenico *et al.*, 1982) of the heat-shock response. It has been known for a long time that transcription of heat-shock genes ceases after a while, even if the heat-shock conditions are maintained (Ashburner and Bonner, 1979). According to the model, this will occur when free ubiquitin becomes available, either through synthesis or when liberated from the denatured proteins following their repair or degradation. We suggested in Section VI that HSP70 recognizes denatured or abnormal proteins, that is, the same kinds of substrates as the degradation system. By binding to them and promoting their refolding, or at least protecting them from the ubiquitinating enzymes, HSP70 would reduce the load on the degradation system and thus modulate the heat-shock response. Alternatively, by solubilizing aggregates of denatured proteins HSP70 may make them more accessible to proteases, thus accelerating their degradation and the concomitant release of ubiquitin, which in turn would lead to inactivation of HSTF. Other forms of regulation are also possible and are not excluded by this model, for example, HSP70 might bind directly to HSTF and prevent HSTF dimerization or disrupt HSTF interactions with other molecules in the transcription complex, as discussed in Section IV,D. However, the most straightforward hypothesis seems to be that denatured proteins form both the signal that causes activation of the heat-shock genes and the substrate for the proteins encoded by them, and thus that the heat-shock response involves a simple closed regulatory loop.

Note Added in Proof

Part of the results quoted as personal communication in Sections III,C; V,B; and VI,A have in the meantime been published or are in press. The references are as follows. Wiederrecht, G., Shuey, D. J., Kibbe, W. A., and Parker, C. S. (1987). The *Saccharomyces* and *Drosophila* heat shock transcription factors are identical in size and DNA binding properties. *Cell* **48**, 507–515. Sorger, P. K., and Pelham, H. R. B. (1987). Purification and characterization of a heat shock element binding protein from yeast. *EMBO J.* (in press). Kingston, R. E., Schuetz, T. J., and Larin, Z. (1987). Heat-inducible human factor that binds to a human hsp70 promoter. *Mol. Cell. Biol.* **7**, 1530–1534. Sorger P. K., and Pelham, H. R. B. (1987). Cloning and expression of a gene encoding hsc73, the major hsp70-like protein in unstressed rat cells. *EMBO J.* **6**, 993–998.

Acknowledgments

We would like to thank Ron Blackman, Leon Browder, Alan Coleman, Robert Kingston, Michael Lewis, Susan Lindquist, Larry Moran, Bill Morgan, Richard Morimoto, Sean Munro, Carl Parker, Guy Riddihough, Bill Segraves, Peter Sorger, Debra Wolgemuth, Barbara Wu, Carl Wu, and Keith Yamamoto for discussions, personal communication of unpublished results, and preprints.

References

Amin, J., Mestril, R., Lawson, R., Klapper, H., and Voellmy, R. (1985). The heat-shock consensus sequence is not sufficient for hsp70 gene expression in *Drosophila melanogaster*. *Mol. Cell. Biol.* **5,** 197–203.

Ananthan, J., Goldberg, A. L., and Voellmy, R. (1986). Abnormal proteins serve as eucaryotic stress signals and trigger the activation of heat shock genes. *Science* **232,** 522–524.

Arrigo, A.-P., Darlix, J.-L, Khandjian, E. W., Simon, M., and Spahr, P.-F. (1985). Characterisation of the prosome from *Drosophila* and its similarity to the cytoplasmic structures formed by the low molecular weight heat-shock proteins. *EMBO J.* **4,** 399–406.

Ashburner, M., and Bonner, J. J. (1979). The induction of gene activity in *Drosophila* by heat shock. *Cell* **17,** 241–254.

Ayme, A., Southgate, R., and Tissieres, A. (1985). Nucleotide sequences responsible for the thermal inducibility of the *Drosophila* small heat-shock protein genes in monkey COS cells. *J. Mol. Biol.* **182,** 469–475.

Bardwell, J. C. A., and Craig, E. A. (1984). Major heat-shock gene of *Drosophila* and the *Escherichia coli* heat-inducible dnaK gene are homologous. *Proc. Natl. Acad. Sci. U.S.A.* **81,** 848–852.

Bensaude, O., Babinet, C., Morange, M., and Jacob, F. (1983). Heat-shock proteins, first major products of zygotic gene activity in mouse embryo. *Nature (London)* **305,** 331–333.

Bienz, M. (1984a). *Xenopus* hsp70 genes are constitutively expressed in injected oocytes. *EMBO J.* **3,** 2477–2483.

Bienz, M. (1984b). Developmental control of the heat-shock response in *Xenopus*. *Proc. Natl. Acad. Sci. U.S.A.* **81,** 3138–3142.

Bienz, M. (1985). Transient and developmental activation of heat-shock genes. *Trends Biochem. Sci.* **10,** 157–161.

Bienz, M. (1986). A CCAAT-box confers cell-type specific regulation on the *Xenopus* hsp70 gene in oocytes. *Cell* **46,** 1037–1042.

Bienz, M., and Gurdon, J. B. (1982). The heat-shock response in *Xenopus* oocytes is controlled at the translational level. *Cell* **29,** 811–819.

Bienz, M., and Pelham, H. R. B. (1982). Expression of a *Drosophila* heat-shock protein in *Xenopus* oocytes: Conserved and divergent regulatory signals. *EMBO J.* **1,** 1583–1588.

Bienz, M., and Pelham, H. R. B. (1986). Heat-shock regulatory elements function as an inducible enhancer in the *Xenopus* hsp70 gene and when linked to a heterologous promoter. *Cell* **45,** 753–760.

Bole, D. G., Hendershot, L. M., and Kearney, J. F. (1986). Post-translational association of immunoglobulin heavy chain binding protein with nascent heavy chains in nonsecreting and secreting hybridomas. *J. Cell Biol.* **102,** 1558–1566.

Bond, U., and Schlesinger, M. J. (1985). Ubiquitin is a heat-shock protein in chicken embryo fibroblasts. *Mol. Cell. Biol.* **5,** 949–956.

Browder, L. W., Heikkila, J. J., Wilkes, J., Wang, T., Pollock, M., Krone, P., Ovsenek, M., and Kloc, M. (1987). Decay of the oocyte-type heat-shock response of *Xenopus laevis*. *Dev. Biol.* (submitted).

Brown, D. D. (1984). The role of stable complexes that repress and activate eucaryotic genes. *Cell* **37,** 359–365.

Brown, D. D., and Schlissel, M. S. (1985). A positive transcription factor controls the differential expression of two 5S RNA genes. *Cell* **42,** 759–767.

Carlson, N., and Rechsteiner, M. (1985). Ubiquitin metabolism following heat shock. *J. Cell Biol.* **101,** 443a (Abstr.).

Cartwright, I. L., and Elgin, S. C. R. (1986). Nucleosomal instability and induction of new upstream protein–DNA associations accompany activation of four small heat-shock protein genes in *Drosophila melanogaster. Mol. Cell. Biol.* **6,** 779–791.

Castano, J. G., Ornberg, R., Koster, J. G., Tobian, J. A., and Zasloff, M. (1986). Eucaryotic pre-tRNA 5′ processing nuclease: Copurification with a complex cylindrical particle. *Cell* **46,** 377–387.

Catelli, M. G., Binart, N., Jung-Testas, I., Renoir, J. M., Baulieu, E. E., Feramisco, J. R., and Welch, W. J. (1985). The common 90-kd protein component of non-transformed "8S" steroid receptors is a heat-shock protein. *EMBO J.* **4,** 3131–3135.

Chappell, T. G., Welch, W. J., Schlossmann, D. M., Palter, K. B., Schlesinger, M. J., and Rothman, J. E. (1986). Uncoating ATPase is a member of the 70 kilodalton family of stress proteins. *Cell* **45,** 3–13.

Ciechanover, A., Finley, D., and Varshavsky, A. (1984). Ubiquitin dependence of selective protein degradation demonstrated in the mammalian cell cycle mutant ts85. *Cell* **37,** 57–66.

Cohen, R. S., and Meselson, M. (1984). Inducible transcription and puffing in *Drosophila melanogaster* transformed with hsp70-phage lambda hybrid heat-shock genes. *Proc. Natl. Acad. Sci. U.S.A.* **81,** 5509–5513.

Cohen, R. S., and Meselson, M. (1985). Separate regulatory elements for the heat-inducible and ovarian expression of the *Drosophila* hsp26 gene. *Cell* **43,** 737–746.

Corces, V., Pellicer, A., Axel, R., and Meselson, M. (1981). Integration, transcription, and control of a *Drosophila* heat-shock gene in mouse cells. *Proc. Natl. Acad. Sci. U.S.A.* **78,** 7038–7042.

Costlow, N., and Lis, J. T. (1984). High-resolution mapping of DNase I-hypersensitive sites of *Drosophila* heat-shock genes in *Drosophila melanogaster* and *Saccharomyces cerevisiae. Mol. Cell. Biol.* **4,** 1853–1863.

Craig, E. A. (1985). The heat shock response. *CRC Crit. Rev. Biochem.* **18,** 239–280.

Craig, E. A., and Jacobsen, K. (1984). Mutations of the heat-inducible 70 kilodalton genes of yeast confer temperature sensitive growth. *Cell* **38,** 841–849.

Craig, E. A., and Jacobsen, K. (1985). Mutations in cognate genes of *Saccharomyces cerevisiae* hsp70 result in reduced growth rates at low temperatures. *Mol. Cell. Biol.* **5,** 3517–3524.

Craig, E. A., Ingolia, T. D., and Manseau, L. J. (1983). Expression of *Drosophila* heat-shock cognate genes during heat shock and development. *Dev. Biol.* **99,** 418–426.

Craig, E. A., Slater, M. R., Boorstein, W. R., and Palter, K. (1985). Expression of the *S. cerevisiae* hsp70 multigene family. *In* "Sequence Specificity in Transcription and Translation" (R. Calendar and L. Gold, eds.), pp. 659–667. Liss, New York.

Czarnecka, E., Gurley, W. B., Nagao, R. T., Mosquera, L. A., and Key, J. L. (1985). DNA sequence and transcript mapping of a soybean gene encoding a small heat-shock protein. *Proc. Natl. Acad. Sci. U.S.A.* **82,** 3726–3730.

DiDomenico, B. J., Bugaisky, G. E., and Lindquist, S. (1982). The heat-shock response is self-regulated at both the transcriptional and posttranscriptional levels. *Cell* **31,** 593–603.

Dierks, P., van Ooyen, A., Cochran, M. D., Dobkin, C., Reiser, J., and Weissmann, C.

(1983). Three regions upstream from the cap site are required for efficient and accurate transcription of the rabbit beta-globin gene in mouse 3T6 cells. *Cell* **32,** 695–706.

Dodson, M., Echols, H., Wickner, S., Alfano, C., Mensa-Wilmot, K., Gomes, B., LeBowitz, J., Roberts, J. D., and McMacken, R. (1986). Specialized nucleoprotein structures at the origin of replication of bacteriophage lambda: Localized unwinding of duplex DNA by a six-protein reaction. *Proc. Natl. Acad. Sci. U.S.A.* **83,** 7638–7642.

Dudler, R., and Travers, A. A. (1984). Upstream elements necessary for optimal function of the hsp70 promoter in transformed flies. *Cell* **38,** 391–398.

Dunn, T. M., Hahn, S., Ogden, S., and Schleif, R. F. (1984). An operator at −280 base pairs that is required for repression of araBD operon promoter: Addition of DNA helical turns between the operator and promoter cyclically hinders repression. *Proc. Natl. Acad. Sci. U.S.A.* **81,** 5017–5020.

Dynan, W. S., and Tjian, R. (1985). Control of eucaryotic messenger RNA synthesis by sequence-specific DNA-binding proteins. *Nature (London)* **316,** 774–778.

Evan, G. I., and Hancock, D. C. (1985). Studies on the interaction of the human c-myc protein with cell nuclei: p62-c-myc as a member of a discrete subset of nuclear proteins. *Cell* **43,** 253–261.

Finley, D., and Varshavsky, A. (1985). The ubiquitin system: Functions and mechanisms. *Trends Biochem. Sci.* **10,** 343–347.

Finley, D., Cierchanover, A., and Varshavsky, A. (1984). Thermolability of ubiquitin-activating enzyme from the mammalian cell cycle mutant ts85. *Cell* **37,** 43–55.

Garabedian, M. J., Shepherd, B. M., and Wensink, P. C. (1986). A tissue-specific transcription enhancer from the *Drosophila* yolk protein 1 gene. *Cell* **45,** 859–867.

Gething, M.-J., McCammon, K., and Sambrook, J. (1986). Expression of wild-type and mutant forms of influenza hemagglutinin: The role of folding in intracellular transport. *Cell* **46,** 939–950.

Glaser, R. L., Wolfner, M. F., and Lis, J. T. (1986). Spatial and temporal pattern of hsp26 expression during normal development. *EMBO J.* **5,** 747–754.

Glover, C. V. C. (1982). Heat-shock effects on protein phosphorylation in *Drosophila*. *In* "Heat Shock: From Bacteria to Man" (M. J. Schlesinger, M. Ashburner, and A. Tissieres, eds.), pp. 227–234. Cold Spring Harbor Laboratory, Cold Spring Harbor, New York.

Goff, S. A., and Goldberg, A. L. (1985). Production of abnormal proteins in *E. coli* stimulates transcription of *lon* and other heat-shock genes. *Cell* **41,** 587–595.

Goodbourn, S., Burstein, H., and Maniatis, T. (1986). The human beta-interferon gene enhancer is under negative control. *Cell* **45,** 601–610.

Griffith, J., Hochschild, A., and Ptashne, M. (1986). DNA loops induced by cooperative binding of lambda repressor. *Nature (London)* **322,** 750–752.

Gurley, W. B., Czarnecka, E., Nagao, R. T., and Key, J. L. (1986). Upstream sequences required for efficient expression of a soybean heat-shock gene. *Mol. Cell. Biol.* **6,** 559–565.

Hackett, R. W., and Lis, J. T. (1983). Localization of the hsp83 transcript within a 3292 nucleotide sequence from the 63B heat-shock locus of *Drosophila melanogaster*. *Nucleic Acids Res.* **11,** 7011–7030.

Han, S., Udvardy, A., and Schedl, P. (1985). *Neurospora crassa* and S1 nuclease cleavage in hsp83 gene chromatin. *J. Mol. Biol.* **184,** 657–665.

Handler, A. M. (1982). Ecdysteroid titers during pupal and adult development in *Drosophila melanogaster*. *Dev. Biol.* **93,** 73–82.

Heikkila, J. J., Miller, J. G. O., Schultz, G. A., Kloc, M., and Browder, L. W. (1985). Heat-shock gene expression during early animal development. *In* "Changes in Eucaryotic Gene Expression in Response to Environmental Stress" (B. G. Atkinson and D. B. Walden, eds.), pp. 135–158. (Cell Biology Monograph). Academic Press, New York.

Hickey, E., Brandon, S. E., Potter, R., Stein, G., Stein, J., and Weber, L. A. (1986). Sequence and organization of genes encoding the human 27 kDa heat-shock protein. *Nucleic Acids Res.* **14**, 4127–4145.

Hightower, L. E. (1980). Cultured animal cells exposed to amino acid analogues or puromycin rapidly synthesize several polypeptides. *J. Cell Physiol.* **102**, 407–427.

Hiromi, Y., and Hotta, Y. (1985). Actin gene mutation in *Drosophila:* Heat-shock activation in indirect flight muscles. *EMBO J.* **4**, 1681–1687.

Hochschild, A. H., and Ptashne, M. (1986). Cooperative binding of lambda repressors to sites separated by integral turns of the DNA helix. *Cell* **44**, 681–687.

Hoffmann, E., and Corces, V. (1986). Sequences involved in temperature and ecdysterone-induced transcription are located in separate regions of a *Drosophila melanogaster* heat-shock gene. *Mol. Cell. Biol.* **6**, 663–673.

Holmgren, R., Corces, V., Morimoto, R., Blackman, R., and Meselson, M. (1981). Sequence homologies in the 5′ regions of four *Drosophila* heat-shock genes. *Proc. Natl. Acad. Sci. U.S.A.* **78**, 3775–3778.

Hunt, C., and Morimoto, R. I. (1985). Conserved features of eucaryotic hsp70 revealed by comparison with the nucleotide sequence of human hsp70. *Proc. Natl. Acad. Sci. U.S.A.* **82**, 6455–6459.

Ingolia, T. D., and Craig, E. A. (1982a). *Drosophila* gene related to the major heat shock-induced gene is transcribed at normal temperatures and is not induced by heat shock. *Proc. Natl. Acad. Sci. U.S.A.* **79**, 525–529.

Ingolia, T. D., and Craig, E. A. (1982b). Four small *Drosophila* heat-shock proteins are related to each other and to mammalian alpha-crystallin. *Proc. Natl. Acad. Sci. U.S.A.* **79**, 2360–2364.

Ireland, R. C., and Berger, E. (1982). Synthesis of low molecular weight heat shock peptides stimulated by ecdysterone in a cultured *Drosophila* cell line. *Proc. Natl. Acad. Sci. U.S.A.* **79**, 855–859.

Ireland, R. C., Berger, E., Sirotkin, K., Yund, M. A., Osterbur, D., and Fristrom, J. (1982). Ecdysterone induces the transcription of four heat-shock genes in *Drosophila* S3 cells and imaginal discs. *Dev. Biol.* **93**, 498–507.

Jones, K. A., Yamamoto, K. R., and Tjian, R. (1985). Two distinct transcription factors bind to the HSV thymidine kinase promoter in vitro. *Cell* **42**, 559–572.

Karch, F., Török, I., and Tissieres, A. (1981). Extensive regions of homology in front of the two hsp70 heat-shock variant genes in *Drosophila melanogaster. J. Mol. Biol.* **148**, 219–230.

Karlik, C. C., Coutu, M. D., and Fyrberg, E. A. (1984). A nonsense mutation within the Act885 actin gene disrupts myofibril formation in *Drosophila* indirect flight muscles. *Cell* **38**, 711–719.

King, M. L., and Davis, R. (1987). Do *Xenopus* oocytes have a heat-shock response? *Dev. Biol.* **119**, 532–539.

Kingston, R. E., Baldin, A. S., Jr., and Sharp, P. A. (1984). Regulation of heat-shock protein 70 gene expression by c-myc. *Nature (London)* **312**, 280–282.

Kingston, R. E., Cowie, A., Morimoto, R. I., and Gwinn, K. A. (1986). Binding of polyomavirus large T antigen to the human hsp70 promoter is not required for trans activation. *Mol. Cell. Biol.* **6**, 3180–3190.

Klemenz, R., and Gehring, W. J. (1986). Sequence requirement for expression of the

Drosophila melanogaster heat-shock protein hsp22 gene during heat shock and normal development. *Mol. Cell. Biol.* **6**, 2011–2019.

Klemenz, R., Hultmark, D., and Gehring, W. J. (1985). Selective translation of heat-shock mRNA in *Drosophila melanogaster* depends on sequence information in the leader. *EMBO J.* **4**, 2053–2060.

Kloetzel, P. M., and Bautz, E. K. F. (1983). Heat-shock proteins are associated with hnRNA in *Drosophila melanogaster* tissue culture cells. *EMBO J.* **2**, 705–710.

Krawczyk, Z., Wisniewski, J., and Biesiada, E. (1987). A hsp70-related gene is constitutively expressed highly in testis of rat and mouse. *Mol. Biol. Rep.* (in press).

Kurtz, S., Rossi, J., Petko, L., and Lindquist, S. (1986). An ancient developmental induction in *Saccharomyces* sporulation and *Drosophila* oogenesis. *Science* **231**, 1154–1157.

Lewis, M. J., and Pelham, H. R. B. (1985). Involvement of ATP in the nuclear and nucleolar functions of the 70 kd heat-shock protein. *EMBO J.* **4**, 3137–3143.

Lin, A. Y., Chang, S. C., and Lee, A. S. (1986). A calcium ionophore-inducible cellular promoter is highly active and has enhancer-like properties. *Mol. Cell. Biol.* **6**, 1235–1243.

Lindquist, S. (1986). The heat shock response. *Annu. Rev. Biochem.* **55**, 1151–1191.

Lowe, D. G., and Moran, L. A. (1984). Proteins related to the mouse L-cell major heat-shock protein are synthesized in the absense of heat-shock gene expression. *Proc. Natl. Acad. Sci. U.S.A.* **81**, 2317–2321.

Mason, J. O., Williams, G. T., and Neuberger, M. S. (1985). Transcription cell-type specificity is conferred by an immunoglobulin VH gene promoter that includes a functional consensus sequence. *Cell* **41**, 479–487.

Mason, P. J., Hall, L. M. C., and Gausz, J. (1984). The expression of heat-shock genes during normal development in *Drosophila melanogaster. Mol. Gen. Genet.* **194**, 73–78.

Mattaj, I. W., Lienhard, S., Jiricny, J., and De Robertis, E. M. (1985). An enhancer-like sequence within the *Xenopus* U2 gene promoter facilitates the formation of stable transcription complexes. *Nature* (*London*) **316**, 163–164.

Mestril, R., Rungger, D., Schiller, P., and Voellmy, R. (1985). Identification of a sequence element in the promoter of the *Drosophila melanogaster* hsp23 gene that is required for its heat activation. *EMBO J.* **4**, 2971–2976.

Mestril, R., Schiller, P., Amin, J., Klapper, H., Ananthan, J., and Voellmy, R. (1986). Heat-shock and ecdysterone activation of the *Drosophila* hsp23 gene: A sequence element implied in developmental regulation. *EMBO J.* **5**, 1667–1673.

Miesfeld, R., Rusconi, S., Godowski, P. J., Maler, B. A., Okret, S., Wikstroem, A.-C., Gustafsson, J. A., and Yamamoto, K. R. (1986). Genetic complementation of a glucocorticoid receptor deficiency by expression of cloned receptor cDNA. *Cell* **46**, 389–399.

Milarski, K. L., and Morimoto, R. I. (1986). Expression of human hsp70 during the synthetic phase of the cell cycle. *Proc. Natl. Acad. Sci. U.S.A.* **83**, 9517–9521.

Mirault, M.-E., Southgate, R., and Delwart, E. (1982). Regulation of heat-shock genes: A DNA sequence upstream of *Drosophila* hsp70 genes is essential for their induction in monkey cells. *EMBO J.* **1**, 1279–1285.

Morgan, W. D., Williams, G. T., Morimoto, R. I., Greene, J., Kingston, R. E., and Tjian, R. (1987). Two transcriptional activators, CCAAT-binding transcription factor and heat shock transcription factor, interact with a human hsp70 gene promoter. *Mol. Cell. Biol.* (submitted).

Morganelli, C. M., Berger, E. M., and Pelham, H. R. B. (1985). Transcription of *Drosophila* small hsp-tk hybrid genes is induced by heat shock and by ecdysterone in transfected *Drosophila* cells. *Proc. Natl. Acad. Sci. U.S.A.* **82**, 5865–5869.

Morimoto, R. I., Hunt, C., Huang, S.-Y., Berg, L., and Banerji, S. S. (1986). Organization, nucleotide sequence and transcription of the chicken hsp70 gene. *J. Biol. Chem.* **261**, 12692–12699.

Munro, S., and Pelham, H. R. B. (1984). Use of peptide tagging to detect proteins expressed from cloned genes: Deletion mapping functional domains of *Drosophila* hsp70. *EMBO J.* **3**, 3087–3093.

Munro, S., and Pelham, H. R. B. (1985). What turns on heat-shock genes? *Nature (London)* **317**, 477–478.

Munro, S., and Pelham, H. R. B. (1986). An hsp70-like protein in the endoplasmic reticulum: Identity with the 78kd glucose-regulated protein and immunoglobulin heavy chain binding protein. *Cell* **46**, 291–300.

Nagao, R. T., Czarnecka, E., Gurley, W. B., Schoeffl, F., and Key, J. L. (1985). Genes for low-molecular-weight heat-shock proteins of soybeans: Sequence analysis of a multigene family. *Mol. Cell. Biol.* **5**, 3417–3428.

Nover, L., ed. (1984). "Heat Shock Response of Eucaryotic Cells." Springer-Verlag, Berlin and New York.

Palter, K. B., Watanabe, M., Stinson, L., Mahowald, A. P., and Craig, E. A. (1986). Expression and localization of *Drosophila melanogaster* hsp70 cognate proteins. *Mol. Cell. Biol.* **6**, 1187–1203.

Parag, H. A., Raboy, B., and Kulka, R. G. (1987). Effects of heat shock on protein degradation in mammalian cells: Involvement of the ubiquitin system. *EMBO J.* **6**, 55–61.

Parker, C. S., and Topol, J. (1984a). A *Drosophila* RNA polymerase II transcription factor binds to the regulatory site of an hsp70 gene. *Cell* **37**, 273–283.

Parker, C. S., and Topol, J. (1984b). A *Drosophila* RNA polymerase II transcription factor contains a promoter region specific DNA binding activity. *Cell* **36**, 357–369.

Pauli, D., Spierer, A., and Tissieres, A. (1986). Several hundred base pairs upstream of *Drosophila* hsp23 and 26 genes are required for their heat induction in transformed flies. *EMBO J.* **5**, 755–761.

Pelham, H. R. B. (1982). A regulatory upstream promoter element in the *Drosophila* hsp70 heat-shock gene. *Cell* **30**, 517–528.

Pelham, H. R. B. (1984). Hsp70 accelerates the recovery of nucleolar morphology after heat shock. *EMBO J.* **3**, 3095–3100.

Pelham, H. R. B. (1985). Activation of heat-shock genes in eucaryotes. *Trends Genet.* **1**, 31–35.

Pelham, H. R. B. (1986). Speculations on the functions of the major heat-shock and glucose-regulated proteins. *Cell* **46**, 959–961.

Pelham, H. R. B., and Bienz, M. (1982). A synthetic heat-shock promoter element confers heat-inducibility on the herpes simplex virus thymidine kinase gene. *EMBO J.* **1**, 1473–1477.

Pelham, H. R. B., and Lewis, M. J. (1983). Assay of natural and synthetic heat-shock promoters in monkey COS cells: Requirements for regulation. *In* "Gene Expression" (D. H. Hamer and M. J. Rosenberg, eds.), pp. 75–85. Liss, New York.

Petko, L., and Lindquist, S. (1986). Hsp26 is not required for growth at high temperatures, nor for thermotolerance, spore development, or germination. *Cell* **45**, 885–894.

Pinhasi-Kimhi, O., Michalovitz, D., Ben-Zeev, A., and Oren, M. (1986). Specific interaction between the p53 cellular tumour antigen and major heat-shock proteins. *Nature (London)* **320**, 182–185.

Ptashne, M. (1986a). "A Genetic Switch." Cell & Blackwell Scientific Press, Cambridge & Palo Alto, California.

Ptashne, M. (1986b). Gene regulation by proteins acting nearby and at a distance. *Nature (London)* **322**, 697–701.

Riddihough, G., and Pelham, H. R. B. (1986). Activation of the *Drosophila* hsp27 promoter by heat shock and by ecdysone involves independent and remote regulatory sequences. *EMBO J.* **5**, 1653–1658.

Rochester, D. E., Winer, J. A., and Shah, D. M. (1986). The structure and expression of maize genes encoding the major heat-shock protein, hsp70. *EMBO J.* **5**, 451–458.

Rothman, J. E., and Schmid, S. L. (1986). Enzymatic recycling of clathrin from coated vesicles. *Cell* **46**, 5–9.

Rotter, V., and Wolf, D. (1985). Biological and molecular analysis of p53 cellular-encoded tumour antigen. *Adv. Cancer Res.* **43**, 113–141.

Rusconi, S., Miesfeld, R., Godowski, P. J., Vanderbilt, J. N., Maler, B. A., and Yamamoto, K. R. (1986). Functional analysis of cloned glucocorticoid receptor sequences. *In* "RNA Polymerase and the Regulation of Transcription" (W. S. Reznikoff, R. R. Burgess, J. E. Dahlberg, C. A. Gross, T. M. Record, and M. P. Wickens, eds.). Elsevier, Amsterdam, in press.

Russnak, R. H., and Candido, E. P. M. (1985). Locus encoding a family of small heat-shock genes in *Caenorhabditis elegans:* Two genes duplicated to form a 3.8-kilobase inverted repeat. *Mol. Cell. Biol.* **5**, 1268–1278.

Sachs, A. B., Bond, M. W., and Kornberg, R. D. (1986). A single gene from yeast for both nuclear and cytoplasmic polyadenylate-binding proteins: Domain structure and expression. *Cell* **45**, 827–835.

Sanchez, E. H., Toft, D. O., Schlesinger, M. J., and Pratt, W. B. (1985). Evidence that the 90-kDa phosphoprotein associated with the untransformed L-cell glucocorticoid receptor is a murine heat-shock protein. *J. Biol. Chem.* **260**, 12398–12401.

Schlesinger, M. J., Aliperti, G., and Kelley, P. M. (1982). The response of cells to heat shock. *Trends Biochem. Sci.* **7**, 222–225.

Schoeffl, F., Raschke, E., and Nagao, R. T. (1984). The DNA sequence analysis of soybean heat-shock genes and identification of possible regulatory promoter elements. *EMBO J.* **3**, 2491–2497.

Schoenfelder, M., Horsch, A., and Schmid, H. P. (1985). Heat shock increases the synthesis of poly(A)-binding protein in HeLa cells. *Proc. Natl. Acad. Sci. U.S.A.* **82**, 6884–6888.

Schuh, S., Yonemoto, W., Brugge, J., Bauer, V. J., Riehl, R. M., Sullivan, W. F., and Toft, D. O. (1985). A 90,000-dalton binding protein common to both steroid receptors and the Rous sarcoma virus transforming protein, pp60 v-sarc. *J. Biol. Chem.* **260**, 14292–14296.

Serfling, E., Jasin, M., and Schaffner, W. (1985). Enhancers and eucaryotic gene transcription. *Trends Genet.* **1**, 224–230.

Sharma, S., Rodgers, L., Brandsma, J., Gething, M.-J., and Sambrook, J. (1985). SV40 T antigen and the exocytic pathway. *EMBO J.* **4**, 1479–1489.

Shuey, D. J., and Parker, C. S. (1986). Binding of *Drosophila* heat-shock gene transcription factor to the hsp70 promoter. *J. Biol. Chem.* **261**, 7934–7940.

Simard, R., and Bernhard, W. (1967). A heat-sensitive cellular function located in the nucleolus. *J. Cell Biol.* **34**, 61–76.

Simon, J. A., Sutton, C. A., Lobell, R. B., Glaser, R. L., and Lis, J. T. (1985). Determinants of heat shock-induced chromosome puffing. *Cell* **40**, 805–817.

Sinibaldi, R. M., and Morris, P. W. (1981). Putative function of *Drosophila melanogaster* heat shock proteins in the nucleoskeleton. *J. Biol. Chem.* **256**, 10735–10738.

Sirotkin, K., and Davidson, N. (1982). Developmentally regulated transcription from *Drosophila melanogaster* chromosomal site 67B. *Dev. Biol.* **89**, 196–210.

Sorger, P. K., and Pelham, H. R. B. (1987). The glucose-regulated protein grp94 is related to heat-shock protein hsp90. *J. Mol. Biol.* **194**, 341–344.

Southgate, R., Ayme, A., and Voellmy, R. (1983). Nucleotide sequence analysis of the *Drosophila* small heat-shock gene cluster at locus 67B. *J. Mol. Biol.* **165**, 35–57.

Takahashi, K., Vigneron, M., Matthes, H., Wildeman, A., Zenke, M., and Chambon, P. (1985). Stereospecific alignments are required for initiation from the SV40 early promoter. *Nature (London)* **319**, 121–126.

Tanaka, K., Ii, K., Ichihara, A., Waxman, L., and Goldberg, A. L. (1986a). A high molecular weight protease in the cytosol of rat liver. I. Purification, enzymological properties and tissue distribution. *J. Biol. Chem.* **261**, 15197–15203.

Tanaka, K., Yoshimura, T., Ichihara, A., Kameyama, K., and Takagi, T. (1986b). A high molecular weight protease in the cytosol of rat liver. II. Properties of the purified protease. *J. Biol. Chem.* **261**, 15204–15207.

Tilly, K., McKittrick, N., Zylicz, M., and Georgopoulos, C. (1983). The dnaK protein modulates the heat-shock response in *Escherichia coli*. *Cell* **34**, 641–646.

Topol, J., Ruden, D. M., and Parker, C. S. (1985). Sequences required for *in vitro* transcriptional activation of a *Drosophila* hsp70 gene. *Cell* **42**, 527–537.

Ungewickell, E. (1985). The 70-kd mammalian heat shock proteins are structurally and functionally related to the uncoating protein that releases clathrin triskelia from coated vesicles. *EMBO J.* **4**, 3385–3391.

Vitek, M. P., and Berger, E. M. (1984). Steroid and high-temperature induction of the small heat-shock proteins in *Drosophila*. *J. Mol. Biol.* **178**, 173–189.

Voellmy, R. (1985). The heat shock genes: A family of highly conserved genes with a superbly complex expression pattern. *BioEssays* **1**, 213–217.

Voellmy, R., Ahmed, A., Schiller, P., Bromley, P., and Rungger, D. (1985). Isolation and functional analysis of a human 70,000-dalton heat-shock protein gene segment. *Proc. Natl. Acad. Sci. U.S.A.* **82**, 4949–4953.

Welch, W. J., and Feramisco, J. R. (1984). Nuclear and nucleolar localization of the 72,000-dalton heat shock protein in heat-shocked mammalian cells. *J. Biol. Chem.* **259**, 4501–4513.

Welch, W. J., and Feramisco, J. R. (1985). Rapid purification of mammalian 70,000-dalton stress proteins: Affinity of the proteins for nucleotides. *Mol. Cell. Biol.* **5**, 1229–1237.

Welch, W. J., and Suhan, J. P. (1986). Cellular and biochemical events in mammalian cells during and after recovery from physiological stress. *J. Cell Biol.* **103**, 2035–2052.

Welch, W. J., Garrels, J. I., Thomas, G. P., Lin, J. J. C., and Feramisco, J. R. (1983). Biochemical characterization of the mammalian stress proteins and identification of two stress proteins as glucose and calcium ionophore-regulated proteins. *J. Biol. Chem.* **258**, 7102–7111.

Wu, B. J., and Morimoto, R. I. (1985). Transcription of the human hsp70 gene is induced by serum stimulation. *Proc. Natl. Acad. Sci. U.S.A.* **82**, 6070–6074.

Wu, B. J., Hurst, H. C., Jones, N. C., and Morimoto, R. I. (1986a). The E1A 13S product of adenovirus-5 activates transcription of the cellular human hsp70 gene. *Mol. Cell. Biol.* **6,** 2994–2999.

Wu, B. J., Kingston, R. E., and Morimoto, R. I. (1986b). Human hsp70 promoter contains at least two distinct regulatory domains. *Proc. Natl. Acad. Sci. U.S.A.* **83,** 629–633.

Wu, B. J., Williams, G. T., and Morimoto, R. I. (1987). Detection of three protein binding sites in the serum regulated promoter of the human gene encoding the 70kDa heat shock protein. *Proc. Natl. Acad. Sci. U.S.A.* **84,** 2203–2207.

Wu, C. (1980). The 5′ ends of *Drosophila* heat-shock genes in chromatin are hypersensitive to DNase I. *Nature (London)* **286,** 854–860.

Wu, C. (1984a). Two protein-binding sites in chromatin implicated in the activation of heat-shock genes. *Nature (London)* **309,** 229–234.

Wu, C. (1984b). Activating protein factor binds in vitro to upstream control sequences in heat-shock gene chromatin. *Nature (London)* **311,** 81–84.

Wu, C. (1985). An exonuclease protection assay reveals heat-shock element and TATA-box DNA-binding proteins in crude nuclear extracts. *Nature (London)* **317,** 84–87.

Yoshinaga, S., Dean, N., Han, M., and Berk, A. J. (1986). Adenovirus stimulation of transcription by RNA polymerase III: Evidence for an E1A-dependent increase in transcription factor IIIC concentration. *EMBO J.* **5,** 343–354.

Zakeri, Z. F., and Wolgemuth, D. J. (1987). Developmental stage-specific expression of the hsp70 gene family during differentiation of the mammalian germ line. *Mol. Cell. Biol.* (submitted).

Zimmerman, J. L., Petri, W., and Meselson, M. (1983). Accumulation of a specific subset of *D. melanogaster* heat-shock mRNAs in normal development without heat shock. *Cell* **32,** 1161–1170.

Zylicz, M., LeBowitz, J. H., McMacken, R., and Georgopoulos, C. (1983). The dnaK protein of *Escherichia coli* possesses an ATPase and autophosphorylating activity and is essential in an in vitro DNA replication system. *Proc. Natl. Acad. Sci. U.S.A.* **80,** 6431–6435.

REGULATORY GENE ACTION DURING EUKARYOTIC DEVELOPMENT

Joel M. Chandlee* and John G. Scandalios

Department of Genetics, North Carolina State University, Raleigh, North Carolina 27695

I. Introduction

It is now a well-established fact that genes are differentially expressed both temporally and spatially during eukaryotic development. It is this differential expression of the structural genes of a multicellular organism that leads to the orderly development and differentiation of the various tissues and organs of an individual. A central concern of developmental biologists is to understand the processes by which the cell regulates the differential activity of its structural genes so that development can be better understood.

Regulation of gene activity can occur at several different levels of control (for review see Brown, 1981; Darnell, 1982; O'Malley *et al.*, 1977). These control levels include (1) availability of the gene, (2) transcriptional control, (3) posttranscriptional control, (4) translational control, and (5) posttranslational control. Various mechanisms of regulation have been identified at each of these levels of control. In order to identify the precise mechanisms used to regulate gene expression one can first analyze enzyme (protein) expression, since this accurately reflects underlying gene activity. Once a system is defined at

* Present address: United States Department of Agriculture, Agricultural Research Service, Plant Genetics and Germplasm Institute, Plant Molecular Genetics Laboratory, Beltsville, Maryland 20705.

the level of enzyme expression it is possible to identify mutants or variants of the normal expression pattern of particular genes. Subsequently, experiments can be designed to determine what specific level of gene control is altered in the variant.

In recent years there has been a considerable advance toward understanding the regulatory processes occurring in higher eukaryotes. Instances of genetically controlled variation in the level of enzyme expression have been identified. These are interesting because they allow for the identification of elements in the genome with a strict regulatory function. Through analysis of these elements it has become increasingly apparent that the activity of structural genes of higher eukaryotes can be influenced by both proximally and distally located regulatory loci whose sole apparent function is to regulate the expression of specific structural genes. Thus, there appears to be information in the genome that programs or controls the expression of particular structural genes. Identification and characterization of these regulatory loci have been made possible through the analysis of well-defined (genetically and biochemically) enzyme systems which exhibit genetically controlled variation of structural gene expression at the protein level. This review attempts to consolidate and summarize the accumulated data on those cases of well-characterized regulatory loci of higher eukaryotes. The organisms included in this review are *Mus musculus* (house mouse), *Drosophila melanogaster* (fruit fly), *Zea mays* (corn), and *Salmo gairdneri* (rainbow trout).

Individual regulatory genes can be classified into one of several different groups depending on how they exert their regulatory function. While in a general sense they are all regulatory (i.e., control the expression of a distinct structural gene), they carry out their regulatory action in different ways. For simplicity, we have tentatively adapted the basic classification scheme of Paigen (1979a) with minor modifications for use in this review. Eukaryotic genes can be classified as follows.

1. *Structural genes.* These code for the amino acid sequence of a protein and are the targets for the action of regulatory genes.

2. *Regulatory genes.* (a) *Processing genes* include those regulatory genes which code for the posttranslational processing machinery of the cell. These processes include protein modification (cleavage, glycosylation, etc.), as well as protein turnover. (b) *Systemic genes* include those regulatory genes which affect the rate of synthesis of a particular protein uniformly throughout all the tissues and organs of the organism. (c) *Effector response genes* regulate the response of a

particular protein to inducer molecules which are necessary to elicit gene activity. (d) *Temporal genes* include those regulatory genes which determine the concentration changes that any given protein undergoes during cellular differentiation. By definition they exhibit both temporal and spatial specificity in their action.

Eukaryotic regulatory genes can be located either proximal or distal to the structural gene under their control, and they can exert their effect in cis or trans. They can be inherited in either an additive manner or in a dominance–recessive manner. Each of these characteristics of a given regulatory locus reveals important information about the nature of its regulatory action. A summary of the properties of well-defined eukaryotic regulatory loci is presented in Table 1. In this review we describe in some detail a number of well-characterized eukaryotic regulatory genes. It may be useful to refer to Table 1 while reading the text so that each individual regulatory gene can be understood from a more general perspective. At the end of the review, we attempt to summarize the current status of our understanding of these regulatory genes. Additional discussions on the regulatory gene systems described herein can be found for the mouse (Felder, 1980; Paigen, 1964, 1979a,b, 1980), *Drosophila* (Laurie-Ahlberg, 1985), maize (Scandalios, 1982; Scandalios and Baum, 1982), and rainbow trout (Allendorf *et al.*, 1983b).

II. Regulatory Genes of the Mouse

A. β-Glucuronidase

One of the most extensively studied examples of complex genetic control over the pattern of expression of a particular enzyme during development is the β-glucuronidase (E.C. 3.2.1.31) gene–enzyme system of the mouse. β-Glucuronidase is a tetrameric glycoprotein present in the lysosomes and microsomes [endoplasmic reticulum (ER)] of several different tissues, but primarily the liver and kidney. It is an acid hydrolase and is likely involved in the hydrolysis of β-glucuronide linkages in mucopolysaccharides and various biological glucuronides such as those of steroid hormones. The structural gene for β-glucuronidase, *Gus*, has been mapped to the distal end of chromosome 5. Much of the known regulatory information affecting expression of the gene also maps to the same position; however, unlinked regulatory sites do exist. The gene encodes a polypeptide of

TABLE 1
Eukaryotic Regulatory Genes

Structural gene affected	Regulatory gene	Organism	Location[a]	Action	Type of gene	Inheritance	Mode of action
β-Glucuronidase (*Gus*) E.C. 3.2.1.31	*Gus-t*	Mouse	Proximate	Trans	Temporal (several tissues)	Additive	Synthesis
	Gus-r	Mouse	Proximate	Cis	Regulatory (several tissues)	Additive	Synthesis
	Gus-u	Mouse	Proximate	Cis	Regulatory (systemic)	—	—
	Eg	Mouse	Distant	Trans	Processing (systemic)	Additive	—
	Tfm	Mouse	Distant	Trans	Regulatory (several tissues)	—	—
α-Galactosidase (*Ags*) E.C. 3.2.1.22	*Agt*	Mouse	Distant	Trans	Temporal (liver)	Additive	Possibly synthesis
β-Galactosidase (*Bgl-e*) E.C. 3.2.1.23	*Bgl-s*	Mouse	Proximate	Cis	Regulatory (systemic)	Additive	Synthesis
	Bgl-t	Mouse	Proximate	—	Temporal (liver)	Additive	Synthesis
	(Interacts with *Bgl-t*)	Mouse	Distant	—	Temporal (liver)	Additive	Synthesis
Arylsulfatase B (*As-1*) E.C. 3.1.6.1	*Asr-1*	Mouse	Possibly proximate	—	Temporal (liver)	Additive	—
H-2 antigen	*Int*	Mouse	Possibly proximate	Cis	Temporal (erythrocytes)	—	—
	Rec	Mouse	Distant	—	Temporal (erythrocytes)	—	—
	Tem	Mouse	Distant	Possibly cis	Temporal (erythrocytes)	—	—

Aminolevulinate dehydratase (*Lv*) E.C. 4.2.1.24	—	Mouse	Proximate	—	Regulatory (systemic)	Additive	Synthesis
Catalase (*Cs-1*) E.C. 1.11.1.6	*Ce*	Mouse	Distant	Trans	Processing (liver)	Dom–rec	Degradation
Alcohol dehydrogenase							
(*Adh1*)	*Adh-1-t*	Mouse	—	—	Temporal (liver)	Additive	Synthesis
(*Adh3*) E.C. 1.1.1.1	*Adh-3-t*	Mouse	Proximate	Cis	Temporal (reproductive tissue)	—	—
Malic enzyme (mitochondrial) (*Mod2*) E.C. 1.1.1.40	*Mdr-1*	Mouse	Proximate	Cis	Temporal (brain)	Additive	Synthesis
Glucose phosphate isomerase (*Gpi-1*) E.C. 5.3.1.9	*Org*	Mouse	Proximate	Cis	Temporal (ovum)	Additive	—
Alcohol dehydrogenase (*Adh*) E.C. 1.1.1.1	—	*Drosophila*	Proximate	Cis	Regulatory (systemic)	Additive	—
α-Amylases (*Amy*) E.C. 3.2.1.1	*Map*	*Drosophila*	Distant (within 2 map units)	Trans	Temporal (PMG)	Possibly additive	—
	—	*Drosophila*	Proximate	Cis	Temporal (PMG)	Possibly additive	—
Molybdenum hydrolases:							
1. Xanthine dehydrogenase (*ry*) E.C. 1.1.1.20	*mal*	*Drosophila*	Distant	Trans (implied)	Processing	Dom–rec	—
2. Aldehyde oxidase (*Aldox*) E.C. 1.2.3.1	*cin*	*Drosophila*	Distant	Trans (implied)	Processing	Dom–rec	—

(*continued*)

TABLE 1 (*Continued*)

Structural gene affected	Regulatory gene	Organism	Location[a]	Action	Type of gene	Inheritance	Mode of action
3. Pyridoxal oxidase (*lpo*) E.C. 1.2.3.8 4. Sulfite oxidase E.C. 1.8.3.1	*lxd*	*Drosophila*	Distant	Trans (implied)	Processing	Dom–rec	—
Aldehyde oxidase (*Aldox*) E.C. 1.2.3.1	—	*Drosophila*	Proximate	Cis	Temporal (pupae)	Additive	Possibly synthesis
	—	*Drosophila*	Possibly proximate	Cis	Temporal (paragonia)	Additive	—
Xanthine dehydrogenase (*ry*)	*i1005*	*Drosophila*	Proximate	Cis	Regulatory (systemic)	Additive	Possibly synthesis
E.C. 1.1.1.20	*i409*	*Drosophila*	Proximate	Cis	Regulatory (fat body)	Additive	Possibly synthesis
Glycerol-3-phosphate dehydrogenase	*Gdt-3*	*Drosophila*	Proximate	Cis	Temporal (fat body, abdomen)	Additive	Possibly synthesis
(*Gpdh*)	—	*Drosophila*	Proximate	Cis	Regulatory (systemic)	Additive	Synthesis
E.C. 1.1.1.8	—	*Drosophila*	Distant	Trans	Temporal (thorax)	Dom–rec	Degradation
Esterase (*est-6*) E.C. 3.1.1.1	*m-est*	*Drosophila*	Distant	Trans	Processing	Dom–rec	—
Dopa decarboxylase (Ddc^{+}) E.C. 4.1.1.28	Ddc^{+4}	*Drosophila*	Proximate	—	Temporal (epidermis)	—	—
Esterase (E_1) E.C. 3.1.1.1	—	Maize	Proximate	Cis	Temporal (endosperm)	Possibly additive	—

Catalase (*Cat1*)	*Car2*	Maize	Possibly proximate	—	Temporal (scutellum)	Additive	Synthesis
E.C. 1.11.1.6 (*Cat2*)	*Car1*	Maize	Distant	Trans	Temporal (scutellum)	Additive	Synthesis
Alcohol dehydrogenase (*Adh2*) E.C. 1.1.1.1	*Adr1*	Maize	Distant	Trans	Temporal (scutellum)	Dom–rec	Degradation
	—	Maize	Proximate	Cis	Temporal (reciprocal effect)	—	Synthesis (in root)
	Adh_r	Maize	Distant	Trans	Regulatory (systemic)	Dom–rec (acts on S allele only)	—
UDPglucose : flavonol 3-*O*-glucosyltransferase (UFGT) (*Bz*) E.C. 2.4.1.91	*C*	Maize	Distant	Trans	—	Dom–rec	—
	R	Maize	Distant	Trans	—	Dom–rec	—
	Vp	Maize	Distant	Trans	—	Dom–rec	—
Zein	*o2*	Maize	Distant and proximate	Trans	—	Dom–rec	—
	o6	Maize	Distant	Trans	—	Dom–rec	—
	o7	Maize	Distant and proximate	Trans	—	Dom–rec	—
	fl2	Maize	Distant and proximate	Trans	—	Dom–rec	—
	fl3	Maize	Distant	Trans	—	Dom–rec	—
	Mc	Maize	Distant	Trans	—	Dom–rec	—
Phosphoglucomutase (*Pgm1*) E.C. 5.4.2.2	*Pgm1-t*	Rainbow trout	—	Possibly cis	Temporal (liver)	Additive	—

[a] Relative to structural gene(s) affected.

molecular weight of 70,000–75,000. This polypeptide can associate into either the L or X form of the active tetrameric enzyme. The X form differs slightly from L in its molecular weight (~2000–3000 less) and has a higher isoelectric point. There is ample evidence that both forms are derived from the same structural gene (Paigen, 1979b). The L form is localized exclusively in the lysosomes while the X form associates with an additional protein factor which acts to anchor the enzyme to microsomal membranes (Swank and Paigen, 1973). This additional protein factor has been called egasyn and is encoded by the *Eg* locus on chromosome 8. It is, therefore, unlinked to the *Gus* locus. Egasyn has a molecular weight of 64,000 and is a glycoprotein (Lusis *et al.*, 1976). One to four molecules can bind to the tetrameric X form of β-glucuronidase. Mouse strains completely lacking any detectable egasyn exist (Ganschow and Paigen, 1967). These lines also lack β-glucuronidase in microsomes. The mutation exhibits additive inheritance such that F_1 heterozygotes have half the normal levels of egasyn (Lusis and Paigen, 1977). The *Eg* locus is regulatory in nature and represents the class of regulatory genes termed processing genes (Paigen, 1979a), since it is involved in the posttranslational localization of a protein within the cellular environment. Recent evidence indicates that egasyn also possesses carboxyl esterase activity and exists in multiple electrophoretic forms (Medda and Swank, 1985). Since only 10% of total liver egasyn is complexed with β-glucuronidase, it was suggested that the remaining 90% is associated with other, as yet unidentified, proteins (Lusis *et al.*, 1976). However, recent studies show this not to be the case (Medda and Swank, 1985). Thus, the interaction between egasyn and β-glucuronidase appears to be more specific than previously thought.

Several regulatory sites map close to *Gus*. The first of these, *Gus-r*, is responsible for the induction of *Gus* gene activity in response to androgenic steroids (e.g., testosterone) (Swank *et al.*, 1973). It represents an example of an effector-response regulatory gene since it regulates the response of a structural gene to an inducer molecule. Mice homozygous for the *Gus-r*^{a} allele can be induced to higher levels than homozygous *Gus-r*^{b} mice. Additional alleles with more extreme degrees of inducibility exist. *Gus-r* acts in a cis manner specifically by controlling the rate of enzyme synthesis (Swank *et al.*, 1973). This control reflects an increase in the activity of β-glucuronidase mRNA (Paigen *et al.*, 1979), but it is not known if the concentration of mRNA is altered or if its accessibility for translation is altered. The alleles of *Gus-r* are inherited in an additive manner. All evidence suggests that *Gus-r* represents a site responsible for the binding of the androgen–

androgen receptor complex which allows for the induction of *Gus* gene activity. Another gene important in the androgen induction of *Gus* then would be the androgen receptor gene. This locus (*Tfm*) has been identified and shown to be unlinked to *Gus*. *Tfm* mutants completely lack androgen inducibility of β-glucuronidase activity due to the absence of the androgen receptor protein (Bullock and Bardin, 1974; DoFuku *et al.*, 1971).

A second regulatory site, *Gus-t*, maps very near to *Gus* and controls the temporal pattern of β-glucuronidase expression during development (Meredith and Ganschow, 1978; Paigen, 1961a,b). This site represents an example of a temporal regulatory gene. The developmental program for β-glucuronidase has been characterized in several different tissues of the mouse (Paigen, 1961a) and has been found to be coordinated with other lysosomal enzymes such as β-galactosidase, α-galactosidase, and arylsulfatase. Despite their coordinate expression, all the genes are regulated independently (Felton *et al.*, 1974; Lusis and Paigen, 1975; Paigen, 1979a). In some mouse strains, β-glucuronidase follows a different developmental profile than that typically observed. In these strains there is a rapid decrease in activity levels at times which vary from tissue to tissue. This decrease in activity is tissue specific since it occurs in some tissues (e.g., liver and kidney), but not in others (e.g., spleen). This decrease in activity is controlled by the *gus-t*^{h} allele at this regulatory site. The alleles of the *gus-t* locus are inherited in an additive manner. In the past it had been difficult to resolve whether *gus-t* acted in a cis or trans manner; however, recent evidence clearly indicates that it acts in trans (Lusis *et al.*, 1983). This trans action is unusual for a proximately located regulatory site which usually exhibits cis action. Trans-acting sites presumably act via diffusible regulatory molecules and are in most cases distantly located relative to the gene being regulated. Immunotitration and pulse-labeling studies have clearly shown that *gus-t* acts to regulate the number of enzyme molecules produced and that this is achieved by controlling the relative rate of enzyme synthesis (Ganschow, 1975; Paigen *et al.*, 1975). Recent evidence for the action of *gus-t* in the liver suggests that it acts only in hepatocytes and not in nonhepatocytic cells (i.e., endothelial, Kupffer, and fat-storing cells) (Paigen and Jakubowski, 1985), so it acts in a cell-type-specific manner within the liver.

A third regulatory site which maps in close proximity to the *Gus* locus is the *gus-u* site. *Gus-u* is a systemic regulator of β-glucuronidase activity and is clearly distinct from *gus-r* and *gus-t*. It acts in a cis manner to uniformly reduce enzyme levels in all tissues to about one-third the normal level of activity (Lusis *et al.*, 1983). Recent evidence for the action of *gus-u* in the liver indicates that it does not exhibit cell-type specificity in its effects among the different cell types of the liver (Paigen and Jakubowski, 1985).

B. α-Galactosidase

The α-galactosidase (E.C. 3.2.1.22) enzyme system of the mouse exhibits many similarities to the β-galactosidase system (Lusis and Paigen, 1976). However, the two enzymes are clearly regulated by independent temporal regulatory genes. The structural gene for α-galactosidase (*Ags*) is located on the X chromosome. A temporal regulatory gene has been identified which effects the developmental pattern of α-galactosidase expression, specifically in the liver after 25 days of age (Lusis and Paigen, 1975). The increase in overall activity reaches a level twice that observed in other typical strains of mice. Other mouse tissues do not exhibit this altered pattern of expression and therefore, the effect is both tissue- and time-specific. Recently, it has been shown (Paigen and Jakubowski, 1985) that the effect of this regulatory gene is also cell-type specific, in that it controls α-galactosidase levels in the hepatocytic cells, but not in the nonhepatocytic cells (i.e., endothelial, Kupffer, and fat-storing cells) of the liver. Elevated levels of enzyme activity in the high-activity strains are due to increased levels of protein, but it is not yet clear if the regulatory mechanism operates at the level of enzyme synthesis. The regulatory gene, *Agt*, is autosomal and, as such, is distant from the structural gene it regulates. F_1 heterozygotes exhibit an additive pattern of inheritance. Such distant, trans-acting regulatory genes are likely to act via some diffusible regulatory molecule. Additive inheritance patterns suggest that they operate by a novel regulatory mechanism, since it cannot be the simple repressor–activator (i.e., ON–OFF switch) model of prokaryotes which would produce a dominance–recessive inheritance pattern.

C. β-Galactosidase

The expression of β-galactosidase (E.C. 3.2.1.23) in the mouse represents a system in which both proximate and distant regulatory sites interact to produce the unique developmental program of expression for this particular enzyme. β-Galactosidase is expressed throughout the different tissues of the mouse and functions in the degradation of glycolipids with a terminal β-galactose. The β-galactosidase gene complex, termed *Bgl*, consists of the structural gene (*Bgl-e*), for which two isozyme alleles exist, i.e., a regulatory gene (*Bgl-s*), which has a systemic effect on the level of β-galactosidase expression (Felton *et al.*, 1974), and a temporal regulatory gene (*Bgl-t*), which specifically affects the level of enzyme expression in the liver only (Paigen *et al.*, 1976).

Bgl-s possesses two alleles: $Bgl\text{-}s^{h}$, which systemically produces approximately twice the activity levels of other mouse strains, which carry the $Bgl\text{-}s^{d}$ allele. These alleles are inherited in an additive manner so that F_1 heterozygotes possess levels of activity intermediate to either parent. *Bgl-s* is tightly linked to *Bgl-e* (Breen *et al.*, 1977). The fact that *Bgl-s* is distinct from *Bgl-e* comes from the lack of a strict association of alleles at one site with that of the other (Breen *et al.*, 1977). Immunological studies have shown that *Bgl-s* regulates the number of enzyme molecules (Meisler, 1976), and pulse-labeling studies have shown that this is controlled at the level of enzyme synthesis (Berger *et al.*, 1978). Electrophoretic analysis has revealed that *Bgl-s* acts in a cis manner. While *Bgl-s* regulates the level of enzyme activity in most tissues in a systemic manner, there are exceptions. For instance, certain mouse strains have high brain activity but also have low salivary gland activity. Other strains possess β-galactosidase activity variation in different cell types within the same tissue (Dewey and Mintz, 1978). The possibility, therefore, exists that additional elements may regulate the expression of the *Bgl-e* structural site and these elements may be unlinked to the β-galactosidase gene complex.

Bgl-t, a temporal regulatory gene, is also contained in the β-galactosidase gene complex. Two alleles exist at this site: $Bgl\text{-}t^{d}$ causes normal (i.e., typical) levels of β-galactosidase in the liver after 25 days of age while $Bgl\text{-}t^{b}$ results in a doubling of the typical levels of β-galactosidase. Genetic analyses place the *Bgl-t* site in close proximity to *Bgl-e* and *Bgl-s*. *Bgl-t* exhibits additive inheritance (Paigen *et al.*, 1976) and controls the overall levels of enzyme synthesis (Berger *et al.*, 1978). Unlinked, distant regulatory sites which interact with *Bgl-t* have been implicated in the regulation of β-galactosidase gene expression (Berger *et al.*, 1979). These distant sites exhibit additive inheritance and control the rate of enzyme synthesis (Berger *et al.*, 1979). It has been shown (Paigen and Jakubowski, 1985) that the effect of the unlinked temporal site occurs specifically in the hepatocytes of the liver and has no effect in nonhepatocytic cell types of the liver.

D. Arylsulfatase B

The predominant arylsulfatase activity in mice is the lysosomal arylsulfatase B (E.C. 3.1.6.1). This enzyme is coded by the structural gene *As-1,* for which two alleles have been identified based on thermolability. This locus is under the control of a temporal regulatory gene *Asr-1,* which specifically affects expression of the enzyme in the liver (Daniel, 1976a,b) by causing a developmental increase in enzyme activity after 35 days of age in some strains of mice. This regulatory locus exhibits additive inheritance and is linked to the *As-1* structural gene. It is not clear at present how close this linkage is. Evidence that the regulatory and structural sites are distinct entities includes the fact that the particular allele carried by a strain at the *As-1* locus is not correlated with any particular allele at the *Asr-1* locus such that all possible allelic combinations occur.

A second regulatory site also affects arylsulfatase B activity; however, this site has not yet been fully characterized (Daniel, 1976a,b). It affects expression of the enzyme in the kidney of males and apparently arises from differences in androgen inducibility of the enzyme. It exhibits additive inheritance and, again, while this site is linked to the *As-1* structural gene, it is not known how close this linkage is.

E. H-2 Antigen

The major histocompatibility antigens of mouse cells (H-2D and H-2K) are encoded in a region near the centromeric end of chromosome 17 termed the *H-2* region (for review, see Klein, 1975) which comprises the major histocompatibility complex of the mouse. Several alternative genetic combinations, or haplotypes, analogous to alleles, have been identified at this complex.

The regulation of the expression of the H-2 antigens in mice involves a complex temporal gene system with both proximate and distant sites (Boubelik *et al.,* 1975; Paigen, 1979b). Almost every cell of the mouse acquires H-2 antigens during early embryonic development. The one exception is erythrocytes, which do not express the antigens until birth. Variation in erythrocyte H-2 antigen expression has been identified (Boubelik *et al.,* 1975). The "early phenotype" expresses H-2 antigens at birth and "late phenotype" does not express H-2 antigens until about 3 days after birth. Evidence indicated that this phenotypic difference was due to a specific alteration in expression of H-2 antigens and not due to nonspecific developmental differences among strains. It was shown that the elements which regulate H-2 antigen expression are distinct from known elements within the *H-2* complex and are not correlated with any specific haplotype. Therefore, mouse strains carrying different *H-2* haplotypes can exhibit the same developmental phenotype (i.e., early or late) and strains carrying the same *H-2* haplotype can exhibit different developmental phenotypes.

The first temporal gene identified which affected H-2 antigen expression in erythrocytes was a locus designated intrinsic (*Int*). This site is linked to the *H-2* complex and is cis acting in F_1 heterozygotes. Despite the cis action of *Int* and the inability to identify recombinants between *Int* and the *H-2* complex in a conventional cross, *Int* is not tightly linked with the *H-2* complex (Boubelik *et al.*, 1975; Paigen, 1979b). Two alleles have been identified at the *Int* locus; one controls the "early phenotype" and the other the "late phenotype."

A second and third gene were subsequently identified which also influenced H-2 antigen expression. Examination of F_2 and backcross data indicated a complex pattern of inheritance. A high frequency of recombination was observed between these sites and the *H-2* complex, indicating that they are not linked. The two regulatory genes are not on chromosome 17 with the *H-2* complex; however, the two sites are linked to each other (about 20 map units apart). These loci have been designated *Tem* and *Rec*. The model for the interaction of the three regulatory genes, including *Int*, is as follows. Two alleles of *Int* exist which condition either early or late expression of the H-2 antigens in the erythrocyte. In the absence of suppression by the *Tem–Rec* system, *Int* regulates the timing of expression in a cis manner. Suppression of the regulatory effects of *Int* gene activity can occur depending on the allele present at the *Rec* locus. Two alleles of *Rec* exist, each matching one of the *Int* alleles. If a *Rec* allele matching the *Int* allele is present, suppression of *Int* function occurs and *Tem* takes over control of the regulation of the timing of H-2 antigen expression. Two alleles exist at *Tem*, one for early and one for late expression. Basically then, the model predicts that the allele at *Int* regulates H-2 antigen expression unless a matching *Rec* allele is present. If so, then *Tem* assumes the control of the timing of H-2 antigen expression. For the model to hold it must be assumed that the *Tem* allele cis to the matching *Rec* allele can take over when the *Rec* allele suppresses *Int*, even though they are not tightly linked. Alternatively, one can assume that *Tem* and *Rec* are on the X chromosome and show haploid expression due to X inactivation. There is, as yet, no clear explanation of how *Rec* carries out its suppressor function.

F. Aminolevulinate Dehydratase

The aminolevulinate dehydratase (E.C. 4.2.1.24) locus (*Lv*) of the mouse is under the control of a systemic regulatory gene which influences enzyme activity levels equally in all tissues (e.g., liver, kidney, and spleen) (Coleman, 1971; Doyle and Schimke, 1969; Russell and Coleman, 1963). The regulatory site is closely linked to the structural gene and is involved in the regulation of enzyme synthesis as shown by pulse-labeling experiments (Doyle and Schimke, 1969). Strains of mice with higher activity levels have a greater rate of synthesis of an immunologically and catalytically identical enzyme to that found in low-activity strains of mice. This regulatory element exhibits additive inheritance, but it is not clear if it is cis or trans acting.

G. Catalase

A regulatory gene has been identified which controls the turnover of catalase (E.C. 1.11.1.6) in the mouse (Ganschow and Schimke, 1969; Heston *et al.*, 1965; Rechcigl and Heston, 1967). Two lines of mice were examined which differed in their overall level of liver catalase (Ganschow and Schimke, 1969; Heston *et al.*, 1965). It was shown that this activity difference was due to a twofold decrease in the rate of catalase degradation in the high-activity line and that this was under genetic control. The regulatory gene involved, *Ce*, specifically controls the rate of catalase degradation in the liver but not the kidney and therefore exhibits tissue specificity. This regulatory gene maps to a site in the mouse genome that is different from that of the catalase structural gene (Hoffman and Grieshaber, 1974, 1976). It is trans acting and exhibits a dominance–recessive pattern of inheritance. The results demonstrated that the turnover of a specific protein in a specific tissue can be genetically controlled. *Ce* is representative of those regulatory elements which have been designated processing genes, since it is involved in the posttranslational processing machinery (i.e., protein degradation).

H. Alcohol Dehydrogenase

Multiple molecular forms of alcohol dehydrogenase (ADH) (E.C. 1.1.1.1) are present in the mouse (Felder *et al.*, 1983). The structural genes *Adh1*, *Adh2*, and *Adh3* code for the isozymes A2, B2, and C2, respectively. The predominant form of ADH in the liver is A2. This form is also found in the kidney, and trace amounts are detectable in adrenal and intestinal tissue. The predominant form of ADH in the stomach is C2, which is also found in the male and female reproductive tissues. Both *Adh1* and *Adh3* genes are located on chromosome 3 and are closely linked. Both loci have been found to be under the independent control of separate and distinct temporal regulatory genes (Balak *et al.*, 1982; Felder *et al.*, 1983; Holmes, 1979).

ADH activity levels in liver extracts from several inbred mouse strains have been found to vary by as much as twofold (Balak *et al.*, 1982). The difference is both tissue and time specific, since kidney extracts do not exhibit the activity difference, and the difference in liver activity levels does not appear until the mice are at least 25 days old. Genetic analysis has demonstrated that the activity difference is controlled by a distinct regulatory site, termed *Adh-1-t*, for which two alleles exist: *Adh-1-t*^{a}, which conditions low liver ADH activity, and *Adh-1-t*^{b}, which conditions high liver ADH activity (Balak *et al.*, 1982). This regulatory locus exhibits additive inheritance such that F_1 heterozygotes possess intermediate levels of activity relative to either parent. At present it is not known whether the regulatory site is proximal or distal to the *Adh1* structural gene. Quantitative immunoprecipitation experiments coupled with biochemical analysis of purified enzyme from high- and low-activity strains indicate that the activity difference is due to an increase in the overall level of liver ADH molecules and not to alterations in the structure of the ADH polypeptides (Balak *et al.*, 1982). Finally, radiolabeling experiments followed by immunoprecipitation have shown that the twofold increase in liver activity in the high-activity strains is the result of a twofold increase in the relative rate of synthesis of liver ADH in these strains (Balak *et al.*, 1982).

The levels of *Adh3* gene expression in the reproductive tissues of the mouse have been shown to be under the control of a temporal regulatory gene termed *Adh-3-t* (Holmes, 1979). Two alleles exist at this locus: *Adh-3-t*^{a}, which allows expression of C2 in the reproductive tissue, and *Adh-3-t*^{b}, which inhibits expression in this tissue. The regulatory gene is cis acting and is tissue and time specific in its effect, as is expected for temporal regulatory genes. The lack of recombination between *Adh-3-t* and *Adh3* suggests that the two sites are closely linked.

I. Malic Enzyme (Mitochondrial)

The *Mod2* nuclear gene of the mouse encodes the mitochondrial malic enzyme (MOD-2) (E.C. 1.1.1.40) for which two isozyme alleles have been identified: *Mod2*^{a} and *Mod2*^{b}. Biochemical and genetic studies have identified a temporal regulatory gene which affects the expression of *Mod2* (Bernstine, 1979; Bernstine and Koh, 1980). Specifically, this regulatory site (*Mdr1*) determines the synthesis rate of the MOD-2 protein. *Mdr1* maps in close proximity to the *Mod2* gene on chromosome 7, is cis acting, and is inherited in an additive manner. Using pulse labeling, immunoprecipitation, and isolectric focusing under denaturing conditions, it was determined that the *Mdr1*^{a} allele conditions a twofold increase in the rate of synthesis of the adjacent *Mod2* gene as compared to the *Mdr1*^{b} allele (Bernstine and Koh, 1980). This experimental approach showed directly that synthesis rates were affected in a cis manner. This method has many advantages over the more commonly used alternative approach to identify cis or trans action based on histochemical staining of gels and looking for an isozyme pattern shift in F_1 hybrids which would indicate a predominance of one allelic form over the other. Obviously, this latter approach is very qualitative. This represents one of the few examples which show definitively that synthesis rates are affected by regulatory gene action. At present, however, it is not known if transcriptional or posttranscriptional processes are affected. The effect is tissue specific, since only the brain enzyme is affected.

J. Glucose Phosphate Isomerase

Genetic variation has been described which affects the expression of glucose phosphate isomerase (GPI) (E.C. 5.3.1.9) during oogenesis in the mouse (Peterson *et al.*, 1978, 1985; McLaren and Buehr, 1981). GPI, encoded within the *Gpi-1* locus, is a dimeric enzyme expressed in the mouse ovum. Several inbred strains of mice vary in the level of expression of this enzyme in the ovum but no difference is found in somatic cells. Genetic analysis has revealed the existence of a regulatory site (*Org*) in close proximity to the structural gene (within 1.1 cM), which is responsible for regulating activity levels (Peterson *et al.*, 1985). In heterozygotes three isozymes are detected on zymograms due to the dimeric nature of GPI. In red blood cells the ratio of the three isozymes is 25% AA : 50% AB : 25% BB, as expected for random association of subunits into a dimer. Ova, however, have a greatly skewed zymogram band ratio of 3.4% AA : 29.9% AB : 66.8% BB when the A subunit comes from the low-activity mouse strain. The ratio of enzyme activity levels in the ova of the low versus high line was 19% A : 81% B and, therefore, the electrophoretic activity distribution of GPI in the F_1 hybrid closely approximates that expected if the monomers were being produced in the ratio predicted from parental activities [i.e., 3.4% + $\frac{1}{2}$(29.9%) ≅ 19% A and 66.8% + $\frac{1}{2}$(29.9%) ≅ 81% B]. These results indicate that the element is cis acting. It is also tissue specific in its action and is inherited in an additive fashion. Evidence that *Org* is distinct from the structural gene comes from the fact that the structural gene can be associated with different activity classes and vice versa.

III. Regulatory Genes of *Drosophila*

A. Alcohol Dehydrogenase

The alcohol dehydrogenase (E.C. 1.1.1.1) gene–enzyme system of *Drosophila* has been studied extensively. A large volume of work has been conducted on the characterization of the structure and function of the protein and on the nature of the posttranslational modification that leads to the production of isozymic forms of the enzyme from a single allele. The gene has been cloned and sequenced (Benyajati *et al.*, 1981; Goldberg, 1980), and a molecular analysis of the gene–enzyme system is now in progress.

Most of the observed quantitative variation of *Drosophila* ADH is associated with structural variants of the *Adh* locus. Lines with the *Adh-S* allele generally have less activity than lines containing the *Adh-F* allele (Gibson, 1972; Grossman, 1980; Lewis and Gibson, 1978; Maroni *et al.*, 1982; McKay, 1981). DNA-sequencing data (Benyajati *et al.*, 1981; Goldberg, 1980) as well as amino acid sequencing data (Fletcher *et al.*, 1978) indicate that the only difference in primary structure between the two alleles is the presence of threonine (*Adh-F*) or lysine (*Adh-S*) at position 192. While it is possible that this single change could be responsible for the activity differences by causing catalytic efficiency differences (Birley *et al.*, 1981; Lewis and Gibson, 1978), other evidence suggests the strong possibility that the activity effect is due to a separate regulatory site which affects the accumulation of enzyme molecules (Birley *et al.*, 1981; Gibson, 1972; Lewis and Gibson, 1978; Maroni *et al.*, 1982; McDonald *et al.*, 1980). It seems likely that the activity difference observed between the F and S forms is due to a closely linked cis-acting regulatory site (Grossman, 1980) which may affect enzyme accumulation through enzyme synthesis, although it could affect degradation rates as well. Strong evidence that the regulatory site can be separated from the structural gene site comes from intragenic class activity variation that has been observed. Occasionally a line is identified with an activity level atypical of its class. For instance, an ADH-S line has been described with half the activity and ADH protein levels of a typical ADH-S line (Thompson and Kaiser, 1977). Again the regulatory site maps very close to the structural gene (Thompson *et al.*, 1977). Another pair of ADH-S lines were shown to have about a twofold difference in activity and, again, the regulatory site mapped close to the *Adh* locus (Grossman, 1980). Similar variation has been described within the F allelic class (McKay, 1981). At present, the information concerning the effect of this regulatory site on tissue and temporal specificity of expression of *Drosophila* ADH is sparse. Overall, it appears that the effects are largely systemic.

B. α-Amylase

The α-amylase (E.C. 3.2.1.1) gene–enzyme system of *Drosophila* exemplifies the role of distant regulatory genes in controlling the expression of a well-defined structural gene. *Drosophila* α-amylase functions in the digestion of starch and glycogen by hydrolyzing internal α-1,4-glucosidic bonds. The enzyme has been purified and fully characterized (Doane *et al.*, 1975). There is much evidence to suggest that the *Amy* structural gene is actually duplicated with the two structural genes tightly linked (about 0.008 map units apart). This has recently been verified through the cloning of the duplicated genes (Gemmill *et al.*, 1985; Levy *et al.*, 1985). Inbred lines exhibit either one or two isozymes since the gene is duplicated and the alleles are codominant. A total of 13 different isozymic banding patterns has been observed (Doane *et al.*, 1975). They are derived from six different electrophoretic mobility variants.

Considerable activity variation has been identified both within and among the various isozyme patterns (Doane *et al.*, 1975, 1983). For example, two isogenic second chromosome lines, Amy^{1a} and Amy^{1c}, exhibit the same isozymic pattern but differ about threefold in activity levels. Rocket immunoelectrophoresis revealed that the activity difference was due to different levels of amylase protein (Hickey, 1981). Among other examples (Doane *et al.*, 1983), one line with the allele Amy^{6} exhibits a threefold increase in amylase activity (Hickey, 1981) as compared to the Amy^{1c} allele.

The amylase system of *Drosophila* is particularly interesting because of the information it provides concerning the developmental regulation of a specific structural gene by a temporal regulatory gene. Amylase is present in the midgut of both larval and adult flies (Doane *et al.*, 1975, 1983). The midgut can be divided into an anterior, a middle, and a posterior region, with amylase only being expressed in the anterior midgut (AMG) and posterior midgut (PMG). The spatial pattern of amylase expression can be determined by dissecting out the midgut and placing it on a thin film of starch gel. With time the cells lyse and release the amylase enzyme, which hydrolyzes the starch in the surrounding gel. Subsequent staining of the starch gel with iodine reveals the pattern of enzyme activity. Three general patterns of enzyme expression in the PMG are revealed by this method: the "A pattern" expresses amylase throughout the PMG, the "B pattern" expresses amylase only in a small anterior region of the PMG, and the "C pattern" lacks amylase throughout the PMG. The important distinction in the expression of these patterns is the time during development when activity appears, since, as the flies age, the B and C patterns come to look like the A pattern; thus, the effect is a temporal one.

The genetic and biochemical bases for the variation of amylase expression in the PMG have been investigated (Abraham and Doane, 1978). The isozyme patterns for *Drosophila* amylase in both the AMG and PMG are always the same. Therefore, the differences between the A pattern and C pattern in the PMG reflect differential regulation of the same structural genes in two tissues and not the expression of distinct structural genes. Genetic studies have shown that the different patterns of expression of amylase in the PMG are controlled by a single gene which is located 2 map units away from the amylase structural gene pair (Abraham and Doane, 1978). This locus has been designated *map* and is a clear example of a temporal regulatory gene, since it regulates the timing of gene expression during development. Two alleles have been identified at this regulatory locus: map^{a} conditions the A pattern in PMG and map^{c} conditions the C pattern of expression. The B pattern is thought to arise from a third allele at this locus, map^{b}. Analysis of F_1 hybrids reveals that *map* acts in trans. For instance, in the parental line map^{a}, $Amy^{1,6}$ isozymes 1 and 6 are expressed in both the AMG and PMG (A pattern here), but in the other parental line map^{c}, $Amy^{2,3}$ isozymes 2 and 3 are expressed in the AMG but not the PMG (C pattern here). The F_1 hybrid expresses all four isozymes in the PMG, so that the map^{a} allele can activate an amylase locus on a separate chromosome. Visual comparisons of staining intensities in homozygous and heterozygous flies suggest that heterozygotes have an intermediate level of enzyme expression, implying that *map* exhibits additive inheritance.

In the heterozygote described above (map^{a}, map^{c}; $Amy^{1,6}$, $Amy^{2,3}$) the four amylase alleles contribute equal amounts of enzyme in the AMG. However, this is not true in the PMG. AMY1, AMY3, and AMY6 are present in about equal amounts but AMY2 levels are much reduced in young flies. This evidence suggests that genetic regulatory variation may exist in some kind of receptor element (possibly for the *map* gene product) in close proximity to the structural gene. Each copy of the duplicated structural gene then has an opportunity for an independent pattern of expression. Other examples of differential expression during development for amylase gene pairs which encode different electrophoretic forms of amylase have been described (Doane, 1970). Thus, it appears that both proximate and distant sites can contribute toward determining the temporal pattern of expression for the duplicated amylase genes of *Drosophila* (Doane *et al.,* 1983).

Variation in the pattern of expression of amylase has also been identified in the AMG. Five different patterns have been observed with respect to amylase gene expression in the AMG (Doane, 1980). Those patterns are independent of the three patterns observed in the PMG. While the results are still preliminary, it appears that the regulatory gene which controls these AMY patterns are located about 3 map units away from the *map* locus (Doane, 1980).

With the recent cloning of the duplicated α-amylase genes (Gemmill *et al.,* 1985; Levy *et al.,* 1985), new insights at the molecular level can be expected soon which will better describe the important regulatory sequences which control amylase gene expression. Due to the proximity of the *map* locus to the structural genes, it may be possible to use the technique of "chromosome walking" to clone this regulatory locus and thus gain a better understanding of its function.

C. Molybdenum Hydrolases

Molybdoenzymes are enzymes that require a molybdenum (Mo)-containing cofactor. Four enzymes of this type have been identified in *Drosophila:* xanthine dehydrogenase (XDH), aldehyde oxidase (AOX), pyridoxal oxidase (PO), and sulfite oxidase (SO). XDH is encoded by the *rosy* (*ry*) locus (Chovnick *et al.,* 1980), AOX by the *Aldox* locus (Dickinson, 1970), and PO by the *low pyridoxal phosphate* (*lpo*) locus (Dickinson and Weisbrod, 1976), while the structural gene for SO has yet to be identified.

The level of expression for these enzymes appears to be controlled by three regulatory loci (Courtwright, 1976; Finnerty, 1976; O'Brien and MacIntyre, 1978). These include *maroonlike* (*mal*), *cinnamon* (*cin*), and *low xanthine dehydrogenase* (*lxd*). A total of 42 recessive *mal* alleles have been identified which fall into six complementation groups (Bentley and Williamson, 1982; Finnerty, 1976). Most of these alleles were artificially induced (by X rays and chemical mutagens) and when present in the homozygous or hemizygous state result in the absence of all XDH, AOX, and PO activity. However, there is still a large amount of XDH and AOX cross-reacting material (CRM) produced (Finnerty *et al.*, 1979). In one study, XDH-CRM was about equal to the wild-type control, while AOX-CRM was about one-half of the control (Browder *et al.*, 1982a,b). Some *mal* alleles produce wild-type levels of PO-CRM while others show none (Warner *et al.*, 1980). The *mal* mutants appear to have little effect on SO activity. Ethyl methanesulfonate (EMS) mutagenesis has produced seven recessive *cin* alleles which comprise four different complementation groups. In the homozygous or hemizygous state, essentially no XDH, AOX, PO, or SO activity is produced; however, again large amounts of XDH-CRM and AOX-CRM are present. Two spontaneous and two EMS-induced recessive *lxd* alleles have been recovered. As with *cin*, the enzyme activities for all four molybdoenzymes are reduced to varying degrees but substantial levels of CRM remain. For example, one *lxd* mutant has an 80 and 90% reduction in XDH and AOX activity levels, respectively, a 90% reduction of SO activity, and no detectable PO activity (Keller and Glassman, 1964; Warner *et al.*, 1980). However, wild-type levels of XDH-CRM and AOX-CRM are present and PO-CRM is not detectable (Browder *et al.*, 1982a,b; Warner *et al.*, 1980).

It would appear from the defects induced by these three mutant loci that they represent regulatory genes which affect the posttranslational modification (Finnerty *et al.*, 1979; Laurie-Ahlberg, 1985) of the molybdoenzymes, making them representative examples of processing genes. A posttranslational level of action would explain the dominance–recessive inheritance pattern. It has been suggested that the gene products of these three regulatory loci may be involved in cofactor synthesis or modification of the Mo cofactor-binding site on the enzyme (O'Brien and MacIntyre, 1978). This idea has been supported by reports which show that *mal* flies have high levels of the Mo cofactor while *cin* and *lxd* mutants have reduced levels (Warner and Finnerty, 1981). It appears that the latter two mutants may be disrupted in cofactor synthesis. For the *mal* mutants the defect appears to be in the modification of the Mo-binding site, since XDH and AOX from *mal* flies lack a sulfur moiety associated with the Mo site. However, a resulfuration procedure can restore the activity of these two enzymes in those flies (Wahl *et al.*, 1982). Additional evidence for this idea comes from the fact that SO lacks the sulfur moiety at the Mo site and its activity is affected by *cin* and *lxd* but not *mal*. Therefore, it appears that the *mal* locus produces a product involved in acquisition of the sulfur moiety while *cin* and *lxd* are involved in other aspects of Mo cofactor synthesis.

D. Aldehyde Oxidase

The first report of a temporal regulatory gene in *Drosophila* came from the analysis of a genetically controlled modification of the developmental program for the expression of aldehyde oxidase (E.C. 1.2.3.1) (Dickinson, 1970, 1975) which is coded by the *Aldox* locus. AOX activity levels of *Drosophila* are low during the larval stages of development, increase to intermediate levels shortly after pupation, rise again shortly before eclosion, and eventually rise to the highest activity levels in newly emerged flies. Among fly lines most activity variation for this enzyme occurs during the pupal stage of development. One line has about 50% of the adult activity level, while another has about 90% of the adult activity level. It is known that this difference is not due to differences in catalytic efficiency of the enzymes. A genetic analysis of pupal stage activity differences was performed using parents that differed in their electrophoretic mobility patterns as well as pupal activity levels. The F_1 hybrids had an activity level intermediate to the two parents and backcrossing F_1 females back to the low-activity line produced two activity classes in the progeny. These results indicate that pupal activity levels are inherited as a single Mendelian gene which exhibits additive inheritance. The regulatory element responsible is tightly linked to the structural gene and is cis acting. It can be shown that the electrophoretic site and regulatory site are distinct because different electrophoretic alleles can be associated with different pupal activity classes and vice versa. This regulatory site was, as mentioned, the first example of a temporal regulatory gene operating in *Drosophila*. Like other temporal genes which control the developmental pattern of expression of an enzyme, this one exhibits both tissue and stage specificity in its action. While not yet resolved, the fact that the element is closely linked to the structural gene and exhibits additive inheritance makes it likely that AOX protein levels are affected at the level of protein synthesis either through altered mRNA population levels or alterations in the translational efficiency of certain mRNAs.

Another tissue-specific variation in the developmental pattern of expression of AOX has been described (Dickinson, 1978). The variant causes a twofold to threefold increase in the amount of AOX in the paragonia (male accessory sex glands) and not in any other tissue. This variation is not associated with any change in enzyme catalytic properties and likely reflects a difference in AOX protein levels, although this has yet to be shown conclusively. Mapping experiments have localized the regulatory site to within 4 map units of the *Aldox* structural gene. It appears to be cis acting and demonstrates additive inheritance. While more work needs to be done, this appears to be a good candidate for a second temporal gene regulating expression of *Drosophila* aldehyde oxidase.

E. Xanthine Dehydrogenase

The *rosy* (*ry*) locus of *Drosophila* encodes the enzyme xanthine dehydrogenase (E.C. 1.1.1.20). It represents one of the most well-defined loci of *Drosophila* (see Chovnick *et al.*, 1977, 1978a,b, 1980) based on the extensive fine-scale genetic mapping performed on this locus (Chovnick, 1973; Chovnick *et al.*, 1971). Through the analysis of naturally occurring enzymatic activity variants, evidence has been obtained which reveals the presence of a cis-acting regulatory element located adjacent to the *ry* structural gene that exhibits additive inheritance (Chovnick *et al.*, 1976; McCarron *et al.*, 1979). Three different activity classes for XDH activity have been identified. Typically, fly lines exhibit normal, intermediate, or high levels of activity. However, one variant (ry^{+4}) has unusually high levels of activity and another variant (ry^{+10}) has unusually low levels of activity (Chovnick *et al.*, 1976; McCarron *et al.*, 1979). Neither variant was found to be associated with an alteration in the kinetic properties or thermolability of the XDH enzyme. Instead, rocket immunoelectrophoresis showed that the variation in the level of activity correlated with a variation in the amount of XDH protein (Chovnick *et al.*, 1976; Edwards *et al.*, 1977; McCarron *et al.*, 1979). Large-scale fine structure recombination analyses indicated that the regulatory site from both ry^{+4} and ry^{+10} could be separated from the structural gene, but both mapped in very close proximity to the left of the structural gene (Chovnick *et al.*, 1976; McCarron *et al.*, 1979). Subsequent experiments have shown that the regulatory sites in these two alleles are also separable in recombination studies (McCarron *et al.*, 1979). The two distinct regulatory sites have been designated *i409* and *i1005*. The ry^{+4} allele possesses *i409H* (high), which conditions overproduction of XDH. The ry^{+10} allele possesses *i1005L* (low), which conditions underproduction of XDH. The normal wild-type alleles carry *i409N* (normal) and *i1005N* sites. Recent evidence has ordered the sites as follows: *i1005–i409–ry* (Clark *et al.*, 1984). It has been shown that the two sites are separated enough to define functionally distinct sequences in the control region. Northern blot analysis has shown that these sites control mRNA population sizes; thus, they likely exert their effect at the transcriptional or posttranscriptional level of control of gene expression (Clark *et al.*, 1984). Finally, evidence has been presented which indicates that *i409*, unlike *i1005*, acts in a tissue-specific manner such that *i409H* of ry^{+4} produces a large increase of XDH in fat body tissue but not malpighian tubules, the two tissues where most XDH activity is confined (Clark *et al.*, 1984). Thus, the detailed examination of the closely linked cis-acting region adjacent to the *ry* locus reveals clearly that genes can exist as a complex containing regulatory information as well as information coding for the primary amino acid sequence of the enzyme. This gene–enzyme system, along with others described in this report, indicate that a cluster of regulatory information can be found in close proximity with structural genes.

F. Glycerol-3-Phosphate Dehydrogenase

The glycerol-3-phosphate dehydrogenase (E.C. 1.1.1.8) (GPDH) gene–enzyme system of *Drosophila* represents another suitable system for analyzing genetically controlled variations in the developmental and tissue-specific patterns of gene expression in *Drosophila* (Wright and Shaw, 1969; MacIntyre and Davis, 1987). Three isozymes of GPDH exist in *Drosophila*, and each exhibits a unique developmental pattern of expression; GPDH-1 and GPDH-2 (a minor isozyme) are found only in late pupal stages and adult flies, while GPDH-3 is found at all stages of development. Each isozyme is also distinct with regard to tissue-specific patterns of expression; GPDH-3 is localized primarily in the fat body and malpighian tubules of the larvae and the abdomen of the adult (includes gut, malpighian tubules, reproductive organs, and cuticle) while GPDH-1 is the predominant isozyme in the thorax of the adult fly. This tissue-specific pattern of expression reflects the unique physiological roles of the two isozymes. Because GPDH-1 plays an important role in flight metabolism, it is primarily localized in the flight muscle of the thorax. On the other hand, because GPDH-3 is important in providing precursors for lipid metabolism, it is associated with the fat body in larvae and the gut in adult flies. Interestingly, genetic and biochemical analyses have shown that both isozymes are encoded by the same structural gene. GPDH-1 and GPDH-3 have an almost identical primary structure, the only difference being that GPDH-3 lacks the terminal three amino acids of GPDH-1 (Niesel *et al.*, 1980, 1982). It is not yet known whether this difference is brought about by a posttranslational modification (cleavage) of the protein or at some pretranslational step (differential transcription or differential processing of mRNA intermediates).

The developmental program for GPDH activity typically increases through each larval instar stage and peaks at about 6 days after hatching of the embryo (i.e., third instar larval stage). This is followed by a rapid decline in activity in early pupae so that by the late pupal stage activity is barely detectable. From this developmental stage, activity levels begin to increase and about 12 days posthatching, the adult fly has and maintains a steady-state level of activity about three to four times that reached in the third instar larvae.

An analysis of coisogenic second chromosome lines revealed variant lines which exhibit alterations in the developmental program of GPDH expression and has led to the identification of a temporal regulatory gene, *Gdt-3*. The developmental profiles of two lines were found to differ to a greater extent in the larval stage (where GPDH-3 predominates) than in the adult stage (where GPDH-1 and GPDH-3 predominate). Analysis of zymogram patterns revealed that the effect was primarily due to a selective reduction of GPDH-3 in both the larvae and adults of the low-activity lines. No differences were found with respect to various biochemical, kinetic, and antigenic properties of GPDH-3 isolated from the high- and low-activity lines. However, CRM level differences were found and appear to account for the activity differences. Genetic analysis showed that *Gdt-3* is closely linked to the *Gpdh* structural gene and represents a cis-acting regulatory element which exhibits additive inheritance. It is responsible for programming the tissue-specific pattern of expression of GPDH-3 activity. The level(s) at which *Gdt-3* acts has yet to be resolved (Bewley, 1983).

Several low-activity lines which uniformly reduce the activity of both GPDH-1 and GPDH-3 have been analyzed. Using a microinjection technique for radiolabeling *Drosophila* proteins *in vivo* in combination with immunoprecipitation of GPDH translation products, it has been possible to estimate turnover parameters of GPDH in these lines. Results indicate that the rate of intracellular degradation for GPDH is the same in the high- and low-activity lines. However, the synthesis rates differ, being much reduced in the low-activity lines. This analysis, combined with genetic studies, has defined a regulatory site which again maps in very close proximity to the *Gpdh* structural gene and follows a pattern of additive inheritance. Analysis of zymogram patterns of F_1 heterozygotes indicates that the element is cis acting. Unlike *Gdt-3*, the element appears to be a systemic regulator since activity levels are reduced uniformly within the different fly tissues (Wilkins *et al.*, 1982).

A third instance of regulatory gene variation affecting *Drosophila* GPDH expression has been described (King and McDonald, 1983). A trans-acting regulatory gene which exhibits a dominance–recessive inheritance pattern was identified on the third chromosome of *Drosophila* and is unlinked to *Gpdh*, which is on the second chromosome. The effect appears to be specifically on activity levels in the adult thorax (where GPDH-1 is localized), since there is no effect on GPDH levels in the abdomen. Turnover studies indicate that this regulatory site does not affect synthesis rates but instead influences the *in vivo* stability of GPDH polypeptides. This effect on GPDH degradation rates along with trans action would suggest a posttranslational mode of operation and would explain the dominance–recessive pattern of inheritance. It should be noted that this regulatory site is not completely specific for GPDH since it was found to affect at least one additional abdominal protein (King and McDonald, 1983).

G. Esterase

The esterases of *Drosophila* represent a heterogeneous group of enzymes which are identified by their ability to hydrolyze ester bonds of several synthetic compounds such as α- and β-naphthyl acetate. Variation in the expression of esterase (E.C. 3.1.1.1) has been studied in some *Drosophila virilis* species (Korochkin, 1980).

The ejaculatory bulb esterase of *D. melanogaster* is encoded by the structural gene *est-6*. Six known electrophoretic mobility variants occur at this locus. Studies have revealed that a second locus exists, *m-est*, which has allele-specific effects on the mobility of the EST-6 isozymes (Cochrane and Richmond, 1979). The *m-est* maps to a different site than *est-6* but the two loci are loosely linked. The modifying effect of this regulatory site is inherited in a dominance–recessive manner and appears to be involved in a posttranslational modification of the *est-6* gene products; however, no biochemical studies have been reported. It appears that *m-est* may also have a modifying effect on leucine aminopeptidase, 1 of 13 other enzymes investigated. Thus, while the action of *m-est* is not restricted to a single enzyme, neither does it act on a broad range of enzymes.

H. Dopa Decarboxylase

Genetic variation in the developmental pattern of expression for dopa decarboxylase (DDC) (E.C. 4.1.1.28) in *Drosophila* has been reported (Estelle and Hodgetts, 1984a,b). DDC is encoded within the *Ddc*$^+$ locus on chromosome 2. Activity is expressed in the epidermal, neural, and ovarian tissues. In the epidermis the majority of DDC activity occurs at six specific times during the life cycle, these being at hatching, the two larval molts, pupariation, pupation, and adult eclosion (Wright *et al.*, 1982). During a screen of coisogenic second chromosome lines, one variant line was found in which the DDC activity was overproduced throughout most of the life cycle, and this activity was correlated to an overproduction of DDC-cross-reacting material (DDC-CRM). Mapping studies revealed that the activity effect was due to an element, termed *Ddc*$^{+4}$, which was closely linked (within 0.15 map units) to the structural gene. Several lines of evidence indicate that the activity effect is not associated with a variant DDC polypeptide (i.e., altered physicochemical or kinetic properties) (Estelle and Hodgetts, 1984a). The most convincing evidence for this is that the same element is responsible for overproduction of enzyme activity (e.g., at hatching and eclosion) and also underproduction of enzyme activity (specifically at pupariation) during the life cycle of the fly. Therefore, the effect of this site exhibits complex temporal specificity. *Ddc*$^{+4}$ also exhibits tissue specificity since only expression in the epidermis and not in neural tissue is altered in the variant fly line. The variant activity levels correlate with altered CRM levels, as well as with altered mRNA population levels, which are overproduced at hatching and eclosion and underproduced at pupariation of the variant fly line. Molecular analyses reveal complex polymorphisms in the 5′-untranslated leader sequence of DDC mRNA and in the DNA region upstream to the transcription start site (Estelle and Hodgetts, 1984b). Eventually, attempts will likely be made to associate particular polymorphisms with DDC overproduction or underproduction with the idea that this will provide greater insight into the molecular mechanisms regulating *Ddc*$^+$ gene expression in *Drosophila*.

Additional variation has also been observed (Marsh and Wright, 1986) at the *Ddc*$^+$ locus of *Drosophila*. In this case, the variants produce elevated levels of DDC-CRM during the two major periods of DDC expression (i.e., eclosion and pupariation), making it distinct from *Ddc*$^{+4}$, which overproduces DDC at eclosion and underproduces DDC at pupariation. This control site(s) was identified in naturally occurring *Drosophila* strains which exhibit both elevated levels of

DDC activity and increased resistance to dietary α-methyldopa and was found to be closely linked to the genes responsible for these two phenotypes (*Ddc*$^+$ and *amd*$^+$; 0.002 cM apart). It is believed that this site(s) leads to the coordinated changes in the expression of the *Ddc*$^+$ and *amd*$^+$ genes in the variant strains.

IV. Regulatory Genes of Maize

A. Esterase

An early description of temporal gene variation in plants involved an esterase ("pH 7.5 esterase") (E.C. 3.1.1.1) mutant of maize (Schwartz, 1962). Unusual alleles at the enzyme structural gene (E_1), called "prime" alleles, caused premature loss of enzyme activity in the developing endosperm. However, there was no loss of activity in other tissues where the esterase was normally expressed (i.e., developing embryo and seedling). Therefore, the effect was tissue specific. Studies revealed that the genetic site responsible for the early loss of activity in the endosperm was closely linked to the structural gene and acted in a cis manner. Evidence that the proximate regulatory site and structural gene site are separate and distinct entities comes from an analysis of several "prime" alleles. The mutant developmental phenotype controlled by the "prime" site can be found in association with different electrophoretically distinct esterase alleles and the same allele can show different developmental phenotypes. Further detailed analysis of this system has not been reported.

B. Catalase

The major catalases (CAT) ($H_2O_2 : H_2O_2$ oxidoreductase) (E.C. 1.11.1.6) of maize, CAT-1, CAT-2, and CAT-3, are encoded in three unlinked structural genes, *Cat1*, *Cat2*, and *Cat3*, respectively (Scandalios, 1979). These genes were mapped in the maize genome using B–A translocation lines (Roupakias *et al.*, 1980): *Cat1* is on the short arm of chromosome 5, *Cat2* is on the short arm of chromosome 1, and *Cat3* is on the long arm of chromosome 1. The catalases of maize represent an interesting system for studying gene expression in a higher eukaryote since the isozymes appear to be regulated by a number of mechanisms during early sporophytic development, including the differential turnover of the *Cat1* and *Cat2* gene products (Quail and Scandalios, 1971) and regulation by an endogenous inhibitor

(Tsaftaris *et al.*, 1980). Catalase activity in the maize scutellum increases to a peak approximately 4 days after soaking the kernels in water for 24 hours and then growing them on moistened germination paper. This time course of activity reflects the differential expression of the CAT-1 and CAT-2 isozymes, both of which are found in the scutellum at this stage of development. CAT-1 activity gradually disappears while CAT-2 activity increases rapidly during the days following seed imbibition. These changes in activity are largely, if not entirely, due to changes in the levels of catalase protein caused by varying rates of synthesis and degradation (Quail and Scandalios, 1971; Scandalios *et al.*, 1980).

As the result of a screening program, several inbred lines that express an altered developmental program for catalase in the scutellum were found. In line R6-67, catalase activity in the scutellum increases rapidly following germination and maintains a level at least twice that observed in the standard inbred line W64A (Scandalios *et al.*, 1980). Using rocket immunoelectrophoresis to measure the amount of catalase protein, it was determined that this elevated catalase activity is concomitant with an increase in CAT-2 protein. Turnover studies showed that these elevated CAT-2 protein levels are a consequence of increased levels of synthesis (Scandalios *et al.*, 1980). Genetic analyses produced results consistent with the hypothesis that the elevated CAT-2 activity (and amount of protein) observed in R6-67 is due to a single locus with additive alleles (Scandalios *et al.*, 1980). By using electrophoretic variants of CAT-2 in genetic crosses, it was determined that the regulatory locus is loosely linked to *Cat2* on the short arm of chromosome 1, approximately 37 map units away. This locus has been designated *Car1* and represents a temporal regulatory gene. Evidence has been obtained which indicates that *Car1* is trans acting and exhibits strict tissue (scutellum), time (4-day postimbibition), and structural gene (*Cat2*) specificity (Scandalios *et al.*, 1980; Chandlee and Scandalios, 1984a) in its action. Although turnover studies (Scandalios *et al.*, 1980) indicate that *Car1* acts by regulating the rate of CAT-2 synthesis, the precise mode of action or level of control of *Car1* over *Cat2* is not yet fully understood. However, it was recently demonstrated (Kopczynski and Scandalios, 1986) that *Car1* acts by regulating the levels of translatable *CAT-2* mRNA in the scutellum.

The mechanism(s) by which *Car1* might regulate the levels of translatable CAT-2 mRNA during early sporophytic development are numerous. Since *Car1* is distally located from the *Cat2* structural gene and is trans acting, it most likely acts through a diffusible regulatory molecule. This molecule could affect the rate of gene transcription,

RNA processing, nuclear transport, or mRNA degradation. The possibility that the regulation may occur at the level of mRNA stability is being given particular attention since preliminary data indicate that the average poly(A) tail length of CAT-2 mRNA in 8-day-old R6-67 scutella may be considerably shorter than that in 8-day-old W64A scutella. Since the poly(A) tail of mRNA is believed to shorten with age (Sheiness and Darnell, 1973), the high levels of CAT-2 mRNA in R6-67 could represent the accumulation of old but functionally stable mRNA. This possibility and others are being explored with the aid of a *Cat2*-specific cDNA clone in order to elucidate further the mechanism(s) responsible for the temporal regulation of *Cat2* gene expression in the maize scutellum.

Attempts were made to determine whether all scutellar cells may be genetically programmed to activate expression of the *Cat2* gene synchronously, or whether there is an asynchronous spatial gradient of *Cat2* gene activation. Utilizing immunofluorescence microscopy and anti-CAT-2 IgG, it was demonstrated that a gradient of *Cat2* gene activation occurs within the scutellar cell mass during postgerminative development. The gradient of *Cat2* gene activation occurs from the outer perimeter of the tissue inward toward the embryonic axis. In an effort to determine a potential site of origin for any putative "triggering signal" for *Cat2* activation, it was demonstrated (Tsaftaris and Scandalios, 1986) that the *Cat2* gene is expressed in the single layer of aleurone cells prior to its expression in any other tissue during kernel development. To our knowledge, this is the first observation of a gradient-type spatial pattern of a eukaryote gene activation occurring in a stable, virtually nondividing tissue such as the maize scutellum.

It is conceivable that the aleurone cell layer, being the outermost layer of the embryonic cells of the kernel, is the first to sense environmental signals (e.g., water and temperature) and responds by furnishing a variety of secretory products to the rest of the kernel tissues to initiate germination and seedling growth. If there is such a response, perhaps a signal (*Car1* product?) for *Cat2* expression is initiated in the aleurone and subsequently diffuses inward to activate *Cat2* in a sequential manner. This notion is supported by the fact that the first groups of scutellar cells to express the *Cat2* gene are those nearest to the aleurone cells. Alternatively, the possibility for the existence of an inhibitory substance(s) in scutellar cells that must be inactivated for CAT-2 synthesis to occur has not been eliminated. Such a mechanism could also explain the gradual pattern of *Cat2* gene expression observed in the scutellum. Irrespective of the mechanism, it is evident that the expression of the *Cat2* structural gene in the scutellum during postgerminative development is not uniform in all scutellar cells but is spatially regulated.

Using approaches and analyses similar to those discussed above, a second regulatory locus has been identified which influences the expression of maize catalase in the scutellum (Chandlee and Scandalios, 1984b). This element, *Car2* acts specifically to decrease the rate of CAT-1 synthesis in the scutellum without affecting CAT-2 synthesis, as shown by density-labeling experiments. This element is genetically independent of *Car1*, but like *Car1*, it is tissue (scutellum), time, and structural gene (*Cat1*) specific in its action. This element exhibits additive inheritance, and the data available suggest that it is either closely linked or contiguous with the *Cat1* structural gene (Chandlee and Scandalios, 1984b). Preliminary experiments designed to determine the level of gene regulation at which *Car2* operates indicate that the overall levels of translatable *CAT-1* mRNA are influenced by *Car2*. It is possible that *Car2* accomplishes this by regulating the transcription rate of the *Cat1* gene. However, alternative possibilities, such as *Car2* affecting *CAT-1* mRNA processing or stability, or affecting the ability of *CAT-1* mRNA to associate with ribosomes, also exist. With the identification of *Car1* and *Car2* it is clear that the catalase developmental program in maize scutellum is attained through an extensive and complex interplay of various genetic sites responsible for the regulation of the expression of the *Cat1* and *Cat2* structural genes.

C. Alcohol Dehydrogenase

Alcohol dehydrogenase (E.C. 1.1.1.1) activity in the scutellum of most maize inbred lines (e.g., W64A) normally decreases following seed imbibition (Scandalios and Felder, 1971). In line R6-67, ADH activity is unusually high in the scutellum of the dry seed and remains at a level nearly two to three times that observed in line W64A during the 10 days following seed inbibition. Using rocket immunoelectrophoresis to measure the amount of ADH protein, it was determined that this variation in ADH activity is primarily due to changes in the amount of ADH-2 protein, the major isozymic form of maize ADH that is encoded in a structural gene (*Adh2*) on the long arm of chromosome 1. The scutella from 10-day-old seedlings were used in subsequent genetic analyses since the differences in ADH activity observed between the two inbred lines is greatest at this point in development. The results indicated that the altered developmental program of ADH-2 in R6-67 is due to a single gene with alleles exhibiting dominance–recessive inheritance (Lai and Scandalios, 1980). At 10 days, heterozygotes had an ADH activity level indistinguishable from that observed in W64A, suggesting that the rapid decrease in ADH-2 activity observed in W64A and in the F_1 progeny is due to a trans-acting dominant allele at this locus. This regulatory gene exhibiting dominance–recessive inheritance was designated *Adr1* and is unlinked to either *Cat2* nor *Adh2* (Lai and Scandalios, 1980).

Turnover studies combining density labeling (deuterium oxide; D_2O) with cesium chloride isopycnic centrifugation were undertaken to determine whether or not the *Adr1* locus controls ADH-2 synthesis and/or degradation. No evidence of significant ADH-2 synthesis in the scutellum of either W64A or R6-67 during early seedling development was found (Lai and Scandalios, 1980). These results confirm those obtained in earlier studies with W64A (Lai and Scandalios, 1977) and suggest that *Adr1* operates by influencing enzyme degradation. This would help to explain the dominance–recessive inheritance pattern since the regulatory action would be occurring posttranslationally.

For several years, it has been known that ADH activity in maize scutellum is negatively correlated with an endogenous inhibitor (Ho and Scandalios, 1975; Lai and Scandalios, 1977). Recent evidence suggests that the inhibitor is a 10,000-Da proteolytic enzyme that specifically inactivates ADH (Lai and Scandalios, 1980). It is tempting to speculate that the gene coding for the ADH inhibitor is *Adr1* and that R6-67 and W64A have alleles that differ in their expression of this inhibitor protein. In any case, the developmental profile of ADH-2 in the scutellum during germination appears to be at least partially determined by the lack of significant ADH-2 synthesis and by the action of an endogenous ADH-specific inhibitor (Scandalios, 1977).

Other instances of regulatory variation within the maize ADH isozyme system have been described. For instance, a cis-acting site that is closely linked with or within the *Adh1* gene [Freeling *Adh1* (Woodman and Freeling, 1981) is the same as Scandalios *Adh2* (Scandalios, 1977)] is responsible for what has been termed the "reciprocal effect" (Woodman and Freeling, 1981). From a study to identify *Adh1* alleles that vary in their quantitative levels of expression it was found that, within a given electrophoretic mobility class (F or S), alleles which were relatively underexpressed in the scutellum were subsequently relatively overexpressed in anaerobically induced roots and vice versa. Genetic analyses demonstrated that the effect on the quantitative levels of enzyme expressed always segregated with the electrophoretic mobility of the enzyme. The reciprocal effect site was estimated to map within 0.45 map units of the structural gene. All evidence indicated that the organ-specific quantitative behavior was not intrinsic to the biochemical properties of the ADH-1 polypeptides but instead was due to rates of polypeptide synthesis, at least in anaerobically grown primary roots (Woodman and Freeling, 1981).

All alleles included in the above study were naturally occurring alleles. An induced mutant, *Adh1-S1951a*, was recovered from progeny of a mutant originally induced with accelerated neon irradiation (Freeling *et al.*, 1982; Freeling and Woodman, 1979). This allele demonstrates an underproduction of ADH-1-S in the scutellum and an overproduction in the anaerobic primary root. The existence of this induced mutant indicates that a single mutational event can alter quantitative expression of *Adh1* simultaneously in different tissues. Together with an analysis of other induced allelic mutations it appears that the *Adh1* locus contains extensive information with regard to tissue specificity of expression and quantitative level of expression as well as the information for the polypeptide primary structure.

Another locus which affects alcohol dehydrogenase expression in the scutellum has been identified (Efron, 1970). Adh_r is linked to the *Adh2* locus, however, it maps 17 map units away. Two alleles exist at this regulatory site: Adh_r^N conditions equal levels of activity of the *Adh2-F* and *Adh2-S* alleles while Adh_r^L conditions lower activity of the *S* allele only. It can exert its effect either cis or trans to the *S* allele, suggesting that a diffusible regulatory molecule is involved. The Adh_r^L allele is dominant to Adh_r^N and only affects the *S* allele.

In summary, the ADH gene–enzyme system in maize will prove to be valuable in analyzing regulatory gene variation within a higher eukaryote. As with the catalase system, it appears that the developmental program of scutellar ADH is highly regulated through a complex interaction of both proximate and distant regulatory sites.

D. UDPGlucose : Flavonol 3-*O*-Glucosyltransferase (UFGT)

Although not fully characterized, there are several putative regulatory genes affecting anthocyanin biosynthesis in maize (Dooner, 1982; Soave and Salamini, 1984). At least nine loci have been implicated in the synthesis of anthocyanins within the aleurone tissue of maize kernels. These include *A*, *A2*, *Bz*, *Bz2*, *C*, *C2*, *Dek1*, *R*, and *Vp*. Four of these loci, *C*, *R*, *Vp*, and *Bz*, are required for the increase in activity of UDPglucose : flavonol 3-*O*-glucosyltransferase (UFGT) (E.C. 2.4.1.91) which is observed during kernel development (Dooner and Nelson, 1979). The *bronze* locus (*Bz*), based on several criteria (Dooner, 1982), appears to be the structural gene for UFGT. This enzyme is responsible for the glucosylation of flavonoids at the 3-OH position, one of the last steps in flavonoid glucoside (anthocyanin) biosynthesis. The *C*, *R*, and *Vp* loci appear to be regulatory in nature.

Alleles at the *C* locus include *C-I*, a dominant inhibitor of anthocyanin pigmentation and UFGT activity; *C*, a color-determining allele; and *c-p* and *c-n*, both recessive colorless (anthocyanin-negative) mutants which also affect UFGT activity. Endosperms with three doses of the *C-I* and *c-p* alleles have only 2–3% of the UFGT activity of normal endosperms (Dooner and Nelson, 1977). Plants homozygous for the *c-p* allele are conditional colored mutants which exhibit colorless aleurone but will accumulate anthocyanin pigments in the aleurone during germination in the light. Individuals homozygous for the *c-n* allele are colorless regardless of light conditions while individuals with the dominant *C* allele accumulate anthocyanin in the aleurone during seed maturation with and without light (Chen and Coe, 1977). If *c-p* kernels are exposed to light while still on the ear, the aleurone tissue will remain colorless until germination, at which time it will accumulate anthocyanin pigments even if the seedling is grown in the dark. Thus, the light stimulus appears to be stored until the onset of germination when induction occurs. A tentative model for the light-induction phenomenon has been proposed (Chen and Coe, 1977); the prominent feature of this model is that light induction and germination induction of anthocyanin pigmentation are two separable events and that the former precedes the latter. Although the regulatory function of *C* is not clear, the *c-p* allele is of major interest since this mutant dramatically affects the timing of gene expression.

The *R* locus is a compound locus associated with a tandem duplication. The *r-g* allele at this locus is a null allele giving rise to plants with no anthocyanin pigmentation in any seed or plant parts. Only 2–3% of the normal UFGT activity found in endosperm can be detected in lines homozygous for this allele (Dooner, 1979; Dooner and Nelson, 1977). Likewise, the *Vp* mutation also causes a similar reduction in UFGT activity in mature endosperm.

Two models have been proposed to explain the interaction between *Bz*, *C*, *R*, and *Vp* (Dooner and Nelson, 1979). According to the first model, *C*, *R*, and *Vp* would be regulatory genes coding for macromolecules which directly turn on the *Bz* gene. In the second model, *C*, *R*, and *Vp* would specify early enzymes in the anthocyanin pathway involved in the synthesis of a flavonoid precursor responsible for UFGT induction. This precursor induction of UFGT could be mediated by way of an activated regulatory protein. Although no clear choice can be made between the two models at this time, it would appear that the precursor-induction model is the more plausible. The evidence which has accumulated to date does suggest that *C* and *R* act before *C2*, *A*, *A2*, *Bz*, and *Bz2* in the anthocyanin biosynthesis pathway (Dooner, 1982). Also, since *Vp* is the only colorless aleurone mutant with a pleiotropic effect (vivipary; i.e., premature germination), it has been suggested that *Vp* may precede *C* and *R* in the pathway. Dooner and Nelson (1979) report that *Vp* aleurones are also deficient in phenylalanine ammonia lyase (PAL) activity. Thus, *Vp* could affect the pathway either through direct control of several genes or by its control of PAL, the enzyme responsible for producing the flavonoid precursor, cinnamic acid.

As mentioned, the *R* locus of maize is one of several genes necessary for anthocyanin pigmentation in the aleurone, possibly through its control of UFGT activity. In addition, the *R* locus appears to control the anthocyanin pigmentation of anthers and other plant parts including the coleoptile, seedling leaf tip, and root. Therefore, it is distinct from the *Bz*, *Bz2*, *A*, *A2*, *C*, and *C2* loci, which are required for pigmentation in the aleurone only. The R : r standard allele is a compound locus composed of a tandem duplication bearing a proximal (to the centromere) plant-pigmenting factor (P) and a distal seed-pigmenting factor (S). Because of this duplication, unequal crossing-over can occur, giving rise to progeny that can lack either member of the duplication. The *r-r* allele has lost (S), but retains (P), and gives rise to plants with colorless aleurone and pigmented plant parts. The *R-g* allele has lost (P), but retains (S), and gives rise to plants with colored aleurone but unpigmented plant parts. The *r-g* allele codes for no pigmentation in any seed or plant part and thus constitutes a null allele for this region. A clear explanation of the *R* locus and the gene symbols used has been presented (Dooner, 1979). An additional plant-pigmenting factor *Lc* has been reported (Bray, 1964) which maps 1.5 map units distal to *R* and appears to be part of a displaced duplicate *R* segment. However, there is no evidence that *Lc* is a compound locus and it does not condition aleurone pigmentation. When *R-g : 1 Lc/R-g : 1 Lc* individuals were crossed as females to *r-g lc/r-g lc* males, progeny were obtained which exhibited a unique tissue specificity in that they lacked both seed and leaf color but had pigmented anthers (*r-r lc*) (Dooner, 1979). This generation of a new tissue-specific function was apparently caused by unequal crossing-over involving two closely linked regions of the *R* locus which control tissue specificity. Based on this and other genetic evidence, it was hypothesized that the cis-acting components of *R* include a proximal region composed of closely linked sites controlling tissue specificity and a distal region necessary for *R* function. Supportive evidence for this has been obtained (Kermicle, 1980) using the controlling elements *Ds* and *Modulator* (*Mp*) to disrupt *R* function in *R-sc* : *124* individuals (presumably through integration at the *R* locus) and then recovering progeny with the original tissue specificity through recombination with an *R* region encoding a different tissue-specific function. These results indicate that *R* function consists of one component which controls tissue-specific expression and another which is common to alleles with different tissue-specific activities.

To explain how *R* may function, a model has been proposed (Dooner, 1979) in which the cis-acting proximal region of the *R* locus controlling tissue specificity is involved in determining in which tissues the *Bz*-activating signal is produced. The signal (e.g., protein or RNA) produced by the distal region of the *R* locus then activates the structural gene (*Bz*) for UFGT.

E. Zein

Several loci have been identified which exert a regulatory effect on the production of maize storage proteins during endosperm development (Salamini and Soave, 1982; Soave *et al.*, 1981; Soave and Salamini, 1984). These storage proteins, collectively termed zein, account for approximately 60% of the endosperm proteins in the mature seed. A thorough characterization of these regulatory loci has been somewhat hindered by the complexity of the zein multigene family, which codes for several different molecular-weight classes of polypeptides that exhibit extensive charge heterogeneity, as shown by isoelectric focusing in polyacrylamide gels (Larkins *et al.*, 1984; Salamini and Soave, 1982). The molecular-weight classes reported in the literature are not entirely consistent; however, two major components are present with molecular weights of about 22K and 19K (Larkins *et al.*, 1984; Salamini and Soave, 1982). It is not known precisely how many structural genes code for zein in the maize genome. The estimates vary depending on the method of analysis. So far, twenty zein genes have been mapped (Soave and Salamini, 1984); of these, eight have been located on the short arm of chromosome 4, two on the long arm of chromosome 4, nine on the short arm of chromosome 7, and one on the long arm of chromosome 10. In general, these locations have been verified by *in situ* hybridization experiments (Viotti *et al.*, 1980).

The first zein regulatory loci to be studied in some detail were the *opaque 2* (*o2*) and *floury 2* (*fl2*) mutants. Subsequently, the *opaque 7* (*o7*), *opaque 6* (*o6*), *floury 3* (*fl3*), *Mc*, and *De*-B30* mutants were identified. The *o2* and *De*-B30* mutants preferentially reduce the level of the 22-kDa zeins, *o7* preferentially reduces the 19-kDa zeins, while *fl2*, *Mc*, and *o6* preferentially reduce the synthesis of all the zein classes to the same extent (Soave and Salamini, 1984). These regulatory genes have been mapped to regions nearby those occupied by some of the zein genes; *o2* is on the short arm of chromosome 7, *fl2* is on the short arm of chromosome 4, and *o7* is on the long arm of chromosome 10.

The mechanism by which these mutations influence zein synthesis is unknown. The genes *o2*, *o7*, and *fl2* decrease zein mRNA populations while *o6*, *Mc*, and *De*-B30* lower the level of translatable zein mRNA (Langridge *et al.*, 1982; Soave and Salamini, 1984). This latter set, however, has not yet been thoroughly probed at the mRNA level to determine if mRNA population sizes are affected (Soave and Salamini, 1984).

The *o2* mutation acts as a simple Mendelian recessive and results in seed with about 50% less zein than normal. Zein synthesis begins several days late, proceeds at a slower rate, and ceases midway through endosperm development. The *o2* locus must act through a diffusible regulatory molecule, because it affects genes not linked to it. It has been shown that a 32-kDa protein is under the control of *o2* (Soave *et al.*, 1981). The b-32 protein is present in the soluble cytoplasm of the endosperm of normal kernels but absent in seven recessive alleles at the *o2* locus (Soave *et al.*, 1981). The *o2* gene does not encode the b-32 protein, however. Instead it appears to be the product of the *o6* locus. It is believed that *o2* activates *o6* which in turn, through the b-32 protein, controls the level of mRNA of a battery of zein genes (Soave *et al.*, 1981; Soave and Salamini, 1984). It is possible that the b-32 protein affects zein mRNA population sizes through transcriptional regulation or, alternatively, reduces the efficiency of translation of zein mRNA.

The *o7* mutation causes a reduction in the synthesis of the 19-kDa zein class. As with *o2*, the mutation acts as a simple Mendelian recessive. The fact that the two regulatory loci affect different zein polypeptides has led to the belief that multiple regulatory pathways are active in zein synthesis (Salamini and Soave, 1982; Soave and Salamini, 1984). Both probably function via diffusible regulatory molecules since they can affect either linked or unlinked zein genes.

The *fl2* mutant appears to act by a different mechanism than the opaque mutants; it reduces overall zein synthesis in the kernel and acts in a dosage-dependent manner (Salamini and Soave, 1982; Soave and Salamini, 1984). The mutant *fl3* shows a similar dosage effect because, as the dose increases, zein production is decreased.

The defective endosperm *B-30* (*De*-B30*) mutation is dominant and causes a 30% reduction in zein synthesis. This mutant is closely linked to *o2* and again preferentially reduces expression of the 22-kDa zein class.

An association has been found between dominant or semidominant zein regulatory loci (*fl2, Mc,* and *De*-B30*) and the overproduction of a 70-kDa protein (Galante *et al.*, 1983; Salamini *et al.*, 1983; Soave and Salamini, 1984). It is possible that this b-70 protein is involved in the zein synthetic–secretory system (Soave and Salamini, 1984). All attempts to explain the regulation of zein production by dominant mutants, as in the case of these three loci, are based on a posttranslational mechanism which is disrupted by mutations at the level of processing or compartmentation of the zein proteins.

An additional factor to be considered in zein regulation is the observation that starch-forming mutants of maize have been reported to reduce zein content (Dalby and Tsai, 1975; Glover *et al.*, 1975; Misra *et al.*, 1975; Tsai *et al.*, 1978). The starch-forming mutants can be divided into two distinct classes: the starch-modified mutants and the starch-deficient mutants. It has been reported (Tsai *et al.*, 1978) that when *o2* is combined with each of the starch-modified mutants, zein content is further reduced while nonzein protein content is not. Their results suggest a cumulative effect between *o2* and the starch-modified genes in reducing zein expression. However, when *o2* is combined with each of the starch-deficient mutants, a synergistic effect is observed with the double mutants accumulating very little zein. In addition, the double-mutant combinations yielded SDS–polyacrylamide gel patterns distinctly different from that characteristic of either parent. In the case of *bt2* (*brittle 2*), *o2*, and *bt2o2* kernels, this reduction in zein content is correlated with a reduced recovery of membrane-bound polysomes capable of directing zein polypeptide synthesis *in vitro* (Tsai *et al.*, 1978). Significantly, the RNase level of *bt2o2* endosperm is roughly three times that observed in either *bt2* or *o2* endosperm and seven times that observed in normal endosperm (from the same genetic background), suggesting that this enzyme plays a role in regulating zein expression (Lee and Tsai, 1984). The activity and timing of expression of RNase I in maize endosperm appears to be under genetic control, but the high variability in enzyme activity observed during development has rendered a thorough genetic analysis difficult (Wilson, 1973, 1978).

V. Regulatory Genes of Rainbow Trout

PHOSPHOGLUCOMUTASE

The *Pgm1* locus of rainbow trout (*Salmo gairdneri*) encodes the enzyme phosphoglucomutase (PGM) (E.C. 5.4.2.2). Two electrophoretically distinct isozyme alleles, as well as a null allele, have been identified at this locus. During a search for intraspecific variation in the tissue-specific pattern of expression of this enzyme (Allendorf *et al.*, 1982, 1983b), a regulatory gene was identified which affects the expression of PGM1 in the liver. This regulatory locus has been designated *Pgm1-t*. Two alleles have been identified; *Pgm1-t(a)* does not express PGM1 in the liver and *Pgm1-t(b)* causes about a 100-fold increase in the level of expression in this tissue. Other tissues (i.e., skeletal muscle, heart, and brain) exhibit equal levels of PGM1 expression regardless of which allele is present at *Pgm1-t*. Therefore, the effect of this regulatory locus is tissue specific. The expression of PGM1 in the liver is likely a recent mutation. Several species closely related to rainbow trout do not express PGM1 in liver tissue. Also, the *Pgm1-t(b)* allele is rare in rainbow trout. It has only been identified in 4 of 30 strains which were screened. Genetic studies indicate that the regulatory locus is inherited in an additive manner; however, it has not been clearly demonstrated whether it acts cis or trans to the structural gene. Also, its location relative to the structural gene has not been determined.

This regulatory system is of special interest since it appears to support the hypothesis that mutations in regulatory genes play an important role in evolutionary processes by causing changes in developmental events that lead to adaptive changes in morphology. Embryos that express PGM1 in the liver hatch earlier than those that do not, possibly due to the faster embryonic developmental rates (Allendorf *et al.*, 1983a,b). They also appear to mature faster sexually. Thus, a single regulatory mutation appears to lead to apparent important phenotypic differences that may be of important adaptive significance.

VI. Discussion

The realization that the majority of the cells of an organism contain the same genetic constitution indicates that differences among differentiated tissues can be accounted for in terms of differential gene expression of the same pool of genetic information. While it is clear that differentiated cell types within an organism express different populations of genes, it is not fully understood how this differential gene activity is achieved. An understanding of the mechanisms involved in the regulation of differential gene expression in higher eukaryotes will eventually lead to a greater understanding of differentiation and development. The identification and characterization of regulatory genes which control different aspects of gene expression will be a valuable contribution toward this end.

There is now available a significant amount of information concerning the nature of higher eukaryotic regulatory genes. Many of these genes have been identified in those organisms that lend themselves well to genetic and biochemical studies (the necessary prerequisites for the identification of genetic regulatory variation); however, it should not be presumed that such regulatory genes are unique to these organisms. On the contrary, they most likely represent general strategies used by eukaryotic cells to regulate the specialized patterns of gene expression necessary for the maintenance of the differentiated states of the different tissues of a multicellular organism. All differentiated organisms must somehow regulate the expression of different genes in different tissues, and one way of doing this is by having programmed "switches" in the genome that turn genes on and off or control their level of expression. Even though a clearer understanding is emerging with regard to the nature of regulatory genes, we are still far from understanding the complex cellular interactions which the regulatory genes respond to so that they know where and when to exert their specific action. Thus, a more complete understanding of the role of regulatory genes in the complex processes of differentiation and development will require a more direct analysis of these regulatory sites.

The majority of the regulatory genes described in this report were identified and characterized using similar methodologies. Initially, it is necessary to work with a gene–enzyme (protein) system which has been well characterized genetically, developmentally, and biochemically. Once this type of information is obtained, variants exhibiting alterations in the "typical" patterns of gene expression within the enzyme system can be analyzed. It is necessary to determine the full extent of the alteration in gene expression (i.e., the tissue and temporal specificity of the effect). It is also important to verify that the altered pattern of gene expression is not due to an alteration in the primary structure of the polypeptide(s) which make up the protein.

Such alterations could conceivably change turnover parameters, subcellular localization, or catalytic efficiencies, which may initially appear to be due to regulatory gene variation. Once this is shown not to be the case it is best to systematically analyze the altered pattern of gene expression, first at the protein level, then at the mRNA level, and finally at the gene level. For instance, it is important to determine if the number of protein molecules is altered in the variant and if this alteration is due to protein synthesis rates. It is important to determine the mode of inheritance of the variation and whether the action is in cis or trans. All these characteristics can be studied by genetic analysis and with basic techniques available for the analysis of proteins (e.g., electrophoresis, density- and radio-labeling techniques, and immunological approaches). Once it clearly appears that a distinct genetic component is involved in controlling the observed protein variation, the next step would be analysis at the mRNA level of gene expression to gain greater insight into the mechanism of action of the regulatory gene. For instance, it is important to determine if the sizes of the mRNA populations are being regulated or if the mRNA availability for translation is regulated. Once again, fairly routine analyses can be performed (e.g., polysome or total mRNA isolation, *in vitro* translation, and Northern blots). Finally, it is important eventually to analyze gene structure and define precisely the DNA sequences that exert the regulatory action. This type of analysis is becoming more amenable with the advances in recombinant DNA technology which have greatly simplified gene cloning. This is evident with the recent cloning of several genes whose expression is controlled by well-characterized regulatory loci (e.g., *Adh* and *Amy* of *Drosophila* and *Adh2*, *Cat2*, *Bz*, *R*, and *A* of maize).

Four general categories of regulatory genes have been discussed in this report. However, it is important to point out that these categories are not mutually exclusive and overlap does occur. These classifications are made to allow for simple generalizations about the properties of eukaryotic regulatory genes and to provide a useful framework in which to organize the information available concerning these regulatory genes. A more complex classification scheme has been presented (Paigen, 1979a) and the one used herein is a simplified modification of that scheme, intended only as a mechanism to facilitate this discussion. In conclusion, it is best to remember that these are regulatory genes, that is, genes that influence the expression of a separate and distinct structural gene. The means by which they achieve this regulation is varied. What follows is a summary of the properties of the different types of eukaryotic regulatory genes:

1. *Processing genes*. This group of regulatory genes affects the expression of structural genes at the posttranslational level. It is, therefore, likely that these genes are less specific in their action than other types of regulatory genes. This is true for most of the processing genes described (i.e., *mal, cin, lxd,* and *m-est*). All these regulatory genes affect the expression of more than one structural gene product. One exception is *Ce*, which appears to specifically degrade only catalase in the liver of the mouse (Heston *et al.*, 1965). However, it is not clear how many other proteins were investigated to determine the overall influence of this degradation process. Since these genes affect posttranslational processes it is expected that they exhibit a dominance–recessive pattern of inheritance and this is observed in all cases except for *Eg* of the β-glucuronidase system of the mouse, which exhibits additive inheritance. However, additive inheritance would be expected for a regulator with the mechanism of action of egasyn (i.e., it binds permanently to β-glucuronidase to carry out its "anchoring" function). All processing genes identified to date map distant to the gene under regulation and also act in trans. Both of these properties might be expected for cellular regulatory mechanisms that act at the posttranslational level to affect the expression of a small set of proteins through the action of a diffusible regulatory molecule (possibly a modifying enzyme or protein in each case).

2. *Systemic regulatory genes*. This group of regulatory genes affects the rates of synthesis of a particular protein uniformly throughout all tissues and organs of an organism. It is likely that these regulators represent sequences adjacent to, or part of, the structural gene under regulation and are important for controlling transcriptional processes which determine the overall mRNA concentration. Alternatively, it is possible that they may act posttranscriptionally and represent sites which function in regulating mRNA processing, stability, or ability to associate with ribosomes. In either case, the regulatory control is exerted pretranslationally. All systemic regulatory genes reported to date map proximate to the structural gene they control and all act in cis. This is consistent with the pattern of additive inheritance. In all cases tested, so far, the synthesis levels of the structural gene product are affected. In the future, it will be necessary to determine precisely at what level these genes operate so that their mechanism of action is better understood. For the purpose of this review *i409* of the *Xdh* system of *Drosophila* and Adh_r of the *Adh* system of maize have been classified as systemic regulatory genes. However, they exhibit properties unlike the other systemic regulators (i.e., *i409* exhibits tissue spe-

cificity and Adh_r exhibits allele specificity while acting in trans). As more information is obtained about their mechanisms of action, they may be classified differently.

3. *Effector response regulatory genes.* This group of regulatory genes regulates the response of particular structural genes to inducer molecules which are required to elicit gene activity. The primary example discussed in this report is *Gus-r* of the β-glucuronidase system of the mouse. This site likely represents a DNA sequence adjacent to the structural gene which is responsible for binding a hormone receptor–hormone complex, which then allows for the induction of gene activity. Like systemic regulatory genes each is located proximate to the structural gene and acts in cis, and the alleles are inherited in an additive manner. Thus, while this type of regulator has similarities to systemic regulators, the primary distinguishing characteristic among the two groups of regulatory genes is the type of signal each group responds to; the signal is known for effector-response regulators (i.e., hormone–hormone receptor complex) and it is not yet known for systemic regulators. As more is learned about the signals that systemic regulators respond to, the distinction between the two groups may become less obvious.

4. *Temporal regulatory genes.* This group of regulatory genes is responsible for controlling the concentration changes that a given protein undergoes during cellular differentiation. In other words, these genes are part of the genetic programming which determines the developmental pattern of expression for a given enzyme or protein. Inherent in this definition is the temporal and spatial specificity of the action of these regulators, since any given protein will commonly exhibit different patterns of developmental expression in different tissues. They are particularly interesting because of their close association with processes of development and differentiation. Temporal regulatory genes appear to fall into two classes. Many, but not all, map proximate to the structural gene being regulated and act in cis. Consequently, they exhibit additive inheritance. In all cases tested so far they appear to influence the rate of synthesis of the regulated protein and, therefore, like systemic regulators, likely act pretranslationally. The second class consists of temporal genes that map distant to the structural gene being regulated and act in trans (i.e., *Agt, Map,* and *Car1*). Again, they appear to influence levels of polypeptide synthesis. What is unusual about this second class is that the genes exhibit additive inheritance even though they presumably act in trans via some diffusible regulatory molecule. This finding suggests that they act

through a unique regulatory mechanism quite different from other distant regulators and that they do not simply activate an ON–OFF switch at the target gene.

Some temporal genes exist which are exceptions to the generalizations of the above two class categories. For instance, *gus-t* of the β-glucuronidase system of the mouse maps proximal to the structural gene, but acts in trans. The significance of this is not yet understood and it is hard to conceive of a reason for the evolutionary selection of a closely linked trans regulator. This likely reflects something integral to the mechanism of action of the *gus-t* site. Other exceptions are *Adr1* of the *Adh* system in maize and a temporal gene affecting the *Gpdh* system of *Drosophila*. In both instances, the temporal gene maps distant to the structural gene and acts in trans; however, both seem to affect degradation rates and exhibit a dominance–recessive pattern of inheritance. In a strict sense these are temporal genes, since they affect developmental patterns of gene expression in specific tissues; however, their mechanism of action is probably similar to processing genes since they must act posttranslationally.

It appears that regulatory gene systems of higher eukaryotes can be and commonly are composed of both proximate and distant regulatory sites which act in concert to bring about the unique pattern of developmental and tissue-specific expression for a given structural gene. The large portion of the genetic regulatory variation which is closely linked to the structural genes probably represents adjacent 5′-regulatory regions to the structural gene. This region may include binding sites for various factors necessary to initiate the process of transcription. However, the verification of this concept awaits future experimental results. In the case of temporal genes, the unlinked regulators may code for the signals which interact with these proximate sites. The high frequency of occurrence of closely linked regulatory information indicates that genes are more than just transcriptional units with simple ON–OFF switches and may have to be redefined to account for the adjacent sequences which contain information encoding both the quantitative and developmental patterns of expression for a given structural gene.

Acknowledgments

Research from the authors' laboratory has been supported by research grants from the National Institutes of Health (GM22733 and GM33817) and the U.S. Environmental Protection Agency (R 812404) to J. G. Scandalios.

REFERENCES

Abraham, I., and Doane, W. W. (1978). Genetic regulation of tissue-specific expression of *Amylase* structural genes in *Drosophila melanogaster. Proc. Natl. Acad. Sci. U.S.A.* **75**, 4446–4450.

Allendorf, F. W., Knudsen, K. L., and Phelps, S. R. (1982). Identification of a gene regulating the tissue expression of a phosphoglucomutase locus in rainbow trout. *Genetics* **102**, 259–268.

Allendorf, F. W., Knudsen, K. L., and Leary, R. F. (1983a). Adaptive significance of differences in the tissue-specific expression of a phosphoglucomutase gene in rainbow trout. *Proc. Natl. Acad. Sci. U.S.A.* **80**, 1397–1400.

Allendorf, F. W., Leary, R. F., and Knudsen, K. L. (1983b). Structural and regulatory variation of phosphoglucomutase in rainbow trout. *In* "Isozymes: Current Topics in Biological and Medical Research" (M. C. Rattazzi, J. G. Scandalios, and G. S. Whitt, eds.), Vol. 9, pp. 123–142. Liss, New York.

Balak, K. J., Keith, R. H., and Felder, M. R. (1982). Genetic and developmental regulation of mouse liver alcohol dehydrogenase. *J. Biol. Chem.* **257**, 15000–15007.

Bentley, M. M., and Williamson, J. H. (1982). The control of aldehyde oxidase and xanthine dehydrogenase activities and crm levels by the *Mal* locus in *Drosophila melanogaster. Can. J. Genet. Cytol.* **24**, 11–17.

Benyajati, C., Place, A. R., Powers, D. A., and Sofer, W. (1981). Alcohol dehydrogenase gene of *Drosophila melanogaster:* Relationship of intervening sequences to functional domains in the protein. *Proc. Natl. Acad. Sci. U.S.A.* **78**, 2717–2721.

Berger, F. G., Paigen, K., and Meisler, M. (1978). Regulation of the rate of β-galactosidase synthesis by the *Bgs* and *Bgt* loci in the mouse. *J. Biol. Chem.* **253**, 5280–5282.

Berger, F. G., Breen, G., and Paigen, K. (1979). Genetic determination of the developmental program for mouse liver β-galactosidase: Involvement of sites proximate to and distant from the structural gene. *Genetics* **92**, 1187–1203.

Bernstine, E. G. (1979). Genetic control of mitochondrial malic enzyme in mouse brain. *J. Biol. Chem.* **254**, 83–87.

Bernstine, E. G., and Koh, C. (1980). A *cis*-active regulatory gene in the mouse: Direct demonstration of *cis*-active control of the rate of enzyme subunit synthesis. *Proc. Natl. Acad. Sci. U.S.A.* **77**, 4193–4195.

Bewley, G. C. (1983). The genetic and epigenetic control of *sn*-glycerol-3-phosphate dehydrogenase isozyme expression during the development of *Drosophila melanogaster. In* "Isozymes: Current Topics in Biological and Medical Research" (M. C. Rattazzi, J. G. Scandalios, and G. S. Whitt, eds.), Vol. 9, pp. 33–62. Liss, New York.

Birley, A. J., Couch, P. A., and Marson, A. (1981). Genetical variation for enzyme activity in a population of *Drosophila melanogaster*—VI. Molecular variation in the control of alcohol dehydrogenase (ADH) activity. *Heredity* **47**, 185–196.

Boubelik, M., Lengerová, A., Bailey, D. W., and Matovsék, V. (1975). A model for genetic analysis of programmed gene expression as reflected in the development of membrane antigens. *Dev. Biol.* **47**, 206–214.

Bray, R. A. (1964). A plant color factor linked to the *R* locus. *Maize Genet. Coop. Newslett.* **38**, 134.

Breen, G., Lusis, A., and Paigen, K. (1977). Linkage of genetic determinants for mouse β-galactosidase electrophoresis and activity. *Genetics* **85**, 73–84.

Browder, L. W., Tucker, L., and Wilkes, J. (1982a). Xanthine dehydrogenase (XDH) cross-reacting material in mutants in *Drosophila melanogaster. Biochem. Genet.* **20**, 125–132.

Browder, L. W., Wilkes, J., and Tucker, L. (1982b). Aldehyde oxidase cross-reacting material in the *Aldox-n, cin, mal* and *lxd* mutants of *Drosophila melanogaster*. *Biochem. Genet.* **20,** 111–124.

Brown, D. D. (1981). Gene expression in eukaryotes. *Science* **211,** 667–674.

Bullock, L., and Bardin, C. (1974). Androgen receptors in mouse kidney: A study of male, female and androgen-insensitive (tfm/y) mice. *Endocrinology* **94,** 746–756.

Chandlee, J. M., and Scandalios, J. G. (1984a). Analysis of variants affecting the catalase developmental program in maize scutellum. *Theor. Appl. Genet.* **69,** 71–77.

Chandlee, J. M., and Scandalios, J. G. (1984b). Regulation of *Cat1* gene expression in the scutellum of maize during early sporophytic development. *Proc. Natl. Acad. Sci. U.S.A.* **81,** 4903–4907.

Chen, S. M., and Coe, E. H. (1977). Control of anthocyanin synthesis by the *C* locus in maize. *Biochem. Genet.* **15,** 333–346.

Chovnick, A. (1973). Gene conversion and transfer of genetic information within the inverted region of inversion heterozygotes. *Genetics* **75,** 123–131.

Chovnick, A., Ballantyne, G. H., and Holm, D. G. (1971). Studies on gene conversion and its relationship to linked exchange in *Drosophila melangaster. Genetics* **69,** 163–178.

Chovnick, A., Gelbart, W., McCarron, M., Osmond, B., Candido, E. P. M., and Baillie, D. L. (1976). Organization of the *rosy* locus in *Drosophila melanogaster:* Evidence for a control element adjacent to the xanthine dehydrogenase structural element. *Genetics* **84,** 233–255.

Chovnick, A., Gelbart, W., and McCarron, M. (1977). Organization of the *rosy* locus in *Drosophila melanogaster. Cell* **11,** 1–10.

Chovnick, A., McCarron, M., Hilliker, A., O'Donnell, J., Gelbart, W., and Clark, S. (1978a). Gene organization in *Drosophila. Cold Spring Harbor Symp. Quant. Biol.* **42,** 1011–1021.

Chovnick, A., McCarron, M., Hilliker, A., O'Donnell, J., Gelbart, W., and Clark, S. (1978b). Organization of a gene in *Drosophila:* A progress report. *Stadler Symp.* **10,** 9–24.

Chovnick, A., McCarron, M., Clark, S. H., Hilliker, A. J., and Rushlow, C. A. (1980). Structural and functional organization of a gene in *Drosophila melanogaster. In* "Development and Neurobiology of *Drosophila*" (O. Siddiqi, P. Babu, L. M. Hall, and S. C. Hall, eds.), pp. 3–23. Plenum, New York.

Clark, S. H., Daniels, S., Rushlow, C. A., Hilliker, A. J., and Chovnick, A. (1984). Tissue-specific and pretranslational character of variants of the *rosy* locus control element in *Drosophila melanogaster. Genetics* **108,** 953–968.

Cochrane, B. J., and Richmond, R. C. (1979). Studies of esterase 6 in *Drosophila melanogaster*. I. The genetics of a post-translational modification. *Biochem. Genet.* **17,** 167–183.

Coleman, D. L. (1971). Linkage of genes controlling the rate of synthesis and structure of aminolevulinate dehydratase. *Science* **173,** 1245–1246.

Courtwright, J. B. (1976). *Drosophila* gene–enzyme systems. *Adv. Genet.* **18,** 249–314.

Dalby, A., and Tsai, C. Y. (1975). Comparisons of lysine and zein and nonzein protein contents in immature and mature maize endosperm mutants. *Crop Sci.* **15,** 513–515.

Daniel, W. L. (1976a). Genetic control of heat sensitivity and activity level of murine arylsulfatase B. *Biochem. Genet.* **14,** 1003–1018.

Daniel, W. L. (1976b). Genetics of murine liver and kidney arylsulfatase B. *Genetics* **82,** 477–491.

Darnell, J. E. (1982). Variety in the level of gene control in eukaryotic cells. *Nature (London)* **297**, 365–371.

Dewey, M., and Mintz, B. (1978). Genetic control of cell-type-specific levels of mouse β-galactosidase. *Dev. Biol.* **66**, 560–563.

Dickinson, W. J. (1970). The genetics of aldehyde oxidase in *Drosophila melanogaster. Genetics* **66**, 487–496.

Dickinson, W. J. (1975). A genetic locus affecting the developmental expression of an enzyme in *Drosophila melanogaster. Dev. Biol.* **42**, 131–140.

Dickinson, W. J. (1978). Genetic control of enzyme expression in *Drosophila:* A locus influencing tissue specificity of aldehyde oxidase. *J. Exp. Zool.* **206**, 333–342.

Dickinson, W. J., and Weisbrod, E. (1976). Gene regulation in *Drosophila:* Independent expression of closely linked, related structural loci. *Biochem. Genet.* **14**, 709–721.

Doane, W. W. (1970). *Drosophila* amylases and problems in cellular differentiation. *In* "Problems in Biology: RNA in Development" (E. W. Hanly, ed.), pp. 73–109. Univ. of Utah Press, Salt Lake City, Utah.

Doane, W. W. (1980). Midgut activity patterns in *Drosophila:* Nomenclature. *Dros. Inf. Serv.* **55**, 36–39.

Doane, W. W., Abraham, I., Koiar, M. M., Martenson, R. E., and Deibler, G. E. (1975). Purified *Drosophila* α-amylase isozymes: Genetical, biochemical and molecular characterization. *In* "Isozymes" (C. L. Markert, ed.), Vol. 4, pp. 585–607. Academic Press, New York.

Doane, W. W., Treat-Clemons, L. G., Gemmill, R. M., Levy, J. N., Hawley, S. A., Buchberg, A. M., and Paigen, K. (1983). Genetic mechanism for tissue-specific control of α-amylase expression in *Drosophila melanogaster. In* "Isozymes: Current Topics in Biological and Medical Research" (M. C. Rattazzi, J. G. Scandalios, and G. S. Whitt, eds.), Vol. 9, pp. 63–90. Liss, New York.

DoFuku, R., Tettenborn, U., and Ohno, S. (1971). Testosterone "Regulon" in the mouse kidney. *Nature (London) New Biol.* **232**, 5–7.

Dooner, H. K. (1979). Identification of an *R*-locus region that controls the tissue specificity of anthocyanin formation in maize. *Genetics* **93**, 703–710.

Dooner, H. (1982). Gene–enzyme relationships in anthocyanin biosynthesis in maize. *In* "Maize for Biological Research" (W. Sheridan, ed.), pp. 123–128. Univ. of North Dakota Press.

Dooner, H., and Nelson, O. (1977). Genetic control of UDP glucose : flavonol 3-*O*-glucosyltransferase in the endosperm of maize. *Biochem. Genet.* **15**, 509–515.

Dooner, H. K., and Nelson, O. E. (1979). Interaction among *C, R,* and *Vp* in the control of the *Bz* glucosyltransferase during endosperm development in maize. *Genetics* **91**, 309–315.

Doyle, D., and Schimke, R. T. (1969). The genetic and developmental regulation of hepatic aminolevulinate dehydratase in mice. *J. Biol. Chem.* **244**, 5449–5459.

Edwards, T. C. R., Candido, E. P. M., and Chovnick, A. (1977). Xanthine dehydrogenase from *Drosophila melanogaster:* A comparison of the kinetic parameters of the pure enzyme from two wild-type isoalleles differing at a putative regulatory site. *Mol. Gen. Genet.* **154**, 1–6.

Efron, Y. (1970). Alcohol dehydrogenase in maize: Genetic control of enzyme activity. *Science* **170**, 751–753.

Estelle, M. A., and Hodgetts, R. B. (1984a). Genetic elements near the structural gene modulate the level of dopa decarboxylase during *Drosophila* development. *Mol. Gen. Genet.* **195**, 434–441.

Estelle, M. A., and Hodgetts, R. B. (1984b). Insertion polymorphisms may cause stage specific variation in mRNA levels for dopa decarboxylase in *Drosophila*. *Mol. Gen. Genet.* **195**, 442–451.

Felder, M. R. (1980). Biochemical and developmental genetics of isozymes in the mouse, *Mus musculus*. *In* "Isozymes: Current Topics in Biological and Medical Research" (M. C. Rattazzi, J. G. Scandalios, and G. S. Whitt, eds.), Vol. 4, pp. 1–68. Liss, New York.

Felder, M. R., Burnett, K. G., and Balak, K. J. (1983). Genetics, biochemistry, and developmental regulation of alcohol dehydrogenase in *Peromyscus* and labortory mice. *In* "Isozymes: Current Topics in Biological and Medical Research" (M. C. Rattazzi, J. G. Scandalios, and G. S. Whitt, eds.), Vol. 9, pp. 143–161. Liss, New York.

Felton, J., Meisler, M., and Paigen, K. (1974). A locus determining β-galactosidase activity in the mouse. *J. Biol. Chem.* **249**, 3267–3272.

Finnerty, V. (1976). Genetic units of *Drosophila*—Simple cistrons. *In* "The Genetics and Biology of *Drosophila*" (M. Ashburner and E. Novitski, eds.), Vol. 1b, pp. 721–766. Academic Press, New York.

Finnerty, V., McCarron, M., and Johnson, G. B. (1979). Gene expression in *Drosophila:* Post-translational modification of aldehyde oxidase and xanthine dehydrogenase. *Mol. Gen. Genet.* **172**, 37–43.

Fletcher, T. S., Ayala, F. J., Thatcher, D. R., and Chambers, G. K. (1978). Structural analysis of the ADH-S electromorph of *Drosophila melanogaster*. *Proc. Natl. Acad. Sci. U.S.A.* **75**, 5609–5612.

Freeling, M., and Woodman, J. C. (1979). Regulatory variant and mutant alleles in higher organisms and their possible origin via chromosomal breaks. *In* "The Plant Seed" (I. Rubenstein, R. L. Phillips, C. E. Green, and B. G. Gengenbach, eds.), pp. 85–111. Academic Press, New York.

Freeling, M., Cheng, D. S., and Alleman, M. (1982). Mutant alleles that are altered in quantitative, organ-specific behavior. *Dev. Genet.* **3**, 179–196.

Galante, E., Vitale, A., Manzocchi, L. A., Soave, C., and Salamini, F. (1983). Genetic control of a membrane component and zein deposition in maize endosperm. *Mol. Gen. Genet.* **192**, 316–321.

Ganschow, R. (1975). Simultaneous genetic control of the structure and rate of synthesis of murine glucuronidase. *In* "Isozymes" (C. L. Markert, ed.), Vol. 4, pp. 633–647. Academic Press, New York.

Ganschow, R., and Paigen, K. (1967). Separate genes determining the structure and intracellular location of hepatic glucuronidase. *Proc. Natl. Acad. Sci. U.S.A.* **58**, 938–945.

Ganschow, R., and Schimke, R. T. (1969). Independent genetic control of the catalytic activity and the rate of degradation of catalase in mice. *J. Biol. Chem.* **244**, 4649–4658.

Gemmill, R. M., Levy, J. N., and Doane, W. W. (1985). Molecular cloning of α-amylase genes from *Drosophila melanogaster*. I. Clone isolation by use of a mouse clone. *Genetics* **110**, 299–312.

Gibson, J. B. (1972). Differences in the number of molecules produced by two allelic electrophoretic enzyme variants in *Drosophila melanogaster*. *Experientia* **28**, 975–976.

Glover, D. V., Crane, P. L., Misra, P. S., and Mertz, E. T. (1975). Genetics of endosperm mutants in maize as related to protein quality and quantity. *In* "High Quality Protein Maize" (Purdue Univ. and the International Maize and Wheat Improvement Center), pp. 228–240. Dowden, Hutchison & Ross, Stroudsburg, Pennsylvania.

Goldberg, D. A. (1980). Isolation and partial characterization of the *Drosophila* alcohol dehydrogenase gene. *Proc. Natl. Acad. Sci. U.S.A.* **77**, 5794–5798.

Grossman, A. (1980). Analysis of genetic variation affecting the relative activities of fast and slow ADH dimers in *Drosophila melanogaster* heterozygotes. *Biochem. Genet.* **18**, 765–780.

Heston, W. E., Hoffman, H. A., and Rechcigl, M. (1965). Genetic analysis of liver catalase activity in two substrains of mice. *Genet. Res.* **6**, 387–397.

Hickey, D. A. (1981). Regulation of amylase activity in *Drosophila melanogaster:* Variation in the number of enzyme molecules produced by different amylase genotypes. *Biochem. Genet.* **19**, 783–796.

Ho, D., and Scandalios, J. G. (1975). Regulation of alcohol dehydrogenase in maize scutellum during germination. *Plant Physiol.* **56**, 56–59.

Hoffman, H. A., and Grieshaber, C. K. (1974). Genetic studies of murine catalase: Liver and erythrocyte catalase controlled by independent loci. *J. Hered.* **65**, 277–279.

Hoffman, H. A., and Grieshaber, C. K. (1976). Genetic studies of murine catalase: Regulation of multiple molecular forms of kidney catalase. *Biochem. Genet.* **14**, 59–66.

Holmes, R. S. (1979). Genetics and ontogeny of alcohol dehydrogenase isozymes in the mouse: Evidence for a *cis*-acting regulatory gene (*Adt-1*) controlling C_2 isozyme expression in reproductive tissues and close linkage of *Adh-3* and *Adt-1* on chromosome 3. *Biochem. Genet.* **17**, 461–472.

Keller, E. C., and Glassman, E. (1964). A third locus (*lxd*) affecting xanthine dehydrogenase in *Drosophila melanogaster. Genetics* **49**, 663–668.

Kermicle, J. L. (1980). Probing the component structure of a maize gene with transposable elements. *Science* **208**, 1457–1459.

King, J. J., and McDonald, J. F. (1983). Genetic localization and biochemical characterization of a *trans*-acting regulatory effect in *Drosophila. Genetics* **105**, 55–69.

Klein, J. (1975). "Biology of the Mouse Histocompatibility-2 Complex." Springer-Verlag, Berlin and New York.

Kopczynski, C. C., and Scandalios, J. G. (1986). *Cat2* gene expression: Developmental control of tanslatable CAT-2 mRNA levels in maize scutellum. *Mol. Gen. Genet.* **203**, 185–189.

Korochkin, L. I. (1980). Genetic regulation of isozyme patterns in *Drosophila* during development. *In* "Isozymes: Current Topics in Biological and Medical Research" (M. C. Rattazzi, J. G. Scandalios, and G. S. Whitt, eds.), Vol. 4, pp. 159–202. Liss, New York.

Lai, Y.-K., and Scandalios, J. G. (1977). Differential expression of *Adh* and its regulation by an endogenous *Adh* specific inhibitor during maize development. *Differentiation* **9**, 111-118.

Lai, Y.-K., and Scandalios, J. G. (1980). Genetic determination of the developmental program for maize scutellar alcohol dehydrogenase: Involvement of a recessive, *trans*-acting, temporal regulatory gene. *Dev. Genet.* **1**, 311–324.

Langridge, P., Pinto-Toro, J. A., and Feix, G. (1982). Transcriptional effects of the opaque-2 mutation of *Zea mays* L. *Planta* **156**, 166–170.

Larkins, B. A., Pedersen, K., Marks, M. D., and Wilson, D. R. (1984). The zein proteins of maize endosperm. *TIBS* **9**, 306–308.

Laurie-Ahlberg, C. C. (1985). Genetic variation affecting the expression of enzyme coding genes in *Drosophila:* An evolutionary perspective. *In* "Isozymes: Current Topics in Biological and Medical Research" (M. C. Rattazzi, J. G. Scandalios, and G. S. Whitt, eds.), Vol. 12, pp. 33–88. Liss, New York.

Lee, L., and Tsai, C. Y. (1984). Effects of RNase on zein synthesis in endosperms of *brittle 2*: *opaque 2* maize double mutant. *Plant Physiol.* **76,** 79–83.

Levy, J. N., Gemmill, R. M., and Doane, W. W. (1985). Molecular cloning of α-amylase genes from *Drosophila melanogaster.* II. Clone organization and verification. *Genetics* **110,** 313–324.

Lewis, N., and Gibson, J. (1978). Variation in amount of enzyme protein in natural populations. *Biochem. Genet.* **16,** 159–170.

Lusis, A. J., and Paigen, K. (1975). Genetic determination of the α-galactosidase developmental program in mice. *Cell* **6,** 371–378.

Lusis, A. J., and Paigen, K. (1976). Properties of mouse α-galactosidase. *Biochim. Biophys. Acta* **437,** 487–497.

Lusis, A., and Paigen, K. (1977). Relationships between levels of membrane bound glucuronidase and the associated protein Egasyn in mouse tissues. *J. Cell Biol.* **73,** 728–735.

Lusis, A., Tomino, S., and Paigen, K. (1976). Isolation, characterization and radioimmunoassay of murine Egasyn, a protein stabilizing glucuronidase membrane binding. *J. Biol. Chem.* **251,** 7753–7760.

Lusis, A., Chapman, V., Wangenstein, R., and Paigen, K. (1973). Transacting temporal locus within the β-glucuronidase gene complex. *Proc. Natl. Acad. Sci. U.S.A.* **80,** 4398–4402.

McCarron, M., O'Donnell, J., Chovnick, A., Bhullar, B. S., Hewitt, J., and Candido, E. P. M. (1979). Organization of the *rosy* locus in *Drosophila melanogaster:* Further evidence of a *cis*-acting control element adjacent to the xanthine dehydrogenase structural element. *Genetics* **91,** 275–293.

McDonald, J. F., Anderson, S. M., and Santos, M. (1980). Biochemical differences between products of the *Adh* locus in *Drosophila. Genetics* **95,** 1013–1022.

MacIntyre, R. J., and Davis, M. B. (1987). A genetic and molecular analysis of the α-glycerophosphate cycle in *Drosophila melanogaster. In* "Isozymes: Current Topics in Biological and Medical Research" (M. C. Rattazzi, J. G. Scandalios, and G. S. Whitt, eds.), Vol. 14, pp. 195–227. Liss, New York.

McKay, J. (1981). Variation in activity and thermostability of alcohol dehydrogenase in *Drosophila melanogaster. Genet. Res.* **37,** 227–237.

McLaren, A., and Buehr, M. (1981). GPI expression in female germ cells of the mouse. *Genet. Res.* **37,** 303–310.

Maroni, G., Laurie-Ahlberg, C. C., Adams, D. A., and Wilton, A. N. (1982). Genetic variation in the expression of *Adh* in *Drosophila melanogaster. Genetics* **101,** 431–446.

Marsh, J. L., and Wright, T. R. F. (1986). Evidence for regulatory variants of the dopa decarboxylase and alpha-methyldopa hypersensitive loci in *Drosophila. Genetics* **112,** 249–265.

Medda, S., and Swank, R. I. (1985). Egasyn, a protein which determines the subcellular distribution of β-glucuronidase, has esterase activity. *J. Biol. Chem.* **260,** 15802–15808.

Meisler, M. (1976). Effects of the *Bgs* locus on mouse β-galactosidase. *Biochem. Genet.* **14,** 921–932.

Meredith, S., and Ganschow, R. (1978). Apparent *trans* control of murine β-glucuronidase synthesis by a temporal genetic elements. *Genetics* **90,** 725–734.

Misra, P. S., Mertz, E. T., and Glover, D. V. (1975). Studies on corn proteins, VI. Endosperm protein changes in single and double endosperm mutants of maize. *Cereal Chem.* **52,** 161–166.

Niesel, D. W., Bewley, G. C., Miller, S. G., Armstrong, F. B., and Lee, C.-Y. (1980). Purification and structural analysis of the soluble *sn*-glycerol-3-phosphate dehydrogenase isozymes in *Drosophila melanogaster. J. Biol. Chem.* **255,** 4073–4080.

Niesel, D. W., Pan, Y. E., Bewley, G. C., Armstrong, F. B., and Li, S. S. (1982). Structural analysis of adult and larval isozymes of *sn*-glycerol-3-phosphate dehydrogenase of *Drosophila melanogaster. J. Biol. Chem.* **257,** 979–983.

O'Brien, S. J., and MacIntyre, R. S. (1978). Genetics and biochemistry of enzymes and specific proteins of *Drosophila. In* "The Genetics and Biology of *Drosophila*" (M. Ashburner and T. R. F. Wright, eds.), Vol. 20, pp. 395–551. Academic Press, New York.

O'Malley, B. W., Towle, H. C., and Schwartz, R. J. (1977). Regulation of gene expression in eukaryotes. *Annu. Rev. Genet.* **11,** 239–275.

Paigen, K. (1961a). The effect of mutation on the intracellular location of β-glucuronidase. *Exp. Cell Res.* **25,** 286–301.

Paigen, K. (1961b). The genetic control of enzyme activity during differentiation. *Proc. Natl. Acad. Sci. U.S.A.* **47,** 1641–1649.

Paigen, K. (1964). *In* "Second International Conference on Congenital Malformations," pp. 181–190.

Paigen, K. (1979a). Acid hydrolases as models of genetic control. *Annu. Rev. Genet.* **13,** 417–466.

Paigen, K. (1979b). Genetic factors in developmental regulation. *In* "Physiological Genetics" (J. G. Scandalios, ed.), pp. 1–61. Academic Press, New York.

Paigen, K. (1980). Temporal genes and other developmental regulators in mammals. *In* "The Molecular Genetics of Development" (T. Leighton and W. F. Loomis, eds.), pp. 419–470.

Paigen, K., and Jakubowski, A. (1985). Cell specificity in the developmental regulation of acid hydrolases by temporal genes. *Dev. Genet.* **5,** 83–91.

Paigen, K., Swank, R., Tomino, S., and Ganschow, R. (1975). The molecular genetics of mammalian glucuronidase. *J. Cell. Physiol.* **85,** 379–392.

Paigen, K., Meisler, M., Felton, J., and Chapman, V. (1976). Genetic determination of the β-galactosidase developmental program in mouse liver. *Cell* **9,** 533–539.

Paigen, K., Laborca, C., and Watson, G. (1979). A regulatory locus for mouse β-glucuronidase induction, *Gur,* controls mRNA activity. *Science* **203,** 554–556.

Peterson, A. C., and Wong, G. G. (1978). Genetic regulation of glucose phosphate isomerase in mouse oocytes. *Nature (London)* **276,** 267–269.

Peterson, A., Choy, F., Wong, G., Clapoff, S., and Friar, P. (1985). Glucosephosphate isomerase (GPI-1) expression in mouse ova: *Cis*-regulation of monomer realization. *Biochem. Genet.* **23,** 827–846.

Quail, P. H., and Scandalios, J. G. (1971). Turnover of genetically defined catalase isozymes in maize. *Proc. Natl. Acad. Sci. U.S.A.* **68,** 1402–1407.

Rechcigl, M., and Heston, W. (1967). Genetic regulation of enzyme activity in a mammalian system by the alteration of the rates of enzyme degradation. *Biochem. Biophys. Res. Commun.* **27,** 119–124.

Roupakias, D. G., McMillin, D. E., and Scandalios, J. G. (1980). Chromosomal location of the catalase structural genes in *Zea mays* using B-A translocations. *Theor. Appl. Genet.* **58,** 211–218.

Russell, R. L., and Coleman, D. L. (1963). Genetic control of hepatic δ-aminolevulinate dehydratase in mice. *Genetics* **48**, 1033–1039.

Salamini, F., and Soave, C. (1982). Zein: Genetics and biochemistry. *In* "Maize for Biological Research" (W. Sheridan, ed.), pp. 155–160. Univ. of North Dakota Press.

Salamini, F., DiFonzo, N., Fornasari, E., Gertinetta, E., Reggiani, R., and Soave, C. (1983). *Mucronate, Mc,* a dominate gene of maize which interacts with *opaque-2* to suppress zein synthesis. *Theor. Appl. Genet.* **65**, 123–128.

Scandalios, J. G. (1977). Isozymes: Genetic and biochemical regulation of alcohol dehydrogenase. *In* "Regulation of Enzyme Synthesis and Activity" (H. Smith, ed.), pp. 129–153. Academic Press, New York.

Scandalios, J. G. (1979). Control of gene expression and enzyme differentiation. *In* "Physiological Genetics" (J. G. Scandalios, ed.), pp. 63–107. Academic Press, New York.

Scandalios, J. G. (1982). Developmental genetics of maize. *Annu. Rev. Genet.* **16,** 85–112.

Scandalios, J. G., and Baum, J. A. (1982). Regulatory gene variation in higher plants. *Adv. Genet.* **21,** 347–370.

Scandalios, J. G., and Felder, M. R. (1971). Developmental expression of alcohol dehydrogenases in maize. *Dev. Biol.* **25,** 641–654.

Scandalios, J. G., Chang, D.-Y., McMillin, D. E., Tsaftaris, A., and Moll, R. H. (1980). Genetic regulation of the catalase developmental program in maize scutellum: Identification of a temporal regulatory gene. *Proc. Natl. Acad. Sci. U.S.A.* **77,** 5360–5364.

Schwartz, D. (1962). Genetics studies on mutant enzymes in maize. III. Control of gene action in the synthesis of pH 7.5 esterase. *Genetics* **47,** 1609–1615.

Sheiness, D., and Darnell, J. E. (1973). Poly(A) segment in mRNA becomes shorter with age. *Nature (London) New Biol.* **241,** 265–268.

Soave, C., and Salamini, F. (1984). The role of structural and regulatory genes in the development of maize endosperm. *Dev. Genet.* **5,** 1–25.

Soave, C., Tardani, L., DiFonzo, N., and Salamini, F. (1981). Zein level in maize endosperm depends on a protein under control of the *opaque-2* and *opaque-6* loci. *Cell* **27,** 403–410.

Swank, R., and Paigen, K. (1973). Biochemical genetic evidence for a macromolecular β-glucuronidase complex in microsomal membranes. *J. Mol. Biol.* **77,** 371–389.

Swank, R., Paigen, K., and Ganschow, R. (1973). Genetic control of glucuronidase induction in mice. *J. Mol. Biol.* **81,** 225–243.

Thompson, J. N., and Kaiser, T. N. (1977). Selection acting upon slow-migrating ADH alleles differing in enzyme activity. *Heredity* **38,** 191–195.

Thompson, J. N., Ashburner, M., and Woodruff, R. C. (1977). Presumptive control mutation for alcohol dehydrogenase in *Drosophila melanogaster. Nature (London)* **270,** 363.

Tsaftaris, A. S., and Scandalios, J. G. (1986). The spatial pattern of catalase (*Cat2*) gene activation in scutella during post-germinative development in maize. *Proc. Natl. Acad. Sci. U.S.A.* **83,** 5549–5553.

Tsaftaris, A. S., Sorenson, J. C., and Scandalios, J. G. (1980). Glycosylation of catalase inhibitor necessary for activity. *Biochem. Biophys. Res. Commun.* **92,** 889–895.

Tsai, C. Y., Larkins, B. A., and Glover, D. V. (1978). Interaction of the *opaque-2* gene with starch-forming mutant genes on the synthesis of zein in maize endosperm. *Biochem. Genet.* **16,** 883–896.

Viotti, A., Pogna, N. E., Balducci, C., and Durante, M. (1980). Chromosomal location of zein genes by *in situ* hybridization in *Zea mays* L. *Mol. Gen. Genet.* **178,** 35–41.

Wahl, R. C., Warner, C. K., Finnerty, V., and Rajagopalan, K. V. (1982). *Drosophila melanogaster mal* mutants are defective in the sulfuration of desulfo Mo hydroxylases. *J. Biol. Chem.* **257**, 3958–3962.

Warner, C. K., and Finnerty, V. (1981). Molybdenum hydroxylases in *Drosophila*. II. Molybdenum cofactor in xanthine dehydrogenase, aldehyde oxidase and pyridoxal oxidase. *Mol. Gen. Genet.* **184**, 92–96.

Warner, C. K., Watts, D. T., and Finnerty, V. (1980). Molybdenum hydroxylases in *Drosophila*. I. Preliminary studies of pyridoxal oxidase. *Mol. Gen. Genet.* **180**, 449–453.

Wilkins, J. R., Shaffer, J. B., and Bewley, G. C. (1982). *In vivo* radiolabeling and turnover of *sn*-glycerol-3-phosphate dehydrogenase in *Drosophila melanogaster:* Gene dosage effects. *Dev. Genet.* **3**, 129–142.

Wilson, C. M. (1973). Plant nucleases. IV. Genetic control of ribonuclease activity in corn endosperm. *Biochem. Genet.* **9**, 53–62.

Wilson, C. M. (1978). Some biochemical indicators of genetic and developmental controls in endosperm. *In* "Maize Breeding and Genetics" (D. B. Walden, ed.), pp. 405–419. Wiley, New York.

Woodman, J. C., and Freeling, M. (1981). Identification of a genetic element that controls the organ-specific expression of *Adh1* in maize. *Genetics* **98**, 357–378.

Wright, D. A., and Shaw, C. (1969). Genetics and ontogeny of α-glycerophosphate dehydrogenase isozymes in *Drosophila melanogaster*. *Biochem. Genet.* **3**, 343–353.

Wright, R. F., Black, B. C., Bishop, C. P., Marsh, J. L., Pentz, R., and Wright, E. Y. (1982). The genetics of dopa-decarboxylase in *Drosophila melanogaster*. V. *Ddc* and 1(2) *Amd* alleles; isolation, characterization and intragenic complementation. *Mol. Gen. Genet.* **188**, 18–26.

THE GENETICS OF BIOGENIC AMINE METABOLISM, SCLEROTIZATION, AND MELANIZATION IN *Drosophila melanogaster**

Theodore R. F. Wright

Department of Biology, University of Virginia, Charlottesville, Virginia 22901

I. Introduction

Biogenic amines are particularly important compounds in insects, for not only do they serve as neurotransmitters and modulaters of adenylate cyclase activity in their neurophysiology, but they are also essential for cross-linking proteins and chitin during sclerotization in the insect cuticle throughout development and especially at times of eclosion (hatching) and molting. A thorough understanding of the regulation of the synthesis, inactivation, and degradation of biogenic amines in the nervous system is mandatory for the elucidation of their role in the neurophysiology and behavior of any organism. Clearly the availability of mutations that interrupt biogenic amine metabolism in an experimentally suitable organism such as *Drosophila* should provide powerful analytical tools for neurochemists, neurophysiologists, and behavioral biologists. In addition, the formation, sclerotization, and pigmentation of cuticle in insects is a complex process which produces a variety of cuticles of different thickness, hardness, flexibil-

* This review is dedicated to Professor Ernst Caspari in recognition of his pioneering research in biochemical genetics.

ity, shape, and color. These different cuticles are major morphological and functional end products of development and differentiation in insects. The primary components of cuticle are lipids, cuticular proteins, chitin, and catecholamines as cross-linking agents. Just how the genes that differentially produce these components are regulated in development is a paradigm for one of the most fundamental questions in genetics and developmental biology. These genes are, in fact, some of the "realisator" genes which must be controlled by "selector" genes (terminology from Garcia-Bellido, 1977), such as the homeotic genes of the Bithorax Complex and the genes responsible for somatic sex determination. Thus, identification and characterization of genes involved in catecholamine metabolism and sclerotization are essential not only to expedite the unraveling of the complex processes of cuticle formation, sclerotization, and pigmentation, but also to facilitate the investigation of how realisator genes are differentially regulated by selector genes at different times in development, in different segments, or in different sexually dimorphic structures.

In order to facilitate and stimulate research on biogenic amines and sclerotization, this review will attempt to summarize what is known about the genetics and biochemistry of biogenic amines and sclerotization in *Drosophila* (see Table 1). In order to do this it is necessary to refer to work done in other insects, particularly in regard to sclerotization and the properties of phenol oxidases. The effects of biogenic amine mutations on the neurophysiology and behavior of *Drosophila* will be indicated, but no attempt will be made to review the neurogenetics and the behavioral genetics of *Drosophila*.

II. General Comments on Biogenic Amine Metabolism

Possible pathways for the synthesis, modification, and degradation of biogenic amines are summarized in Figs. 1–3. Omitted from these figures are steps involving the conjugation of biogenic amines with sulfates and the formation of glucosides. In addition, only the degradation of dopamine (DA) and norepinephrine (NE) is depicted in Fig. 3, whereas other biogenic amines such as dihydroxyphenylalanine (DOPA), epinephrine (E), and octopamine (OA) may be degraded by similar enzymatic reactions.

The metabolism of biogenic amines in insects has been reviewed by Brown and Nestler (1985), Pichon and Manaranche (1985), and Brunet (1980). Not all the reactions depicted in Figs. 1–3 have been reported to occur in insects and by no means have all these reactions

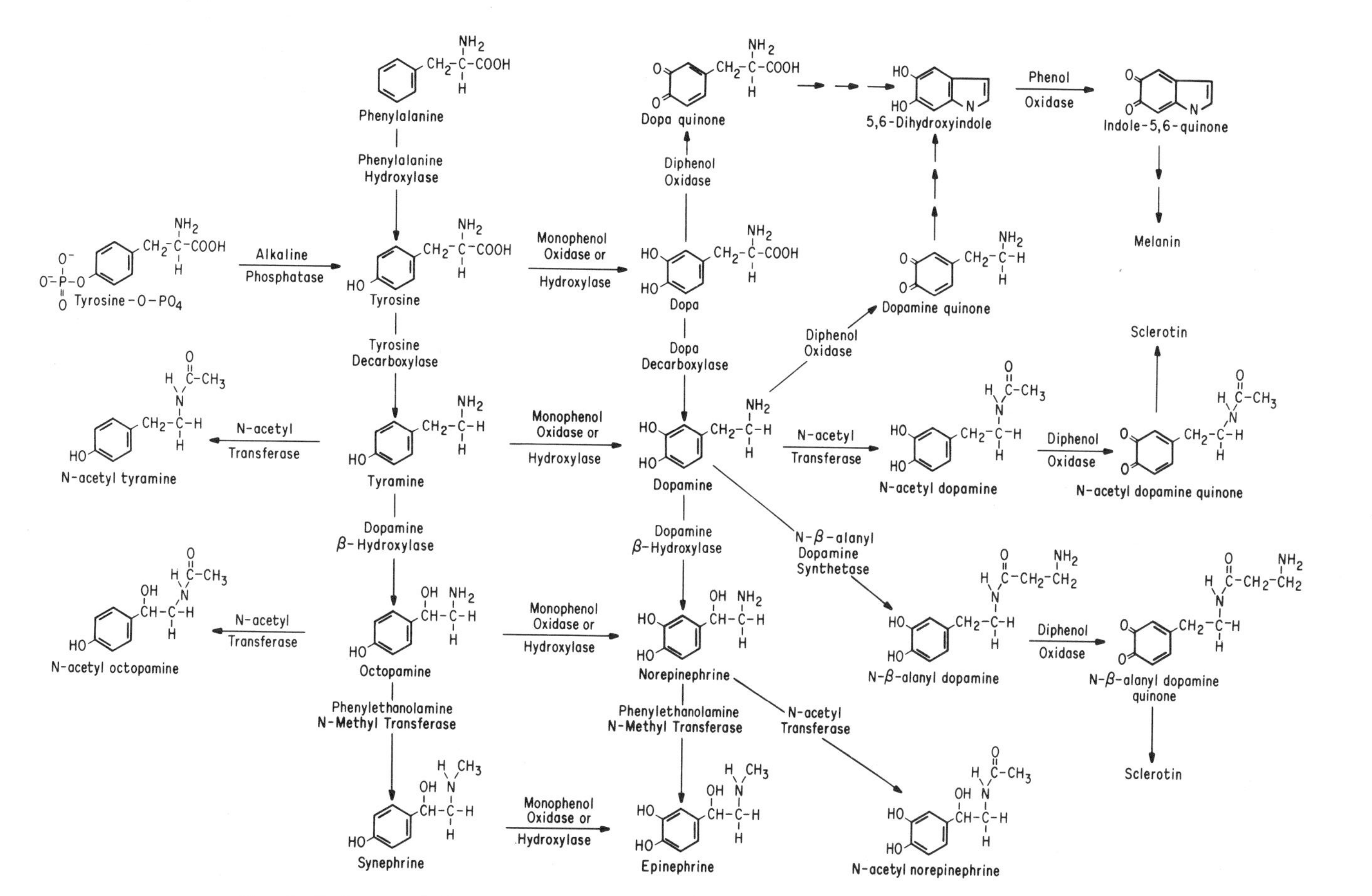

FIG. 1. Pathways in the synthesis and modification of monophenolamines and catecholamines.

TABLE 1

Genes Involved in Biogenic Amine Metabolism, Sclerotization, or Melanization in *Drosophila melanogaster*

Gene[a]	Locus[b]	Enzyme, protein, or process affected	Product affected[c]	Section[d]	Reference[e]
Aph-1, Alkaline phosphatase-1	3–47.3	Alkaline phosphatase-1	Tyrosine-*O*-phosphate	VI	Harper and Armstrong (1974)
b, black	2–48.5; 34E5–35D1	β-Alanine synthesis, reduced diphenol oxidase	β-Alanine, NBAD ↓	XIII, A and D XI, F, 3	Pedersen (1982)
Bc, Black cells	2–80.6	Diphenol oxidase, black crystal cells	—	XI,E,2	Rizki *et al.* (1980)
Bld, Blond	T(1; 2)Bld = T(1;2)1C3–4; 60B12–13	Diphenol oxidase	—	XI,F,3	Mitchell *et al.* (1967)
ca, claret	3–100.7; 99C–E		4-*O*-Glucosyl-*N*-acetyldopamine	III,E	Okubo (1958)
Dat, Dopamine acetyltransferase	2–107.0; 60B1–10	Dopamine acetyltransferase		XII	Huntley (1978)
* *Ddc, Dopa decarboxylase*	2–53.9+; 37C1,2	Dopa decarboxylase	DOPA ↑ ;DA, NADA, NBAD, serotonin ↓	VII,B; VIII	Wright *et al.* (1982); Livingstone and Tempel (1984)
* *Dox-A2, Diphenol oxidase-A2*	2–53.9; 37B10–13	Diphenol oxidase A2 component	DOPA, DA, NADA ↑	XI,C,1	Pentz *et al.* (1986)
Dox-3, Diphenol oxidase-3	Chromosome 2	Diphenol oxidase A3 component	—	XI,C,2	Rizki *et al.* (1985)
dy, dusky	1–36.2	Dark body color	β-Alanine ↓	XIII,D	Lindsley and Grell (1968)

e, ebony	3–70.3; 93D2–3	β-Alanyldopamine synthetase	β-Alanine ↑; NBAD ↓	XIII,A–C	Black *et al.* (1987)
		Diphenol oxidase		XI,F,3	Mitchell *et al.* (1967)
* *hk, hook*	2–53.9; 37B10–13	Bristle development	—	X,A	Wright (1987)
* *l(2)amd, alpha methyl dopa*	2–53.9+; 37C1,2	α-Methyldopa hypersensitive; codes for a DDC homologous protein	Catecholamine X ↓	IX	Black *et al.* (1987b); Eveleth and Marsh (1986a)
* *l(2)37Ba*	2–53.9; 37B10–13	Cuticle development, melanotic tumors	—	X,A	Wright (1987)
* *l(2)37Bc*	2–53.9; 37B10–13	Cuticle development, melanotic tumors	—	X,A	Wright (1987)
* *l(2)37Bd*	2–53.9+; 37B13–C1,2	Cuticle development	—	X,A	Wright (1987)
* *l(2)37Be*	2–53.9; 37B10–13	Cuticle development	—	X,A	Wright (1987)
* *l(2)37Bg*	2–53.9; 37B13–C1,2	Cuticle development, melanotic tumors	—	X,A	Wright (1987)
* *l(2)37Ca*	2–53.9+; 37C1,2–4	Cuticle development	—	X,A	Wright (1987)
* *l(2)37Cc*	2–53.9+; 37C1,2–4	Cuticle development, melanotic tumors	—	X,A	Eveleth and Marsh (1986b)
* *l(2)37Cd*	2–53.9+; 37C1,2–4	Cuticle development, melanotic tumors	—	X,A	Wright (1987)

(continued)

TABLE 1 (*Continued*)

Gene[a]	Locus[b]	Enzyme, protein, or process affected	Product affected[c]	Section[d]	Reference[e]
* *l(2)37Ce*	2–53.9+; 37C1,2–4	Cuticle development	—	X,A	Wright (1987)
* *l(2)37Cf*	2–53.9+; 37C3–7	Brain tumor, cuticle development	—	X,A	Wright (1987)
l(3)tr, lethal translucida	3–20.0	Amino acid metabolism	Tyrosine-*O*-phosphate ↑; tyrosine ↓	VI	Mitchell and Simmons (1961)
* *lz, lozenge*	1–27.7	Diphenol oxidase, crystal cells	—	XI,E,3	Rizki and Rizki (1981)
* *per, period*	1; 3B1,2	Tyrosine decarboxylase; codes for a proteoglycan	Octopamine ↓	VIII	Livingstone and Tempel (1983); Jackson *et al.* (1986); Reddy *et al.* (1986)
Phox, Phenol oxidase	2–80.6	Diphenol oxidase; A1, A2, and A3 components not affected	—	XI,F,1	Batterham and Mackechnie (1980)
ple, pale	3–18; 65A–E	Tyrosine hydroxylase?	Underpigmented; catecholamines ↓	III,A	K. White (personal communication)
Pu, Punch	1–97; 57C4–6	GTP cyclohydrolase	Pteridine cofactor ↓ ?; unpigmented, weak cuticle	III,A	Mackay and O'Donnell (1983); Reynolds and O'Donnell (1987)
qs, quicksilver	1–39.5; 10F1–10	Diphenol oxidase A1, A2, and A3 components	DOPA, NADA, NBAD ↑ ↑; DA ↑	XI,D,1	Pentz *et al.* (1987)

t, tan	1–27.5; 8C2–3; C14–D1	β-Alanyldopamine hydrolase	NBAD↑↑; DA↓	XIII,A and B	B. C. Black *et al.* (personal communication)
Tcr, Third chromosome resistance	3–39.6	α-Methyldopa resistant	—	IX,C,2	C. P. Bishop *et al.* (personal communication)
* *Tyrosine hydroxylase*	3; 65B	Tyrosine hydroxylase (cloned gene)	—	III,A	Neckameyer *et al.* (1986)
tyr-1, tyrosinase-1	2–54.5; 37F6–38B6	Diphenol oxidase A1, A2, and A3 components	—	XI,D,2	Pentz *et al.* (1987)
Tyr-2, Tyrosinase-2	2–57	Diphenol oxidase	—	XI,C,2	Lindsley and Grell (1968)
Tyr-3, Tyrosinase-3	3–right arm	Diphenol oxidase	—		Lindsley and Grell (1968)
s, sable	1–43.0	Dark body color	β-Alanine↓	XIII,D	Lindsley and Grell (1968)
sp, speck	2–107.0; 60B13–C5	Diphenol oxidase A2 component?	Melanization↑	XI,F,2	Warner *et al.* (1975)
str, straw	2–55.1; 41B–C	Diphenol oxidase	—	XI,F,3	Mitchell (1966)
su(b), suppressor of black	1–0.0	Suppresses *black*	β-Alanine↑	XIII,D	Sherald (1981)
Su(b), Suppressor of black	1–55.5	Suppresses *black*	—	XIII,D	Pedersen (1982)
su(f), suppressor of forked	1–65.9	Diphenol oxidase, suppressor of *lz*	—	XI,E,3	Snyder and Smith (1976)
* *su(s), suppressor of sable*	1–0	Diphenol oxidase, suppressor of *sp*	—	XI,F,2	Warner *et al.* (1975)

(*continued*)

TABLE 1 (*Continued*)

Gene[a]	Locus[b]	Enzyme, protein, or process affected	Product affected[c]	Section[d]	Reference[e]
su(r), suppressor of rudimentary	1–27.7	Partial *black* phenotype; enhancer of *black*	Melanization ↑	XIII,D	Pedersen (1982)
svr, silver	1–0.0; 1B6–7	Light body color	NADA ↑ ↑	XIV,A	B. C. Black *et al.* (personal communication)
* *y, yellow*	1–0.0; 1B1	Membrane or secreted structural protein	Melanization ↓	XIV,B	Geyer *et al.* (1986)
		Diphenol oxidase		XI,F,3	Mitchell *et al.* (1967)

[a] An asterisk (*) indicates gene has been cloned.
[b] Locus: chromosome–genetic locus; polytene chromosome bands.
[c] Product affected: ↓, decreased; ↑, increased.
[d] Section: location of discussion in this review.
[e] Reference: most recent key reference.

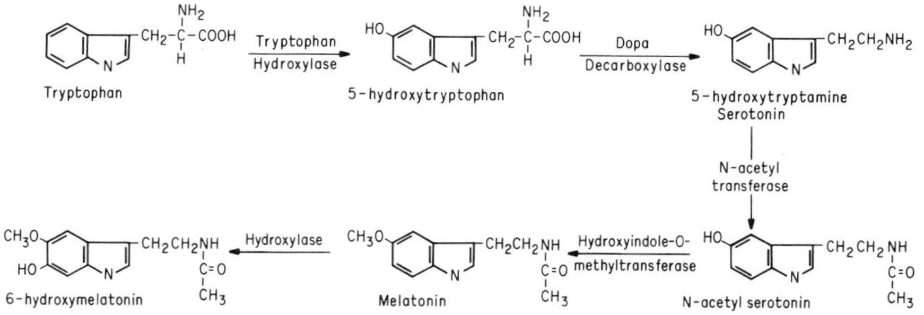

FIG. 2. The synthesis of serotonin and its derivatives. Melatonin and 6-hydroxymelatonin have not been identified in insects.

been assayed in *Drosophila*. An inspection of Figs. 1 and 2 should make it clear that the different classes of biogenic amines, monophenolamines, catecholamines (diphenolamines), and indoleamines, are capable of undergoing similar reactions, i.e., decarboxylation, β-hydroxylation, N-acetylation, ring oxidation, N-methylation, O-methylation, and oxidative deamination with the resulting aldehyde reduced to an alcohol by aldehyde reductase or converted to an acid by aldehyde dehydrogenase. *In vitro* many of the enzymes that carry out these reactions are quite versatile, being capable of using either a monophenolamine, a catecholamine, or an indolamine as a substrate. Whether or not all three classes of biogenic amines are used effectively as substrates *in vivo* in both the epidermis–cuticle and the nervous system has to be determined by assaying precursor and product pools, by injecting radioactive precursors into whole organisms or by incubating whole tissue explants in radioactive precursors, with the subsequent assay of radioactive products. These experiments are incapable of determining if genetically different, more specific, enzymes exist *in vivo*, e.g., a decarboxylase with high affinity for tyrosine but low affinity for DOPA and 5-hydroxytryptophan, in contrast to another, different decarboxylase with opposite substrate affinities (see Section VIII below). Except by separately purifying and characterizing the two enzymes, the most effective way to discriminate between them is to isolate null mutations of least one of the enzymes, a procedure almost limited to *Drosophila*. A similar and particularly relevant question in insects is whether the same specific reaction, e.g., decarboxylation of DOPA, is carried out by two different enzymes, one in the epidermis–cuticle and another in the nervous system. Again, except for isolating and characterizing the enzyme from the two different tissues, the most reliable method for making the judgment is to

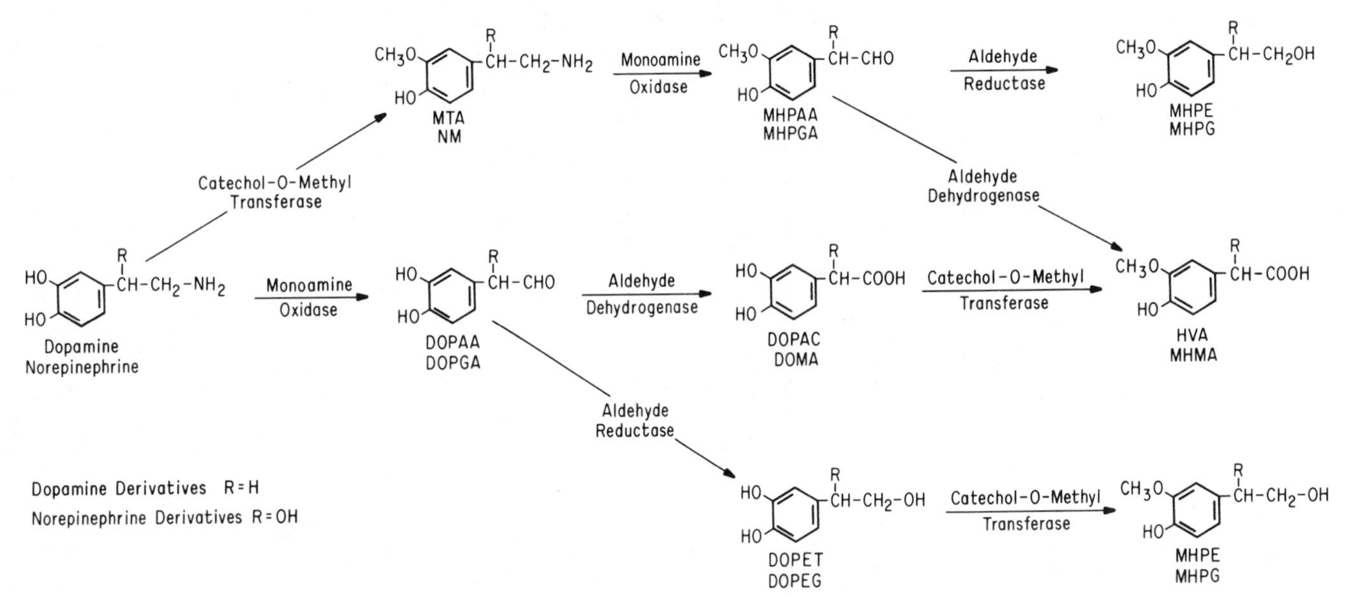

FIG. 3. The oxidative deamination and O-methylation of dopamine and norepinephrine and their derivatives. The designations of the dopamine derivatives are given above the designations of the norepinephrine derivatives: DOMA, 3,4-dihydroxymandelic acid; DOPAA, 3,4-dihydroxyphenylacetaldehyde; DOPAC, 3,4-dihydroxyphenylacetic acid; DOPEG, 3,4-dihydroxyphenylglycol; DOPET, 3,4-dihydroxyphenylethanol; DOPGA, 3,4-dihydroxyphenylglycolaldehyde; HVA, homovanillic acid; MHMA, 3-methyoxy-4-hydroxymandelic acid; MHPAA, 3-methoxy-4-hydroxyphenylacetaldehyde; MHPE, 3-methoxy-4-hydroxyphenylethanol; MHPG, 3-methoxy-4-hydroxyphenylglycol; MHPGA, 3-methoxy-4-hydroxphenylglycolaldehyde; MTA, 3-methoxytyramine; NM, normetanephrine.

have a null mutation for the enzyme (see discussion of dopa decarboxylase in Section VII).

III. Biogenic Amine Metabolism in *Drosophila*

The purpose of this section is to provide a catalog of those reactions in biogenic amine metabolism that have been shown to occur in *Drosophila* and for which there is little or no genetics. Those reactions which are mediated by enzymes for which more genetic information is available are discussed in other sections of this review.

A. Ring Hydroxylation

Both phenylalanine and tryptophan, precursors in the synthesis of biogenic amines, are essential metabolites in the diet of *Drosophila* (Geer, 1966). On the other hand, tyrosine is not a dietary essential, being synthesized from phenylalanine by hydroxylation.

Phenylalanine hydroxylase activity has been assayed and characterized in extracts of whole *Drosophila melanogaster* and a developmental profile has been published, with maximum activity found at pupariation and pupation (Geltosky and Mitchell, 1980). Similar to the rat liver enzyme (Kaufman, 1971), the *Drosophila* enzyme requires a cofactor for which synthetic 6,7-dimethyl-4-aminotetrahydropteridine could be successfully substituted (Geltosky and Mitchell, 1980). In the rat liver system, the oxidized pteridine is converted back to the active reduced cofactor by a NADPH-linked regeneration system, i.e., a NADPH-requiring dihydropteridine reductase which can be replaced by dithiothreitol in the incubation mixture (Bublitz, 1969). The *Drosophila* enzyme was assayed with dithiothreitol present. The effects of specific inhibitors were not monitored. There is no reported genetics on phenylalanine hydroxylase in *Drosophila,* but it is probable that some of the eye pigment mutations that are involved in pteridine metabolism may affect the synthesis or reduction of tetrahydropteridine cofactors required for the activity of phenylalanine hydroxylase, tyrosine hydroxylase (see below), and the molybdenum hydroxylases, including xanthine dehydrogenase, aldehyde oxidase, and pyridoxal oxidase (Fan and Brown, 1979; Phillips and Forrest, 1980; Wahl *et al.,* 1982).

One of these eye pigment loci, *Punch* (*Pu,* 2–97; 57C4–6) codes for the enzyme guanosine triphosphate cyclohydrolase I (GTP CH; E.C. 3.5.4.16) (Mackay and O'Donnell, 1983; Mackay *et al.,* 1985; Weisberg and O'Donnell, 1986), which catalyzes the first reaction in pteridine biosynthesis in *Drosophila* (Brown *et al.*, 1979), converting GTP to dihydroneopterin triphosphate. Presumably GTP CH activity is required for the synthesis of any pteridine cofactor, and therefore the fact that null mutations and deletions of the *Punch* locus are recessive lethals is not unexpected. In their screen of the second chromosome of *Drosophila* for embryonic lethals that affected larval cuticle, Nüsslein-Volhard *et al.* (1984) recovered four *Punch* alleles which produced unhatched larvae with unpigmented cuticle and mouthparts. (The locus for these alleles was initially designated *unpigmented, upi.*) Subsequently Reynolds and O'Donnell (1987) have confirmed the unpigmented embryonic lethal phenotype for additional *Pu* alleles and for a homozygous deletion of the *Punch* locus. In addition to having unpigmented mouthparts and denticle belts, they found that the cuticle was weak and ruptured easily when manipulated. This unpigmented phenotype of *Punch* embryos is very similar, if not identical, to the embryonic lethal phenotypes of dopa decarboxylase-deficient mutations in *Ddc* (see Section VII,B), of phenol oxidase-deficient mutations in *quicksilver* (see Section XI,D,1), and of putative tyrosine hydroxylase-deficient mutations (*pale;* see below). This suggests that in the embryonic hypoderm GTP CH activity is necessary for the synthesis of a pteridine cofactor required for the activity of one or more enzymes involved in biogenic amine metabolism, perhaps tyrosine hydroxylase or even possibly phenylalanine hydroxylase or both.

In mammals there are at least two enzymatic reactions capable of hydroxylating tyrosine to dihydroxyphenylalanine. One is the monophenol oxidase activity catalyzed by the copper enzyme tyrosinase (E.C. 1.14.18.1), which can be inhibited by phenylthiourea. The other is catalyzed by tyrosine 3-hydroxylase (TH) (E.C. 1.14.16.2), which requires the cofactor 2-amino-4-hydroxy-6,7-dimethyl-5,6,7,8-tetrahydropterine ($DMPH_4$). In mammals TH is found in the nervous system and in the adrenal medulla. One might, therefore, expect to find TH in the nervous system of insects also. If the oxidation of tyrosine to DOPA is the rate-limiting step in catecholamine synthesis, and if this step is exclusively mediated by the monophenol oxidase activity of tyrosinase in the hemolymph and cuticle, and, on the other hand, exclusively mediated by TH in the nervous system, it would provide a means for differentially regulating the pathway to neurotransmitter synthesis in the nervous system vis-à-vis the pathway to sclerotization and melanization in the hemolymph and cuticle. It has been tacitly assumed that TH exists in *Drosophila*, but no enzymatic evidence to support this assumption has been published. Livingstone and Tempel (1983), however, did demonstrate that dissected intact adult *Drosophila* brains, when incubated with tritiated tyrosine or tritiated tryptophan, produced tritiated dopamine and tritiated serotonin, respectively. These results, supplemented by the analysis of the effects of the *per*^{o} and *Ddc* mutations (see Section VIII), showed that *Drosophila* is capable of hydroxylating tyrosine and tryptophan. However, they do not discriminate between hydroxylation by pteridine-dependent tyrosine or tryptophan hydroxylases or by one or more copper-requiring monophenol oxidases. No one has reported attempting specifically to assay TH activity in dissected *Drosophila* central nervous systems, but B. C. Black (personal communication), using the assay of Nagatsu *et al.* (1964), failed to find TH activity in extracts of whole adult *Drosophila* even though high levels of TH activity were measured in extracts of rat adrenal medulla. As a corollary to this negative result, no one has reported attempting to assay monophenol oxidase activity in dissected *Drosophila* brains.

However, using a molecular approach rather than an enzymatic one, Neckameyer *et al.* (1986) have very recently recovered *Drosophila* genomic clones and cDNAs by reduced stringency hybridization to a partial cDNA probe containing coding sequences for rat tyrosine hydroxylase (Chikaraishi *et al.*, 1983). Sequencing of one of the *Drosophila* cDNAs shows the deduced amino acid sequences are very well conserved between the rat and *Drosophila*. Results showed 72% of the amino acids are identical among 140 residues apparently located in the catalytic region of the enzyme (Joh *et al.*, 1983; Grima *et al.*, 1985). A 3.2-kb transcript was found to be 20-fold more abundant in *Drosophila* head tissues than in bodies. The *Drosophila* cDNA hybridizes *in situ* to region *65B* of the third chromosome. Embryos homozygous for the embryonic lethal mutation *pale* (*ple*, 3–18; 65A–E) have underpigmented mouthparts and denticle belts (Jürgens *et al.*, 1984) similar to dopa decarboxylase-deficient embryos (see Section VII,B). Using glyoxylic acid-induced histofluorescence to identify catecholamine-containing cells and a monoclonal antibody against serotonin to identify serotonin-containing cells, K. White (personal communication) has determined that in contrast to the wild type, *pale* embryonic nervous systems exhibit no catecholamine signal whereas both wild type and *pale* exhibit a serotonin signal. The interpretation is that *pale* may be a lethal mutation in the tyrosine hydroxylase gene which is required for the synthesis of catecholamines in the nervous system but not serotonin. If these preliminary results are confirmed, the *pale* mouthparts and denticle belts suggest that tyrosine hydroxylase activity is required in the cuticle for complete pigmentation (and sclerotization?).

Figure 1 depicts the possible hydroxylation of tyramine to dopamine, octopamine to norepinephrine, and synephrine to epinephrine. The effect of the mutants *per*^{o} and *Ddc*^{ts2} on tyrosine decarboxylase and dopa decarboxylase (Livingstone and Tempel, 1983) (see Section VIII) suggests that in intact adult brains the synthesis of dopamine by hydroxylating tyramine occurs at very low levels, if at all. There is some question whether norepinephrine occurs in *Drosophila,* but if it does it might arise via hydroxylation of octopamine, but there is no evidence on this (see discussion of β-hydroxylation below). Finally, the presence of neither synephrine nor epinephrine has been reported in *Drosophila* (see Section III,C on N-methylation).

B. β-Hydroxylation

Tunnicliff *et al.* (1969) measured levels of biogenic amines in *D. melanogaster* strains selected for locomotor activity. Norepinephrine (noradrenaline) levels were reported to be maximal in the active strain in adult females and lowest in the inactive strain. The reverse was found for dopamine levels, and serotonin did not vary among the strains. The control strain was intermediate for both norepinephrine and dopamine. This report of measuring various levels of norepinephrine in *Drosophila* stands in contrast to the report of Black *et al.* (1987a), who were unable to identify norepinephrine in adult *Drosophila* among the alumina-extracted, electrochemically active compounds separated by high-performance liquid chromatography (HPLC) (see Section V). Although unidentified compounds were observed and isolated, none showed the same retention time as the norepinephrine standard with all the running buffers used nor did fast atom bombardment of isolated compounds identify any compound as norepinephrine. Although norepinephrine has been reported to be present in other insect species (see review by Brown and Nestler, 1985), no additional reports of noradrenaline in *Drosophila* are known, and no one reports measuring dopamine β-hydroxylase in extracts of *Drosophila.*

On the other hand, when dissected, intact adult *Drosophila* brains were incubated in tritiated tyrosine, radiolabeled dopamine, tyramine, and octopamine were found (Livingstone and Tempel, 1983). Since the last compound is the β-hydroxylated derivative of tyramine (see Fig. 1), this indicates the presence of tyramine β-hydroxylase in adult *Drosophila* brains. Livingstone and Tempel (1983) determined that one-third as much [^{3}H]octopamine was accumulated in brains of *per*° homozygotes as in wild type. Although *per*° brains had wild-type levels of tyramine β-hydroxylase activity, tyrosine decarboxylase activity was reduced to 35% of wild type (see Section VIII). Thus *per*° mutants, which lack normal circadian rhythms (Knopka and Benzer, 1971), do not have reduced or altered tyramine β-hydroxylase activity.

It would be interesting to determine if the same enzyme is capable of β-hydroxylating both tyramine and dopamine *in vivo* or if octopamine can be ring hydroxylated *in vivo* to noradrenaline. A mutation inactivating tyramine β-hydroxylase might be required to answer the question definitively.

C. N-Methylation

The addition of a methyl group to the amine group of octopamine and norepinephrine converts the former into synephrine and the latter into epinephrine (Fig. 1). Epinephrine is found in some insects, but is rare and its function is unknown (Brown and Nestler, 1985). Pinchon and Manaranche (1985) report that epinephrine has not been demonstrated to be present in the central nervous system of insects and neither review (Brown and Nestler, 1985; Pinchon and Manaranche, 1985) even mentions synephrine. In their survey of catecholamines throughout development, Black *et al.* (1987a) found no epinephrine in *D. melanogaster*. Thus the enzyme phenylethanolamine *N*-methyltransferase and its genetics are unknown in *Drosophila*.

D. Oxidative Deamination, O-Methylation, and N-Acetylation

Possible pathways for the degradation and inactivation of catecholamines are presented in Fig. 3. Monoamine oxidase (MAO) activity, which effects the oxidative deamination of catecholamines, is found in some insect tissues, including the brain, at very low levels in comparison to activities in mammals; using a variety of assays, it has not been found at all in numerous insects (Brown and Nestler, 1985). However, very low levels of MAO activity have been measured in *D. melanogaster* whole fly and whole head preparations (Dewhurst *et al.*, 1972). Tyramine, dopamine, and serotonin were used as substrates, with the last two giving much lower activities. The activities are so low that in assays for dopa decarboxylase activity the use of MAO inhibitors is not necessary. No genetics on MAO has been done in *Drosophila*.

Dewhurst *et al.* (1972) report measuring the activity of catecholamine-*O*-methyltransferase (COMT) in *Drosophila* whole fly and whole head homogenates. The activities are approximately an order of magnitude lower than the very low MAO activities found. Brown and Nestler (1985) state the only evidence for COMT activity in insects is that of Dewhurst *et al.* (1972) in *Drosophila*. There is no genetics of COMT in *Drosophila*.

In this same study on MAO and COMT Dewhurst *et al.* (1972) found that in *Drosophila* nervous tissue *N*-acetyltransferase (NAT) activity (see Fig. 1) is 50-fold greater than MAO activity and 500-fold greater than COMT activity. These results suggested to them that NAT activity provides the major route for biogenic amine regulation in *Drosophila,* rather than MAO or COMT. The biochemistry and genetics of NAT or DAT (dopamine acetyltransferase) are presented in Section XII.

As indicated in Fig. 3, in mammals the aldehyde products of the oxidative deamination of catecholamines by MAO can be further metabolized to alcohols by a $NADP^+$-dependent aldehyde reductase (E.C. 1.1.1.2) or reduced to an acid by a NAD^+-dependent aldehyde dehydrogenase (E.C. 1.2.1.3). There is no information on the existence of these enzymatic activities or their products in *Drosophila* and, of course, no genetics.

E. Biogenic Amine Conjugates

Inert, highly soluble conjugates of tyrosine or phenylalanine are found in the hemolymph of numerous insects. Concentrations of these conjugates are high in dipteran larvae just prior to pupariation, indicating that these conjugates provide a mechanism for storing high levels of tyrosine prior to its mobilization into the puparium at the time of pupariation. These conjugates can be dipeptides, particularly β-alanyltyrosine and γ-glutamylphenylalanine (Bodnaryk, 1972), β-glucosides (Chen *et al.*, 1978; Zomer and Lipke, 1981; Hopkins *et al.*, 1984; Bodnaryk *et al.*, 1974), sulfates (Bodnaryk and Brunet, 1974; Bodnaryk *et al.*, 1974), and phosphates (Seligman *et al.*, 1969; Bodnaryk, 1972; Bodnaryk *et al.*, 1974), e.g., tyrosine-O-phosphate, particularly in *Drosophila* (Mitchell *et al.*, 1960; Mitchell and Lunan, 1964; Chen *et al.*, 1966).

In a survey of fully grown larvae of 46 species of Diptera, Bodnaryk (1972) found that all members of the genus *Sarcophaga* contained β-alanyltyrosine, all the species of *Musca* contained γ-glutamylphenylalanine, and all the *Drosophila* species surveyed contained tyrosine-

O-phosphate. In this survey this last compound was found only in *Drosophila*, β-alanyltyrosine was found in only one non-*Sarcophaga* species, and γ-glutamylphenylalamine was found only in one non-*Musca* species. In a survey of 11 *Drosophila* species belonging to three different subgenera, tyrosine-*O*-phosphate was present in all but one species, *Drosophila busckii*, which contained another phenolic compound (Chen, 1975). This phenolic compound was subsequently identified as β-glucosyl-*O*-tyrosine and was determined to be present maximally at pupariation, disappearing almost completely immediately thereafter (Chen *et al.*, 1978). Another phenolic glucoside, 4-*O*-glucosyl-*N*-acetyldopamine, has been reported to be present in the mutant *claret* (~3–100.7) in *D. melanogaster* (Okubo, 1958). It is thought to be a detoxification product of dopamine.

The presence of glucosides of other catechols has been reported in a number of insects, including the 4-*O*-β-glucoside of *N*-acetyldopamine (NADA) in *Calliphora* (Karlson and Sekeris, 1964), protocatechuic acid 4-*O*-β-glucoside in the colleterial glands of cockroaches (Brunet and Kent, 1955), a metabolite of tyramine in nervous tissue of *Manduca sexta*, metabolites of NADA, *N*-acetylnorepinephrine and their acid derivatives in *Periplaneta* hemolymph (Koeppe and Mills, 1974, 1975), and probably the conjugates of NADA and *N*-β-alanyldopamine (NBAD) in the hemolymph and cuticle of *M. sexta* (Hopkins *et al.*, 1984). In many of these studies the identity of the isolated conjugate as a glucoside has been based on susceptibility of the compound to hydrolysis by a β-glucosidase. Bodnaryk *et al.* (1974) point out that β-glucosidase preparations can be contaminated with a highly active acid phosphatase, an observation which brings doubt to the validity of some of the above-listed identifications.

The presence in *Drosophila* of sulfate conjugates of catecholamines has not been reported, and it is not apparent that anyone has looked for such derivatives. Bodarnyk and Brunet (1974) have identified dopamine 3-*O*-sulfate in extracts of newly eclosed *Periplaneta americana* and subsequently it was shown that 3-*O*-sulfates of dopamine and *N*-acetyldopamine are synthesized *in vivo* from injected radioactive precursors. It is apparent that the former sulfate can serve as a precursor for the latter sulfate, indicating a capability for acetylating the sulfate conjugate.

It is clear that most of the biogenic amine conjugates mentioned above are not directly incorporated into the cuticle intact, but rather the conjugates are first hydrolyzed and the biogenic amine moiety directly incorporated into the cuticle, e.g., as NADA and NBAD (Bodnaryk *et al.*, 1974; Hopkins *et al.*, 1984), or the biogenic amine moieties, e.g., tyrosine, β-alanine, and dopamine, are subject to further metabolism prior to incorporation into cuticle, e.g., as NADA or NBAD (Bodnaryk, 1970, 1971; Bodnaryk and Brunet, 1974; Bodnaryk *et al.*, 1974). Since these highly soluble conjugates are found in large amounts primarily in the hemolymph just prior to pupariation or at the time of ecdysis, and since the biogenic amine moieties are subsequently found in the cuticle, it is apparent that one of the primary roles of these conjugates is to provide an innocuous way of building up the concentration of highly toxic biogenic amino components of the cuticle prior to their use during pupariation and ecdysis. In addition, these conjugates may play a key role in the nervous system by providing a mechanism of inactivating biogenic amine neurotransmitters or modulators. There is very little information on the presence of these conjugates in the nervous system of insects, although sulfate conjugates of octopamine, dopamine, and serotonin have been found in nerves of the crustacean *Homarus americanus* (lobster) (Kennedy, 1978).

Mitchell and co-workers (Mitchell and Simmons, 1961; Simmons and Mitchell, 1961; Mitchell *et al.*, 1960) were the first to report the natural occurrence of tyrosine-*O*-phosphate, which was found in extracts of larvae of *D. melanogaster*. It starts to accumulate early in the third larval instar prior to 65 hours after egg laying and reaches a maximum concentration of 6 μmol/g of wet weight at 90 hours after egg laying, where it remains until pupariation, when its concentration drops abruptly (Mitchell and Lunan, 1964; Lunan and Mitchell, 1969). Of the tyrosine-*O*-phosphate recovered, 96% was recovered from hemolymph plasma with barely a trace in the blood cells, and only 4% was found in the remaining larval tissues, which included the cuticle (Lunan and Mitchell, 1969). Tyrosine-*O*-phosphate is neither a substrate for oxidation by phenol oxidase nor is it an inhibitor of the oxidation of tyrosine or DOPA by this enzyme. After pupariation there is an equimolar decrease in tyrosine-*O*-phosphate from 6.4 ± 0.9 to 0.2 ± 0.2 μmol/g of tissue correlated with the increase in inorganic phosphate from 12.1 ± 1.4 to 18.7 ± 1.2 μmol/g of tissue. Incorporation studies using radioactive tyrosine-*O*-phosphate, labeled either in the tyrosine moiety or the phosphate moiety, have not been done for *Drosophila* cuticle. However, Bodnaryk *et al.* (1974), investigating the metabolism of *N*-acetyldopamine 3-*O*-phosphate in *P. americana*, have demonstrated that the NADA moiety is incorporated into the cuticle whereas the phosphate moiety is not, similar to the fate of *N*-acetyldopamine 3-*O*-sulfate (see above). They suggest these conjugates may be the form in which NADA is transported from the blood into the epidermis by a putative carrier protein.

Compelling circumstantial evidence (Schneiderman *et al.*, 1966; Harper and Armstrong, 1972, 1973, 1974) indicates that the enzyme which hydrolyzes tyrosine-*O*-phosphate at pupariation in *D. melanogaster* is the alkaline phosphatase specified by the *Aph-1* locus at 47.3 on the third chromosome (denoted 3–47.3) (Beckman and Johnson, 1964a,b; Wallis and Fox, 1968). This evidence and the genetics and biology of *Aph-1* are presented in Section VI.

IV. Melanization, Quinone Sclerotization, and β-Sclerotization in Insects

A. Melanization

Using cyclic voltammetry to investigate both electrochemically and enzymatically catalyzed reactions in hemolymph and cuticle of *M. sexta*, Kramer *et al.* (1983; Aso *et al.*, 1984) established the presence of the pathway of reactions depicted in Fig. 4. Except for minor differences in terminology, the reaction sequence is identical to that found in mammalian cells (Mason, 1955; Pawelek *et al.*, 1980; Körner and Pawelek, 1982). The reactions include the diphenol oxidase (tyrosinase)-mediated oxidation of catecholamines to their respective *o*-quinones and their spontaneous cyclization to leucoaminochromes, followed by their reoxidation by diphenol oxidase to their respective *p*-quinone imines. Partially purified cuticular diphenol oxidase (tyrosinase) was also shown to hydroxylate tyrosine and tyramine to DOPA and DA, respectively, and to oxidize 5,6-dihydroxyindole (DHI) to indole-5,6-quinone (Aso *et al.*, 1984). The order of substrate preferences was NBAD, NADA, DA, DHI, DOPA. Although NBAD and NADA are the preferred substrates for oxidation, they were found to cyclize at a rate 40-fold lower than DA (Kramer *et al.*, 1983). If these properties are maintained *in vivo*, they suggest that NBAD and NADA subserve different physiological roles vis-à-vis DA (see below).

A dopa quinone imine conversion factor (QICF) has also been partially purified from *M. sexta* extracts by Aso *et al.* (1984) and is apparently equivalent to the mammalian dopachrome conversion factor (Pawelek *et al.*, 1980). The *Manduca* QICF is a heat-labile and protease-susceptible substance that facilitates the conversion of dopa quinone imine to 5,6-dihydroxyindole but does not accelerate the conversion of DA quinone imine to DHI. Aso *et al.* (1984) suggest

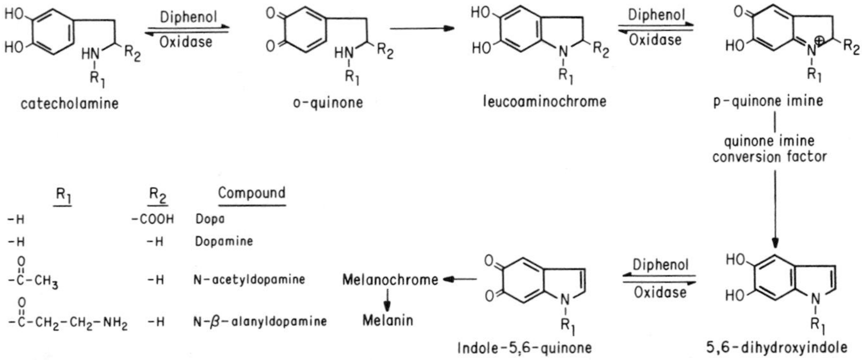

R_1	R_2	Compound
-H	-COOH	Dopa
-H	-H	Dopamine
$-\overset{O}{\overset{\Vert}{C}}-CH_3$	-H	N-acetyldopamine
$-\overset{O}{\overset{\Vert}{C}}-CH_2-CH_2-NH_2$	-H	N-β-alanyldopamine

FIG. 4. The oxidation of catecholamines by diphenol oxidase in the synthesis of their derivatives involved in sclerotization and melanization. N-acetyldopamine and *N*-β-alanyldopamine quinone imines apparently do not undergo indolization (Aso *et al.*, 1984). The specific dopa quinone imine conversion factor in *M. sexta* may be dopa quinone imine decarboxylase, which is not a DDC (Aso *et al.*, 1984). {Figure adapted from Aso *et al.* (1984). Reprinted with permission from [*Insect Biochem.*, **14**, 463–472, Y. Aso, K. J. Kramer, T. L. Hopkins, and S. Z. Whetzel, Properties of tyrosine and dopa quinone imine conversion factor from pharate pupal cuticle of *Manduca sexta* (*L*)], Copyright (1984), Pergamon Journals Ltd.}

that it may be a specific dopa quinone imine decarboxylase since the partially purified preparation did not convert DOPA to DA. Apparently because of the presence of the acyl substituents on the nitrogen, the *p*-quinone imines derived from NADA and NBAD do not undergo indolization (Aso *et al.*, 1984). This observation, along with the reduced rate of cyclization of NADA and NBAD *o*-quinones, suggests that at least in *M. sexta*, NADA and NBAD derivatives are not involved in melanization. These observations are consistent with NADA and NBAD derivatives being primarily involved in cross-linking proteins and/or chitin in sclerotization and that the DOPA and DA derivatives are the primary precursors for polymerization into melanin (Hopkins *et al.*, 1982, 1984; Brunet, 1980). A scheme showing the role of these compounds in sclerotization and melanization is presented in Fig. 5 (Fig. 9 in Aso *et al.*, 1984). This scheme indicates that DOPA and DA derivatives may be directly involved in sclerotization as well as in melanization, but that NADA and NBAD derivatives are only involved in the former. This scheme emphasizes the many different ways catecholamine derivatives can cross-link protein (and chitin). If numerous different cuticular proteins with different properties can be cross-linked by this array of different catecholamine deriva-

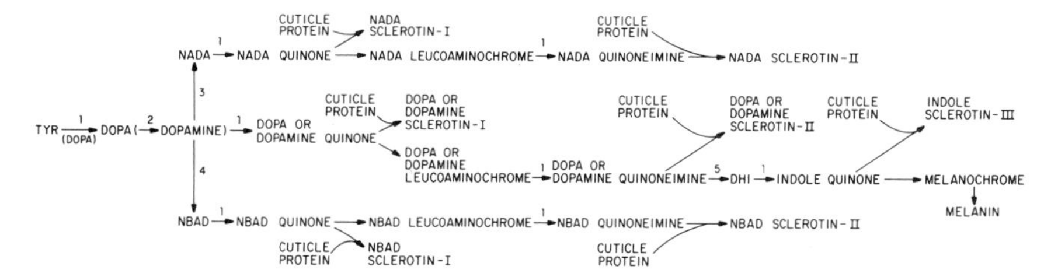

FIG. 5. Hypothetical pathway for metabolism of tyrosine and catecholamines to form sclerotin and melanin in insect cuticle. Numbered reactions catalyzed by enzymes: (1) tyrosinase, (2) dopa decarboxylase (parentheses indicate DOPA may enter Mason–Raper pathway directly), (3) NADA synthase, (4) NBAD synthase, (5) dopa quinone imine conversion factor. Other reactions probably occur spontaneously. Sclerotin-I, -II, and -III denote quinone, quinone imine, and indole quinone ring systems, respectively, used as protein cross-linking agents. {After Aso *et al.* (1984). Reprinted with permission from [*Insect Biochem.* **14**, 463–472, Y. Aso, K. J. Kramer, T. L. Hopkins, and S. Z. Whetzel, Properties of tyrosine and dopa quinone imine conversion factor from pharate pupal cuticle of *Manduca sexta* (L)], Copyright (1984), Pergamon Journals, Ltd.}

tives, one can begin to understand how so many different cuticles with such different properties can be generated.

The extensive genetics of diphenol oxidase (tyrosinase) activity in *D. melanogaster* is summarized in Section XI. There is no information on the dopa quinone imine conversion factor (dopa quinone imine decarboxylase) in *Drosophila.* However, there are two reports of dopa decarboxylase activities in *D. melanogaster* which are not precipitated by antibodies to the prevalent dopa decarboxylase coded for by the *Ddc* locus (Spencer *et al.*, 1983; Bishop and Wright, 1987). These activities could be due to dopa quinone imine decarboxylase, tyrosine decarboxylase, or yet another enzyme with dopa decarboxylase activity. If either of the observed activities in *Drosophila* turn out to be a dopa quinone imine decarboxylase, then the *Drosophila* enzyme would differ from the *M. sexta* enzyme, for in *Drosophila* DOPA is decarboxylated to DA, but in *M. sexta* it is not (see above).

B. QUINONE SCLEROTIZATION AND β-SCLEROTIZATION

The biochemical aspects of sclerotization in insects have been reviewed by Sherald (1980), Brunet (1980), Lipke *et al.* (1983), and Andersen (1985). Other topics relevant to sclerotization in insects have also been reviewed, including the structure of the integument (Hepburn, 1985), chitin metabolism (Kramer *et al.*, 1985), and cuticular proteins (Silvert, 1985) and their genetics (Fristrom *et al.*, 1986).

Sclerotization (or tanning, a term used so loosely that it will be avoided here) primarily involves the incorporation of catecholamines into the cuticle, where they are oxidized to their respective quinones, which then covalently interact with amino groups of different protein molecules or with chitin, thereby effecting cross-links between different protein molecules and/or chitin. This results in the stabilization and hardening of the cuticle. Although Pryor (1940a,b) first suggested this scheme for sclerotization in insects, it was Karlson *et al.* (1962) who established that *N*-acetyldopamine was a primary sclerotizing agent in *Calliphora,* in contrast to *N*-acetyltyramine and the 4-*O*-β-glucoside of NADA which also were found to be present. It took yet another 20 years before Hopkins *et al.* (1982) established that *N*-β-alanyldopamine also is a major sclerotizing agent like NADA, at least in *M. sexta* (and *Drosophila;* see Sections V and XIII).

Quinone sclerotization (ring sclerotization, quinone tanning) occurs when a catecholamine is oxidized to its corresponding quinone by cuticular diphenol oxidase (tyrosinase, polyphenol oxidase), which then reacts with the ε-amino or N-terminus of a cuticular protein to give the protein–catechol ring adduct (Fig. 6, upper pathway). Further oxidation of this adduct by diphenol oxidase yields a further reaction with the ε-amino or N-terminus of another protein, producing protein ring cross-links which are stabilized by tautomerization. Apparently catecholamine derivatives react with proteins only through the two sites on the ring depicted in Fig. 6 (the second and fifth carbons of a four-substituted catechol) (Lipke *et al.*, 1983).

β-Sclerotization, whereby protein molecules are cross-linked through the β-carbon of the side chain of catecholamines (see the side-chain cross-links of proteins R_3 and R_4 in the last compound of the lower pathway in Fig. 6), was initially proposed by Andersen (1970, 1974; Andersen and Barrett, 1971). Evidence that such side-chain cross-links occur in cuticle has most recently been summarized and evaluated by Lipke *et al.* (1983) and Andersen (1985). It seems clear that β-sclerotization does in fact occur in insect cuticles, but two different mechanisms have been proposed by these authors by which these side-chain cross-links are established. It is not clear that the two mechanisms are mutually exclusive, i.e., that both could not coexist in the cuticle of a single insect species or that the different mechanism may be found in different insect species.

Andersen (1985) proposes that a desaturase present in the cuticle removes a hydrogen from the side chain of NADA, which introduces a double bond in the side chain between the β- and α-carbons. The

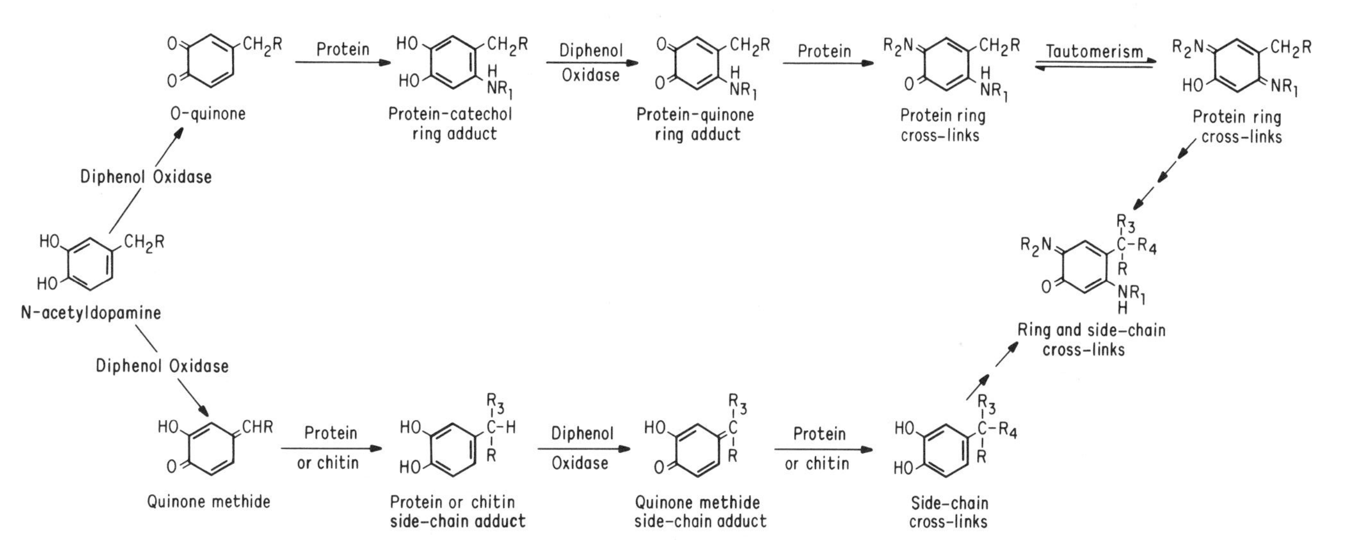

FIG. 6. Two proposed pathways for cross-linking catecholamines and proteins during sclerotization in insect cuticle (adapted from Lipke *et al.*, 1983). Upper pathway: quinone sclerotization. Lower pathway: quinone methide sclerotization, β-sclerotization, or side-chain sclerotization; R = $CH_2NHCOCH_3$ if *N*-acetyldopamine is the sclerotizing agent or R = $CH_2NHCOCH_2CH_2NH_2$ if it is *N*-β-alanyldopamine; R_1 and R_2 = proteins, R_3 and R_4 = proteins or chitin. The mixed ring and side-chain cross-linked adduct is included to indicate that various mixtures of the two types of cross-linking may occur sequentially (see Lipke *et al.*, 1983).

resulting dehydro-*N*-acetyldopamine would be oxidized to its quinone by diphenol oxidase, which would then presumably cross-link with proteins through either or both of the side-chain carbons, producing a colorless sclerotin. This scheme also permits quinone ring sclerotization to take place at the same time. The desaturase necessary for this scheme has not been purified or characterized but its activity has been demonstrated in locust femur (Andersen and Roepstorff, 1982).

Lipke *et al.* (1983) proposed a pathway for β-sclerotization based on reactive quinone methides as intermediates (Fig. 6, lower pathway). They suggest such a pathway is consistent with isolates from biological systems and the known chemical behavior of aromatic and quinonoid reagents. The quinone methide derivative could arise directly by means of diphenol oxidase oxidation of NADA or indirectly by tautomerization of the *o*-quinone of NADA. The NADA quinone methide intermediate would then react with protein or chitin through the β-carbon of the side chain and would be reoxidized to a quinone methide intermediate, which would again react with protein or chitin through the β-carbon, providing for protein–protein, protein–chitin, and chitin–chitin cross-links. They propose that the reaction with the β-carbon of the NADA quinone methide intermediate would occur through amino acid side chains of proteins: imidazolyl of histidine, amino of lysine and N-terminals, phenolic of tyrosine, hydroxyls of serine and threonine, and hydroxyls in the third and sixth positions of *N*-acetylglucosaminyl units of chitin. They also propose that all possible combinations of mixed ring and side-chain cross-links should occur. One possible combination with two ring and two side-chain cross-links is depicted in Fig. 6.

Although the genetics of diphenol oxidase activity is extensive (Section XI), so far it has not shed much light on mechanisms of sclerotization. It should be noted that protein can be cross-linked by other mechanisms which do not involve catecholamines. In particular, in the elastic protein resilin, dityrosine and tertyrosine cross-links formed from endogenous tyrosine groups can be effected by diphenol oxidase (tyrosinase) or peroxidase (Andersen, 1966). Furthermore, it has been established that an endogenous peroxidase activity in chorions of *Drosophila* oocytes is responsible for cross-linking chorion proteins via di- and tertyrosine bridges during the last stage of oogenesis (Mindrinos *et al.*, 1980). As yet no genes responsible for this peroxidase activity have been identified.

For a thorough, historical, critical evaluation of the metabolism of aromatic amino acids involved in sclerotization in insects see the review by Brunet (1980).

V. Developmental Profiles of Catecholamine Pools during Embryogenesis and after Adult Eclosion in *Drosophila*

Black *et al.* (1987a) have isolated and identified catecholamines extracted from 70 g of adult *Drosophila* by absorption to alumina. Subsequent fractionation on a BioGel P-2 column was monitored for electrochemically active compounds by injection of samples of the fractions into a high-performance liquid chromatography apparatus fitted with an electrochemical (EC) detector. Selected regions of the eluate were concentrated and rechromatographed individually in Sephadex LH-20 columns with fractions again monitored by HPLC-EC. Individual peaks identified by OD_{280} and electrochemical activity were collected, providing pure compounds in most cases. DOPA, DA, NADA, and NBAD were recovered and their structural identity confirmed by both high- and low-resolution mass spectrometry. Four additional compounds were recovered that have properties consistent with their being unidentified catecholamines, i.e., low-resolution fast atom bombardment mass spectra, electrochemical activity, UV–VIS spectra, and absorption to aluminum oxide. None of the unidentified catecholamines was epinephrine, norepinephrine, dihydroxybenzoic acid, or α-methyldopa based on their HPLC retention times with different running buffers or by fast atom bombardment mass spectrometry.

In order to investigate the role of known catecholamines as well as of two of the unidentified catecholamines mentioned above, Black *et al.* (1987a) have carried out a detailed developmental analysis comparing catecholamine pool alterations during the formation of unpigmented cuticle during embryogenesis (Fig. 7) with pool changes during the formation of pigmented cuticle following adult eclosion (Fig. 8). Pools of DOPA, DA, NADA, and the two unidentified catecholamines designated VIII and X were measured at hourly intervals during embryogenesis (Fig. 7). Prior to egg laying, small pools of these five catecholamines are present in ovarian tissue. No NBAD was detected during embryogenesis. Of particular interest is the observation that it is the pool size of catecholamine X that starts to increase when unpigmented cuticle formation begins at 10 hours of embryogenesis (Poulson, 1950; Hillman and Lesnick, 1970), and not DA or NADA, which are not found until after 15 hours of development after the initial appearance of dopa decarboxylase activity (Beall and Hirsh, 1984; Gietz and Hodgetts, 1985). Furthermore, catecholamine X is the predominant catecholamine present throughout the remainder of embryogenesis. Its possible role in the sclerotization of unpigmented cuticle is discussed in relation to the genetics of the *l(2)amd* gene

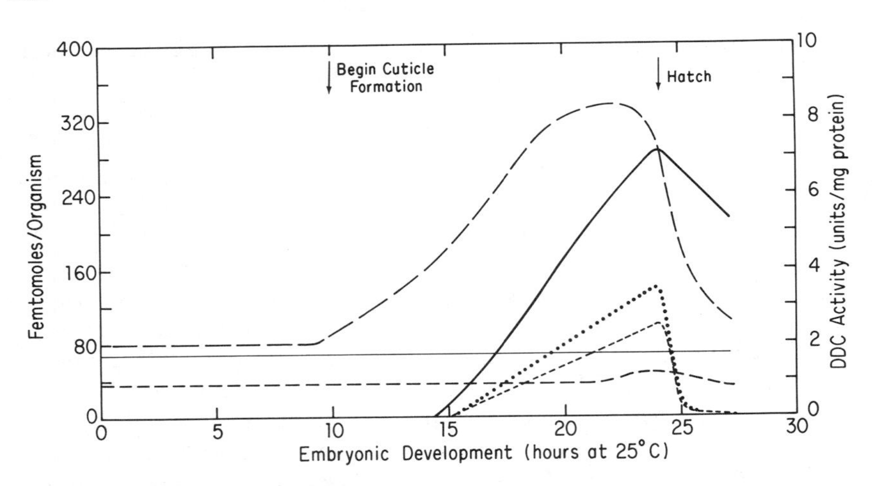

FIG. 7. Developmental profiles of catecholamine pools throughout embryogenesis when unpigmented sclerotization of cuticle takes place. The dopa decarboxylase developmental profile of Kraminsky *et al.* (1980) is included for reference (after Black *et al.*, 1987a). (– – –) DOPA, (- - - -) dopamine, (· · · ·) NADA, (———) compound VIII, (— — —) compound X, (———) DDC activity.

(Section IX,B). At the time of adult eclosion, when probably both unpigmented and pigmented sclerotization of the cuticle are taking place, the pool of catecholamine X is already elevated, whereas the pools of NADA and particularly NBAD begin to increase in size very rapidly immediately after eclosion. For all three of these catecholamines, maximum pools sizes are achieved at 18 hours after eclosion at 22°C, when the pools of all the catecholamines except catecholamine VIII begin to decrease in size, with NBAD pools subsequently leveling off at a concentration of approximately 100 pmol/fly greater than NADA. (Note in Fig. 8 the scale for NBAD concentration is fourfold that for the other catecholamines.) Why adult NBAD pools are sexually dimorphic is not known, although heavy melanization in the male abdominal cuticle may divert significant amounts of DOPA and/or DA into melanin rather than NBAD.

A comparison of the developmental profiles of the catecholamine pools at embryonic eclosion (Fig. 7) and after adult eclosion (Fig. 8) obviously shows that they are completely different. Particularly notable are the absence of NBAD in embryos and the coordinate changes in the levels of DA and NADA pools in embryos, in contrast to completely independent changes in DA and NADA pools in young adults.

Although developmental profiles of catecholamine pools in *Dro-*

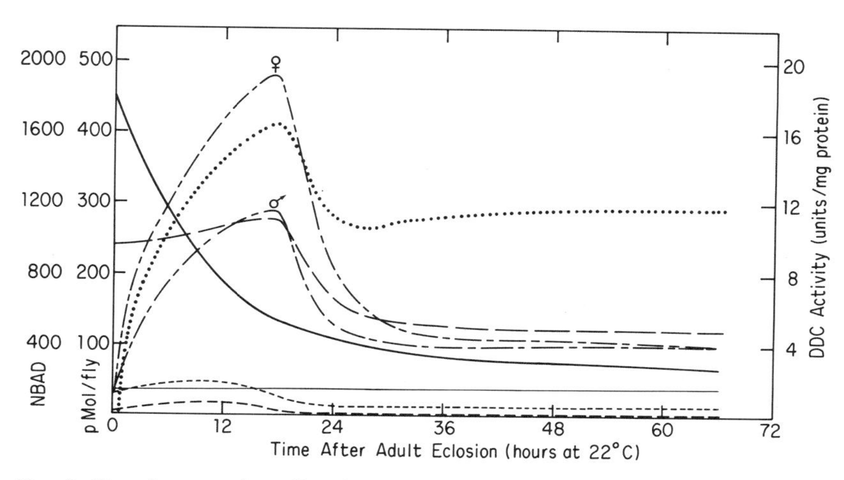

FIG. 8. Developmental profile of catecholamine pools in adults from 0 to 3 days after eclosion when pigmented and unpigmented sclerotization take place. The sexual dimorphism of *N*-*β*-alanyldopamine pools is distinguished by ♀ for females and ♂ for males. The developmental profile of dopa decarboxylase activity from Kraminsky *et al.* (1980) is included for reference (after Black *et al.*, 1987a). (— - —) NBAD, (– – –) DOPA, (- - - -) dopamine, (· · · ·) NADA, (———) compound VIII, (— — —) compound X, (———) DDC activity.

sophila prior to and after pupariation and during metamorphosis have yet to be reported, a catecholamine pool profile of sessile third instar larvae prior to spiracle eversion is presented by Black *et al.* (1987a). At this time in development DOPA, DA, and NADA pools are very small, and there is no measurable NBAD peak. By far the largest pool is found in the catecholamine X peak with significant amounts in catecholamine VIII also present. The levels of DA and NADA found are consistent with those reported earlier (Hodgetts and Knopka, 1973).

The above studies were done using extracts of whole flies, embryos, or larvae. Developmental profiles of biogenic amine pools in dissected nervous systems have not been reported. This information would be very useful along with complete developmental pool profiles on whole organisms throughout the three larval instars of *Drosophila,* as well as between pupariation and adult eclosion. The larval data would supplement the work of Kaznowski *et al.* (1985, 1986), who have provided a thorough morphological analysis and description of cuticle secretion during the second and third larval instars (including ecdysis) of *D. melanogaster* and, by means of electron microscope autoradiography, have followed the incorporation of tritiated leucine, tyrosine, and glucosamine into the integument of *Drosophila.*

Whether there are mutations in *Drosophila* which will elucidate cuticle secretion remains to be seen.

VI. The Genetics and Biology of Alkaline Phosphatase-1 and the Metabolism of Tyrosine-*O*-Phosphate

That alkaline phosphatase-1 (E.C. 3.1.3.1) in *Drosophila* is the enzyme *in vivo* which mobilizes tyrosine by hydrolyzing tyrosine-*O*-phosphate to tyrosine and inorganic phosphate has not been unequivocally established. However, the circumstantial evidence is strong. The enzyme is found primarily (but not exclusively; see below) in late third instar larvae and early pupae (Beckman and Johnson, 1964b; Schneiderman *et al.*, 1966; Sena, 1966; Wallis and Fox, 1968) and has been histochemically localized by two different methods in the epidermis of the late third instar larva (Yao, 1950; Schneiderman *et al.*, 1966). Schneiderman *et al.* (1966) also extracted the expected two different electrophoretic forms of the enzyme from isolated third instar larval hulls (muscle and epidermal preparations) from the two strains of *Drosophila* homozygous for two different electrophoretic variant alleles of the *Alkaline phosphatase-1* locus (see below). Furthermore, using partially purified preparations of alkaline phosphatase-1 produced by four different naturally occurring alleles (see below), Harper and Armstrong (1972, 1973, 1974) have demonstrated that tyrosine-*O*-phosphate is clearly the most effective natural substrate for all four of the allelic forms of the enzyme, and that it is bound as effectively as the nonphysiological substrate *p*-nitrophenylphosphate. Alkaline phosphatase-1 purified 100-fold was not inhibited by inorganic phosphate, phenylalanine, or tyrosine, but was inhibited by high concentrations of the substrate, tyrosine-*O*-phosphate (≥ 7 m*M*) and by KCN and L-cysteine (Harper and Armstrong, 1974). The *Drosophila* alkaline phosphatase-1 enzyme has been extensively characterized by Harper and Armstrong (1972, 1973, 1974).

Alkaline phosphatase-1 (*Aph-1*, 3–47.3) was the second locus in *Drosophila* established on the basis of the segregation of electrophoretic variant alleles (Beckman and Johnson, 1964a,b). The initial two alleles discovered, Aph^4 (= Aph^F) and Aph^6 (= Aph^S), are the two most common naturally occurring alleles (F. M. Johnson, unpublished; cited in Harper and Armstrong, 1973). Alkaline phosphatase-1 from Aph^4/Aph^4 homozygous larvae migrates faster in starch gels than enzyme from Aph^6/Aph^6 larvae. Heterozygous Aph^4/Aph^6 larvae produce the two parental types of alkaline phosphatase-1 plus a hybrid type with a mobility intermediate between the two parental types, indicating that the enzyme is at least a homodimer. Three (perhaps four) more alleles have been found (Wallis and Fox, 1968; Harper and Armstrong, 1973).

Individuals homozygous for the Aph^0 allele are completely viable and fertile even though extracts of Aph^0/Aph^0 larvae yield no active alkaline phosphatase-1 in gels (Johnson, 1966). Apparently this allele does produce an enzymatically inactive subunit capable of dimerizing with enzymatically active subunits, for Aph^0/Aph^6 and Aph^0/Aph^A heterozygous larval extracts exhibit not only the expected parental forms characteristic of Aph^6/Aph^6 and Aph^A/Aph^A, respectively, but also hybrid forms. In both cases the hybrid forms have mobilities which suggest that the inactive subunit has a mobility similar to or identical with the active Aph^4 fast-migrating subunit. Consistent with this interpretation is that fact that Aph^0/Aph^4 heterozygotes produce only a parental-type fast-migrating form of the enzyme and no distinguishable hybrid form.

To date no one has measured tyrosine-*O*-phosphate levels in Aph^0/Aph^0 late third instar larvae and early pupae to determine whether or not the compound is hydrolyzed in these individuals. Without further data, at least three interpretations are possible to explain the apparently normal phenotype of Aph^0/Aph^0 homozygotes. First, sufficient tyrosine may be mobilized from other sources to effect normal sclerotization. If so, tyrosine-*O*-phosphate levels may remain very high throughout the pupal stage in Aph^0 homozygotes. Second, other enzymes, e.g., acid phosphatase (see discussion below), can hydrolyze sufficient tyrosine-*O*-phosphate. Third, the mutant alkaline phosphatase-1 in Aph^0/Aph^0 homozygotes may be active *in vivo* and may only be inactivated when the larva is homogenized. Combinations of these three possibilities could, of course, obtain.

One should note that tyrosine-*O*-phosphate is also an effective *in vitro* substrate for the lysosomal enzyme acid phosphatase (E.C. 3.1.3.2) (Yasbin *et al.*, 1978) specified by the *Acph-1* locus (3–101.1) in *Drosophila* (MacIntyre, 1966). Although the cellular and histological distribution and developmental profile of acid phosphatase (Yasbin *et al.*, 1978) suggest that it is not responsible for the *in vivo* mobilization of tyrosine from tyrosine-*O*-phosphate at pupariation, the possibility is not completely excluded. Not only are homozygotes for null alleles of *Acph-1* completely viable and fertile like $Aph\text{-}1^0$, but double-null homozygotes for both the alkaline and acid phosphatases, $Aph\text{-}1^0$ $Acph\text{-}1^{n5}/Aph\text{-}1^0$ $Acph\text{-}1^{n5}$, are also perfectly viable and fertile (Sawicki and MacIntyre, 1978).

In their survey of alkaline phosphatases in *Drosophila*, Schneiderman *et al.* (1966) report the appearance in late prepupae and young pupae of an alkaline phosphatase (band 5) which migrates faster than alkaline phosphatase-1 (band 2, specified by *Aph-1*). As band 5 activity increases, alkaline phosphatase-1 (band 2) decreases and disappears. Band 5 activity increases considerably in late pupae and was demonstrated (Schneiderman *et al.*, 1966) to derive from the yellow body, a pellet of sloughed larval midgut cells formed in the newly developed imaginal gut and excreted as the meconium after emergence (Bodenstein, 1950). It was assumed that the yellow-body phosphatase was larval gut alkaline phosphatase (bands 3, 4a, and 4b) with its mobility modified during histolysis. However, attempts to modify larval gut phosphatase were unsuccessful (Schneiderman *et al.*, 1966). In contrast, Wallis and Fox (1968) determined that the electrophoretic mobility of the pupal yellow-body alkaline phosphatase is modified by each of the four *Aph-1* alleles investigated. Except for the one specified by *Aph-1*^{A}, the electrophoretic mobilities of the pupal alkaline phosphate variants parallel those of the larval alkaline phosphatase-1 variants. The segregation and careful mapping of the pupal variants exactly correlate with the segregation and mapping of the larval variants (Wallis and Fox, 1968), indicating that the larval and pupal variants are specified by the same locus. Schneiderman (1967) reports that a number of proteolytic enzymes are capable of transforming the larval alkaline phosphatase-1 into an active alkaline phosphatase with the electrophoretic mobility of the pupal yellow-body enzyme. The same transformation can be produced by heat-labile extracts of larval gut. These observations demonstrate the relationship between the larval alkaline phosphatase-1 and the pupal yellow-body enzyme but do not elucidate the *in vivo* origin of the modified pupal alkaline phosphatase-1.

Not including allelic variant forms, Schneiderman *et al.* (1966) demonstrated the existence of seven electrophoretically distinguishable forms of alkaline phosphatase in various tissues and at various stages of development in *Drosophila*. Other than *Aph-1*, only the genetics of the adult hind gut alkaline phosphatase (band 7) has been investigated (see *Aph-2*, Schneiderman *et al.*, 1966; Lindsley and Zimm, 1985).

Finally, for tyrosine-*O*-phosphate, the mutation *lethal translucida* *l(3)tr* (3–20.0), when homozygous, produces larvae which become bloated and transparent from accumulation of abnormal amounts of hemolymph. The mutation apparently causes some sort of abnormality in protein synthesis since in the hemolymph concentrations of amino acids are higher than normal and concentrations of proteins are lower (Hadorn, 1956). In contrast to wild-type larvae which contain approximately equal amounts of tyrosine and tyrosine-*O*-phosphate, *l(3)tr* larvae contain essentially no free tyrosine but contain abnormally large amounts of tyrosine-*O*-phosphate (Mitchell and Simmons, 1961). Why tyrosine and hemolymph protein levels are depressed and tyrosine-*O*-phosphate and other amino acid levels are elevated is not known.

VII. Dopa Decarboxylase in *D. melanogaster*

The control of dopa decarboxylase (DDC) in insects has been reviewed (Sekeris and Fragoulis, 1985). The summary in this section will consider only DDC in *D. melanogaster*.

A. Biology and Biochemistry

The enzyme (E.C. 4.1.1.28) 3,4-dihydroxy-L-phenylalanine carboxylyase, in *Drosophila*, catalyzes the decarboxylation of DOPA to dopamine (Fig. 1) and 5-hydroxytryptophan to serotonin (5-hydroxytryptamine) (Fig. 2) but not tyrosine to tyramine (Livingstone and Tempel, 1983). It was initially assayed by Lunan and Mitchell (1969) and McCaman *et al.* (1972), both of whom provided partial developmental profiles. Greater than 90% of the activity is found in the epidermis–cuticle (Scholnick *et al.*, 1983). DDC activity has been assayed in the central nervous system (CNS) of adults and white prepupae (Dewhurst *et al.*, 1972; Wright, 1977; Livingstone and Tempel, 1983; Bray and Hirsh, 1986), in the proventriculus dissected from late third instar larvae (Wright and Wright, 1978), and in ovaries (Wright *et al.*, 1981b; B. C. Black, unpublished). That the same gene codes for DDC activity in the epidermis, CNS, and proventriculus was demonstrated by the assay of less than 50% wild-type (+/+) activity in these tissues in *Ddc* mutant heterozygotes (*Ddc*^{n}/+) (Wright, 1977; E. Y. Wright and T. R. F. Wright, unpublished). Five peaks of DDC activity are evident when whole organisms are assayed throughout development. These are found at the end of embryogenesis (Gietz and Hodgetts, 1985), at each larval molt, at pupariation, and at adult eclosion (Marsh and Wright, 1980; Kraminsky *et al.*, 1980). The largest peak at pupariation is induced by a coincident peak of the molting hormone, ecdysone (Marsh and Wright, 1980), and has been shown to be due to a rapid increase in translatable DDC mRNA following the administration of 20-OH-ecdysone (Kraminsky *et al.*, 1980; Clark *et al.*, 1986). The induction of DDC transcripts in isolated imaginal disks (presumptive adult epidermis) occurs only by exposure to a pulse of 20-OH-ecdysone and not when the hormone is present continuously (Clark *et al.*, 1986). It is not clear whether or not DDC in the CNS responds to ecdysone. The developmental profile of DDC activity in dissected CNSs from midthird larval instar through pupariation and metamorphosis resembles the whole organism profile and therefore suggests that ecdysone may be regulating DDC levels in the CNS also (E. Y. Wright and T. R. F. Wright, unpublished). Hormonal regulation of DDC in *Drosophila* is extensively discussed in Hodgetts *et al.* (1986a,b).

Drosophila DDC has been purified and shown to be a homodimer with a subunit molecular weight of 54,000 (Clark *et al.*, 1978). The enzyme requires pyridoxal 5′-phosphate for activity. A thorough kinetic analysis including product inhibition studies has been carried out by Black and Smarrelli (1986), who showed that since endogenously bound pyridoxal 5′-phosphate is always present, the addition of exogenous cofactor is not an absolute requirement for activity but does provide for a threefold increase in activity. The amino acid sequence of *Drosophila* DDC has not been directly determined, but has been derived from the nucleotide sequence of the gene (Eveleth *et al.*, 1986), providing evidence for two RNA splicing alternatives which appear to encode two different protein isoforms. That different isoforms of the enzyme do exist is indicated by the data of Bishop and Wright (1987), who demonstrate that adult DDC is more thermolabile than white prepupal DDC. Although the genetic data (see below) indicate that there is only one gene coding for DDC in the *Drosophila* genome, DDC activity which is not precipitated by polyclonal antibodies to the prevalent DDC has been found in *Drosophila* Kc tissue-culture cells (Spencer *et al.*, 1983) and in extracts of whole white prepupae and newly eclosed adults (Bishop and Wright, 1987). Whether these activities may be the result of tyrosine decarboxylase activity (Livingstone and Tempel, 1983), dopa quinone imine decarboxylase (see Section IV,A), or some other enzyme with DDC activity is not known.

B. Genetics

Hodgetts (1975), using a series of Y;autosome translocations produced by Lindsley *et al.* (1972) found that the only DDC dosage-sensitive region in the whole genome is on the left arm of the second chromosome between bands 36E–F and 37D. The subsequent use of overlapping deficiencies (Wright *et al.*, 1976b; Gilbert *et al.*, 1984) and *in situ* hybridization with cloned *Ddc* DNA (Hirsh and Davidson, 1981) located the *Ddc* gene in the band 37C1,2. Most of the 50 "point" mutations in the *Ddc* gene (Wright *et al.*, 1976a, 1981a, 1982) were isolated as lethals over deficiencies which delete band 37C1,2 and were then identified as *Ddc* alleles by assaying for reduced DDC activity in heterozygotes over the balancer chromosome, CyO, carrying a wild-type *Ddc* allele. Lethal alleles of *Ddc* have been mapped 0.025 map units to the right of *hook* (*hk*, 2–53.9) and 0.002 map units to the right of *l(2)amd* (see Section IX,B). Except for five temperature-sensitive lethal alleles, Ddc^{ts1}–Ddc^{ts5}, only three nonlethal alleles, Ddc^{lo1}, Ddc^{+4}, and Ddc^{DE1}, have been isolated (Wright *et al.*, 1982; Estelle and Hodgetts, 1984a,b; Bishop and Wright, 1987).

Consistent with the enzyme being a homodimer, some intragenic heteroallelic heterozygotes exhibit both positive and negative complementation (Wright *et al.*, 1982). Different Ddc^{nx}/Ddc^{ny} heterozygous combinations are differentially viable and produce different amounts of DDC activity, which is usually thermolabile. On the other hand, some $Ddc^{n}/+$ heterozygotes produce DDC activities that are significantly lower than the expected 50% (24–40%), indicating that the mutant : wild-type heterodimers are inactive. Most zygotes hemizygous for *Ddc* lethal alleles over *Df(2L)TW130*, an 8- to 12-band deficiency for the region, die as active larvae unable to eclose from the embryonic membranes, with clearly underpigmented (incompletely sclerotized?) mouthparts and denticle belts. A few hemizygotes manage to hatch, with most of them dying at pupariation, but some continue development to die as pharate adults. This is true even for Ddc^{n27}/*Df(2L)TW130* hemizygotes where Ddc^{n27} is a 2.3-kb intragenic deletion. Thus a few individuals with absolutely no DDC activity can develop to the pharate adult stage. The number of individuals that do this can be increased by artificially releasing Ddc^{n27}/*Df(2L)TW130* hemizygotes from their embryonic membranes (Valles and White, 1986). Individuals with extremely low levels of DDC activity (0.5–2% of wild type) will eclose as adults exhibiting an incomplete sclerotization phenotype ("escaper" phenotype) (see colored illustration in Pentz *et al.*, 1986). These adults will often die or get stuck in the food within 24 hours of eclosion. Their macrochaetae are often very thin, long, and straw colored or colorless. The whole adult integument remains light, not taking on its normal pigmentation. Abdominal markings are apparent but do not darken. After aging a few hours, wing axillae and leg joints become melanized, perhaps as a result of the phenol oxidase wound reaction to the rupture of weakened cuticle along with elevated DOPA pools. Flies with an extreme phenotype, perhaps due to weakened cuticle, appear to walk on their tibiae rather than on their tarsi but their leg movements appear to be coordinated (Wright *et al.*, 1976a). Zygotes with these very low DDC activities will have their developmental time prolonged for as many as 4 or 5 days and mutant puparia are easily identified by melanization at both ends of abnormally colored greenish-gray pupa cases. Genotypes that produce DDC levels low enough to effect these mutant phenotypes are intragenic heteroallelic complementing heterozygotes which produce less than 5% of the expected number of survivors (e.g., Ddc^{n8}/Ddc^{n6}, Ddc^{n5}/Ddc^{n6}, and Ddc^{n5}/Ddc^{n4}) (Wright *et al.*, 1976a), Ddc^{ts2} hemizygotes raised continuously at 22 or 25°C, or Ddc^{ts1} or Ddc^{ts2} homozygotes heat treated at the restrictive temperature of 30°C for a 24- or 48-hour pulse prior to adult eclosion (T. R. F. Wright, unpublished).

Two *Ddc* mutant alleles, Ddc^{DE1} and Ddc^{+4}, alter the developmental expression of DDC. Adults homozygous for Ddc^{DE1}, a γ-ray-induced allele with DDC activity of <5% of wild type, express the usual incomplete sclerotization phenotype except the macrochaetae are morpholocally normal and normally pigmented, suggesting that *Ddc* gene expression is elevated in the bristle-forming cells vis-à-vis the other epidermal cells. DDC activities in Ddc^{DE1} homozygous embryos and CNSs dissected from white prepupae are also at this reduced level (<5%). However, activity in Ddc^{DE1} whole wandering-stage late third instar larvae (predominantly epidermal activity) is 5- to 10-fold higher, with levels at least 20% of wild type (Bishop and Wright, 1987). Ddc^{DE1} DDC from both whole adults and whole white prepupae is more thermolabile than wild-type DDC from these stages, with adult Ddc^{DE1} DDC being more thermolabile than white prepupal Ddc^{DE1} DDC. Although the precise intragenic location of the Ddc^{DE1} lesion has yet to be determined, the data indicate that somehow the lesion has a more drastic effect on the adult, embryonic, and white prepupal CNS DDC isoform(s) than on the white prepupal epidermal and adult bristle DDC isoform(s) (Bishop and Wright, 1987). On the other hand, Ddc^{+4}, an allele isolated from a natural population, overproduces epidermal DDC at all stages except pupariation: larval eclosion—141%; second instar/third instar molt—150%; adult eclosion—118%; versus pupariation—50% (Estelle and Hodgetts, 1984a). The mutant does not affect DDC activity in brains dissected from wandering third instar larvae and 0- to 2-hour-old adult females. The epidermal DDC activity differences are directly correlated with differences in the amount of epidermal DDC cross-reacting material. Seven differences between cloned *Ddc*$^{+4}$ DNA and cloned Canton S wild-type DNA were found (Estelle and Hodgetts, 1984b), which consisted of six small (<100 bp) restriction-length (insertion–deletion) polymorphisms and one restriction site polymorphism. Five of the seven differences are in the 5′-untranslated leader sequence of the DDC mRNA or in the 4.5 kb of DNA upstream of the transcription start site. Recently Hodgetts *et al.* (1986b) reported that one of the differences is a 12-bp duplication of a poly(A–T)-rich region at −280 bp upstream from the transcription start site, and another difference is an A to G transition at −220 in a sequence which bears considerable homology to the canonical enhancer sequence of mammalian viruses. The G substitution in *Ddc*$^{+4}$ makes the sequence more homologous to the viral canonical sequence than the Canton S wild-type sequence and could account for the overproduction of DDC at three of the four stages of development. DDC at the other stage, pupariation, is probably regulated differently by ecdysone (Hodgetts *et al.*, 1986a).

C. Mutant Neurophysiological Phenotypes

Tempel *et al.* (1984) have demonstrated that *Ddc* mutations reduce learning acquisition approximately in proportion to their effect on enzymatic activity. Tempel and co-workers shifted 3-day-old adult *Ddc*^{ts2} homozygotes and *Ddc*^{ts1}/*Df(2L)TW130* hemizygotes, which had been raised at a permissive temperature (20°C), to a restrictive temperature (29°C) for 3 days prior to performing behavioral assays at 25°C. These assays showed that associative learning was reduced significantly, and that experience-dependent male courtship depression was absent. Memory retention, however, was unaltered. Of the mutant population, 5% showed little positive phototaxis although electroretinograms were normal. Negative geotaxis, strength, coordination, walking ability, and olfactory acuity were also normal. However, the threshold for proboscis extension in response to sucrose was raised significantly from 0.004 to 0.025 *M* (Tempel *et al.*, 1984). Thus, reduced DDC activity in mature adults seriously affects limited, but definite, aspects of the behavior of the fly.

In the dissected whole CNS of wild-type third instar larvae, discrete groups of serotonin-like immunoreactive cells are arranged in a bilaterally symmetrical fashion in the brain and ventral nerve cord (White and Valles, 1985). This immunoreactivity is completely abolished in CNSs dissected from *Ddc*^{n27}/*Df(2L)TW130* larvae (Valles and White, 1986). However, in these mutant larvae the same CNS cells, which if wild type would presumably show immunoreactivity, are capable of serotonin uptake from the medium (Valles and White, 1986). These results suggest that serotonin synthesis is not required for the differentiation of these particular specialized nerve cells.

Budnik *et al.* (1986) used glyoxylic-induced histofluorescence to identify catecholamine-containing neurons in dissected central nervous systems from wild-type and DDC-deficient larvae [Ddc^{n27}/*Df(2L)TW130* or Ddc^{n27}/Ddc^{n27}]. The same specific cells displayed catecholamine histofluorescence in both wild-type and mutant CNSs, the latter probably because of the accumulation of DOPA. In the mutant, additional "novel" neuronal cells, not apparent in wild type, became fluorogenic earlier in development than the normal fluorogenic cells. Incubation of dissected wild-type CNSs in DOPA or dopamine showed that normally nonfluorogenic neurons would sequester the catecholamines and that these neurons included the novel fluorogenic subset of neurons observed in the mutant. Presumably the very high DOPA pools in the mutant are being phenocopied in the wild type by incubating the CNSs in exogenous catecholamines.

Gailey *et al.* (1987) have contructed an unstable Ring-X chromosome from a Rod-X chromosome carrying a P-element-mediated Ddc^{+} insert. This facilitates the production of mosaic clones of cells deficient in DDC. The analysis of such clones showed that there is no absolute requirement for Ddc^{+} expression in either the epidermis or the nervous system, but that very large clones do reduce the viability of mosaic individuals (Gailey *et al.*, 1987).

D. Molecular Biology

The *Ddc* gene has been cloned (Hirsh and Davidson, 1981), and the genomic sequence has been completely determined along with the partial sequence of cDNAs (Eveleth *et al.*, 1986; Morgan *et al.*, 1986). Four species of RNA are found which hybridize to *Ddc* DNA (Beall and Hirsh, 1984; Gietz and Hodgetts, 1985; Clark *et al.*, 1986). These are 4.0, 3.0, 2.3, and 2.0 kb in length. The 4.0- and 3.0-kb RNAs are unprocessed and partially processed message precursors, respectively. The 2.3- and 2.0-kb RNAs have both been found on polysomes (Gietz and Hodgetts, 1985) and both are postulated to be mature messages that arise as a result of alternate processing events and are proposed to produce two different isoforms of DDC (Eveleth *et al.*, 1986). During development, the 2.3- and 2.0-kb *Ddc* mRNAs along with the 4.0- and 3.0-kb precursors are found in late embryos, isolated imaginal disks evaginating *in vitro,* and eclosing adults, but only the 2.0-kb *Ddc* mRNA along with the 4.0- and 3.0-kb precursors are found at pupariation (Beall and Hirsh, 1984; Gietz and Hodgetts, 1985; Clark *et al.*, 1986). The interpretation is that at pupariation, in response to a rising ecdysone titer, only the 2.0-kb mRNA is produced, whereas at the other three stages, in response to a falling ecdysone titer, both the 2.0- and 2.3-kb mRNAs are produced. Since *Ddc* responds to ecdysone, Eveleth *et al.* (1986) searched for upstream sequences similar to upstream sequences in other hormonally regulated genes. They found an upstream sequence ATGAAAAATAATGCCTTT in *Ddc* resembling a similarly located sequence, ATGGAAA--TACCTTT, in the ecdysone-regulated *Sgs-4* gene (Muskavitch and Hogness, 1982), and similar to the ATGGAAC sequence in the *74EF* gene (Moritz *et al.*, 1984). This sequence also is homologous to the canonical mammalian virus enhancer sequence (G)TGGAAA(G) (Weiher *et al.*, 1983), and is the one mutated by the Ddc^{+4} overproducer mutation.

Morgan *et al.* (1986) have very recently presented evidence that the unique primary transcript of the *Ddc* gene is differentially spliced in the hypoderm and CNS to mediate the synthesis of two tissue-specific isoforms of DDC. The CNS isoform has an additional 33–35 amino acids on the N-terminus not found in the hypodermal isoform. These additional N-terminal amino acids are encoded in the second exon, which is present in the 2.3-kb mRNA found in the CNS, but is spliced out of the 2.1-kb (2.0-kb) mRNA. The untranslated first, small third, and large fourth exons are present in both the 2.1- (2.0-) and the 2.3-kb mRNAs and both have the same transcription start site. See Morgan *et al.* (1986) for experimental evidence that supports these conclusions and for data on the separation of two DDC isoforms by column chromatography.

E. Germ-Line Transformation and the Analysis of Developmental Regulation

The germ-line transformation of Ddc^{+} DNA has been effected by Scholnick *et al.* (1983) and Marsh *et al.* (1985) using different P-element vector constructs of the same 7.5-kb *Pst*I restriction enzyme fragment. Of the transformed strains, 14 of 16 exhibited approximately normal levels of DDC and normal tissue and temporal expression of the transposed *Ddc* genes. One strain had unexpectedly low levels of DDC in both sexes of newly eclosed adults but had the normal amount of activity at pupariation, and the other strain had increased levels of DDC activity at all stages (Marsh *et al.*, 1985). Of the 16 transformants, 2 were X-linked, with one being dosage compensated (Scholnick *et al.*, 1983) and the other not (Marsh *et al.*, 1985).

The 5′-flanking sequences necessary for the developmentally regulated expression of *Ddc* have begun to be defined by Hirsh and coworkers (Hirsh, 1986; Hirsh *et al.*, 1986; Scholnick *et al.*, 1986) by deleting cloned *Ddc* DNA *in vitro* and reintroducing different deletion constructs back into the genome by P-element-mediated germline transformation. Subsequently the *in vivo* developmental expression of these *Ddc* constructs is monitored. Those *Ddc* genes containing 208 or more basepairs of 5′-flanking DNA appear to have normally regulated hypodermal activity, but those with less than 24 bp do not. However, the latter are expressed using the usual wild-type *Ddc* RNA start site even though the "TATA"-box sequences are deleted. The speculation is that the use of the normal start site is regulated by the adjacent *Adh* gene also present in the P-element construct (Hirsh, 1986).

P-Element-transformed constructs with 5′-flanking sequences deleted from −208 to −106 show nearly normal expression of *Ddc* but are quite variable (Scholnick *et al.*, 1986); deletion of −208 to −83 reduces whole animal expression (primarily hypodermal activity) both at pupariation and at adult eclosion, but the former more than the latter. The −208 to −59 deletion did not reduce activity at pupariation any further but led to a twofold to threefold decrease in expression at adult eclosion. The −208 to −38 deletion depressed expression further and the −208 to −33 deletion extending just into the TATA box essentially abolished expression. All −208 deletions extending as far as −83 had normal CNS *Ddc* expression within a factor of two of wild type. Deleting to −59 (−208 to −59, −208 to −38, and −208 to −33) abolished all CNS activity, both larval and adult. The −208 to −59 deletion does not decrease whole animal expression at pupariation but leads to a twofold to threefold loss in adults. In contrast, it decreases CNS expression 50-fold. Thus Scholnick *et al.* (1986) conclude that this region contains a CNS-specific regulator element. Since a 16-bp sequence, TGAACCGGTCCTGCGG, in this region of *D. melanogaster* shows perfect homology with the similar 16-bp sequence in the 5′-flanking region of the *Drosophila virilis Ddc* gene, Scholnick *et al.* (1986) infer that this 16-bp sequence is the CNS-specific regulator element.

Drosophila virilis Ddc DNA transformed into *D. melanogaster* exhibits normal temporal and tissue-specific expression (Bray and Hirsh, 1986). This indicates that the *Ddc* gene from *D. virilis,* which is separated from *D. melanogaster* by approximately 60 million years (Throckmorten, 1975), is capable of responding to *D. melanogaster* transregulatory factors. Although the two species exhibit qualitatively similar developmental profiles and tissue-specific expression, quantitatively they are distinctly different. The quantitative expression of the *D. virilis* transformed genes is characteristic of *D. virilis*, indicating that the transformed genes are still regulated by their own cis-acting elements. A comparison of the DNA sequences from the 5′-flanking regions of *Ddc* from the two species revealed the presence of a cluster of five small conserved sequences 8–16 bp long within 150 bp upstream of the RNA start point in the region required for normal developmental expression of the *D. melanogaster* gene. Although the arrangement of the clusters relative to each is somewhat different in the two species, they exhibit between 80 and 100% sequence identity (Bray and Hirsh, 1986). It is proposed that these conserved sequences identify regulatory sequences within the regulatory regions established functionally by the deletion constructs in *D. melanogaster*, as discussed above.

VIII. Tyrosine and 5-Hydroxytryptophan Decarboxylase Activities

Livingstone and Tempel (1983) investigated tyrosine decarboxylase (TDC) activity, dopa decarboxylase activity, and 5-hydroxytryptophan decarboxylase (5HTDC) activity in *Drosophila* by incubating dissected adult and larval brains in the tritiated precursors tyrosine and tryptophan, followed by the separation of precursors and products by high-voltage electrophoresis and by assaying the decarboxylase activities directly in homogenates of whole adult heads. Wild-type brains, when incubated in [^{3}H]tyrosine, produced [^{3}H]dopamine, [^{3}H]tyramine, and [^{3}H]octopamine, and when incubated in [^{3}H]tryptophan produced tritiated serotonin. Incubation in [^{3}H]tyrosine of *Ddc* mutant brains [brains dissected from *Ddc*^{ts1}/*Df(2L)TW130* adults after exposure to the restrictive temperature for 3 days] produced [^{3}H]tyramine and [^{3}H]octopamine but no [^{3}H]dopamine, and after incubation in [^{3}H]tryptophan produced no tritiated serotonin. That the *Ddc* mutation affects DDC activity and 5HTDC activity, but not TDC activity, was confirmed by the direct decarboxylase assays of wild-type and *Ddc*^{ts1}/*Df(2L)TW130* whole head homogenates. From these results one can conclude that the *Ddc* locus codes for an enzyme with both dopa decarboxylase and 5-hydroxytryptophan decarboxylase activities but no tyrosine decarboxylase activity. Furthermore, one can infer that another locus codes for an enzyme with tyrosine decarboxylase activity.

This conclusion is supported by the effects of the mutation *per*^{o}

(1; 3B1,2) on these decarboxylase activities (Livingstone and Tempel, 1983). The *per*^{o} allele completely eliminates circadian rhythms in *Drosophila* and the alleles *per*^{long} and *per*^{short} lengthen and shorten circadian rhythms, respectively. When whole brains dissected from *per*^{o} adults and larvae are incubated in [^{3}H]tyrosine the levels of both [^{3}H]tyramine and [^{3}H]octopamine produced are reduced to one-third of wild type, but [^{3}H]dopamine levels are normal. Direct measurements of enzymatic activities in whole head homogenates showed that *per*^{o} reduced TDC activity to 35% of wild type, but had no effect on DDC or 5HTDG activities nor on tyramine β-hydroxylase activity. The *per*^{long} and *per*^{short} alleles also reduced TDC activity to 50 and 60% of wild type, respectively. Whole head homogenates from doubly mutant individuals, *per*^{o}/*per*^{o}; *Ddc*^{ts1}/*Df(2L)TW130*, had 35% TDC activity like *per*^{o}/*per*^{o}, but also had almost no DDC activity. These results with *per*^{o} are consistent with TDC and DDC being two separate activities coded by separate genes. Since deficiencies and duplications of the *per*$^{+}$ region do not produce a dosage effect on tyrosine decarboxylase activity, Livingstone and Tempel (1983) conclude that *per* is not a structural gene for tyrosine decarboxylase. Apparently *per* mutations reduce TDC activity indirectly via some unknown mechanism.

The *per* locus has been the subject of intensive investigation at the molecular level, including cloning, sequencing, and the derivation of an amino sequence from the nucleotide sequence (Bargiello *et al.*, 1984; Reddy *et al.*, 1984). The derived amino acid sequence indicates the *per*$^{+}$ protein is a chondroitin sulfate proteoglycan and not an enzyme subunit (Jackson *et al.*, 1986; Reddy *et al.*, 1986).

IX. The Genetics of α-Methyldopa Sensitivity and Resistance

A. α-Methyldopa and Dopa Decarboxylase *in Vitro* and *in Vivo*

The DOPA analog, α-methyldopa (αMD), is a competitive inhibitor of *Drosophila* DDC at saturating pyridoxal 5′-phosphate concentrations with respect to DOPA (K_I = 1400 *M*), but is a noncompetitive inhibitor when pyridoxal 5′-phosphate is nonsaturating (K_I = 5 *M*; $K_{m, app}$ = 142 *M*) (Black and Smarrelli, 1986). This behavior of *Drosophila* DDC is similar to that of mammalian enzymes (for comparisons see Black and Smarrelli, 1986). It is evident the compound is not a very effective inhibitor of *Drosophila* DDC *in vitro*. Using HPLC with an electrochemical detector, Black and Smarrelli (1986) did not recover any α-methyldopamine when αMD was incubated with DDC, leading them to conclude that *Drosophila* DDC does not decarboxylate αMD. However, using tritiated αMD as a substrate in the microradioassay of McCaman *et al.* (1972), L. Hendrickson and T. R. F. Wright (unpublished) measured significant amounts of radioactivity (30% of that with [^{3}H]DOPA as a substrate) in the organic phase, indicating that αMD is processed to some sort of an amine product. The product was not further characterized, but in this assay the amine products dopamine, tyramine, 5-hydroxytryptamine, and histamine are quantitatively extracted into the organic phase with respect to the amino acid substrates DOPA, tyrosine, 5-hydroxytryptophan, and histidine (McCaman *et al.*, 1972).

The initial rationale for the use of αMD in the genetic analysis of DDC in *Drosophila* was to use the dietary administration of the analog inhibitor as a specific *in vivo* discriminator of abnormally high levels of DDC activity (resistant strains) and abnormally low levels of DDC activity (hypersensitive strains) (Sherald, 1973; Sparrow and Wright, 1974; Sherald and Wright, 1974). When fed to adult *Drosophila*, αMD has no effect except to sterilize females, which lay large numbers of eggs that turn quite black, perhaps from the oxidation of αMD incorporated into the eggs. This αMD-induced female sterility is reversible (C. P. Bishop and T. R. F. Wright, unpublished). Growing larvae fed sufficient concentrations of αMD die at the next molt, and those puparia formed by larvae fed lower concentrations are quite flexible, in contrast to the very rigid structures normally formed. These observations suggest that αMD interferes with normal sclerotization in some way (Sparrow and Wright, 1974). This inference has been substantiated by electron microscopic studies on the effects of dietary αMD on the cuticle of sheep blowfly larvae (*Lucilia cuprina*) (Turnbull *et al.*, 1980). Electron micrographs of newly synthesized cuticle of larvae fed αMD (and other DDC inhibitors) show that the ultrastructure of the lipid-rich epicuticle layer is clearly abnormal. The newly formed cuticle of αMD-fed larvae permits the free movement of water in both directions and, therefore, is defective as a water permeability barrier. *Lucilia* larvae fed toxic levels of αMD (or other DDC inhibitors) can be rescued by simultaneous addition of *N*-acetyldopamine to the food. The outer epicuticle of rescued larvae appears to be morphologically normal (Turnbull and Howells, 1980; Turnbull *et al.*, 1980).

Since the DDC inhibitors used in these experiments may be inhibiting other biogenic amine-metabolizing enzymes *in vivo*, one cannot conclude that the defective cuticle produced is primarily due to the *in vivo* inhibition of DDC. Quite to the contrary, studies of *D. melanogaster* with genetically altered levels of DDC show that strains with three doses of the *Ddc* gene and 150% levels of DDC activity are not more resistant to dietary αMD than strains with two gene doses and 100% levels of DDC activity (Wright *et al.*, 1976b; Marsh and Wright, 1986). In fact, the reverse may be true, with high-level DDC activity strains being more sensitive. Furthermore, decreasing DDC activity to approximately 5% of normal by the use of partially complementary heteroallelic heterozygotes, e.g., Ddc^{n1}/Ddc^{n8}, does not result in greater sensitivity to dietary αMD (Marsh and Wright, 1986). Thus one can conclude that DDC activity does not strongly influence the resistance to dietary αMD, and that dietary αMD can not be used as a discriminator of *in vivo* levels of DDC activity in *D. melanogaster* as originally proposed.

B. α-Methyldopa-Hypersensitive Locus, *l(2)amd*

The *amd* [*l(2)amd; α-methyldopa*, 2–53.9+] locus is 0.002 cM distal to *Ddc* (Wright *et al.*, 1981a). To date 34 *amd* point mutations have been isolated (Wright *et al.*, 1982). Heterozygotes of amorphic mutations (*amd*/+) die on levels of dietary αMD on which wild type live. Thus *amd* mutations are dominantly hypersensitive to αMD and also to another analog inhibitor of DDC, N^1-(DL-seryl)-N^2-(2,3,4-trihydroxybenzl)hydrazine (Hoffman–LaRoche No. 4-4602/1) (Wright *et al.*, 1976a). Resistance to dietary αMD is correlated directly with amd^+ gene dosage; individuals with more doses of amd^+ are more resistant than individuals with fewer doses and vice versa (Wright *et al.*, 1976b; Marsh and Wright, 1986). From this one can infer that *in vivo* the protein product of the amd^+ gene binds this DOPA analog, α-methyldopa, and either eventually effects its inactivation or is itself directly inhibited by αMD.

Homo- and hemizygotes of amorphic *amd* alleles die as normally pigmented larvae prior to eclosion from the egg membranes. These larvae have necrotic, extruded anal organs and burst very easily when manipulated, indicating the incomplete sclerotization of the colorless body wall cuticle (Wright, 1977). Electron micrographs indicate that the anal organ defect is a result of the incomplete formation or sclerotization of the cuticular suture between the anal organ cells and the normal cells of the epidermis (J. C. Sparrow, personal communication). Some hemizygotes of *amd* hypomorphic alleles hatch and complete larval development, forming normally pigmented pseudopupae that are abnormally flexible (T. R. F. Wright, unpublished). The *amd* heterozygotes (*amd*/+) and *amd* intragenic complementing heteroallelic heterozygotes (amd^{H1}/amd^{H89}) do not alter DDC, diphenol oxidase, or dopamine acetyltransferase activities in any way whatsoever, neither as adults nor as white prepupae and neither in the epidermis nor in the central nervous system of white prepupae (Wright *et al.*, 1976a, 1982; Wright, 1977; Huntley, 1978; E. Y. Wright and T. R. F. Wright, unpublished).

The inference made from the *amd* mutant phenotype described above, that amd^{+} gene activity is necessary for colorless sclerotization, is strongly supported by the effect *amd* mutations have on catecholamine metabolism. Black *et al.* (1987b) have found that in catecholamine pools of intragenic complementary heteroallelic heterozygous adults (amd^{H1}/amd^{H89}) a prominent electroactive compound is missing. This compound has been identified as a catecholamine by low-resolution mass spectroscopy and by other criteria (see Section V) and has been designated catecholamine X (Black *et al.*, 1987b). Its complete structure has yet to be solved. Catecholamine X pools occur in embryos at the time of initial colorless sclerotization which begins prior to the appearance of DDC activity and dopamine (Fig. 7). Ddc^{ts2} homozygous adults at 22°C have markedly reduced levels of dopamine, *N*-acetyldopamine, and *N*-β-alanyldopamine, but levels of catecholamine X are slightly elevated. These facts indicate that the *amd* gene codes for an enzyme which converts DOPA to catecholamine X in a separate branch of the pathway (see Fig. 9). This *amd* enzyme would bind αMD more efficiently than DDC.

One of the 17 known transcription units in the *Ddc* cluster (Section X) located between DNA coordinates −5.7 and −3.75 less than 2.5 kb distal to *Ddc* has been identified as *amd* by the determination that two alleles, amd^{37} and amd^{40}, are small aberrations in the DNA located between −4.5 and −3.5 within this transcriptional region (Black *et al.*, 1987b). This transcription unit has been characterized and sequenced by Marsh *et al.* (1986). A 2-kb transcript is most abundant at about 12 hours of embryogenesis and lower levels are detected throughout most of embryogenesis and in adult females but not in males. Unique stage-specific transcripts of 1 and 0.6 kb are produced in midthird instar and late third instar larvae, respectively (Marsh *et al.*, 1986).

The structures and sequences of *Ddc* and *amd* have been reported by Eveleth *et al.* (1986) and Marsh *et al.* (1986), respectively. The two sequences have been examined by Eveleth and Marsh (1986a) for homologies using computerized dot matrix analysis. Although *amd* has only one intron whereas *Ddc* has two, striking homology was

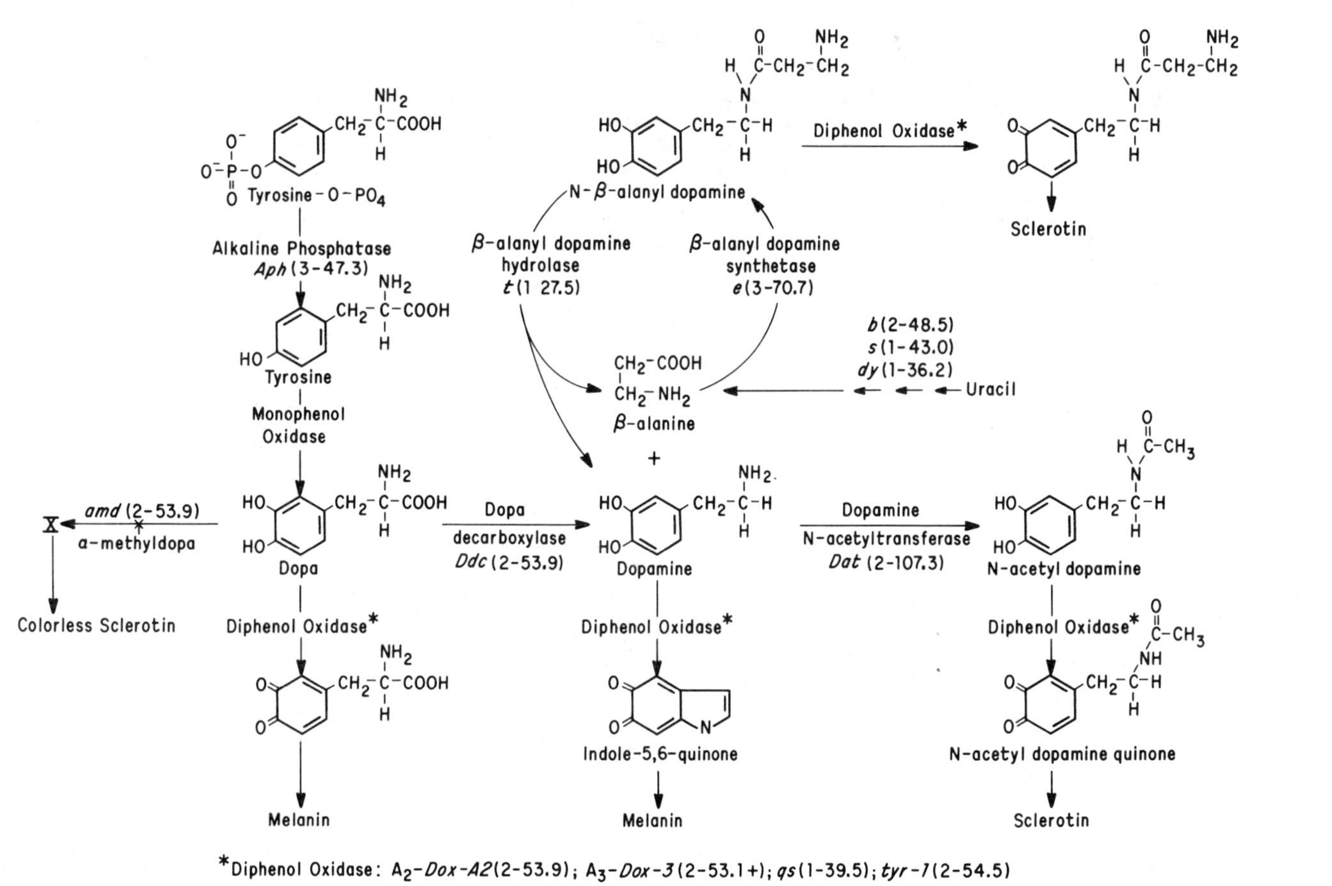

FIG. 9. Proposed pathway for catecholamine metabolism in *D. melanogaster* indicating the reactions affected by specific mutations. The numbers in parentheses after the gene symbols specify chromosome and locus. Mutations affecting diphenol oxidase are listed at the bottom. See Table 1 for more information on the mutations (after Wright, 1987).

found between exon II and exon III of *Ddc* and the two exons of *amd*. "Overall approximately 55% of the bases match between the two sequences (excluding intron regions), however two areas of more extensive homology are apparent. One area beginning near the second exon of *amd* is over 80% homologous over 100 bp and a second run of 124 bp (700 bp from the 3′ end) is approximately 90% homologous" (Eveleth and Marsh, 1986a). Subsequently, D. D. Eveleth and J. L. Marsh (personal communication) have found a third region of very high homology which, in turn, is homologous to the pyridoxal-binding peptide of porcine DDC (Bossa *et al.*, 1977). They suggest the other two regions of very high homology may be the substrate (DOPA)-binding site and the subunit interaction site. Comparison of the deduced amino acid sequences from *Ddc* and *amd* (Eveleth and Marsh, 1986a) shows two regions of considerable amino acid sequence conservation, and analysis of deduced secondary structure indicates the conservation of considerable structural similarity between the two proteins. Eveleth and Marsh (1986a) conclude that *Ddc* and *amd* are products of a gene duplication event with subsequent divergence to related subspecialties in catecholamine metabolism. Since the transcripts of the two genes are not found at the same time in development, Eveleth and Marsh (1986a) concluded that they are not coordinately regulated.

C. α-Methyldopa-Resistant Strains

1. *Dominant Second Chromosome-Resistant Strains*

In screening for dominant resistance to dietary αMD, Sherald and Wright (1974) isolated three strains which are now designated as R_2, Ddc^{RE} (R^E, RM1), and Ddc^{RS} (R^S, S) [in parentheses are the designations used previously, the first designation being from Marsh and Wright (1979, 1986), followed by the designation used in Sherald and Wright (1974)]. The ethyl methanesulfonate-induced mutation in R_2, a very resistant strain, was mapped to the vicinity of both *l(2)amd* and *Ddc* and has been shown not to affect DDC activity at all (Sherald and Wright, 1974). Except that B. C. Black (unpublished) found that R_2 did not alter catecholamine pools of newly eclosed adults, no further work has been done on this mutation. When the *amd* protein is eventually identified, it will be interesting to determine if this mutation alters the K_I of the *amd* protein to αMD.

Both Ddc^{RS} and Ddc^{RE} strains exhibit the dual phenotype of increased resistance to dietary αMD and elevated DDC activities relative to Oregon-R controls (Sherald and Wright, 1974; Marsh and Wright, 1986). Both phenotypes are genetically located very close

(<0.05 cM) to *Ddc* and *l(2)amd*. Relative to the control, *Ddc*^{RS} increases DDC activity 41% and DDC cross-reacting material 37%, and similarly *Ddc*^{RE} increases DDC activity 58% and DDC cross-reacting material 56% (Marsh and Wright, 1986). Since increased DDC activity does not lead to an increase in dietary αMD resistance (Section IX,A) and since additional doses of the *amd*$^+$ gene do lead to an increase in resistance (Section IX,A), it is unlikely that in the *Ddc*^{RE} and *Ddc*^{RS} strains the increased resistance is a result of the increased DDC activity but rather is due to increased *amd*$^+$ activity. Both cytogenetic and molecular analyses indicated that these overproduction variants are not the result of small duplications of the *Ddc* and *amd* genes and are not associated with small (≤100 bp) insertions or deletions. Since both strains probably acquired these traits spontaneously, Marsh and Wright (1986) conclude that accumulated changes in a genetic element (or elements) in close proximity to the *Ddc*$^+$ and *amd*$^+$ genes lead to coordinated changes in the expression of the *Ddc* and *amd* genes in these strains.

2. *Third Chromosome-Resistant Mutations*

In *D. melanogaster* two alleles at the *Third chromosome resistance* locus (*Tcr*, 3–39.6) were isolated in a screen of ethyl methanesulfonate-mutagenized third chromosomes for dominant resistance to dietary αMD (C. P. Bishop, A. F. Sherald, and T. R. F. Wright, unpublished). Both *Tcr* alleles are recessive lethals, and they partially complement each other. Almost half (48.3%) of the *Tcr*^{40}/*Tcr*^{45} heterozygotes die as embryos, but some complete development to eclose as adults. The *Tcr*^{40}/*Tcr*^{45} heterozygotes that fail to hatch from the egg membranes appear to be mature larvae except that their mouthparts are usually underpigmented (incompletely sclerotized?) like lethal *Ddc* homozygous embryos, and after they die, the unhatched larvae darken very slowly compared to other embryonic lethals. Approximately 7.5% of expected *Tcr*^{40}/*Tcr*^{45} heterozygotes eclose as adults and are usually found dead or dying in the food. Adults which were not in the food usually were unable to walk on their tarsi but walked instead on their metatarsi or tibiae with uncoordinated leg movements. Both pharate and eclosed adults had a variable pleiotropic, phenotype, including dark, rough, often reduced eyes with black necrotic spots in them; missing bristles, particularly on the head; one or all the ocelli missing; notched wings; reduced scutellum; *Minute*-like bristles mixed with wild-type bristles; unexpanded ptilinum; reduced sex combs; and a grayish abdomen. *Tcr*^{40}/*Tcr*^{45} puparia were normally colored and sclerotized. Thus, reduced *Tcr* function adversely affects the development of many different adult cuticular structures.

Of particular interest is the fact that the presence of either of the *Tcr* alleles in the genome suppresses the lethality of partially complementing heteroallelic *amd* heterozygotes. For example, 1.2% of expected $l(2)amd^{H82}/l(2)amd^{H1}$ heterozygotes ecloses as adults whereas 111% of expected $l(2)amd^{H82}/l(2)amd^{H1};Tcr^{40}/+$ adults survives. This is true for most, but not all, of the different heteroallelic heterozygotes tested, and Tcr^{40} did not rescue $l(2)amd^{H1}/Df(2L)TW130$ heterozygotes, which have no amd^{+} function at all. Therefore, it appears that at least a partially functional amd protein is required before a *Tcr* allele can suppress lethality caused by *l(2)amd.*

No suppression was found for the converse genotype, i.e., *amd* lethal alleles as heterozygotes over wild type did not suppress Tcr^{40}/Tcr^{45} lethality. In fact, *l(2)amd* alleles have an adverse maternal effect. For example, when present in the maternal genome, $l(2)amd^{H26}/+$ increases the embryonic mortality of Tcr^{40}/Tcr^{45} heterozygotes from 19 to 64%.

Three attributes of the *Tcr* mutation, resistance to dietary αMD, mutant cuticular phenotypes, and suppression of *amd* lethality, indicate the wild-type *Tcr* product is necessary for normal cuticle formation and sclerotization and is in some way involved in catecholamine metabolism. Not unexpectedly a preliminary assay of catecholamine pools in adults showed that Tcr^{40}, when heterozygous over the wild-type allele, does not perturb the pools noticeably. Tcr^{40}/Tcr^{45} adults have yet to be assayed (C. P. Bishop, A. F. Sherald, and T. R. F. Wright, unpublished).

It appears that the presence of one *Tcr* dominant mutant allele in the genome somehow increases the production of catecholamine X (Fig. 9) in partially complementing *amd* heteroallelic heterozygotes. That this could be accomplished by a variety of mechanisms is particularly evident when rescue of these partially complementing *amd* heterozygotes is reported to be effected by feeding DOPA, tyramine, octopamine, and pyridoxal 5′-phosphate (P. D. L. Gibbs and J. L. Marsh, unpublished results, cited in Eveleth and Marsh, 1986a).

X. The *Ddc* Cluster of Functionally Related Genes

In the immediate vicinity of the *Ddc* gene a dense cluster of functionally related genes has been identified which are involved in catecholamine metabolism, in the formation, sclerotization, and pigmentation of cuticle, and in female fertility. Recent evidence on sequence homologies indicates that the genes are evolutionarily related as are the genes in most of the other dense clusters in *Drosophila* (Wright,

1987). Unlike most of the other clustered genes in *Drosophila,* the genes in the *Ddc* cluster have evolved far enough from each other to have acquired separate, vital functions, some of which involve the specification of related enzymes in the same pathway (Wright, 1987).

A. Genetic, Cytological, and Molecular Organization of the *Ddc* Cluster

The *Ddc* cluster has been arbitrarily delimited by the small 8- to 12-band deficiency *Df(2L)TW130* (37B9–C1; 37D1–2) on the left arm of the second chromosome. Approximately 315 point mutations have been isolated in this small region, most, but not all, by Wright's lab (Wright *et al.,* 1981a, 1982; Wright, 1987). Almost all of these mutations have been assigned to 18 genes, 16 of which are vital genes (lethal genes), plus 1 nonvital gene with a mutant adult phenotype, *hook* (*hk,* 2–53.9), and 1 female sterile gene, *fs(2)TW1.* These 18 genes are nonrandomly distributed relative to the salivary gland chromosome bands in the region, with, for example, as many as 7 genes in a 1- to 3-band region, seriously violating the one gene : one band hypothesis (see Wright, 1987, for data and discussion). The use of 21 deficiencies (deletions) with breakpoints within the *Df(2L)TW130* region has permitted the assignment of the 18 *Ddc* region genes to 10 genetic regions.

Approximately 160 kb of contiguous DNA has been cloned from the *Ddc* region (Hirsh and Davidson, 1981; Gilbert and Hirsh, 1981; Gilbert *et al.,* 1984; Steward *et al.,* 1984; E. S. Pentz, G. R. Hankins, and T. R. F. Wright, unpublished), and 17 of the above 21 deficiency breakpoints have been physically located in the DNA (11 depicted in Fig. 10), permitting the assignment of all 18 *Ddc* cluster genes to segments of the cloned DNA (Gilbert *et al.,* 1984; Pentz and Wright, 1986; E. S. Pentz, G. R. Hankins, and T. R. F. Wright, unpublished). Furthermore, of the 315 mutations in the region, 149 have been screened for physical differences in the DNA by Southern blotting restriction digests of genomic DNA. Most of the 12 mutations which have been located in the cloned DNA in this way are small deletions (Gilbert *et al.,* 1984; Pentz and Wright, 1986; Black *et al.,* 1987b; G. R. Hankins, J. Kullman, E. S. Pentz, and T. R. F. Wright, unpublished). The sites of six of these small aberrations are designated in Fig. 10 by triangles immediately above the cloned DNA line.

It is clearly evident from an inspection of Fig. 10 that the 18 genes are also nonrandomly distributed at the DNA level, with 13 of the 18 genes located in two distinct subclusters. Included in the proximal

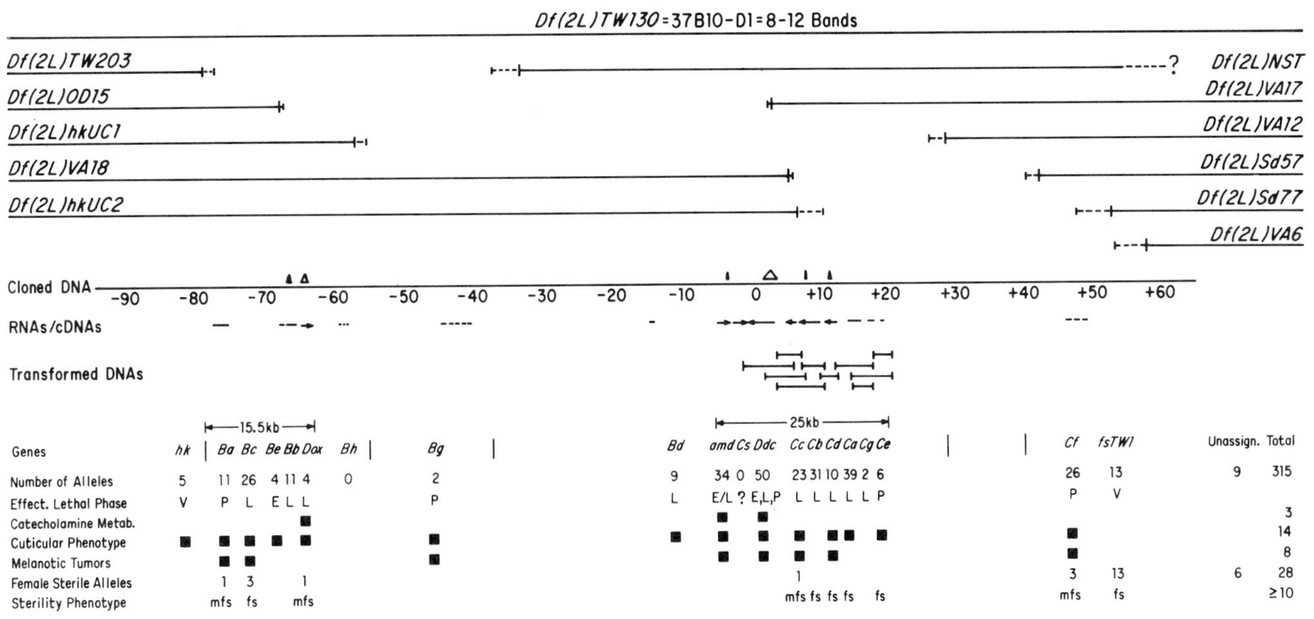

Df(2L)TW130=37B10-D1=8-12 Bands

Df(2L)TW203
Df(2L)OD15
Df(2L)hkUC1
Df(2L)VA18
Df(2L)hkUC2
Df(2L)NST
Df(2L)VA17
Df(2L)VA12
Df(2L)Sd57
Df(2L)Sd77
Df(2L)VA6

Cloned DNA -90 -80 -70 -60 -50 -40 -30 -20 -10 0 +10 +20 +30 +40 +50 +60

RNAs/cDNAs

Transformed DNAs

15.5kb
25kb

Genes hk Ba Bc Be Bb Dox Bh Bg Bd amd Cs Ddc Cc Cb Cd Ca Cg Ce Cf fsTW1 Unassign. Total

Number of Alleles 5 11 26 4 11 4 0 2 9 34 0 50 23 31 10 39 2 6 26 13 9 315

Effect. Lethal Phase V P L E L L P L E/L ? E,L,P L L L L L P P V

Catecholamine Metab. 3

Cuticular Phenotype 14

Melanotic Tumors 8

Female Sterile Alleles 1 3 1 1 3 13 6 28

Sterility Phenotype mfs fs mfs mfs fs fs fs fs mfs fs ≥10

FIG. 10. The genetic and molecular organization of the *Ddc* region. Deficiencies: solid lines represent deleted DNA, with dashed lines indicating uncertainty of the position of the breakpoint. Cloned DNA coordinates given in kilobases, from Gilbert *et al.* (1984). Small triangles above the cloned DNA line physically locate small deletion mutations and short lines underneath designate regions which hybridize to mRNAs or cDNAs, with arrowheads representing direction of transcription. Transformed DNA lines indicate the segments of DNA that have been transformed by P-elements. All the gene symbols except *hk*, *Dox* (*Dox-A2*), *Bh*, *amd* [*l(2)amd*], *Cs*, *Ddc*, and *fsTW1* [*fs(2)TW1*] should be preceded by *l(2)37*, e.g., *l(2)37Ba*. Effective lethal phase designations: E, embryonic; L, larval; P, pupal; and V, viable. Solid squares underneath a gene symbol indicate that mutant alleles of that gene alter catecholamine metabolism, express a mutant cuticular phenotype, or produce melanotic tumors. Sterility phenotype: individuals hemizygous for female-sterile, *ts*, or hypomorphic alleles or heterozygous for complementing heteroalleles are female sterile (fs) or both male and female sterile (mfs). See text for the sources of the information included in this figure (after Wright, 1987).

subcluster are the 8 mutually exclusive lethal genes, *l(2)amd, Ddc, l(2)37Cc, l(2)37Cb, l(2)37Cd, l(2)37Ca, l(2)37Cg,* and *l(2)37Ce,* along with one transcription unit, *Cs,* in which no mutations have been isolated. These 9 genes are located in a maximum of 25 kb of DNA of which at least 70% is transcribed. The distal subcluster includes 5 mutually exclusive lethal genes, *l(2)37Ba, l(2)37Bc, l(2)37Be, l(2)37Bb,* and the diphenol oxidase gene, *Dox-A2,* in a maximum of 15.5 kb of DNA. The location of a small *hook* insertion mutation in a 7-kb fragment of DNA immediately distal to the distal subcluster suggests that this nonvital gene may also belong in that subcluster. However, *hook* and *l(2)37Bg, l(2)37Bd, l(2)37Cf,* and *fs(2)TW1* have been designated as "scattered" genes within the *Ddc* cluster. In addition to *hook, l(2)37Cf* and *fs(2)TW1* have been physically localized fairly accurately, but *l(2)37Bg* and *l(2)37Bd* have not (Fig. 10).

In the *Ddc* region, 17 segments of DNA have been shown to hybridize with RNAs or cDNAs (Hirsh and Davidson, 1981; Gilbert and Hirsh, 1981; Gilbert, 1984; Pentz and Wright, 1986; Spencer *et al.*, 1986a; Marsh *et al.*, 1986; M. E. Freeman, G. R. Hankins, J. Kullman, E. S. Pentz, and T. R. F. Wright, unpublished) and of these, 11 have been assigned to mutant complementation groups either by mutant alleles that physically alter the DNA or by the rescue of mutant alleles by P-element-mediated transformed segments of DNA, or both. The former method identified the transcriptional units of *amd, Ddc, l(2)37Cc, l(2)37Cb,* and *l(2)37Cd* in the proximal subcluster, *l(2)37Bb* and *Dox-A2* in the distal subcluster, and the scattered gene *l(2)37Cf* (Pentz and Wright, 1986; Black *et al.*, 1987b; G. R. Hankins, J. Kullman, E. S. Pentz, and T. R. F. Wright, unpublished). Rescue of mutants by transformed segments of DNA identified transcription units for *l(2)37Ca, l(2)37Cg,* and *l(2)37Ce* and in addition confirmed the identity of the four transcription units, *Ddc, l(2)37Cc, l(2)37Cb,* and *l(2)37Cd* (Scholnick *et al.*, 1983; J. Kullman and T. R. F. Wright, unpublished).

Two of the six unassigned transcription units will probably be identified with *l(2)37Ba, l(2)37Bc,* or *l(2)37Be* in the distal subcluster. The two unassigned units located between DNA coordinates −45 to −41 and −15.4 to −14.6 may eventually be assigned to *l(2)37Bg* and *l(2)37Bd,* respectively, or, like *Bh* and *Cs,* they may be genes in which no mutations have been recovered. The *Bh* transcription unit is defined by two cDNA clones which hybridize to DNA between coordinates −59.4 and −58.6 just proximal to *Dox-A2* (E. S. Pentz and T. R. F. Wright, unpublished), and the *Cs* transcription unit at −3.7 to −1.5 was characterized by Spencer *et al.* (1986a) as being transcribed off the strand opposite from *Ddc* and actually overlapping *Ddc* by 88 bp at the 3′ terminus (Spencer *et al.*, 1986b). Although the possibility has not been completely eliminated, it is highly unlikely that *Cs*, which is barely expressed in the third larval instar and is maximally expressed in male testes, is the *l(2)37Bd* transcript, since *l(2)37Bd* is a larval lethal. When complete tissue- and time-specific transcription mapping is eventually completed for the *Ddc* region, additional transcription units for genes with no mutants may be found in this very densely populated region of the Drosophila genome. Of the 18 genes identified by mutations, 5 remain unordered relative to each other. They are *l(2)37Ba*, *l(2)37Bc*, and *l(2)37Be* in the distal subcluster and *l(2)37Ca* and *l(2)37Cg* in the proximal subcluster.

B. The Functional Relatedness of the Genes in the *Ddc* Cluster

Although the function of several of the genes in the *Ddc* region, *Ddc*, *amd*, and *Dox-A2*, has been investigated extensively, evidence on the function of most of the other 15 genes is far from complete. It is, however, known that the 3 well-studied genes function in catecholamine metabolism, and on the basis of mutant phenotypes, it is inferred that the activity of at least 14 of the genes in the *Ddc* region, including *Ddc*, *amd*, and *Dox-A2*, is required for the normal formation, sclerotization, and/or pigmentation of cuticle. Mutations in many of these genes effect suprisingly similar mutant phenotypes (see below), suggesting that they are involved in the same physiological process, but since most of the mutations are lethal, the function of each is indispensable. Furthermore, melanotic tumors are present in mutant larvae, pupae, or adults of 8 of the 18 genes including *Ddc* and *amd* and may be symptomatic of altered catecholamine pools. Finally, it has been determined that the activity of at least 10 of the genes in the region are required for normal female fertility, a high concentration of female fertility genes in the *Drosophila* genome. Whether the sterility of mutant females arises directly or indirectly from defects in catecholamine metabolism or not has not been determined. Further information on the function of the genes in the *Ddc* region is presented below.

1. *Catecholamine Metabolism*

In addition to the *dopa decarboxylase* gene, *Ddc*, two other genes in the *Ddc* region have been demonstrated to be involved in catecholamine metabolism. All three of these genes are within 65 kb of each other. One is *Dox-A2,* which specifies the A2 component of the diphenol oxidase enzyme and is located approximately 63 kb distal to *Ddc* in the distal subcluster (Pentz *et al.,* 1986; Pentz and Wright, 1986). The other gene is the α-methyldopa-hypersensitive gene, *l(2)amd,* located approximately 2.5 kb distal to *Ddc,* the activity of which is required for the accumulation by catecholamine X, an unidentified catecholamine in *Drosophila* (Black *et al.,* 1987b). Detailed discussions of the genetics of these three genes are found elsewhere in this review: *Ddc* in Section VII; *l(2)amd* in Section IX,B; and *Dox-A2* in Section XI,C,1.

2. *Formation, Sclerotization, and Pigmentation of Cuticle*

Mutant phenotypes observed at various stages of development but particularly at the pharate adult stage have led to the conclusion that the function of at least 14 of the 18 genes in the *Ddc* cluster is required for the normal formation, sclerotization, and/or pigmentation of the cuticle (see Fig. 10) (Wright, 1987; T. R. F. Wright, unpublished). A variety of different genotypes including hypomorphic and *ts* alleles and partially complementing heteroallelic heterozygotes were used in order to investigate the effects of inadequate gene activity on the cuticle of pharate adults for the genes with effective lethal phases prior to this very late stage in development.

Amorphic alleles of *l(2)37Ba* and *l(2)37Ce* and hypomorphic alleles of *l(2)37Be, l(2)37Bc, l(2)37Bg,* and *l(2)37Bd* effect very similar mutant phenotypes in pharate adults incapable of eclosion, with *l(2)37Bg* and *l(2)37Bc* exhibiting the most extreme manifestation of the mutant phenotype. In individuals which are less abnormal, the cuticle is incompletely formed over most of the abdomen, and in those areas where complete cuticle is absent no bristles are formed and there is no evidence of the normal segmentation or pattern of melanization. In these regions the epidermis is complete and may have laid down one or more of the layers of the cuticle, e.g., a thin procuticle, but since no sectioned material has been examined these inferences may not be valid. The expression of this phenotype can be quite variable even for the amorphic pupal lethal alleles of *l(2)37Ba* and *l(2)37Ce.* For instance, small or large areas of the abdomen may be invested with normal cuticle including normal bristles and indications of segmentation and patterns of melanization. On the other hand, cuticle formation may be incomplete not only over the abdomen but also in the thorax and even parts of the head and eyes [e.g., *l(2)37Bd* alleles]. Puparia of *l(2)37Bg*^{1} hemizygotes are larger and darker than normal and when dissected yield pupae with incomplete cuticle formation over the entire head, thorax, and abdomen, except for the genital disk area where the cuticle is complete and normally tanned. Head eversion is complete in these *l(2)37Bg* pupae, but there is no evidence of legs, wings, or ommatidia. Dark puparia are also produced by *l(2)37Bc* hemizygotes and evidence of regions of internal melanization are usually apparent. These arise from the rupture of pharate adult epidermis. Occasionally *l(2)37Bc* individuals can be dissected from puparia which have only minor rupturing but exhibit the incomplete cuticle formation phenotype described above. Except for *l(2)37Bg*, for all of the above mutations the thoracic, leg, and head cuticle usually appear to be formed, but the thoracic cuticle in particular appears to be incompletely sclerotized (T. R. F. Wright, unpublished).

Unlike the above mutations, eclosed adults homozygous for *ts* alleles of *l(2)37Ca* and *l(2)37Cc* at the permissive temperature have more or less normal abdominal cuticle but have abnormal thoracic cuticle. The latter appears to be normally formed but is incompletely sclerotized. Deformations in the thorax are apparent, including very prominent thoracic sutures and indentations probably caused by contractions of the indirect flight muscles. Perhaps due to partially collapsed thoraces, wing expansion never occurs in $l(2)37Cc^{ts}$ individuals and only occasionally occurs normally in $l(2)37Ca^{ts}$ adults. The *l(2)37Cd* hemizygotes that pupate and the *l(2)37Cf* pharate adults show a similar but less extreme phenotype than *l(2)37Cc* and *l(2)37Ca*. Various genotypes such as $l(2)37Ca^{ts}$ and $l(2)37Cc^{ts}$ hemizygotes and partially complementing intragenic heteroallelic heterozygotes of *l(2)37Ca*, *l(2)37Cc*, and *l(2)37Cf* eclose after an extended development time, often are small, often are a darker tan color than wild type, and have moderate *Minute*-sized bristles. Somatic mosaic patches homozygous for an *l(2)37Ca* amorphic allele exhibit abnormally tiny bristles (C. P. Bishop, unpublished), suggesting the production of limited amounts of bristle proteins (T. R. F. Wright, unpublished). It is interesting that the one nonvital, "visible" mutation *hook* produces flies with morphologically abnormal bristles which are hooked at the tip or are blunted. Also the eyes may be slightly roughened and wings somewhat divergent (Lindsley and Grell, 1968; Mitchell and Lipps, 1978).

The obvious mutant cuticular phenotypes of *Ddc*, *l(2)amd*, and *Dox-A2* are described in other sections: *Ddc* in Section VII; *l(2)amd* in Section IX,B; and *Dox-A2* in Section XI,C,1. Except for *Ddc*, *l(2)amd*, and some mutations with melanotic pseudotumors, amorphic mutations with effective lethal phases prior to pupation exhibit no obvious morphological differences in whole or dissected homo- or hemizygous individuals. Many take significantly longer in terms of days to complete development and a few are small or have undersized imaginal disks or other organs (T. R. F. Wright, unpublished).

3. *Melanotic Pseudotumors*

In *D. melanogaster* melanotic tumors are formed by aggregations of hemolymph cells around foreign substances such as bacteria and parasitic wasp larvae (Nappi and Streams, 1969), and in numerous mutant strains, around various different endogenous tissues (Rizki and Rizki, 1984). These aggregations then melanize. With a couple of exceptions (Gateff, 1978; Hanratty and Ryerse, 1981), the melanotic masses are not true tumors (for discussion see Gateff, 1978; Rizki and Rizki, 1984; for a review of their genetics, see Sparrow, 1978).

Of the eight genes in the *Ddc* cluster which produce melanotic tumors, at least one allele, usually many more, produces melanotic tumors in hemizygous or homozygous larvae, pupae, and/or adults (see Fig. 10). Although many major melanotic tumor genes have only been localized to chromosomes and not to specific loci (Sparrow, 1978), the genes in *Ddc* cluster producing melanotic tumors are the highest concentration of such genes known in the *Drosophila* genome. Very little work has been done on the melanotic tumor phenotype of these genes, and therefore it is not known why these mutations effect the production of melanotic tumors. The possibility exists they may result from unusual catecholamine pools resulting from blocks in catecholamine metabolism. For example, *Ddc* mutants may be prone to produce melanotic tumors because of very high pools of DOPA, which could be metabolized to melanin very easily. Perhaps this is an indication that the genes that produce melanotic tumors, *l(2)37Ba*, *l(2)37Bc*, *l(2)37Bg*, *l(2)37Cc*, *l(2)37Cd*, and *l(2)37Cf*, also function in catecholamine metabolism in some way.

4. *Female Sterility*

The normal activity of at least 10 of the 18 genes in the *Ddc* cluster has been shown to be necessary for normal female fertility and the activity of at least 4 of these is also required for male fertility (see Fig. 10) (P. O. Cecil and T. R. F. Wright, unpublished). Since most of the genes in this region are vital genes, hypomorphic and *ts* alleles, intragenic complementing heteroallelic heterozygotes, and female- and male-sterile alleles were used to produce adults that could be tested for sterility. Since this is a high concentration of female fertility genes, it is possible that their coordinate activity in the ovary might provide a reason for why these genes remain clustered. Three of the remaining genes, *hk, l(2)amd,* and *Ddc,* have been tested and have been shown not to be required for either male or female fertility (P. O. Cecil and T. R. F. Wright, unpublished). This in spite of the fact that DDC activity (Wright *et al.*, 1981b) and *l(2)amd* transcripts (K. Konrad and J. L. Marsh, personal communication) have been found in ovaries. The other five genes in the *Ddc* cluster, *l(2)37Be, l(2)37Bb, l(2)37Bg, l(2)37Bd,* and *l(2)37Cg,* have not been adequately tested for female or male fertility effects (T. R. F. Wright, unpublished).

A total of 28 female-sterile (fs) and male- and female-sterile (mfs) mutations have been isolated in the *Ddc* region (T. R. F. Wright, unpublished; D. L. Lindsley, personal communication; T. Schupbach, personal communication). Of these, 9 are *fs* or *mfs* alleles of five lethal loci, *l(2)37Ba, l(2)37Bc, Dox-A2, l(2)37Cc,* and *l(2)37Cf* (Fig. 10), and 13 are alleles of the only nonvital, female-sterile locus, *fs(2)TW1*, established in the *Ddc* region. The initial allele of *fs(2)TW1* was established when the female-sterile lesion in the Ddc^{ts1} chromosome (Wright *et al.*, 1981b) was subsequently separated from Ddc^{ts1} by recombination. The ovary autonomous phenotype of Ddc^{ts1} chromosome established by ovary transplants (Wright *et al.*, 1981b) was due to the *fs(2)TW1* lesion in the chromosome and not to the Ddc^{ts1} lesion as reported in that paper. The *fs(2)TW1* females lay many eggs which do not develop at all and therefore remain white. The *fs(2)TW1* gene is the most proximal gene in the *Ddc* cluster, being located in the vicinity of cloned DNA coordinates +49 and +58. No transcript for this gene has been identified yet (G. R. Hankins and T. R. F. Wright, unpublished).

The *fs* alleles of *l(2)37Cf* and one *fs* alleles of *l(2)37Cc* show reduced viability particularly at higher temperatures, and, therefore, probably are not ovary-specific lesions. The one *fs* allele of *l(2)37Ba*, the three *l(2)37Bc fs* alleles, and the one *Dox-A2 fs* allele do not reduce viability significantly (T. R. F. Wright, unpublished), and, therefore, may be good candidates for being lesions in ovary-specific control regions.

Six of the twenty-eight fs mutations in the *Ddc* region have not been assigned to known loci primarily because they produce high levels of female sterility as heterozygotes over nonoverlapping deficiencies and lethal alleles of noncontiguous vital genes (T. R. F. Wright, unpublished). It would be most interesting if this unconventional behavior is indicative of lesions in regional ovary-specific enhancers.

5. *Mutations of l(2)37Cf Produce Malignant Brain Tumors*

Homozygous third instar larvae for 13 of the 14 alleles of *l(2)37Cf* examined have enlarged brains (G. R. Hankins and T. R. F. Wright, unpublished). Some mutant alleles produce brains which are more than eight times the normal volume. Although the eye-antennal imaginal disks may be enlarged in some cases, the ventral nerve cord remains uniformily unaffected. Following transplantation of pieces of mutant [$l(2)37Cf^{14}$] third instar brains into abdomens of wild-type adult hosts, the transplanted tissue proliferates tremendously, filling and bloating the entire abdominal and thoracic cavities with neurons (E. Gateff, personal communication). These are malignant cells with, for example, metastases in the indirect flight muscle of the thorax. It is evident that the loss of $l(2)37Cf^{+}$ function results in excessive (precocious?) growth of the brain hemispheres *in vivo* and in the malignant behavior of this tissue when transplanted in the adult hemocoel.

Mutations in the *l(2)37Cf* locus produce primarily pupal lethality and the numerous hypomorphic alleles permit many *l(2)37Cf* homo- and hemizygotes to eclose as adults. Thus most of the larvae with these significantly enlarged brains are capable of pupating and developing the pharate adult stage and even eclosing as adults. These adults have greatly reduced female and male fertility (G. R. Hankins, unpublished). The latter results primarily from abnormal behavior since the males are so lethargic they will only mate successfully when wild-type females bump into them (A. C. Kenyon and G. R. Hankins, unpublished). Whether female sterility results from a behavioral abnormality or not has yet to be determined. Catecholamine pools have yet to be assayed for *l(2)37Cf* mutations, but the cuticle of pharate adults is not completely normal, appearing to be somewhat incompletely sclerotized. It would be most interesting if the abnormal development of the brain results from a defect in catecholamine metabolism. This enlarged brain phenotype has not been observed for any of the other mutations in the *Ddc* cluster (T. R. F. Wright, unpublished).

C. SEQUENCE HOMOLOGIES WITHIN THE *Ddc* CLUSTER

The striking sequence homologies between *Ddc* and *l(2)amd* suggest that the two genes are products of a duplication event (Eveleth and Marsh, 1986a), as discussed in Section X,B. In addition to the *Ddc–l(2)amd* homologies, recent preliminary results using low-stringency hybridization conditions demonstrate sequence homology in the proximal subcluster between *l(2)37Cb* and *l(2)amd* but not *Ddc*

(J. Kullman and T. R. F. Wright, unpublished). Furthermore, low-stringency homologies have been found between sequences from the proximal subcluster and sequences from the distal subcluster and from scattered gene regions, e.g., the transcribed region of *Cs* cross-hybridizes to genomic DNA from the putative location of *hk* and to genomic DNA from the distal subcluster (J. Kullman and T. R. F. Wright, unpublished). Also under high-stringency conditions (>80%), the 0.84-kb *Bh* cDNA from the −60 to −58.5 region immediately proximal to the distal subcluster hybridizes to cloned genomic DNA from the +40 to +46.1 region containing the coding region for *l(2)37Cf* (E. S. Pentz, G. R. Hankins, and T. R. F. Wright, unpublished). Complementary to this result, a cDNA from the *l(2)37Cf* region cross-hybridizes with genomic sequences in the *l(2)37Bb-Dox-A2-Bh* region, and a cDNA from *l(2)37Bb* cross-hybridizes with genomic DNA from the *l(2)37Cf-fs(2)TW1* region at low (~50%) stringency. Delimiting precisely which sequences cross-hybridize is still being determined. It is important to note that there are both intra- and intersubcluster homologies and also homologies to the scattered genes, suggesting that numerous genes in the *Ddc* region are evolutionarily related. In this regard it is significant that in vertebrates a number of catecholamine-metabolizing enzymes show sequence homologies (Joh *et al.*, 1984).

A detailed comparison of the *Ddc* cluster with other clusters of functionally related genes in the *Drosophila* genome is presented elsewhere (Wright, 1987) along with a consideration of why the genes in the *Ddc* region remain clustered rather than being dispersed. This includes preliminary experimental data on the function of clustered genes from the proximal *Ddc* subcluster when individually transposed to other sites in the genome (J. Kullman, unpublished).

XI. The Genetics and Biology of Phenol Oxidase Activities

A. Different Types of Phenol Oxidases in Insects

In insects phenol oxidases have several physiological roles, being involved in cuticle sclerotization, melanization, wound healing, and encapsulation of parasites (see reviews by Brunet, 1980; Anderson, 1985). Tyrosinases, a subset of phenol oxidases, are bifunctional copper enzymes with monophenol oxidase and *o*-diphenol oxidase activities which are inhibited by phenylthiourea and carbon monoxide. Physiologically, the former oxidizes tyrosine to DOPA and the latter oxidizes catecholamines (*o*-diphenols), DOPA, dopamine, *N*-acetyldopamine, and *N*-β-alanyldopamine to their respective quinones (Figs. 1 and 9), leucoaminochrome to *p*-quinone imine, 5,6-dihydroxyindole to indole-5,6-quinone (Fig. 4), and catecholamines to quinone methides (Fig. 6). The monophenol oxidase activity is characterized by oxidizing tyrosine to DOPA in the absence of a pterin cofactor, in contrast to tyrosine-*O*-hydroxylase (Section III,A). Diphenol oxidase (E.C. 1.10.3.1) activity is limited to *o*-diphenols, i.e., *p*-diphenols are not oxidized by tyrosinases. The recent *Drosophila* literature uses the more general term phenol oxidase in the place of tyrosinase. Although the more general term should include laccases (see below), with two exceptions (Ohnishi, 1954a,b; Yamazaki, 1969) most of the *Drosophila* work has been done on tyrosinase. Laccases (E.C. 1.10.3.2), also copper proteins, are distinguished from tyrosinases by their ability to oxidize *p*-diphenols in addition to *o*-diphenols and their inability to oxidize monophenols. They are not inhibited by carbon monoxide (Brunet, 1980).

In many insects a protyrosinase (prophenol oxidase) that must undergo an activation process is found in the hemolymph. For some insects, activation is carried out in a proteolytic process by an enzyme present in cuticle. This enzyme has been identified as a serine protease in the cuticle of the silkworm, *Bombyx mori* (Dohke, 1973), which upon activation releases a 5000-Da peptide from the prophenol oxidase (Ashida and Dohke, 1980). Similar serine protease activators have been identified in hemolymph of *Sarcophaga bullata* and *M. sexta* (Sugumaran *et al.*, 1985) and in cuticle of *M. sexta* (Aso *et al.*, 1985). Massive amounts of two endogenous protease inhibitors have been isolated from the hemolymph of *M. sexta* and one from *Sarcophaga* (Sugumaran *et al.*, 1985) which prevent *in vivo* melanization in spite of the simultaneous presence of prophenol oxidase, serine protease activator, and catecholamine precursors.

There is still uncertainty whether endogenous tyrosinase in the cuticle and cuticular laccases are derived from the hemolymph protyrosinase. The data of Hackman and Goldberg (1967) on the dipteran *L. cuprina* and that of Ishaaya (1972) on *Spodoptera littoralis* larvae are cited by Anderson (1985) as indicating that the enzymes from the two sources are different. On the other hand, the activated protyrosinase from the hemolymph and the endogenous cuticular tyrosinase of *M. sexta* have very similar physical, chemical, and kinetic properties, indicating a relationship between them (Aso *et al.*, 1984, 1985). It would not be surprising if in some insects the hemolymph and cuticular enzymes are genetically related and in others they are not. Genetic data suggest that in *Drosophila,* hemolymph and cuticular phenol oxidases are coded for by the same genes (see below).

Barrett and Anderson (1981) have characterized three cuticular phenol oxidases from *Calliphora vicina* third instar wandering larvae. Enzyme A, pH optimum 7.0, is a typical tyrosinase, oxidizing both monophenols and *o*-diphenols but not *p*-diphenols, and being inhibited by thiourea and phenylthiourea. Enzymes B and C are both laccases, oxidizing both *o*- and *p*-diphenols but not monophenols. The former has a pH optimum of 4.5 and the latter, 7.0. All three activities were separated by gel electrophoresis. None of these three cuticular enzymes needs to be activated. Andersen speculates that enzyme A is the same enzyme as the hemolymph tyrosinase incorporated into the cuticle in its active, aggregated form and is thought to be involved in wound healing on the basis of its similarity to the wound-healing phenol oxidase of *Calpodes ethlius* (Lai-Fook, 1966) which has now been purified and characterized (Barrett, 1984a,b). The laccases, enzymes B and C, from *Calliphora* are thought to function in sclerotization, with enzyme B being most similar to the locust (Andersen, 1978) and the *B. mori* (Yamazaki, 1972) cuticular phenol oxidases.

B. Phenol Oxidase Activity in *Drosophila*

In *Drosophila,* biochemical and genetic investigations of soluble tyrosinase activities have been under way a long time (Graubard, 1933; Ohnishi, 1953; Horowitz and Fling, 1955). With the important exception of the work by Ohnishi (1954a,b) and Yamazaki (1969) on an insoluble cuticular laccase in *D. virilis,* almost all the work has been concentrated on the mono- and diphenol oxidase activities which develop in supernatants of whole larval, pupal, or imaginal extracts. Neither monophenol oxidase (tyrosine oxidized to DOPA) nor diphenol oxidase (DOPA oxidized to dopaquinone) activity is present in fresh supernatants. Both activities appear with the same sigmoid kinetics when supernatants are incubated at 0°C. The lag period varies considerably depending on the individual strain, stage, or sex used. Both activities are inhibited by the same inhibitors, e.g., phenylthiocarbimide. After the enzyme is activated it can be removed from the supernatant by centrifugation at 35,000 *g* (Lewis and Lewis, 1963). Likewise, when a supernatant is allowed to activate prior to electrophoresis all activity remains at the origin (Mitchell and Weber, 1965). This indicates that a high degree of aggregation accompanies the appearance of these enzyme activities. When a supernatant is subjected to electrophoresis prior to activation, three components, A1, A2, and

A3, are resolvable after treatment of the gel with an activation solution and incubation with substrate (Mitchell and Weber, 1965). A1 possesses both monophenol and diphenol oxidase activities while A2 and A3 have only diphenol oxidase activity. Mutations in two different genes clearly indicate that the A2 and A3 components are separately mutable (Rizki *et al.*, 1985; Pentz *et al.*, 1986).

Several components in the activation process have been partially purified and their interactions investigated (Seybold *et al.*, 1975). Pre-S interacts with S-activator to yield S, and then S acts on P to produce P′, which interacts with the A components to yield active phenol oxidase. No work has been done to determine if a serine protease or its inhibitors are somehow involved in this cascade.

Levels of activatable diphenol oxidase activity have been assayed throughout development in *Drosophila*. Small but significant amounts of phenol oxidase potential activity have been found late in embryogenesis (Pentz *et al.*, 1986) and in first and second instar larvae (Geiger and Mitchell, 1966). Activity begins to rise 12 hours after the second molt and continues to rise steadily to a maximum just before pupariation (Geiger and Mitchell, 1966). Mitchell (1966) has demonstrated that the maximum rate of phenol oxidase activity achieved by activation in extracts drops precipitously after pupariation until between 4 and 8 hours after pupariation. No phenol oxidase activity can be activated in extracts even after 3-hour incubations. At 12 hours postpupariation (i.e., at pupation) the level of activation rises just as rapidly to a rate approximately two-thirds as high as the maximum found at pupariation. The levels then stay constant throughout metamorphosis until pigmentation commences at 75 hours, when the rate of activation decreases slowly to one-third the pupariation level at adult eclosion. The 4-hour prepupal hiatus in the ability to activate is due to the absence of the S component (Geiger and Mitchell, 1966), but all other components necessary for high levels of activity are present. This hiatus was exploited by Seybold *et al.* (1975) in the extraction and purification of most of the phenol oxidase components. (See also developmental profile in Pentz *et al.*, 1986.)

Yamazaki (1969) investigated a cuticular phenol oxidase from white prepupae of *D. virilis* which could not be solubilized from the cuticle. Using acetone-treated cuticular preparations, she showed the laccase present could oxidize phenylenediamine and *N*-acetyldopamine but not tyrosine or *p*-cresol. Besides this absence of monophenol oxidase activity, this enzyme differs from the soluble-hemolymph enzyme by its insensitivity to CO inhibition, but inhibition by KCN and sodium diethyldithiocarbamate indicates that it is a copper protein. Its activity peaks at pupariation and remains high throughout metamorphosis and subsequently for at least 2 imaginal days. No one has reported work on this enzyme in *D. melanogaster*.

C. Structural Loci

It is highly likely that two of the protein components of phenol oxidase in *D. melanogaster*, A2 and A3, are specified by the structural loci *Dox-A2* and *Dox-3*, respectively.

1. Dox-A2 (Diphenol oxidase-A2, 2–53.9; 37B10-13)

Dox-A2 is located in the *Ddc* gene cluster (Section X) 63 kb distal to *Ddc* in the distal subcluster. *Dox-A2* function is required for the production of the A2 component of the complex phenol oxidase enzyme and is probably its structural gene (Pentz *et al.*, 1986). *Dox-A2* lethal alleles as heterozygotes (*Dox-A2/+*) reduce diphenol oxidase activities to 47–79% of wild type (+/+) but do not affect monophenol oxidase activity. In 1 : 2 : 1 mixtures of *Dox-A2*^{1}/*Dox-A2*^{1} : *Dox-A2*^{1}/*CyO* : *CyO*/*CyO* 20- to 24-hour embryos, pool sizes of DOPA, dopamine, and *N*-acetyldopamine are elevated, indicating that *Dox-A2* function is necessary to oxidize these compounds to their respective quinones (Fig. 9) (Pentz *et al.*, 1986). Only the A2 component is reduced by *Dox-A2* alleles in polyacrylamide gels.

Hemizygotes, *Dox-A2*^{n}/*Df(2L)TW130*, of all three lethal alleles die during the first larval instar, and although their mouthparts and denticle belts are normally pigmented, they do not turn black after dying. A very rare *Dox-A2*^{1} homozygous mutant individual survived to the pharate adult stage and was released alive from the puparium by dissection. The mutant was completely unpigmented, with its bristles and cuticle being completely colorless (see colored illustration in Pentz *et al.*, 1986). This phenotype is identical to individuals with extremely low DDC activity except this *Dox-A2* homozygote never developed melanin in the joints of the legs and axillae of the wings typical of *Ddc* mutants. Even after the weak cuticle of the abdomen ruptured and this mutant died, it never melanized or tanned in any way. This suggests that normal A2 component must be present to have any functional phenol oxidase activity at all, i.e., the A1 and A3 components are not active *in vivo* in the absence of A2 activity. Furthermore, it indicates that *Dox-A2* activity is required for both hemolymph and cuticular phenol oxidase activity. The pigment deposited during embryogenesis in the mouthparts and denticle belts of *Dox-A2* homozygotes may be due to a maternal component, protein or mRNA, but no experiments have been done to verify this possibility.

A nonlethal mutant allele of *Dox-A2* has been isolated which is completely female sterile when heterozygous over *Dox-A2* lethal alleles and deficiencies for the *Dox-A2* region, yet viabilities are only marginally depressed to 60% of expected (P. O. Cecil and T. R. F. Wright, unpublished) and phenol oxidase activity in whole fly extracts is not depressed (E. S. Pentz, unpublished). Although activities have not yet been assayed in dissected ovaries, the data suggest that this may be an ovary-specific mutation with little or no effect on epidermal expression.

After irradiating a variety of stages of development to induce somatic recombination, no *Dox-A2*^{1}, *Dox-A2*^{2}, or *Dox-A2*^{3} hemizygous [*Dox-A2*^{n}/*Df(2L)TW130*] somatic clones have been recovered, indicating that in the epidermis the *Dox-A2* lethal mutations are cell lethals (E. S. Pentz, unpublished). This is a surprising result since epidermal cells hemizygous for both *Ddc* and *l(2)amd* lethal alleles are cell viable. The former produce colorless cuticle patches while the latter do not have a mutant phenotype (P. Ripoll and A. Garcia-Bellido, unpublished; C. P. Bishop, unpublished).

The *Dox-A*^{1} and *Dox-A*^{2} alleles are 0.1- and 1.1-kb deletions, respectively, located within 3.5–4.8 kb of the proximal breakpoint of *Df(2L)OD15* (Fig. 10) (Pentz and Wright, 1986). A 1.7-kb *Dox-A2* mRNA has been identified in 15- to 17-hour-old embryos, crawling third instar larvae, and 1- to 4-day-old adults. cDNA clones indicate that the 3′ end is centromere proximal and that the coding region contains at least one small intron (Pentz and Wright, 1986).

2. *Dox-3 (Diphenol oxidase-3, Second Chromosome) and Tyr-2 (Tyrosinase-2, 2–57)*

A fast-migrating electrophoretic mobility variant of the A3 component of *Drosophila* phenol oxidase was identified and isolated by Rizki *et al.* (1985). Extracts from individuals heterozygous for the fast variant and the ubiquitous slow variant exhibit three A3 bands in gels: light-staining parental-type fast and slow bands and a heavier staining hybrid band with a mobility intermediate between the two parental types. The gene responsible for these differences was determined to be on the second chromosome and then was mapped by means of recombination and overlapping deficiencies to a locus between *rdo* (2–53.1) and *M(2)m* in the region of bands 36E4–F5 (Rizki *et al.*, 1985). E. S. Pentz and T. R. F. Wright (unpublished) have been unable to confirm this location since the serially overlapping deficiencies *Df(2L)H20* (36A7–10; 36E4-F1), *Df(2L)TW203* (36E4–F1; 37B9–C1), and *Df(2L)TW130* (37B9–C1; 37D1–2) do not delete the *Dox-3*^{S} allele. If only one of the recombinant progeny derived from the mapping crosses of Rizki *et al.* (1985) is designated as a double crossover, *Dox-3*^{F} and *Dox-3*^{S} can be interpreted to map to the right of *pr* (2–54.5), perhaps even as electrophoretic variant alleles of *Tyr-2*.

Tyr-2 (*Tyrosinase-2*, 2–57) is a naturally occurring allele found in an *In(2L)Cy* + *In(2R)Cy* chromosome by Lewis and Lewis (1963). It reduces phenol oxidase activity about 50% in combination with some modifying genes. The effect is dominant (Lindsley and Grell, 1968).

D. Other Loci with Major Effects on Phenol Oxidase Activity

1. *qs (quicksilver, 1–39.5; 10F1–10)*

Nine mutant alleles have been isolated at the *qs* locus in three different laboratories on the basis of three different criteria. Craymer (1984) reported an allele of *qs* (designated as *qs*^{1} by Pentz *et al.*, 1987) that is a homozygous zygotic lethal, but not a cell lethal. In gynandromorphs, *qs* autonomously depigments cuticular tissue, including chaetae with "viability reduced as a result of weakened cuticle" (Craymer, 1984). Sherald (1981) isolated *qs*^{2} (*su*^{18}) originally as a nonlethal suppressor of *black* (*b*) and mapped it to 1–39.5. Wieschaus *et al.* (1984) isolated seven sex-linked embryonic lethal alleles with unpigmented mouthparts and denticle belts at a locus designated by them as *faintoid* (*ftd*). The one *ftd* allele tested has been shown to be noncomplementary with *qs*^{2} and therefore has been designated as *qs*^{3} (Pentz *et al.*, 1987), giving precedence in the name of the locus to Craymer (1984). Apparently *qs*^{1}, *qs*^{2}, and *qs*^{3} are all hypomorphic alleles whose expression can be varied by temperature, nutrition, and the accumulation of modifiers. For example, viabilities on food with different sources of yeast can be very different. Values for viabilities expressed as a percentage of expected on food 1 versus food 2 are as follows: *qs*^{1}/*Y*, 0% versus 6%; *qs*^{2}/*Y*, 1.2% versus 107%; *qs*^{1}/*qs*^{2}, 0% versus 20%; *Df(1)RA47*/*qs*^{2}, 0% versus 11% (Pentz *et al.*, 1987). All *qs* hemi- or homozygotes or heteroallelic heterozygotes that hatch are smaller than normal and exhibit an incompletely pigmented phenotype with a yellowish tinge to it. Some are extremely pale with bristles with very little pigmentation and with wings which, when expanded, are "glassy" clear and very fragile. The overall phenotype is very similar to the *Ddc* and *Dox-A2* incomplete sclerotization phenotypes. The extremely pale *qs*^{2}/*Y* flies never exhibit a wound reaction, i.e., they never turn black. Mutant puparia are also clearly underpigmented. As homozygotes, two of the four small deficiencies that delete *qs*$^{+}$ exhibit the faintoid phenotype of underpigmented mouthparts and denticle belts in unhatched lethal embryos (Wieschaus *et al.,* 1984). The mutant phenotypes of the other two deficiencies exhibit earlier developmental defects.

Although also quite variable, phenol oxidase activities in activated extracts of any genotype carrying any of the three *qs* alleles are always significantly depressed (Pentz *et al.,* 1987). Under poor nutritional conditions qs^2/qs^3 heterozygous females produced no measurable levels of phenol oxidase activity, and qs^2/Y males had only 4.5% of control activity. Upon separation in gels, it is apparent that the activities of all three components, A1, A2, and A3, are coordinately reduced in qs^2/Y males that are producing measurable levels of phenol oxidase activity (Pentz *et al.,* 1987). Assays for activator activity show that the initial rates of activation with control and qs^2/Y activators are identical. Catecholamine pools of qs^2/Y males 60–80 minutes after eclosion show dramatic increases (threefold to eightfold) in DOPA, *N*-β-alanyldopamine, and *N*-acetyldopamine pools in comparison to controls and a significant twofold increase in the dopamine pool (Pentz *et al.,* 1987). These results are consistent with the requirement for phenol oxidase activity to oxidize these four catecholamines to their respective quinones prior to melanization and tanning (Fig. 9).

So far there is no evidence to indicate that *qs* is the structural locus for any of the phenol oxidase A components or for any of the proteins in the activation cascade. Nevertheless, similar to *Dox-A2* mutations, *qs* mutations demonstrate that the phenol oxidase activity is vital, that both the hemolymph prophenol oxidase and the cuticular enzyme require *qs* gene function for activity, suggesting they may be the same enzyme. One would predict that, also like *Dox-A2,* an amorphic mutation of *qs* would be a cell lethal. The determination of the specific function of qs^+ may involve cloning and sequencing the gene to provide a derived amino acid sequence which may have a recognizable function. Recombinant DNA clones of *qs* have not yet been identified.

2. *tyr-1 (tyrosinase-1, 2–54.5; 37F6–38B6)*

The mutant *tyr-1*, originally designated *alpha-1* (α^1), was isolated as a spontaneous low-activity variant (Lewis and Lewis, 1961, 1963) which when homozygous reduces phenol oxidase activity to approximately 10% of most normal strains. This residual 10% activity is stable, not being subject to the usual phenol oxidase variation in different genetic backgrounds or under different environmental conditions (Lewis and Lewis, 1963). Lewis and Lewis (1963) mapped the gene to a locus at 2–52.4 (4.2 map units to the right of *black*). Subsequently, the *tyr-1* mutation has been more accurately mapped to a locus which so far is inseparable from *pr* (2–54.5) and is uncovered by the small deficiency, *Df(2L)TW150* (37F5–38A1; 38B2–C1), which also uncovers *pr* (Huntley, 1978; K. Theissen and T. R. F. Wright, unpublished; Pentz *et al.*, 1987). Lewis and Lewis (1963) reported that *tyr-1* phenol oxidase is more thermolabile than wild-type phenol oxidase and, therefore, suggest that *tyr-1* is a structural gene for the enzyme. Using the original strain in which *tyr-1* was isolated, Pentz *et al.* (1987) have been unable to show that *tyr-1* phenol oxidase is differentially thermolabile. It has been reported (Warner *et al.*, 1974) that *tyr-1* pupae have all three A components, A1, A2, and A3, resolvable on gels but in reduced amounts compared to the Samarkand wild-type control. This result has been confirmed by Pentz *et al.* (1987), who, furthermore, show that activator preparations made from the *tyr-1* strain are equally efficient as activator preparations from the wild-type control.

Although careful inspection of *tyr-1* homozygous adults reveals that they are slightly, but noticeably, underpigmented (RK3), with only 10% phenol oxidase activity they are much more normally pigmented than *qs* mutants with as much as 30% wild-type phenol oxidase activity. Furthermore, the hemolymph of *tyr-1* homozygous larvae will turn black, yet *tyr-1* adults do not, and *qs* mutant larvae and adults show no wound reaction. These results suggest that the two mutations, *tyr-1* and *qs*, are affecting different aspects of the compartmentalization of phenol oxidase in *Drosophila.*

Only one allele of *tyr-1* has been identified to date. As with *Dox-A2* and *qs*, one would expect amorphic alleles of *tyr-1* to be lethal. A number of lethal genes have been identified in the *Df(2L)TW150* region by P. Gay and D. Contamine (personal communication), but none has as yet been assayed for phenol oxidase activity. As with *qs*, recombinant DNA clones of *tyr-1* have not yet been identified.

E. Mutations Affecting Crystal Cells and Phenol Oxidase Activity

1. *Crystal Cells*

In *D. melanogaster* larvae 10% of the circulating hemocytes are crystal cells, which contain very large cytoplasmic paracrystalline inclusions (Rizki, 1957). The crystal cells rupture very readily when the body wall is broken or the hemolymph is removed from the hemocoel. Any experimental manipulation that disrupts the paracrystalline inclusions leads to a blackening of the crystal cell, and the pigment formed will spread throughout the hemolymph. Rizki and Rizki (1959) demonstrated the incorporation of [^{14}C]tyrosine into the paracrystalline structures and the presence of phenol oxidase activity in the surrounding cytoplasm in crystal cells. At that time they proposed "that separation of enzyme and substrate (DOPA) within the crystal cell is achieved by packaging the latter in paracrystalline form" (Rizki *et al.*, 1980). Recently, based on the use of monoclonal antibodies, it has been demonstrated that the paracrystalline inclusions are made up of one or more components of the hemolymph phenol oxidase and not semicrystalline DOPA (Rizki *et al.*, 1985; Rizki and Rizki, 1986).

2. Bc (Black cells, 2–80.6)

In homozygous larvae of the mutation *Black cells* (*Bc* 2–80.6) the crystal cells are blackened and lack paracrystalline inclusions, and their hemolymph does not turn black when exposed to air (Rizki *et al.*, 1980). *Bc*/*Bc* larvae have no activatable phenol oxidase activity, and *Bc*/+ larvae, which develop black crystal cells later in development than *Bc* homozygotes (late first instar versus midembryogenesis), have significantly reduced levels of phenol oxidase activity. Although no activity is measurable in *Bc*/*Bc* homozygotes, the larvae form normally pigmented puparia and pharate adult cuticle. This suggests that the phenol oxidase activity present in crystal cells may not be required for sclerotization, and that the presumed activity of one or more cuticular enzymes can function without a contribution from the crystal cells or some other hemolymph source. Dead phenocopies of *Bc*/*Bc* larvae can be made by treating the larvae with methanol, ethanol, or hot water.

Note that *Bc* (*Black cells*) should not be confused with *l(2)37Bc*, a vital gene in the *Ddc* cluster. The *l(2)37Bc* designation has often been abbreviated to *Bc* in the literature.

3. lz (lozenge, 1–27.7)

One locus with a visible phenotype which has been studied with respect to phenol oxidase activity is *lozenge* (*lz*), (1–27.7). A series of mutant alleles at this locus produce flies with oval-shaped eyes of varying widths and rough surface and with abnormal female accessory sex organs and claws. Peeples and co-workers (1968, 1969a,b), using some coisogenic strains, reported a clear-cut correlation between the severity of the visible phenotype and reduction in phenol oxidase activity. Using tyrosine and DOPA as separate substrates, they showed that the severest mutations have no A1 activity and reduced levels of A2–A3 activity (A2 and A3 were not resolved on their gels). Warner *et al.* (1974), working with noncoisogenic strains, were unable to duplicate Peeples' results with two alleles, yet a third, new, very severe allele, not included in Peeples' survey, had no phenol oxidase activity at all, i.e., no A1, A2, or A3. Other data which link the *lz* locus with phenol oxidase relate to suppression of *lz* mutations. The visible phenotype of some *lz* alleles can be suppressed by the *suppressor of forked* [*su(f)*, 1–65.9]. Snyder and Smith (1976) have correlated this suppression of the visible phenotype with increases in phenol oxidase activity.

Rizki and Rizki (1981) have demonstrated that 5 of the 15 *lz* alleles tested suppress the Bc phenotype, i.e., *lz/Y*; *Bc/Bc* males have no black cells in the hemolymph or lymph glands, nor do they have any crystal cells. The 10 *lz* alleles that do not suppress *Bc* do have crystal cells which can be blackened by heat. There are, however, morphological differences in the crystal cells of some of these Bc-nonsuppressor *lz* alleles (Rizki and Rizki, 1981). This suggests that the spectrum of phenol oxidase activities found among the *lz* alleles arises from a spectrum of crystal cell defects. Whether any or all of the other pleiotropic defects of the *lz* alleles arise as a result of reduced or no phenol oxidase activity or from other unknown causes is problematical.

F. Miscellaneous Loci Which Affect Phenol Oxidase Activity

1. *Phox (Phenol oxidase, 2–80.6)*

It is of some interest to note that the genetic location of *Bc*, 2–80.6, and the locus of *Phox*, 2–80.6, are identical. Batterham and Mackechnie (1980) have demonstrated an additional phenol oxidase, PHOX, in gels in addition to the A1, A2, and A3 components. PHOX activity is activated in gels by 50% propan-2-ol and other "synthetic" activators but not by natural pupal activator (see Rizki *et al.*, 1985, and Pentz *et al.*, 1986, for additional information on natural and synthetic activators). It is a tyrosinase, oxidizing only monophenols and *o*-diphenols with a native molecular weight of 108,000 (Batterham and Chambers, 1981). The mutants *tyr-1* and lz^3 exhibit no PHOX activity, lz^g has reduced activity, lz^{37h} has wild-type levels of activity, and other *lz* alleles show intermediate activities (all assayed on gels). Using three electrophoretic variant alleles, $Phox^F$, $Phox^S$, and $Phox^I$, Batterham and Mackechnie (1980) mapped the locus with high precision to 2–80.6. No *Phox* alleles affecting activity levels have been established (perhaps *Bc*?).

2. *sp (speck, 2–107.0; 60B13–C5)*

Homozygotes for *sp* and sp^2 have heavily melanized areas (specks) in wing axillae, dark body color, and darkened pupal anal papillae

(Lindsley and Grell, 1968). The speck phenotype and the dark anal papillae phenotype are characteristic phenotypes of *Ddc* mutant zygotes with less than 2–3% wild-type DDC activity. The dark body color is, of course, reminiscent of *ebony* and *black*, both of which affect catecholamine metabolism (see below). DDC and dopamine acetyltransferase activities have been assayed in *sp* homozygotes and no differences in activities were found (A. F. Sherald, M. D. Huntley, and T. R. F. Wright, unpublished). On the other hand, Warner *et al.* (1975) report a drastic reduction in the A2 component of phenol oxidase which is completely restored in $su(s)^2;sp$ individuals. The visible mutant phenotypes of *sp* are also suppressed by $su(s)^2$ (Lindsley and Grell, 1968). This result suggests that another gene besides *Dox-A2* is somehow involved with the A2 component. Pentz *et al.* (1987) have been unable to confirm reduced phenol oxidase activity levels in extracts of the *sp* and sp^2 stocks maintained at the University of Virginia, nor have they been able to demonstrate reduced amounts of the A2 component in gels. Since Batterham and Mackechnie (1980) also report the absence of what is probably the A2 component in a *sp* stock, it is possible that the disparate results may be due to differences in nutrition or to the accumulation of modifiers. It would be interesting to know if lethal alleles of *sp* like those for *Dox-A2* and *qs* could be recovered. Finally, for *sp*, in the light of the *Ddc* gene cluster, it is interesting to note that no recombinants have been obtained between *sp* and the low-dopamine acetyltransferase mutant allele, Dat^{lo} (see below).

3. *Phenol Oxidase Effects of the Body and Bristle Color Mutations y (yellow), e (ebony), b (black), stw (straw), and Bld (Blond)*

Mitchell's work (Mitchell, 1966; Mitchell *et al.*, 1967) on the levels of phenol oxidase activity available for activation in supernatants at different times in development for the mutations *y*, *e*, *b*, stw^5, and *Bld* [segregant from T(1;2)Bld] and heat-induced phenocopies of *Bld* suggest the following scenario (see below for mutant phenotypes and loci of the first three mutations and Lindsley and Grell, 1968, for the last two). Normally when the soluble phenol oxidase is activated it forms aggregates, oxidizes its substrates to their respective quinones, and is inactivated as a result of its enzymatic activity while it is incorporated with its quinone product into cuticle, cross-linking with cuticular proteins. As this process proceeds *in vivo* fewer and fewer soluble precursor molecules capable of activation remain. Mutations that prevent or decrease the rate of activation (stw^5, *Bld*, and *Bld* phenocopies) or prevent the synthesis of key substrates, e.g., β-alanyldopamine by *b* and *e*, particularly at pupariation, prevent the normal depletion of the phenol oxidase precursors from the hemolymph. Thus if mutant extracts are assayed for levels of activatable precursors, higher than wild-type levels of phenol oxidase activity will be found in the activated mutant extracts at specific times in development. The higher level of activatable phenol oxidase and visible phenotype (straw-colored bristles, reduced wing pigmentation, and impaired wing formation) of stw^5 are thought to arise from a reduced rate of activation 70–80 hours postpupariation prior to eclosion. The striking increase in activatable phenol oxidase and visible mutant phenotype of *Bld* and heat-induced *Bld* phenocopies may also be due to reduced rates of activation. The *Bld* phenol oxidase and visible phenotypes, "bristles gleaming yellow at tips and for varying lengths of more basal regions," are associated with the 2^DX^P aneuploid segregant of T(1;2)Bld = T(1;2)1C3–4; 60B12–13. Other deficiencies and aneuploid segregants indicate that this is not a dosage effect, i.e., *Bld* is a dominant gain-of-function mutation, an antimorph. Thus the breakpoint may be altering the gene product of a gene in the X chromosome in 1C3–4 right next to *svr* (see below) or on 2R in 60B12–13 precisely in the *Dat–sp* region suspected to contain a cluster of functionally related genes (see below).

4. *PPF, an in Vivo Pigment-Producing Factor*

Pigment-producing factor, PPF, is a heat-stable macromolecule existing in multiple polymeric forms which when injected into *Drosophila* prepupae and newly eclosed adults (but not eggs or larvae) induces the formation of black pigmentation (melanin) throughout the hemolymph of the recipient individual (Henderson and Glassman, 1969). Although eggs and larvae do not respond to the injection of PPF, it can be extracted from all developmental stages. The activity of PPF purified from third instar larvae is resistant to nucleases, proteases, amylase, and lipase and does not contain sulfur or phosphorus. The molecule can be hydrolyzed to compounds that react with Ehrlich's reagent, suggesting the presence of indoles. Henderson and Glassman (1969) suggest that PPF may be a storage form of the tyrosinase substrates. Whether PPF can serve as a substrate for tyrosinase activity is not clear, but when extracts of Oregon-R larvae were incubated with PPF, the extracts turned black within an hour, a reaction which did not occur in extracts to which PPF was not added. PPF was found to be associated with cellular particles 6 μm in diameter.

Henderson and Glassman (1969) report finding two strains, designated ppf^- strains, which do not turn black when injected with PPF.

All other stocks examined responded to PPF injection and were designated as ppf$^+$ strains. Although, the two ppf$^-$ strains carry recessive visible mutations, *w ec* and *rc pr,* the inability to respond to PPF is determined by multigenic factors and not by these recessive visible mutations. No tyrosinase activity was detected in third instar larval extracts of either of the ppf$^-$ strains, but, in extracts from adults less than 4 hours old, the *w ec* stock had two-thirds the Oregon-R tyrosinase activity and the *rc pr* had less than one-half. These results suggest that injected PPF, no longer in its normal biochemical compartment, interacts in some stage-specific way with the phenol oxidase activation cascade, permitting it to proceed. Whether the recipient responds by turning black (ppf$^+$) or not (ppf$^-$) depends on the presence or absence of tyrosinase activity at that stage. There is no report of any further work on the PPF particle of Henderson and Glassman (1969).

XII. Dopamine Acetyltransferase

Drosophila dopamine acetyltransferase (DAT) (E.C. 2.3.1.5), an *N*-acetyltransferase, has been characterized by Maranda and Hodgetts (1977). It uses acetyl CoA to acetylate dopamine to *N*-acetyldopamine, which has been identified as the classic "sclerotization compound" in insects (Sekeris and Karlson, 1962). Marsh and Wright (1980) determined the developmental profile of DAT in *Drosophila* using whole organisms. The peaks of DAT activity do not correlate with the peaks of DDC activity nor with levels of the molting hormone, ecdysone. Using the temperature-sensitive ecdysone-less mutation, ecd^1, Marsh and Wright (1980) found minimal evidence that the absence of ecdysone perturbs the normal regression of DAT activity at the normal time for pupariation. The data permit one to conclude that DDC and DAT are not coordinately regulated throughout the life cycle of *Drosophila,* and that DAT is not obviously controlled by ecdysone.

The exact tissue-specific localization of DAT in *Drosophila* has yet to be determined. Maranda and Hodgetts (1977) were unable to demonstrate significant levels of activity in their epidermal preparations, yet found a lot of activity associated with a muscle fraction in larvae. From this they speculate that DAT activity is localized in the oenocytes, segmentally distributed cells located between muscle and epidermis, and that dopamine must be transported to the oenocytes from the epidermis by carrier proteins, and NADA is then transported back to the epidermals. There is evidence that such carrier proteins do exist in Lepidoptera (Koeppe and Gilbert, 1974). One should note that

DAT activity has been localized in the epidermis and trachea of other insects, e.g., *Periplaneta americana* (Murdock and Omar, 1981). Maranda and Hodgetts (1977), Dewhurst *et al.* (1972), and E. Y. Wright (unpublished) have assayed relatively high DAT activities in *Drosophila* CNS, i.e., the ratio of CNS activity to whole white prepupal activity is much higher for DAT than DDC. In addition, Dewhurst *et al.* (1972) demonstrated that supernatants of adult brain and thoracic ganglia acetylate not only dopamine but also tyramine and serotonin, and that the most effective substrate was tyramine. This observation has been confirmed by Maranda and Hodgetts (1977), but they also find that dopamine is the most effective substrate for *N*-acetyltransferase in extracts of isolated abdomens. Whether a single enzymatic entity is responsible for the acetylation of dopamine, tyramine, and serotonin has not been determined.

Using segmental aneuploids (Lindsley *et al.*, 1972) and "free" deletions, Huntley (1978) located the single DAT dosage-sensitive region in the genome close to the tip of 2R in 60B1–10. From an unmutagenized lab stock, she also isolated a mutation Dat^{lo}, which when homozygous reduces DAT levels to 25–30% of controls and also produced a qualitatively mutant DAT enzyme which is thermolabile vis-à-vis wild-type DAT. This strongly suggests that Dat^{lo} is a mutation in the structural gene for DAT, a 29,000-Da monomeric enzyme (Maranda and Hodgetts, 1977). Huntley (1978) and T. R. F. Wright (unpublished) have mapped this mutation to within 0.17 cM of *sp* (2–107.0), reported to be located in 60B13–C5 (Lindsley and Grell, 1968), i.e., no recombinants between *Dat* and *sp* have been recovered.

That there may be a cluster of functionally related genes affecting catecholamine metabolism and sclerotization surrounding the *Dat* locus is suggested by the mutant effects of *sp* (see Section XI,F,2) and other adjacent mutations, including *Forkoid (Fo), l(2)Nova Scotia* [*l(2)NS*], and *Pin* and the breakpoint of T(1;2)Bld (see Section XI,F,3). Amorphic, probably lethal, alleles of *Dat* have not yet been sought and recombinant DNA clones for *Dat* have not been identified.

XIII. The Genetics of β-Alanine, Dopamine, and *N*-β-Alanyldopamine Metabolism in *Drosophila*

A. Mutant Phenotype of *black*, *ebony*, and *tan*

In *Drosophila* the role of β-alanine has been defined through the study of two dark color mutations, *black* (*b*, 2–48.5; 34E5–35D1) and *ebony* (*e*, 3–70.3; 93D2–3). Both produce adults that are extensively melanized and puparia that remain white. It has been shown that *black* homozygous flies and pupae have low levels of β-alanine and that injection of β-alanine into white prepupae and pharate adults results in normal coloration of puparia and adults, respectively (Hodgetts, 1972; Hodgetts and Choi, 1974; Sherald, 1981), and β-alanine injection into newly eclosed *black* adults increases the stiffness and puncture resistance of wings (Jacobs, 1985). Although *ebony* homozygotes synthesize β-alanine, they are incapable of incorporating it into cuticle (Jacobs and Brubaker, 1963; Hodgetts and Knopka, 1973). Dopamine and β-alanine pools in *ebony* flies just prior to and following eclosion are twice wild-type levels, but levels of DOPA, DAT, and DDC activities are the same as wild type (Hodgetts, 1972; Hodgetts and Knopka, 1973). That β-alanine utilization is somehow involved in sclerotization is emphasized by electron microscope studies which demonstrate that the pupal cuticles of both *black* and *ebony* have not compacted normally (Jacobs, 1978, 1980, 1985; Sherald, 1981). The mutant *tan* (*t*, 1–27.5; 8C2–3;C14–D1), with a body color more tan than wild type and with light larval mouthparts, has dopamine levels in late pupae and newly eclosed adults that are approximately 50% of wild-type controls (Knopka, 1972).

Using HPLC and electrochemical detection, B. C. Black, W. E. McIvor, and T. R. F. Wright (unpublished) have looked at catecholamine pool sizes in several wild-type and numerous mutant *Drosophila* strains, including *b, e,* and *t*. The catecholamine pools found 40 minutes after eclosion show that in both *ebony* and *black* the *N*-β-alanyldopamine pool is reduced drastically and there is a real increase in the dopamine pool. In *tan* the NBAD pool is huge, with some reduction in the dopamine pool. These results are consistent with earlier determinations of DA pools summarized above. From these data one can infer that the activity of β-alanyldopamine synthetase (Fig. 9) is defective in *ebony* and that *tan* affects the further metabolism of NBAD.

B. β-Alanyldopamine Synthetase : *ebony* and β-Alanyldopamine Hydrolase : *tan* Gene–Enzyme Systems

An assay of β-alanyldopamine synthetase (BAS) activity has been developed based on electrochemically measuring NBAD levels after HPLC, and the enzyme in *Drosophila* has been characterized (Black *et al.*, 1987c). The native enzyme is 90,000 Da and requires ATP and $MgCl_2$ for maximum activity *in vitro* and must be quite pure before linear reactions are obtained. Nonlinear reactions occur because the product, NBAD, is further rapidly metabolized enzymatically in crude homogenates (see below). Since *Drosophila* BAS is a relatively labile enzyme which must be partially purified to obtain linearity, it is not a very convenient enzyme for genetic and developmental studies. However, it has been determined that in *ebony* (*e/e*) BAS activity is reduced to levels below the limits of detection. The *ebony* allele, e^1, used for this work may be a hypomorph, since very infrequently a small NBAD pool in e^1/e^1 homozygotes was observed.

In crude homogenates, the enzyme that rapidly depletes the NBAD being produced by BAS has been partially purified and characterized (B. C. Black, W. E. McIvor, and T. R. F. Wright, unpublished) and has been designated as β-alanyldopamine hydrolase (BAH) (Fig. 9). Its native molecular weight is 70,000, and it is stable and has no special cofactor requirements. It uses NBAD as substrate, hydrolyzing it to dopamine and β-alanine. It is not inhibited by trypsin, chymotrypsin, or pepsin inhibitors. Consistent with the catecholamine pools, homo- or hemizygous *tan* flies (*t/t* and *t/Y*) have no BAH activity.

Since heterozygotes for both *ebony* (*e/+*) and *tan* (*t/+*) have reduced BAS and BAH activities, respectively, vis-à-vis, homozygous wild type, i.e., show dosage effects, it is likely that these two genes are structural loci for the two enzymes. Presumably these two enzymes, which alternatively synthesize and hydrolyze NBAD, will be localized in different physiological compartments within the organism.

The above information, both old and new, permits the following interpretation of the phenotypes of *black* and *ebony*. The reason for the excessive melanization of *black* and *ebony* adults must be due to the shunting of the excessive levels of dopamine in both mutants to melanin via indole-5,6-quinone (Fig. 9) (Hodgetts and Choi, 1974). One would infer that since the puparia of both *black* and *ebony* remain white, that in wild-type puparia β-alanine, dopamine, and BAS activity are required for quinone sclerotization at this time. Hodgetts (1972; Hodgetts and Knopka, 1973) reports that at pupariation *ebony* does accumulate levels of both dopamine and β-alanine that are twice as high as in wild type. The puparia of both *ebony* and *black*, although white, are hard, i.e., sclerotized, therefore pathways toward colorless sclerotization must be active at this time. Since there are high levels of DAT prior to pupariation, perhaps NADA is involved in Andersen's colorless β-sclerotization (Andersen, 1974) or Sugumaran and Lipke's (1983) colorless quinone methide sclerotization (Fig. 6). Alternatively, or in addition, the putative *l(2)amd* pathway to colorless sclerotization (Section IV,B) is probably operative at this time.

C. Neurological Effects of *ebony* and *tan*

The electroretinograms of flies homozygous for either *e* or *t* are abnormal, lacking ON and OFF transients (Hotta and Benzer, 1969; Pak *et al.*, 1969). Phototaxis is reduced by *ebony* (Durrwachter, 1957), and *tan* has been characterized as being nonphototactic (McEwen, 1918; Hotta and Benzer, 1969; Pak *et al.*, 1969) and, on the other hand, as having a normal photoresponse (Hadler, 1964). Mating success of *ebony* is reduced in comparison to wild type (Rendel, 1951), perhaps due to frequent aborted courtship because of mismounting by males (Crossley and Zuill, 1970). B. E. Zamsky and L. Tompkins (personal communication) report that *tan, ebony,* and the double-mutant *tan; ebony* males move around the mating chamber aimlessly, oblivious to the presence of the female, suggesting they are unable to respond to either visual or olfactory stimuli from the female. Mutant males court only when they accidentally bump into females. This erratic behavior resembles that of *sbl;gl* mutant flies (Tompkins, 1984), suggesting olfactory blindness in *tan, ebony,* and *tan;ebony* flies also. Explanations of these apparently similar effects of *ebony* and *tan* are not obvious, particularly since NBAD is missing in *ebony* and *tan;ebony* catecholamine pools, and abnormal levels of NBAD accumulate in *tan* catecholamine pools.

Apparently the electroretinograms of homozygous *black* flies are normal, but their locomotor activity is reduced (Hotta and Benzer, 1969).

D. Mutations Affecting β-Alanine Levels and the *black* Phenotype

It is clear that the *black* mutation in *Drosophila* partially blocks the synthesis of β-alanine; however, just what step in the pathways to β-alanine synthesis is affected is not known. β-Alanine is synthesized by several different pathways, including the decarboxylation of aspartate, only reduced 6% in *black* homozygotes, and by the synthesis and degradation of uracil (Jacobs, 1974). In *Musca* 56% of β-alanine is synthesized from uracil and 24% from aspartic acid, but whether the 24% derives from the direct α-decarboxylation of aspartate or indirectly via uracil is not known (Ross and Monroe, 1972). Mutations in the uracil synthetic pathway in *Drosophila* which do *not* produce a *black* phenotype are *rudimentary* (*r*, 1–55.3), defective in carbamoyl phosphate synthase, aspartate transcarbamoylase, and/or dihydroorotase (Jarry and Falk, 1974; Norby, 1973; Rawls and Fristrom, 1975);

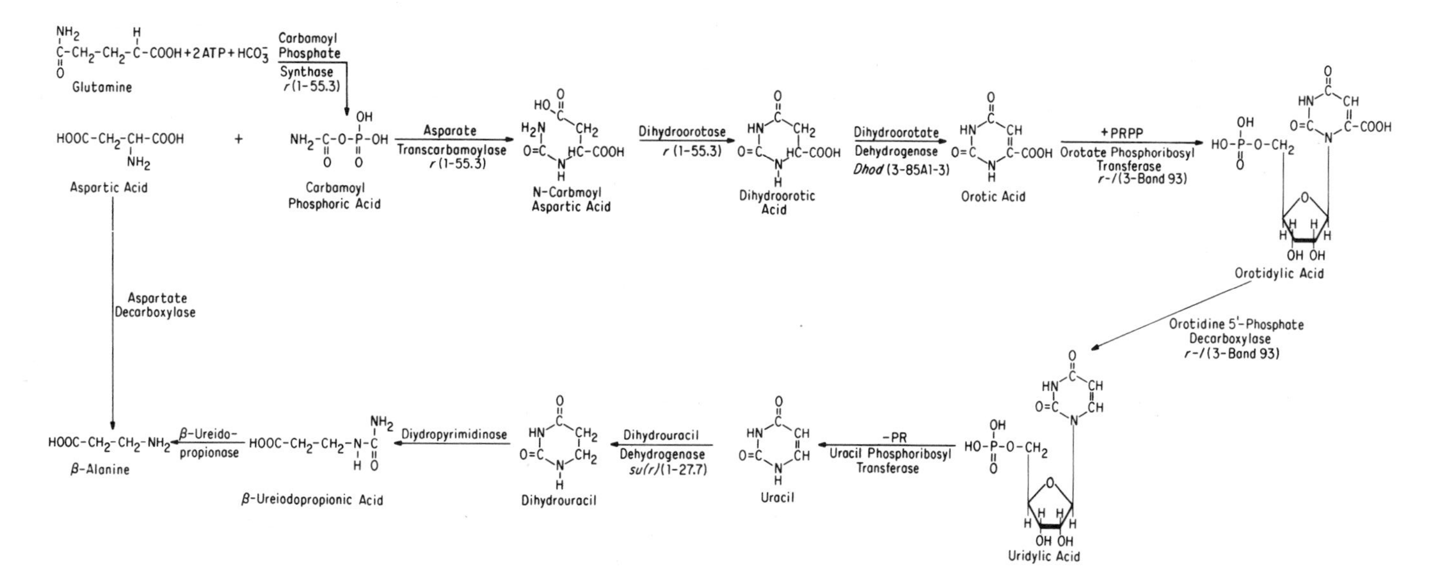

FIG. 11. Two possible routes by which β-alanine could be synthesized with reactions affected by specific mutations in *D. melanogaster* indicated. The numbers in parentheses after the gene symbols specify chromosome and locus. See text for the phenotypic effects of the various mutations.

Dhod (3–85A1–3), defective in dihydroorotate dehydrogenase (Rawls *et al.*, 1981); and *rudimentary-like* (*r-l*, 3–band 93), defective in orotate phosphoribosyltransferase and orotidine-5′-phosphate decarboxylase (Connor and Rawls, 1982; Rawls, 1980) (Fig. 11). On the other hand, the mutation *suppressor of rudimentary* [*su(r)*, 1–27.7], which is defective in dihydrouracil dehydrogenase activity, makes flies slightly darker than normal and definitely enhances the mutant expression of *black* in *su(r);b* flies. Similarly, feeding the analog inhibitor of dihydrouracil dehydrogenase, 6-azathymine, produces a partial black phenotype (Pedersen, 1982). Dihydrouracil dehydrogenase activity has not been measured in *black*, but the next enzyme in the uracil degradative pathway, dihydropyrimidinase, has been found to be present at wild-type levels in *black* (Hankins and Sherald, 1981). The semidominant *Suppressor of black* [*Su(b)*, 1–55.5], which maps adjacent to the *rudimentary* mutants (1–55.3), may suppress the black phenotype by up-regulating r^+, resulting in the overproduction of uracil, an inference which is consistent with *su(r)* being epistatic to *Su(b)*, i.e., *su(r) Su(b);b* flies are black. It is not known where the recessive *suppressor of black* [*su(b)*, 1–0.0] fits into β-alanine metabolism except that in *su(b);b* β-alanine levels are wild type and in *su(b);b*$^+$ β-alanine levels are supranormal (Sherald, 1981). B. C. Black, G. R. Hankins, and T. R. F. Wright (unpublished) have determined that while the uridine pool in *black* is depressed (66%), the uracil pool is normal. They also determined that β-alanine pools are not only depressed in newly eclosed *black* flies (b^3/b^3, 26%) but also in all the dark body color mutations tested: *dusky* (*dy*, 1–36.2, 32%), *sable* (*s*, 1–43.0, 39%), and even *ebony* (48%) and *tan* (39%) had reduced levels.

The role of β-alanine in the sclerotization of pigmentation of the cuticle has been variously discussed by Hodgetts and Choi (1974), Sherald (1980, 1981), Jacobs (1985), and Pedersen (1982). The last source concisely summarizes the genetics and biochemistry of β-alanine metabolism in *D. melanogaster*. Until the effects of the *black* mutation on uracil phosphoribosyltransferase, dihydrouracil dehydrogenase, and β-ureidopropionase activities are measured, one can only speculate about the specific defect in *black*. Since aspartate decarboxylase activity in *black* is only 6% lower than in wild type (Jacobs, 1974), it seems unlikely that this is the basic defect. Although perturbations in uracil degradation and perhaps in uracil synthesis can enhance, suppress, and even partially phenocopy the black phenotype, one cannot yet conclude that the mutant *black* affects this pathway.

XIV. Two Additional Body Color Mutations

A. *silver* (*svr*, 1–0.0; 1B6–7)

The allele svr^1, a hypomorphic allele of the X-linked locus *silver* (*svr*, 1–0.0; 1B6–7), causes the color of the legs, wings, veins, and integument to be pale and silvery and the wings of all males and some of the females to be pointed (Lindsley and Grell, 1968). Amorphic alleles are lethal, dying at pupation. Although the DA and NBAD pools in 40-minute-old svr^1/svr^1 adults are essentially normal, there is a striking increase (5- to 10-fold) in the NADA pool accompanied by a significant increase in three unidentified peaks with retention times between NADA and NBAD, with the slowest of these being quite large (B. C. Black, W. E. McIvor, and T. R. F. Wright, unpublished). The compounds represented in these peaks are unknown (they are not catecholamine VIII or X; see Section V), but in the light of how the materials are prepared for HPLC fractionation and their electrochemical reactivity, they are probably unidentified catecholamine conjugates. The three peaks are barely noticeable, if present at all, in pools from wild-type and most other mutant strains. Thus even this hypomorphic *svr* mutation drastically alters the catecholamine pool in the adult, probably blocking an important unknown branch pathway of catecholamine metabolism originating from NADA.

Mutant svr^1 flies have normal electroretinograms (Hotta and Benzer, 1969), but are extremely photonegative, avoiding light if at all possible (M. Seiger, personal communication). Males lose the following response in the light and, therefore, are unable to mate in the light but do so in the dark. Considering the effect of *silver* on catecholamine pool sizes and phototropism, its chromosomal location is particularly noteworthy. Its cytological location is *1B5–6* immediately proximal to the *scute* (*sc*) complex in *1B3*. This region, *1B1,2–1B13–14,* has been studied extensively by Garcia-Bellido and Santamaria (1978), Ripoll and Garcia-Bellido (1979), Jiménez and Campos-Ortega (1979), and White (1980). It is clear that there are genes in this region that are necessary for the differentiation (maintenance) of the embryonic central nervous system and the differentiation of peripheral nervous elements of chaetae and sensillae of adult cuticle. From her embryological analysis of overlapping deficiencies and duplications and a single embryonic lethal, *l(1)vnd,* White concludes that there must be at least two genes in this region required for neural development, one mapping between *1B2–3* and *1B6–7* and the other mapping between *1B6–7* and *1B9–10*. The cytological location of *svr* has been designated as *1B6–7* right in the middle of this region (Lindsley and Grell, 1968). In addition, Ripoll and Garcia-Bellido (1979) report that imaginal epidermal cells homozygous for *Df(1)260-1* (1A1–1B4–6; not deficient for *svr*) are viable whereas cells homozygous for *Df(1)svr* (1A1–1B10–13; deficient for *svr*) are lethal, implicating the requirement for a gene in this region for the maintenance of epidermal cells as well as neural cells. The relationship of *svr* to these vital functions has not yet been explored.

B. *yellow* (*y*, 1–0.0; 1B1)

The mutant alleles of the *yellow* locus (*y*, 1–0.0; 1B1) can generally be classified into two types: (1) the y^1 type, the amorphic or deficient type which totally removes black pigment from adult flies and larvae, leaving the adult cuticle and its derived structures, i.e., bristles, aristae, wings, sex combs, tarsal claws, genitalia, larval mouthparts, and denticle belts, uniformly yellowish brown and (2) the y^2 type, which exhibit different abnormal patterns of cuticular coloration with some parts of the cuticle and its derivatives showing a wild-type pigmentation and other parts being yellowish brown. The actual pattern of pigmentation being produced by the y^2 type is allele specific (Lindsley and Grell, 1968; Brehme, 1941; Nash and Yankin, 1974). Even though y^1-type alleles do not deposit melanin in cuticular structures, their whole fly catecholamine pool profiles are not perturbed (B. C. Black, unpublished), and they do exhibit the wound reaction, producing normally activatable prophenol oxidase.

Males hemizygous for *y* show a reduction in success in fertilizing wild-type females. This is related to a change in mating behavior involving abnormally long courtship latency and reduced orientation time and not to mechanical changes in the cuticle (Bastock, 1956; Burnet and Wilson, 1980). By feeding α-dimethyltyrosine to wild-type *D. melanogaster* larvae Burnet *et al.* (1973) were able to phenocopy both the mutant yellow body color and the mutant courtship defect.

The *yellow* gene has been cloned in several laboratories and a 1.9-kb polyadenylated RNA was identified as the *yellow*-encoded transcript (Biessmann, 1985; Campuzano *et al.*, 1985; Parkhurst and Corces, 1986; Chia *et al.*, 1986). This transcription is present at high levels in mid to late pupae when melanization of the adult cuticle occurs, but is also found at very high levels in late embryos and larvae (Biessmann, 1985). Geyer *et al.* (1986) have determined the complete nucleotide sequence of the *yellow* gene and have derived a conceptualized amino acid sequence for the *yellow*-encoded protein from it. The N-terminus is hydrophobic and contains many features common to signal peptides, suggesting that it is a membrane or secreted protein. Also suggestive of membrane or signal proteins, the yellow protein has two sites which could act in glucosylation. The calculated secondary structure of the protein suggests that it is unable to fold into any long α-helical or β-pleated sheet structures, indicating that it is not globular and, therefore, probably not an enzymatic protein but a structural protein. No proteins homologous to the yellow protein were found in the Bionet computer facility. Geyer *et al.* (1986) suggest that the yellow protein functions to cross-link indole-5,6-quinone moieties during melanization. They suggest that when it is secreted it reacts with indole-5-quinone through its cysteine and methionine residues (residues which are absent in the abundant cuticle proteins analyzed to date). Further, they suggest that mutations in the *yellow* gene would not inhibit melanin formation but would result in the production of an altered form of melanin (see Waddington, 1942) or in the accumulation of indole-5,6-quinone, which, if it self-polymerizes in the presence of β-alanine, would result in tan pigment rather than a black pigment (Sherald, 1980). The fact that the expression of *yellow* is cell autonomous in mosaics would be explained by it being immobilized in the membrane or, if secreted, immobilized by being cross-linked together. Many aspects of this hypothesis need to be confirmed and strengthened and an explanation of the behavioral effects of *yellow* sought.

XV. Concluding Remarks

It is hoped that even a superficial perusal of the above compendium will emphasize that even though it looks like we may know a lot about the genetics and biochemistry of biogenic amine metabolism in *Drosophila,* we really do not. Even though more than 40 genes have been identified as being involved in some way, there is a very large amount of work still to be done. There are numerous "loose ends" to be tied up and major aspects of biogenic amine metabolism which have yet to be seriously approached. The fact that we know with any precision the actual function of only 11 of these 40 genes leaves a rather large number of loose ends. However, with the genetic and molecular biological techniques now available in *Drosophila,* given the resources, the analysis of any one of these genes could be driven, in a relatively short time, to the point where a nucleotide sequence and derived amino acid sequence of the product of that gene are available. Three such examples summarized above are *per*, *yellow*, and *l(2)amd*. Even though we still do not know exactly what they do, the amount of information gained from the amino acid sequences is tremendous. Just the fact that we now know the former two are nonenzymatic proteins and that the last is clearly an enzyme probably involved in catecholamine metabolism has a fundamental bearing on what experiments will be done next.

A survey of the biochemistry, genetics, and molecular biology of cuticular proteins in *Drosophila* has not been included in this review, although they are, of course, involved in sclerotization. The two reasons for their omission are that this review is already long and that their genetics, biochemistry, and molecular biology have been thoroughly reviewed elsewhere (Silvert, 1985; Fristrom *et al.*, 1986). Chihara and Kimbrell (1986) report the most recent results on the genetics of cuticular proteins.

The genetics of *Drosophila* provides a number of different approaches toward initiating the investigation of areas of biogenic amine metabolism which have previously not been studied or are completely novel to any organism. If, for example, an enzyme such as tyrosine-β-hydroxylase can conveniently be assayed in whole *Drosophila* adults, a dosage-sensitive region (activity in one dose versus two doses versus three doses) in the genome for this enzyme can be found by surveying the whole genome with just 30 crosses (Lindsley *et al.*, 1972; see also Hodgetts, 1975, for an elegant example involving DDC). Once a dosage-sensitive region is found, it can be delineated with overlapping deletions and then point mutations can be recovered in the critical region (for DDC, see Wright *et al.*, 1976a,b). This essentially brute-force process can involve a significant investment of resources and time. However, once *any* gene is found in *Drosophila* it can be cloned and sequenced and mutants induced and used as tools in subsequent biochemical, developmental, neurological, and behavioral analyses. A far easier way of finding the location of the gene of interest in *Drosophila* is to obtain cloned DNA or large amounts of gene-specific mRNA for the gene from another organism and, hoping it has sufficient homology, use it to carry out *in situ* hybridization to *Drosophila* polytene chromosomes, or directly screen a *Drosophila* genomic recombinant DNA library for the *Drosophila* gene, e.g., tyrosine hydroxylase (Section III,A). Once the location is known, deletions and point mutations can be induced.

B. C. Black's unpublished work on *silver*, as summarized above, exemplifies another approach, almost unique to *Drosophila*, for discovering completely novel aspects of biogenic amine metabolism. This involves taking mutants with interesting phenotypes [such as body color mutants (*silver*) or unhatched lethal larvae with underpigmented mouthparts and denticle belts [*Ddc* or qs^3 (faintoid)] (see above)] and looking for differences in biogenic amine pools. Once differences are found, the work can begin perhaps on a branch pathway completely unknown in any organism. It is interesting to note just how many of the mutants discussed above that have effects on biogenic amine metabolism have body color phenotypes. B. C. Black has already surveyed the adult stages, but not other stages, of most of the obvious body color mutations in *Drosophila* (see table in Black *et al.*, 1987b). There are, in addition, numerous embryonic lethal mutations with suggestive mutant phenotypes described by Nüsslein-Volhard, Wieschaus, and co-workers (Nüsslein-Volhard *et al.*, 1984; Jürgens *et al.*, 1984; Wieschaus *et al.*, 1984) which would be most interesting to look at for perturbations in biogenic amine pools, even in mixed samples where only one-quarter of the embryos are lethal homozygotes [e.g., see *Dox-A2*[1] in Section XI,C,1 and Pentz *et al.* (1986)]. On the other hand, to avoid losing a mutant phenotype in a mixture of embryos of different genotypes, CNSs from individual mutant embryos with underpigmented mouthparts could be surveyed for the absence of serotonin, catecholamines, or other biogenic amine histofluorescence (see *pale*, Section III,A, and *Ddc*, Section VII,C).

Still another approach to finding genes involved in biogenic amine metabolism is to screen for mutations that make *Drosophila* hypersensitive or resistant to analog inhibitors of the enzyme of interest, as exemplified by the use of dietary α-methyldopa (Section X) (Sherald and Wright, 1974; Sparrow and Wright, 1974) (see also Howard *et al.*, 1975). Since these studies were initiated almost 15 years ago, the drug companies have produced many inhibitors that might be productive.

Another approach to identifying genes involved in novel aspects of biogenic amine metabolism in *Drosophila* is to examine genes which are very closely linked to known biogenic amine genes, as exemplified by the *Ddc* cluster of functionally related genes (Section X). My prediction is that as more groups of closely linked genes are carefully examined in *Drosophila*, many more clusters of functionally related genes will be identified, in contrast to the traditionally held view that functionally related genes are usually scattered throughout the genome. Besides the *Ddc* cluster, there is probably a cluster of functionally related genes in the *Dat–sp* region near the tip of the right arm of the second chromosome, and there is already abundant evidence that *silver* is located in a cluster of closely linked genes involved in the development of the nervous system (Section XIV,A).

Finally, the effect of the mutations described in this review on the neurophysiology and behavior of *Drosophila* have been underemphasized for a number of reasons, not the least of which is the author's lack of expertise in these areas. The neurogenetics and behavioral genetics of *Drosophila* are extensive and very active areas of research that have been thoroughly reviewed (Hall, 1982, 1985; Dudai, 1985; Tanouye *et al.*, 1986; Tompkins, 1986).

For those interested in more information on the neurophysiological and behavioral effects of mutations discussed here, refer to these reviews or, of course, go to the original literature. However, one final, general, very trite observation is, perhaps, in order, and that is that the effects that mutant genes have on metabolism or development are almost always extensively pleiotropic. Mutations affecting biogenic amine metabolism are certainly no exception, for it is obvious that blocking a single reaction can have multiple, drastic effects on biogenic amine pools measured in whole organism extracts, and have multiple and sometimes drastic effects on the neurophysiology and behavior of the organism. To suggest that the biochemical basis of various behaviors can be determined from the analysis of perturbed pools in whole organism mutant extracts does not seem reasonable, particularly if one considers the massive involvement of biogenic amines in cuticular sclerotization. Certainly the initial step in the analysis must be to determine the effects that these mutations have on isolated nervous system preparations (Livingstone and Tempel, 1983) and on specific groups of cells in the central nervous system (Valles and White, 1985). It seems evident that these last two types of studies must be followed by more sophisticated experiments such as the analysis of genetically produced mosaics (Gailey *et al.*, 1987) as well as many more experiments demanding even more ingenuity. At this point in time the genetics and, to a certain extent, the biochemistry are easy and have and will continue to produce valuable mutations which can be very useful tools for the judicious use of neurophysiologists and behavioral biologists in their very much more difficult investigations.

Acknowledgments

I thank my wife, Eileen Y. Wright, for her encouragement and forbearance during the writing of this review. The contributions, many of them unpublished, of present and past members of my laboratory at the University of Virginia are gratefully acknowledged. The contributors are Clifton P. Bishop, Bruce C. Black, Patricia O. Cecil, Pamela N. Fornili, Mark E. Freeman, Gerald R. Hankins, Marianne D. Huntley, John Kullman,

Lee Litvinas, J. Lawerence Marsh, Ellen S. Pentz, Allen F. Sherald, Ruth Steward, and Eileen Y. Wright. The author appreciates the courtesies extended by Dr. J. Lawrence Marsh and his co-workers at the University of California, Irvine; Dr. Ross B. Hodgetts and his co-workers at the University of Alberta; Dr. Jay Hirsh and his co-workers at Harvard University; and Dr. Kalpana White and her co-workers at Brandeis University by providing preprints of papers submitted for publication and papers in press. The author's resarch is supported by the National Institutes of Health Research Grant GM 19242.

References

Andersen, S. O. (1966). Covalent cross-links in a structural protein, resilin. *Acta Physiol. Scand.* **66** (Suppl. 263), 1–81.

Andersen, S. O. (1970). Isolation of arternone (2-amino-3′,4′-dihydroxyacetophenone) from hydrolysates of sclerotized insect cuticle. *J. Insect Physiol.* **16**, 1951–1959.

Andersen, S. O. (1974). Evidence for two mechanisms of sclerotization in insect cuticle. *Nature (London)* **251**, 507–508.

Andersen, S. O. (1978). Characterization of a trypsin-solubilized phenoloxidase from locust cuticle. *Insect Biochem.* **8**, 143–148.

Andersen, S. O. (1985). Sclerotization and tanning of the cuticle. *In* "Comparative Insect Physiology, Biochemistry, and Pharmacology" (G. A. Kerkut and L. I. Gilbert, eds.), Vol. 3, pp. 59–74. Pergamon, New York.

Andersen, S. O., and Barrett, F. M. (1971). The isolation of ketocatechols from insect cuticle and their possible role in sclerotization. *J. Insect Physiol.* **17**, 69–83.

Andersen, S. O., and Roepstorff, P. (1982). Sclerotization of insect cuticle. III. An unsaturated derivative of N-acetyldopamine and its role in sclerotization. *Insect Biochem.* **12**, 269–276.

Ashida, M., and Dohke, K. (1980). Activation of prophenoloxidase by the activating enzyme of the silkworm, *Bombyx mori*. *Insect Biochem.* **10**, 37–47.

Aso, Y., Kramer, K. J., Hopkins, T. L., and Whetzel, S. Z. (1984). Properties of tyrosine and dopa quinone imine conversion factor from pharate pupal cuticle of *Manduca sexta* (L). *Insect Biochem.* **14**, 463–472.

Aso, Y., Kramer, K. J., Hopkins, T. L., and Lookhart, G. L. (1985). Characterization of haemolymph protyrosinase and a cuticular activator from *Manduca sexta* (L). *Insect Biochem.* **15**, 9–17.

Bargiello, T. A., Jackson, F. R., and Young, M. W. (1984). Restoration of circadian behaviorial rhythms by gene transfer in *Drosophila*. *Nature (London)* **312**, 752–754.

Barrett, F. M. (1984a). Purification of phenolic compounds and a phenoloxidase from larval cuticle of the red-humped oakworm, *Symmerista cannicosta* Francl. *Arch. Insect Biochem. Physiol.* **1**, 213–223.

Barrett, F. M. (1984b). Wound healing phenoloxidase in larval cuticle of *Calpodes ethlius* (Lepidoptera:Hesperiidae) *Can. J. Zool.* **62**, 834–838.

Barrett, F. M., and Andersen, S. O. (1981). Phenoloxidases in larval cuticle of the blowfly, *Calliphora vicina*. *Insect Biochem.* **11**, 17–23.

Bastock, M. (1956). A gene mutation which changes a behavior pattern. *Evolution* **10**, 421–439.

Batterham, P., and Chambers, G. K. (1981). The molecular weight of a novel phenol oxidase in *D. melanogaster*. *Dros. Inf. Serv.* **56**, 18.

Batterham, P., and MacKechnie, S. W. (1980). A phenol oxidase polymorphism in *Drosophila melanogaster*. *Genetica* **54**, 121–125.

Beall, C. J., and Hirsh, J. (1984). High levels of intron-containing RNAs are associated with expression of the *Drosophila* DOPA decarboxylase gene. *Mol. Cell. Biol.* **4,** 1669–1674.

Beckman, L., and Johnson, F. M. (1964a). Genetic variations of phosphatases in larvae of *Drosophila melanogaster. Nature (London)* **201,** 321.

Beckman, L., and Johnson, F. M. (1964b). Variations in larval alkaline phosphatase controlled by *Aph* alleles in *Drosophila melanogaster. Genetics* **49,** 829–835.

Biessmann, H. (1985). Molecular analysis of the *yellow* gene region of *Drosophila melanogaster. Proc. Natl. Acad. Sci. U.S.A.* **82,** 7369–7373.

Bishop, C. P., and Wright, T. R. F. (1987). Ddc^{DE1}, a mutant differentially affecting both stage and tissue specific expression of DOPA decarboxylase in *Drosophila. Genetics* **115,** 477–491.

Black, B. C., and Smarrelli, J., Jr. (1986). A kinetic analysis of *Drosophila melanogaster* carboxylase. *Biochim. Biophys. Acta* **870,** 31–40.

Black, B. C., McIvor, W. E., and Wright, T. R. F. (1987a). Isolation and identification of catecholamines from adult *Drosophila* and developmental profiles during pigmented and unpigmented sclerotization. *Insect Biochem.,* in press.

Black, B. C., Pentz, E. S., and Wright, T. R. F. (1987b). The alpha methyl dopa hypersensitive gene, *1(2)amd,* and two adjacent genes in *Drosophila melanogaster.* Physical location and direct effects of *amd* on catecholamine metabolism. *Mol. Gen. Genet.* (in press).

Black, B. C., McIvor, W. E., and Wright, T. R. F. (1987c). A new enzyme, β-alanylamine synthetase, ligating β-alanine and dopamine in *Drosophila. Insect Biochem.* (in press).

Bodenstein, D. (1950). The postembryonic development of *Drosophila. In* "The Biology of *Drosophila*" (M. Demerec, ed.), pp. 275–367. Wiley, New York.

Bodnaryk, R. P. (1970). Effect of DOPA-decarboxylase inhibition on the metabolism of β-alanyl-L-tyrosine during puparium formation in the fleshfly *Sarcophaga bullata.* Parker. *Comp. Biochem. Physiol.* **35,** 221–227.

Bodnaryk, R. P. (1971). Studies on the incorporation of β-alanine into the puparium of the fly, *Sarcophaga bullata. J. Insect Physiol.* **17,** 1201–1210.

Bodnaryk, R. P. (1972). A survey of the occurrence of β-alanyltyrosine, γ-glutamylphenylalanine and tyrosine-*O*-phosphate in the larval stages of flies (Diptera). *Comp. Biochem. Physiol.* **43B,** 587–592.

Bodnaryk, R. P., and Brunet, P. C. J. (1974). 3-*O*-Hydrosulphato-4-hydroxyphenethyamine (dopamine 3-*O*-sulphate), a metabolite in the sclerotization of insect cuticle. *Biochem. J.* **138,** 463–469.

Bodnaryk, R. P., Brunet, P. C. J., and Koeppe, J. K. (1974). On the metabolism of *N*-acetyl dopamine in *Periplaneta americana. J. Insect Physiol.* **20,** 911–923.

Bossa, F., Martini, F., Barra, D., Borri Voltattorni, C., Minelli, A., and Turano, C. (1977). The chymotryptic phosphopyridoxyl peptide of dopa decarboxylase from pig kidney. *Biochem. Biophys. Res. Commun.* **78,** 177–184.

Bray, S. J., and Hirsh, J. (1986). The *Drosophila virilis* dopa decarboxylase gene is developmentally regulated when integrated into *Drosophila melanogaster. EMBO J.* **5,** 2305–2311.

Brehme, K. (1941). The effect of adult body color mutations upon the larvae of *Drosophila melanogaster. Proc. Natl. Acad. Sci. U.S.A.* **27,** 254–261.

Brown, C. S., and Nestler, C. (1985). Catecholamines and indolalkylamines. *In* "Comparative Insect Physiology, Biochemistry and Pharmacology" (G. A. Kerkut and L. I. Gilbert, eds.), Vol. 11, pp. 435–496. Pergamon, New York.

Brown, G. M., Krivi, G. G., Fan, C. L., and Unnasch, T. R. (1979). The biosynthesis of pteridines in *Drosophila melanogaster*. *In* "Chemistry and Biology of Pteridines" (R. L. Kislick and G. M. Brown, eds.), pp. 81–86. Elsevier, New York.

Brunet, P. C. J. (1980). The metabolism of the aromatic amino acids concerned in the cross-linking of insect cuticle. *Insect Biochem.* **10**, 467–500.

Brunet, P. C. J., and Kent, P. W. (1955). Observations on the mechanisms of a tanning reaction in *Periplaneta* and *Blatta*. *Proc. R. Soc. London Ser. B* **144**, 259–274.

Bublitz, C. (1969). A direct assay for liver phenylalanine hydroxylase. *Biochim. Biophys. Acta* **191**, 249.

Budnik, V., Martin-Morris, L., and White, K. (1986). Perturbed pattern of catecholamine-containing neurons in mutant *Drosophila* deficient in the enzyme dopa decarboxylase. *J. Neurosci.* **6**, 3682–3691.

Burnet, B., and Wilson, R. (1980). Pattern mosaicism for behavior controlled by the *yellow* locus in *Drosophila melanogaster*. *Genet. Res.* **36**, 235–247.

Burnet, B., Connolly, K., and Harrison, B. (1973). Phenocopies of pigmentary and behavioral effects of the *yellow* mutant in *Drosophila* induced by α-dimethyltyrosine. *Science* **181**, 1059–1060.

Campuzano, S., Carramolino, L., Cabrera, C. V., Ruiz-Gomez, M., Villares, R., Boronat, A., and Modolell, J. (1985). Molecular genetics of the *achaete-scute* gene complex of *D. melanogaster*. *Cell* **40**, 327–338.

Chen, P. S. (1975). Ein neues Tyrosinderivat bei *Drosophila busckii*. *Rev. Suisse Zool.* **82**, 673–675.

Chen, P. S., Hanimann, F., and Roeder-Guanella, C. (1966). Phosphatester der Aminosäuren Serin und Tyrosin sowie des Äthanolamins in *Drosophila melanogaster*. *Rev. Suisse Zool.* **73**, 219–228.

Chen, P. S., Mitchell, H. K., and Neuweg, M. (1978). Tyrosine glucoside in *Drosophila busckii*. *Insect Biochem.* **8**, 279–286.

Chia, W., Howes, G., Martin, M., Meng, Y. B., Moses, K., and Tsubota, S. (1986). Molecular analysis of the *yellow* locus of *Drosophila*. *EMBO J.* **5**, 3597–3605.

Chihara, C. J., and Kimbrell, D. A. (1986). The cuticle proteins of *Drosophila melanogaster:* Genetic localization of a second cluster of third-instar genes. *Genetics* **114**, 393–404.

Chikaraishi, D. M., Brilliant, M. H., and Lewis, E. J. (1983). Cloning and characterization of rat-brain-specific transcripts: Rare brain-specific transcripts and tyrosine hydroxylase. *Cold Spring Harbor Symp. Quant. Biol.* **48**, 309–318.

Clark, W. C., Pass, P. S., Venkataraman, B., and Hodgetts, R. B. (1978). Dopa decarboxylase from *Drosophila melanogaster*, purification, characterization and an analysis of mutants. *Mol. Gen. Genet.* **162**, 287–297.

Clark, W. C., Doctor, J., Fristrom, J. W., and Hodgetts, R. B. (1986). Differential response of the dopa decarboxylase gene to 20-OH-ecdysone in *Drosophila melanogaster*. *Dev. Biol.* **114**, 141–150.

Connor, T. W., and Rawls, J. M., Jr. (1982). Analysis of the phenotypes exhibited by rudimentary-like mutants of *Drosophila melanogaster*. *Biochem. Genet.* **20**, 607–619.

Crossley, S., and Zuill, E. (1970). Courtship behavior of some *Drosophila melanogaster* mutants. *Nature (London)* **225**, 1064–1065.

Craymer, L. (1984). Report of L. Craymer. *Dros. Inf. Serv.* **60**, 234–236.

Dewhurst, S. A., Croker, S. G., Ikeda K., and McCaman, R. E. (1972). Metabolism of biogenic amines in *Drosophila* nervous tissue. *Comp. Biochem. Physiol.* **43B**, 975–981.

Dohke, K. (1973). Studies on prephenoloxidase-activating enzyme from cuticle of the silkworm, *Bombyx mori*. II. Purification and characterization of the enzyme. *Arch. Biochem. Biophys.* **157**, 210–221.

Dudai, Y. (1985). Genes, enzymes and learning in *Drosophila*. *Trends Neurosci.* **8**, 18–21.

Dürrwächter, G. (1957). Untersuchungen über Phototaxis and Geotaxis einiger *Drosophila*-Mutanten nach Aufzucht in verschiedenen Lichtbedingungen. *Z. Tierpsychol.* **14**, 1–28.

Estelle, M. A., and Hodgetts, R. B. (1984a). Genetic elements near the structural gene modulate the level of dopa decarboxylase during *Drosophila* development. *Mol. Gen. Genet.* **195**, 434–441.

Estelle, M. A., and Hodgetts, R. B. (1984b). Insertion polymorphisms may cause stage specific variation in mRNA levels for dopa decarboxylase in *Drosophila*. *Mol. Gen. Genet.* **195**, 442–451.

Eveleth, D. D., and Marsh, J. L. (1986a). Evidence for evolutionary duplication of genes in the dopa decarboxylase region of *Drosophila*. *Genetics* **114**, 469–483.

Eveleth, D. D., Jr., and Marsh, J. L. (1986b). Sequence and characterization of the *Cc* gene, a member of the dopa decarboxylase gene cluster of *Drosophila*. *Nucleic Acids Res.* **14**, 6169–6183.

Eveleth, D. D., Gietz, R. D., Spencer, C. A., Nargang, F. E., Hodgetts, R. B., and Marsh, J. L. (1986). Sequence and structure of the dopa decarboxylase gene of *Drosophila:* Evidence for novel RNA splicing variants. *EMBO J.* **5**, 2663–2672.

Fan, C. L., and Brown, G. M. (1979). Partial purification and some properties of biopterin synthetase and dihydrobiopterin oxidase from *Drosophila melanogaster*. *Biochem. Genet.* **17**, 351–369.

Fristrom, J. W., Alexander, S., Brown, E., Doctor, J., Fechtel, K., Fristrom, D., Kimbrell, D., King, D., and Wolfgang, W. (1986). Ecdysone regulation of cuticle protein gene expression in *Drosophila*. *Arch. Insect Biochem. Physiol. Suppl.* **1**, 119–132.

Gailey, D. A., Bordne, D. L., Vallés, A. M., Hall, J. C., and White, K. (1987). Construction of an unstable *Ring X* chromosome bearing the autosomal gene *Dopa decarboxylase* in *Drosophila melanogaster* and analysis of *Ddc* mosaics. *Genetics* **115**, 305–311.

Garcia-Bellido, A. (1977). Homeotic and atavic mutations in insects. *Am. Zool.* **17**, 613–629.

Garcia-Bellido, A., and Santamaria, P. (1978). Developmental analysis of the achaete-scute system of *Drosophila melanogaster*. *Genetics* **88**, 469–486.

Gateff, E. (1978). Malignant neoplasms of genetic origin in *Drosophila melanogaster*. *Science* **200**, 1448–1459.

Geer, B. W. (1966). Utilization of D-amino acids for growth by *Drosophila melanogaster* larvae. *J. Nutr.* **90**, 31–39.

Geiger, H. R., and Mitchell, H. K. (1966). Salivary gland function in phenol oxidase production in *Drosophila melanogaster*. *J. Insect Physiol.* **12**, 749–754.

Geltosky, J. E., and Mitchell, H. K. (1980). Developmental regulation of phenylalanine hydroxylase activity in *Drosophila melanogaster*. *Biochem. Genet.* **18**, 781–791.

Geyer, P. K., Spana, C., and Corces, V. G. (1986). On the molecular mechanism of gypsy-induced mutations at the *yellow* locus of *Drosophila melanogaster*. *EMBO J.* **5**, 2657–2662.

Gietz, R. D., and Hodgetts, R. B. (1985). An analysis of dopa decarboxylase expression during embryogenesis in *Drosophila melanogaster*. *Dev. Biol.* **107**, 142–155.

Gilbert, D. M. (1984). The isolation and characterization of a gene cluster flanking the *Drosophila* dopa decarboxylase gene. Ph.D. thesis, Harvard University, Cambridge.

Gilbert, D., and Hirsh, J. (1981). The dopa decarboxylase gene locus of *Drosophila melanogaster:* Orientation of the gene and preliminary mapping of genetic markers. *In* "Developmental Biology of Purified Genes" (D. D. Brown, ed.), pp. 11–16. Academic Press, New York.

Gilbert, D., Hirsh, J., and Wright, T. R. F. (1984). Molecular mapping of a gene cluster flanking the *Drosophila* dopa decarboxylase gene. *Genetics* **106**, 679–694.

Graubard, M. A. (1933). Tyrosinase in mutants of *Drosophila melanogaster. J. Genet.* **27**, 119–218.

Grima, B., Lamouroux, A., Blanot, F., Biguet, N. F., and Mallet, J. (1985). Complete coding sequence of rat tyrosine hydroxylase mRNA. *Proc. Natl. Acad. Sci. U.S.A.* **82,** 617–621.

Hackman, R. H., and Goldberg, M. (1967). The *o*-diphenoloxidases of fly larvae. *J. Insect Physiol.* **13,** 531–544.

Hadler, N. M. (1964). Genetic influence on phototaxis in *Drosophila melanogaster. Biol. Bull.* **126,** 264–273.

Hadorn, E. (1956). Patterns of biochemical and developmental pleiotropy. *Cold Spring Harbor Symp. Quant. Biol.* **21,** 363–373.

Hall, J. C. (1982). Genetics of the nervous system in *Drosophila. Q. Rev. Biophys.* **15,** 223–479.

Hall, J. C. (1985). Genetic analysis of behavior in insects. *In* "Comparative Insect Physiology, Biochemistry and Pharmacology" (G. A. Kerkut and L. I. Gilbert, eds.), Vol. 9, pp. 287–373. Pergamon, New York.

Hankins, G. R., and Sherald, A. F. (1981). Hydropyrimidine hydrase in *D. melanogaster. Dros. Inf. Serv.* **56,** 57–58.

Hanratty, W. P., and Ryerse, J. S. (1981). A genetic melanotic neoplasm of *Drosophila melanogaster. Dev. Biol.* **83,** 238–249.

Harper, R. A., and Armstrong, F. B. (1972). Alkaline phosphatase of *Drosophila melanogaster.* I. Partial purification and characterization. *Biochem. Genet.* **6,** 75–82.

Harper, R. A., and Armstrong, F. B. (1973). Alkaline phosphatase of *Drosophila melanogaster.* II. Biochemical comparison among four allelic forms. *Biochem. Genet.* **10,** 29–38.

Harper, R. A., and Armstrong, F. B. (1974). Alkaline phosphatase of *Drosophila melanogaster.* III. Tyrosine-*O*-phosphate as substrate. *Biochem. Genet.* **11,** 177–180.

Henderson, A. S., and Glassman, E. (1969). A possible storage form for tyrosinase substrates in *Drosophila melanogaster. J. Insect Physiol.* **15,** 2345–2355.

Hepburn, H. R. (1985). Structure of the integument. *In* "Comparative Insect Physiology, Biochemistry and Pharmacology" (G. A. Kerkut and L. I. Gilbert, eds.), Vol. 3, pp. 1–58. Pergamon, New York.

Hillman, R., and Lesnick, L. H. (1970). Cuticle formation in the embryo of *Drosophila melanogaster. J. Morphol.* **131,** 383–396.

Hirsh, J. (1986). The *Drosophila melanogaster* dopa decarboxylase gene: Progress in understanding the *in vivo* regulation of a higher eukaryotic gene. *In* "Molecular and Developmental Biology" (L. Bogorad, ed.), pp. 103–144. Liss, New York.

Hirsh, J., and Davidson, N. (1981). Isolation and characterization of the dopa decarboxylase gene of *Drosophila melanogaster. Mol. Cell. Biol.* **1,** 475–485.

Hirsh, J., Morgan, B. A., and Scholnick, S. B. (1986). Delimiting regulatory sequences of the *Drosophila melanogaster Ddc* gene. *Mol. Cell. Biol.* **6,** 4548–4557.

Hodgetts, R. B. (1972). Biochemical characterization of mutants affecting the metabolism of β-alanine in *Drosophila. J. Insect Physiol.* **18,** 937–947.
Hodgetts, R. B. (1975). The response of dopa decarboxylase activity to variations in gene dosage in *Drosophila:* A possible location of the structural gene. *Genetics* **79,** 45–54.
Hodgetts, R. B., and Choi, A. (1974). β-alanine and cuticle maturation in *Drosophila. Nature* (*London*) **252,** 710–711.
Hodgetts, R. B., and Konopka, R. J. (1973). Tyrosine and catecholamine metabolism in a wild-type *Drosophila melanogaster* and a mutant, *ebony. J. Insect Physiol.* **19,** 1211–1220.
Hodgetts, R. B., Clark, W. C., Eveleth, D. D., Jr., Gietz, R. D., Spencer, C. A., and Marsh, J. L. (1986a). Hormonal aspects of the regulation of dopa decarboxylase in *Drosophila melanogaster. Prog. Dev. Biol. A,* 221–234.
Hodgetts, R. B., Clark, W. C., Gietz, R. D., Sage, B. A., and O'Connor, J. D. (1986b). Stage-specific mechanisms regulate the expression of the dopa decarboxylase gene during *Drosophila* development. *Arch. Insect Biochem. Physiol. Suppl.* **1,** 97–104.
Hopkins, T. L., Morgan, T. D., Aso, Y., and Kramer, K. J. (1982). *N*-β-alanyldopamine: Major role in insect cuticle tanning. *Science* **217,** 364–366.
Hopkins, T. L., Morgan, T. D., and Kramer, K. J. (1984). Catecholamines in haemolymph and cuticle during larval, pupal and adult development of *Manduca sexta* (L.). *Insect Biochem.* **14,** 533–540.
Horowitz, N. H., and Fling, N. (1955). The autocatalytic production of tyrosinase in extracts of *Drosophila melanogaster. In* "Amino Acid Metabolism" (W.D. McElroy and B. Glass, eds.), pp. 207–218. Johns Hopkins Press, Baltimore.
Hotta, Y., and Benzer, S. (1969). Abnormal electroretinograms in visual mutants of *Drosophila. Nature* (*London*) **222,** 354–356.
Howard, B. D., Merriam, J. R., and Meshul, C. (1975). Effects of neurotropic drugs on *Drosophila melanogaster. J. Insect Physiol.* **21,** 1397–1405.
Huntley, M. D. (1978). Genetics of catecholamine metabolizing enzymes in *Drosophila melanogaster.* Ph.D. dissertation, University of Virginia.
Ishaaya, I. (1972). Studies on the haemolymph and cuticular phenoloxidase in *Spodoptera littoralis* larvae. *Insect Biochem.* **2,** 409–419.
Jackson, F. R., Bargiello, T. A., Yun, S.-H., and Young, M. W. (1986). Product of *per* locus of *Drosophila* shares homology with proteoglycans. *Nature* (*London*) **320,** 185–188.
Jacobs, M. E. (1974). β-Alanine and adaptation in *Drosophila. J. Insect Physiol.* **20,** 859–866.
Jacobs, M. E. (1978). β-Alanine tanning of *Drosophila* cuticles and chitin. *Insect Biochem.* **8,** 37–41.
Jacobs, M. E. (1980). Influence of β-alanine on ultrastructure, tanning, and melanization of *Drosophila melanogaster. Biochem. Genet.* **18,** 65–76.
Jacobs, M. E. (1985). Role of β-alanine in cuticular tanning, sclerotization, and temperature regulation in *Drosophila melanogaster. J. Insect Physiol.* **31,** 509–515.
Jacobs, M. E., and Brubaker, K. K. (1963). β-alanine utilization of ebony and non-ebony *Drosophila melanogaster. Science* **139,** 1282–1283.
Jarry, B., and Falk, D. (1974). Functional diversity within the *rudimentary* locus of *Drosophila melanogaster. Mol. Gen. Genet.* **135,** 113–122.
Jimenez, F., and Campos-Ortega, J. A. (1979). A region of the *Drosophila* genome necessary for CNS development. *Nature* (*London*) **282,** 310–312.

Joh, T. H., Baetge, E. E., Ross, M. E., and Reis, D. J. (1983). Evidence for the existence of homologous gene coding regions for the catecholamine biosynthetic enzymes. *Cold Spring Harbor Symp. Quant. Biol.* **48**, 327–335.

Joh, T. H., Baetge, E. E., and Reis, D. J. (1984). Molecular biology of catecholamine neurons: Similar gene hypothesis. *Hypertension* **6**, II-1–II-6.

Johnson, F. M. (1966). *Drosophila melanogaster:* Inheritance of a deficiency of alkaline phosphatase in larvae. *Science* **152**, 361–362.

Jürgens, G., Wieschaus, E., Nüsslein-Volhard, C., and Kluding, H. (1984). Mutations affecting the pattern of larval cuticle in *Drosophila melanogaster.* II. Zygotic loci on the third chromosome. *Wilhelm Roux's Arch. Dev. Biol.* **193**, 283–295.

Karlson, P., and Sekeris, C. E. (1962). *N*-Acetyldopamine as sclerotizing agent of the insect cuticle. *Nature (London)* **195**, 183–184.

Karlson, P., and Sekeris, C. E. (1964). Biochemistry of insect metamorphosis. *In* "Comparative Biochemistry" (M. Florkin and H. S. Mason, eds.), Vol. 6, pp. 221–243. Academic Press, New York.

Karlson, P., Sekeris, C. E., and Sekeri, K. E. (1962). Zum Tyrosinstoffwechsel der Insekten. VI. Identifizierung von *N*-acetyl-3,4-dihydroxy-β-phenäthylamin (*N*-acetyl-dopamin) als Tyrosinmetabolit. *Hoppe-Seyler's Z. Physiol. Chem.* **327**, 86–94.

Kaufman, S. (1971). The phenylalanine hydroxylating system from mammalian liver. *Adv. Enzymol.* **35**, 245–319.

Kaznowski, C. E., Schneiderman, H. A., and Bryant, P. J. (1985). Cuticle secretion during larval growth in *Drosophila melanogaster. J. Insect Physiol.* **31**, 801–813.

Kaznowski, C. E., Schneiderman, H. A., and Bryant, P. J. (1986). The incorporation of precursors into *Drosophila* larval cuticle. *J. Insect Physiol.* **32**, 133–142.

Kennedy, M. B. (1978). Products of biogenic amine metabolism in the lobster: Sulfate conjugates. *J. Neurochem.* **30**, 315–320.

Knopka, R. J. (1972). Abnormal concentrations of dopamine in a *Drosophila* mutant. *Nature (London)* **239**, 281–282.

Knopka, R. J., and Benzer, S. (1971). Clock mutants of *Drosophila melanogaster. Proc. Natl. Acad. Sci. U.S.A.* **68**, 2112–2116.

Koeppe, J. K., and Gilbert, L. I. (1974). Metabolism and protein transport of a possible pupal cuticle tanning agent in *Manduca sexta. J. Insect Physiol.* **20**, 981–992.

Koeppe, J. K., and Mills, R. R. (1974). Possible involvement of 3,4-dihydroxyphenylacetic acid as a sclerotization agent in the American cockroach. *J. Insect Physiol.* **20**, 1603–1609.

Koeppe, J. K., and Mills, R. R. (1975). Metabolism of noradrenalin and dopamine during ecdysis by the American cockroach. *Insect Biochem.* **5**, 399–408.

Körner, A. M., and Pawelek, J. (1982). Mammalian tyrosinase catalyzes three reactions in the biosynthesis of melanin. *Science* **217**, 1163–1165.

Kramer, K. J., Nuntnarumit, C., Aso, Y., Hawley, M. D., and Hopkins, T. L. (1983). Electrochemical and enzymatic oxidation of catecholamines involved in sclerotization and melanization of insect cuticle. *Insect Biochem.* **13**, 475–479.

Kramer, K. J., Dziadik-Turner, C., and Koga, D. (1985). Chitin metabolism in insects. *In* "Comparative Insect Physiology, Biochemistry and Pharmacology" (G. A. Kerkut and L. I. Gilbert, eds.), Vol. 3, pp. 75–116. Pergamon, New York.

Kraminsky, G. P., Clark, W. C., Estelle, M. A., Gietz, R. B., Sage, B. A., O'Connor, J. D., and Hodgetts, R. B. (1980). Induction of translatable mRNA for dopa decarboxylase in *Drosophila:* An early response to ecdysterone. *Proc. Natl. Acad. Sci. U.S.A.* **77**, 4175–4179.

Lai-Fook, J. (1966). The repair of wounds in the integument of insects. *J. Insect Physiol.* **12**, 195–226.

Lewis, H. W., and Lewis, H. S. (1961). Genetic control of dopa oxidase activity in *Drosophila melanogaster*. II. Regulating mechanism and inter- and intra-strain heterogeneity. *Proc. Natl. Acad. Sci. U.S.A.* **47**, 78–86.

Lewis, H. W., and Lewis, H. S. (1963). Genetic regulation of dopa oxidase activity in *Drosophila*. *Ann. N.Y. Acad. Sci.* **100**, 827–839.

Lindsley, D. L., and Grell, E. H. (1968). Genetic variations of *Drosophila*. Inst. of Wash. Publ. No. 627.

Lindsley, D., and Zimm, G. (1985). The genome of *Drosophila melanogaster*. Part 1: Genes A-K. *Dros. Inf. Serv.* **62**, 1–227.

Lindsley, D. L., Sandler, L., Baker, B. S., Carpenter, A. T. C., Dennell, R. E., Hall, J. C., Jacobs, P. A., Miklos, G. L. C., Davis, B. K., Gethmann, R. C., Hardy, R. W., Hessler, A., Miller, S. M., Nozawa, H., Parry, D. M., and Gould-Somero, M. (1972). Segmental aneuploidy and the genetic gross structure of the *Drosophila* genome. *Genetics* **71**, 157–184.

Lipke, H., Sugumaran, M., and Henzel, W. (1983). Mechanisms of sclerotization in Dipterans. *Adv. Insect Physiol.* **17**, 1–84.

Livingstone, M. S., and Tempel, B. L. (1983). Genetic dissection of monoamine neurotransmitter synthesis in *Drosophila*. *Nature (London)* **303**, 67–70.

Lunan, K. D., and Mitchell, H. K. (1969). The metabolism of tyrosine-*O*-phosphate in *Drosophila*. *Arch. Biochem. Biophys.* **132**, 450–456.

McCaman, M. W., McCaman, R. E., and Lees, G. J. (1972). Liquid cation exchange—A basis for sensitive radiometric assays for aromatic amino acid decarboxylases. *Anal. Biochem.* **45**, 242–252.

McEwen, R. S. (1918). The reaction to light and to gravity in *Drosophila* and its mutants. *J. Exp. Zool.* **25**, 49–106.

MacIntyre, R. J. (1966). Genetics of acid phosphatase in *Drosophila melanogaster* and *Drosophila simulans*. *Genetics* **53**, 461–474.

Mackay, W. J., and O'Donnell, J. M. (1983). A genetic analysis of the pteridine biosynthetic enzyme, guanosine triphosphate cyclohydrolase, in *Drosophila melanogaster*. *Genetics* **105**, 35–53.

Mackay, W. J., Reynolds, E. R., and O'Donnell, J. M. (1985). Tissue-specific and complex complementation patterns in the *Punch* locus of *Drosophila melanogaster*. *Genetics* **111**, 885–904.

Maranda, B., and Hodgetts, R. B. (1977). A characterization of dopamine acetyltransferase in *Drosophila melanogaster*. *Insect Biochem.* **7**, 33–43.

Marsh, J. L., and Wright, T. R. F. (1980). Developmental relationships between dopa decarboxylase, dopamine acetyltransferase and ecdysone in *Drosophila*. *Dev. Biol.* **80**, 379–387.

Marsh, J. L., and Wright, T. R. F. (1986). Evidence for regulatory variants of the dopa decarboxylase and alpha-methyldopa hypersensitive loci in *Drosophila*. *Genetics* **112**, 249–265.

Marsh, J. L., Gibbs, P. D. L., and Timmons, P. M. (1985). Developmental control of transduced dopa decarboxylase genes in *D. melanogaster*. *Mol. Gen. Genet.* **198**, 393–403.

Marsh, J. L., Erfle, M. P., and Leeds, C. A. (1986). Molecular localization, developmental expression and nucleotide sequence of the alpha methyldopa hypersensitive gene of *Drosophila*. *Genetics* **114**, 453–467.

Mason, H. S. (1955). Comparative biochemistry of the phenolase complex. *Adv. Enzymol.* **16**, 105–184.
Mindrinos, M. N., Petri, W. H., Galanopoulos, V. K., Lombard, M. F., and Margaritis, L. H. (1980). Crosslinking of the *Drosophila* chorion involves a peroxidase. *Wilhelm Roux's Arch. Dev. Biol.* **189**, 187–196.
Mitchell, H. K. (1966). Phenol oxidases and *Drosophila* development. *J. Insect Physiol.* **12**, 755–766.
Mitchell, H. K., and Lipps, L. S. (1978). Heatshock and phenocopy induction in *Drosophila. Cell* **15**, 907–918.
Mitchell, H. K., and Lunan, K. D. (1964). Tyrosine-*O*-phosphate in *Drosophila. Arch. Biochem. Biophys.* **106**, 219–222.
Mitchell, H. K., and Simmons, J. R. (1961). Amino acids and derivatives in *Drosophila. In* "Amino Acid Pools" (J. T. Holden, ed.), pp. 136–146. Elsevier, Amsterdam.
Mitchell, H. K., and Weber, U. M. (1965). *Drosophila* phenol oxidases. *Science* **148**, 964–965.
Mitchell, H. K., Chen, P. S., and Hadorn, E. (1960). Tyrosine phosphate on paper chromatograms of *Drosophila melanogaster. Experientia* **14**, 410.
Mitchell, H. K., Weber, U. M., and Schaar, G. (1967). Phenol oxidase characteristics in mutants of *Drosophila melanogaster. Genetics* **57**, 357–368.
Morgan, B. A., Johnson, W. A., and Hirsh, J. (1986). Regulated splicing produces different forms of dopa decarboxylase in the central nervous system and hypoderm of *Drosophila melanogaster. EMBO J.* **5**, 3335–3342.
Moritz, T. H., Edstrom, J. E., and Pongs, O. (1984). Cloning of a gene localized and expressed at the ecdysteroid regulated puff 74EF in salivary glands of *Drosophila* larvae. *EMBO J.* **3**, 289–295.
Murdock, L. L., and Omar, D. (1981). *N*-Acetyldopamine in insect nervous tissue. *Insect Biochem.* **11**, 161–166.
Muskavitch, M. A. T., and Hogness, D. S. (1982). An expandable gene that encodes a *Drosophila* glue protein is not expressed in variants lacking remote upstream sequences. *Cell* **29**, 1041–1051.
Nagatsu, T., Levitt, M., and Udenfriend, S. (1964). Tyrosine hydroxylase: The initial step in norepinephrine biosynthesis. *J. Biol. Chem.* **239**, 2910.
Nappi, A. J., and Streams, F. A. (1969). Haemocytic reactions of *Drosophila melanogaster* to the parasites *Pseudocoila millipes* and *P. bochei. J. Insect Physiol.* **15**, 1551–1566.
Nash, W. G., and Yarkin, R. J. (1974). Genetic regulation and pattern formation: A study of the *yellow* locus in *Drosophila melanogaster. Genet. Res.* **24**, 19–26.
Neckameyer, W. S., Chikaraishi, D., and Quinn, W. G. (1986). Molecular cloning of a tyrosine hydroxylase gene in *Drosophila melanogaster. Soc. Neurosci. Abstr.* **12**, 949.
Norby, S. (1973). The biochemical genetics of *rudimentary* mutants in *Drosophila melanogaster.* I. Aspartate carbamoyltransferase levels in complementing and non-complementing strains. *Hereditas* **73**, 11–16.
Nüsslein-Volhard, C., Wieschaus, E., and Kluding, H. (1984). Mutations affecting the pattern of the larval cuticle in *Drosophila melanogaster.* I. Zygotic loci on the second chromosome. *Wilhelm Roux's Arch Dev. Biol.* **193**, 267–282.
Ohnishi, E. (1953). Tyrosinase activity during puparium formation in *Drosphila melanogaster. Jpn. J. Zool.* **11**, 69–74.

Ohnishi, E. (1954a). Tyrosinase in *Drosophila virilis*. *Annot. Zool. Jpn.* **27**, 33–39.
Ohnishi, E. (1954b). Activation of tyrosinase in *Drosophila virilis*. *Annot. Zool. Jpn.* **27**, 188–193.
Okubo, S. (1958). Occurrence of *N*-acetylhydroxytyramine glucoside in *Drosophila melanogaster*. *Med. J. Osaka Univ.* **9**, 327–337.
Pak, W. L., Grossfield, J., and White, N. V. (1969). Nonphototactic mutants in a study of vision of *Drosophila*. *Nature (London)* **222**, 351–354.
Parkhurst, S. M., and Corces, V. G. (1986). Interactions among the gypsy transposable element and the *yellow* and the *Suppressor of Hairy-Wing loci in Drosophila melanogaster. Mol. Cell. Biol.* **6**, *47–53.*
Pawelek, J., Korner, A., Bergstrom, A., and Bologna, J. (1980). New regulators of melanin biosynthesis and the autodestruction of melanoma cells. *Nature (London)* **286**, 617–619.
Pedersen, M. B. (1982). Characterization of an X-linked semi-dominant suppressor of *black,Su(b)* (1-55.5) in *Drosophila melanogaster*. *Carlsberg Res. Commun.* **47**, 391–400.
Peeples, E. E., Barnett, D. R., and Oliver, C. P. (1968). Phenol oxidase of a *lozenge* mutant of *Drosophila*. *Science* **159**, 550–552.
Peeples, E. E., Geisler, A., Whitcraft, C. J., and Oliver, C. P. (1969a). Activity of phenol oxidase at the puparium formation stage in development of nineteen *lozenge* mutants of *Drosophila melanogaster*. *Biochem. Genet.* **3**, 563–569.
Peeples, E. E., Geisler, A., Whitcraft, C. J., and Oliver, C. P. (1969b). Comparative studies of phenol oxidase activity during pupal development of three lozenge mutants (lz^S, lz, lz^k) of *Drosophila melanogaster*. *Genetics* **62**, 161–170.
Pentz, E. S., and Wright, T. R. F. (1986). A diphenol oxidase gene is part of a cluster of genes involved in catecholamine metabolism and sclerotization in *Drosophila*. II. Molecular localization of the *Dox-A2* coding region. *Genetics* **112**, 843–859.
Pentz, E. S., Black, B. C., and Wright, T. R. F. (1986). A diphenol oxidase gene is part of a cluster of genes involved in catecholamine metabolism and sclerotization in *Drosophila*. I. Identification of the biochemical defect in *Dox-A2* [*l(2)37Bf*] mutants. *Genetics* **112**, 823–841.
Pentz, E. S., Black, B. C., and Wright, T. R. F. (1987). Phenol oxidase in *Drosophila:* Effects of *quicksilver* and *tyrosinase-1*. In preparation.
Phillips, J. P., and Forrest, H. S. (1980). Ommochromes and pteridines. *In* "The Genetics and Biology of *Drosophila*" (M. Ashburner and T. R. F. Wright, eds.), Vol. 2d, pp. 541–623. Academic Press, London.
Pichon, Y., and Manaranche, R. (1985). Biochemistry of the nervous system. *In* "Comparative Insect Physiology, Biochemistry and Pharmacology" (G. A. Kerkut and L. I. Gilbert, eds.), Vol. 10, pp. 417–450. Pergamon, New York.
Poulson, D. F. (1950). Histogenesis, organogenesis, differentiation in the embryo of *Drosophila melanogaster* Meigen. *In* "Biology of *Drosophila*" (M. Demerec, ed.), pp. 168–274. Wiley, New York.
Pryor, M. G. M. (1940a). On the hardening of the ootheca of *Blatta orientalis*. *Proc. R. Soc. London Ser. B* **128**, 378–393.
Pryor, M. G. M. (1940b). On the hardening of the cuticle of insects. *Proc. R. Soc. London Ser. B* **128**, 393–407.
Rawls, J. M. (1980). Identification of a small region that encodes orotate phosphoribosyltransferase and orotidylate decarboxylase in *Drosophila melanogaster*. *Biochem. Genet.* **18**, 43–49.

Rawls, J., and Fristrom, J. (1975). A complex genetic locus that controls the three first steps of pyrimidine biosynthesis in *Drosophila*. *Nature* (*London*) **255**, 738–740.

Rawls, J. M., Chambers, C. L., and Cohen, W. S. (1981). A small genetic region that controls dihydroorotate dehydrogenase in *Drosophila melanogaster*. *Biochem. Genet.* **19**, 115–127.

Reddy, P., Zehring, W. A., Wheeler, D. A., Pirrotta, V., Hadfield, C., Hall, J. C., and Rosbash, M. (1984). Molecular analysis of the *period* locus in *Drosophila melanogaster* and identification of a transcript involved in biological rhythms. *Cell* **38**, 701–710.

Reddy, P., Jacquier, A. C., Abovich, N., Petersen, G., and Rosbash, M. (1986). The *period* clock locus of *D. melanogaster* codes for a proteoglycan. *Cell* **46**, 53–61.

Rendel, J. (1951). Mating of *ebony*, *vestigial* and wild type *D. melanogaster* in light and dark. *Evolution* **5**, 226–230.

Reynolds, E. R., and O'Donnell, J. M. (1987). An analysis of the embryonic defects in *Punch* mutants of *Drosophila melanogaster*. *Dev. Biol.* (in press).

Ripoll, P., and Garcia-Bellido, A. (1979). Viability of homozygous deficiencies in somatic cells of *Drosophila melanogaster*. *Genetics* **91**, 443–453.

Rizki, T. M. (1957). Alterations in the hemocyte population of *Drosophila melanogaster*. *J. Morphol.* **100**, 437–458.

Rizki, T. M., and Rizki, R. M. (1959). Functional significance of the crystal cell in the larva of *Drosophila melanogaster*. *J. Biophys. Biochem. Cytol.* **5**, 235–240.

Rizki, T. M., and Rizki, R. M. (1981). Alleles of *lz* as suppressors of the *Bc*-phene in *Drosophila melanogaster*. *Genetics* **97**, s90.

Rizki, T. M., and Rizki, R. M. (1984). The cellular defense system of *Drosophila melanogaster*. *In* "Insect Ultrastructure" (R. C. King and H. Akai, eds.), Vol. 2, pp. 579–604. Plenum, New York.

Rizki, T. M., and Rizki, R. M. (1986). Surface changes on hemocytes during encapsulation in *Drosophila melanogaster* Meigen. *In* "Hemocytic and Humoral Immunity in Arthropods" (A. P. Gupta, ed.), Chap. 6, pp. 157–190. Wiley, New York.

Rizki, T. M., Rizki. R. M., and Grell, E. H. (1980). A mutant affecting the crystal cells in *Drosophila melanogaster*. *Wilhelm Roux's Arch. Dev. Biol.* **188**, 91–99.

Rizki, T. M., Rizki, R. M., and Bellotti, R. A. (1985). Genetics of a *Drosophila* phenoloxidase. *Mol. Gen. Genet.* **201**, 7–13.

Ross, R. H., and Monroe, R. E. (1972). β-Alanine metabolism in the housefly, *Musca domestica:* Studies on anabolism in the early puparium. *J. Insect Physiol.* **18**, 1593–1597.

Sawicki, J. A., and MacIntyre, R. J. (1978). Localization at the ultrastructural level of maternally derived enzyme and determination of the time of paternal gene expression for acid phosphatase-1 in *Drosophila melanogaster*. *Dev. Biol.* **63**, 47–58.

Schneiderman, H. (1967). Alkaline phosphatase relationships in *Drosophila*. *Nature* (*London*) **216**, 604–605.

Schneiderman, H., Young, W. J., and Childs, B. (1966). Patterns of alkaline phosphatase in developing *Drosophila*. *Science* **151**, 461–463.

Scholnick, S. B., Morgan, B. A., and Hirsh, J. (1983). The cloned dopa decarboxylase gene is developmentally regulated when reintegrated into the *Drosophila* genome. *Cell* **34**, 37–45.

Scholnick, S. B., Bray, S. J., Morgan, B. A., McCormick, C. A., and Hirsh, J. (1986). CNS and hypoderm regulatory elements of the *Drosophila melanogaster* dopa decarboxylase gene. *Science* **234**, 998–1002.

Sekeris, C. E., and Fragoulis, E. G. (1985). Control of DOPA-decarboxylase. *In* "Comparative Insect Physiology, Biochemistry and Pharmacology" (G. A. Kerkut and L. I. Gilbert, eds.), Vol. 8, pp. 147–164. Pergamon, New York.

Sekeris, C. E., and Karlson, P. (1962). Zum Tyrosinstoffwechsel der Insekten. VII. Der katabolische Abbau des Tyrosins und die Biogenese der Sklerotisierungssubstanz, *N*-acetyl-dopamin. *Biochim. Biophys. Acta* **62,** 103–113.

Seligman, M., Friedman, S., and Frankel, G. (1969). Hormonal control of turnover of tyrosine and tyrosine phosphate during tanning of the adult cuticle in the fly, *Sarcophaga bullata. J. Insect Physiol.* **15,** 1085–1101.

Sena, E. P. (1966). Developmental variation of alkaline phosphatases in *Drosophila melanogaster.* M.S. thesis, Cornell University.

Seybold, W. D., Meltzer, P. S., and Mitchell, H. K. (1975). Phenol oxidase activation in *Drosophila:* A cascade of reactions. *Biochem. Genet.* **13,** 85–108.

Sherald, A. F. (1973). Genetic control of dopa decarboxylase in *Drosophila.* Ph.D. thesis, University of Virginia.

Sherald, A. F. (1980). Sclerotization and coloration of the insect cuticle. *Experientia* **36,** 143–146.

Sherald, A. F. (1981). Intergenic suppression of the black mutation of *Drosophila melanogaster. Mol. Gen. Genet.* **183,** 102–106.

Sherald, A. F., and Wright, T. R. F. (1974). The analog inhibitor, α-methyl dopa, as a screening agent for mutants elevating levels of dopa decarboxylase activity in *Drosophila melanogaster. Mol. Gen. Genet.* **133,** 25–36.

Silvert, D. J. (1985). Cuticular proteins during postembryonic development. *In* "Comparative Physiology, Biochemistry and Pharmacology" (G. A. Kerkut and L. I. Gilbert, eds.), Vol. 2, pp. 239–254. Pergamon, New York.

Simmons, J. R., and Mitchell, H. K. (1961). Metabolism of peptides in *Drosophila. In* "Amino Acid Pools" (J. T. Holden, ed.), pp. 147–155. Elsevier, Amsterdam.

Snyder, R. W., and Smith, P. D. (1976). The *suppressor of forked* mutation in *Drosophila melanogaster:* Interactions with the *lozenge* gene. *Biochem. Genet.* **14,** 611–617.

Sparrow, J. C. (1978). Melanotic "tumours." *In* "The Genetics and Biology of Drosophila" (M. Ashburner and T. R. F. Wright, eds.), Vol. 2B, pp. 277–313. Academic Press, London.

Sparrow, J. C., and Wright, T. R. F. (1974). The selection for mutants in *Drosophila melanogaster* hypersensitive to α-methyl dopa, a dopa decarboxylase inhibitor. *Mol. Gen. Genet.* **130,** 127–141.

Spencer, C. A., Stevens, B., O'Connor, J. D., and Hodgetts, R. B. (1983). A novel form of DOPA decarboxylase produced in *Drosophila* cells in response to 20-hydroxyecdysone. *Can. J. Biochem. Cell. Biol.* **61,** 818–825.

Spencer, C., Gietz, R. D., and Hodgetts, R. B. (1986a). Analysis of the transcription unit adjacent to the 3′-end of the DOPA decarboxylase gene in *Drosophila melanogaster. Dev. Biol.* **114,** 260–264.

Spencer, C., Gietz, R. D., and Hodgetts, R. B. (1986b). Overlapping transcription units in the dopa decarboxylase region of *Drosophila. Nature (London)* **322,** 279–281.

Steward, R., McNally, F. J., and Schedl, P. (1984). Isolation of the *dorsal* locus of *Drosophila. Nature (London)* **311,** 262–265.

Sugumaran, M., and Lipke, H. (1983). Quinone methide formation from 4-alkylcatechols: A novel reaction catalyzed by cuticular phenoloxidase. *FEBS Lett.* **155,** 65–68.

Sugumaran, M., Saul, S. J., and Ramesh, N. (1985). Endogenous protease inhibitors prevent undesired activation of prophenolase in insect haemolymph. *Biochem. Biophys. Res. Commun.* **3**, 1124–1129.

Tanouye, M. A., Kamb, C. A., Iverson, L. E., and Salkoff, L. (1986). Genetics and molecular biology of ionic channels in *Drosophila. Annu. Rev. Neurosci.* **9**, 255–276.

Tempel, B. L., Livingstone, M. S., and Quinn, W. G. (1984). Mutations in the dopa decarboxylase gene affect learning in *Drosophila. Proc. Natl. Acad. Sci. U.S.A.* **81**, 3577–3581.

Throckmorton, L. H. (1975). The phylogeny, ecology, and geography of *Drosophila. In* "Handbook of Genetics" (R. C. King, ed.), Vol. 3, pp. 421–469. Plenum, New York.

Tompkins, L. (1984). Genetic analysis of sex appeal in *Drosophila melanogaster. Behav. Genet.* **14**, 411–440.

Tompkins, L. (1986). Genetic control of sexual behavior in *Drosophila melanogaster. Trends Genet.* **2**, 14–17.

Tunnicliff, G., Rick, J. T., and Connolly, K. (1969). Locomotor activity in *Drosophila*—V. A comparative biochemical study of selectivity bred populations. *Comp. Biochem. Physiol.* **29**, 1239–1245.

Turnbull, I. F., and Howells, A. J. (1980). Larvicidal activity of inhibitors of dopa decarboxylase on the Australian sheep blowfly, *Lucilia cuprina. Aust. J. Biol. Sci.* **33**, 169–181.

Turnbull, I. F., Pyliotis, N. A., and Howells, A. J. (1980). The effects of dopa decarboxylase inhibitors on the permeability and ultrastructure of the larval cuticle of the Australian sheep blowfly, *Lucilia cuprina. J. Insect Physiol.* **26**, 525–532.

Vallés, A. M., and White, K. (1986). Development of the neurons committed to serotonin differentiation in mutant *Drosophila* unable to synthesize serotonin. *J. Neurosci.* **6**, 1482–1491.

Waddington, C. H. (1942). Body color genes in Drosophila. *Proc. Zool. Soc. London Ser. A* **III**, 173–180.

Wahl, R. C., Warner, C. K., Finnerty, V., and Rajagopalan, K. V. (1982). *Drosophila melanogaster ma-1* mutants are defective in sulfuration of desulfo-molybdenum hydroxylases. *J. Biol. Chem.* **257**, 3958–3962.

Wallis, B. B., and Fox, A. S. (1968). Genetic and developmental relationships between two alkaline phosphatases in *Drosophila melanogaster. Biochem. Genet.* **2**, 141–158.

Warner, C. K., Grell, E. H., and Jacobson, K. B. (1974). Phenoloxidase activity and the *lozenge* locus of *Drosophila melanogaster. Biochem. Genet.* **11**, 359–365.

Warner, C. K., Grell, E. H., and Jacobson, K. B. (1975). Mechanism of suppression in Drosophila. III. Phenoloxidase activity and the *speck* locus. *Biochem. Genet.* **13**, 353–357.

Weiher, H., Konig, M., and Gruss, P. (1983). Multiple point mutations affecting the Simian Virus 40 enhancer. *Science* **219**, 626–631.

Weisberg, E. P., and O'Donnell, J. M. (1986). Purification and characterization of GTP cyclohydrolase I from *Drosophila melanogaster. J. Biol. Chem.* **261**, 1453–1458.

White, K. (1980). Defective neural development in *Drosophila melanogaster* embryos deficient for the tip of the X chromosome. *Dev. Biol.* **80**, 332–344.

White, K., and Vallés, A. M. (1985). Immunohistochemical and genetic studies of serotonin and neuropeptides in *Drosophila. In* "Molecular Basis of Neural Development" (G. M. Edelman, W. E. Gall, and W. M. Cowan, eds.), pp. 547–564. Neurosciences Research Foundation Series. Wiley, New York.

Wieschaus, E., Nüsslein-Volhard, C., and Jürgens, G. (1984). Mutations affecting the pattern of the larval cuticle in *Drosophila melanogaster.* III. Zygotic loci on the X-chromosome and fourth chromosome. *Wilhelm Roux's Arch. Dev. Biol.* **193,** 296–307.

Wright, T. R. F. (1977). The genetics of dopa decarboxylase and α-methyl dopa sensitivity in *Drosophila melanogaster. Am. Zool.* **17,** 707–721.

Wright, T. R. F. (1987). The genetic and molecular organization of the dense cluster of functionally related, vital genes in the dopa decarboxylase region of the *Drosophila melanogaster* genome. *In* "Eukaryotic Chromosomes: Structure and Function" (W. Hennig, ed.), Vol. 14: Results and Problems in Cell Differentiation, pp. 95–120. Springer-Verlag, Heidelberg.

Wright, T. R. F., and Wright, E. Y. (1978). Developmental effects of dopa decarboxylase deficient mutants, Ddc^{n}, in *Drosophila melanogaster. Int. Congr. Genet., 14th, Moscow* (Abstracts, Part I, Continuation Section 13–20), p. 615.

Wright, T. R. F., Bewley, G. C., and Sherald, A. F. (1976a). The genetics of dopa decarboxylase in *Drosophila melanogaster.* II. Isolation and characterization of dopa decarboxylase deficient mutant and their relationship to the α-methyl dopa hypersensitive mutants. *Genetics* **84,** 287–310.

Wright, T. R. F., Hodgetts, R. B., and Sherald, A. F. (1976b). The genetics of dopa decarboxylase in *Drosophila melanogaster.* I. Isolation and characterization of deficiencies that delete the dopa-decarboxylase-dosage-sensitive region and the α-methyl-dopa-hypersensitive locus. *Genetics* **84,** 267–285.

Wright, T. R. F., Beermann, W., Marsh, J. L., Bishop, C. P., Steward, R., Black, B. C., Tomsett, A. D., and Wright, E. Y. (1981a). The genetics of dopa decarboxylase in *Drosophila melanogaster.* IV. The genetics and cytology of the 37B10-37D1 region. *Chromosoma* **83,** 45–58.

Wright, T. R. F., Steward, R., Bentley, K. W., and Adler, P. N. (1981b). The genetics of dopa decarboxylase in *Drosophila melanogaster.* III. Effects of a temperature sensitive dopa decarboxylase deficient mutation on female fertility. *Dev. Genet.* **2,** 223–235.

Wright, T. R. F., Black, B. C., Bishop, C. P., Marsh, J. L., Pentz, E. S., Steward, R., and Wright, E. Y. (1982). The genetics of dopa decarboxylase in *Drosophila melanogaster.* V. *Ddc* and *l(2)amd* alleles: Isolation, characterization and intragenic complementation. *Mol. Gen. Genet.* **188,** 18–26.

Yamazaki, H. I. (1969). The cuticular phenoloxidase in *Drosophila virilis. J. Insect Physiol.* **15,** 2203–2211.

Yamazaki, H. I. (1972). Cuticular phenoloxidase from the silkworm, *Bombyx mori:* Properties, solubilization, and purification. *Insect Biochem.* **2,** 431–444.

Yao, T. (1950). Cytochemical studies on the embryonic development of *Drosophila melanogaster.* II. Alkaline and acid phosphatases. *Q. J. Microsc. Sci.* **91,** 79–106.

Yasbin, R., Sawicki, J., and MacIntyre, R. J. (1978). A developmental study of acid phosphatase-1 in *Drosophila melanogaster. Dev. Biol.* **63,** 35–46.

Zomer, E., and Lipke, H. (1981). Tyrosine metabolism in *Aedes aegypti.* II. Arrest of sclerotization by MON 0585 and Diflubenzuron. *Pestic. Biochem. Physiol.* **16,** 28–37.

DEVELOPMENTAL CONTROL AND EVOLUTION IN THE CHORION GENE FAMILIES OF INSECTS

F. C. Kafatos,*,† N. Spoerel,* S. A. Mitsialis,* H. T. Nguyen,* C. Romano,* J. R. Lingappa,* B. D. Mariani,* G. C. Rodakis,‡ R. Lecanidou,‡ and S. G. Tsitilou‡

*Department of Cellular and Developmental Biology, The Biological Laboratories, Harvard University, Cambridge, Massachusetts 02138

†Institute of Molecular Biology and Biotechnology and Department of Biology, University of Crete, Crete, Greece

‡Department of Biochemistry, Cell and Molecular Biology and Genetics, University of Athens, Panepistimiopolis, Kouponia, Athens 15701, Greece

I. Introduction

The insect eggshell, or chorion, is a complex and physiologically important extracellular assembly composed of multiple proteins. Its structure and physiology have been reviewed (Hinton, 1981; Margaritis, 1985). Its biochemistry is well understood in two groups, fruit flies and silk moths (Regier and Kafatos, 1985).

The follicular epithelial cells are responsible for elaborating the chorion around each oocyte during a brief period at the end of oogenesis. In both flies and moths, groups of chorion proteins are deposited by these cells in overlapping succession, according to a developmental program which is presumably responsible for the intricate morphogenesis of the chorion (Regier *et al.*, 1982; Mazur *et al.*, 1982). Accordingly, groups of proteins have been classified as developmentally early, early middle, late middle, late, or very late, although the developmental kinetics are more complex than these terms imply (Paul and Kafatos, 1975; Bock *et al.*, 1983, 1986).

Different orders of insects have eggshells of widely different composition and structure, presumably reflecting different physiological needs during embryonic development. Figure 1 contrasts chorion structure in two orders separated by more than 200 million years of evolution, or nearly as much as that between birds and mammals: in moths a "laminated plywood" lamellar structure of horizontal fibers predominates, whereas in flies the dominant structure is a "trabecular," vaulted endochorion. As will be reviewed below, the genes responsible for these structures also vary widely in sequence and chromosomal organization. However, they do share stringent regulatory

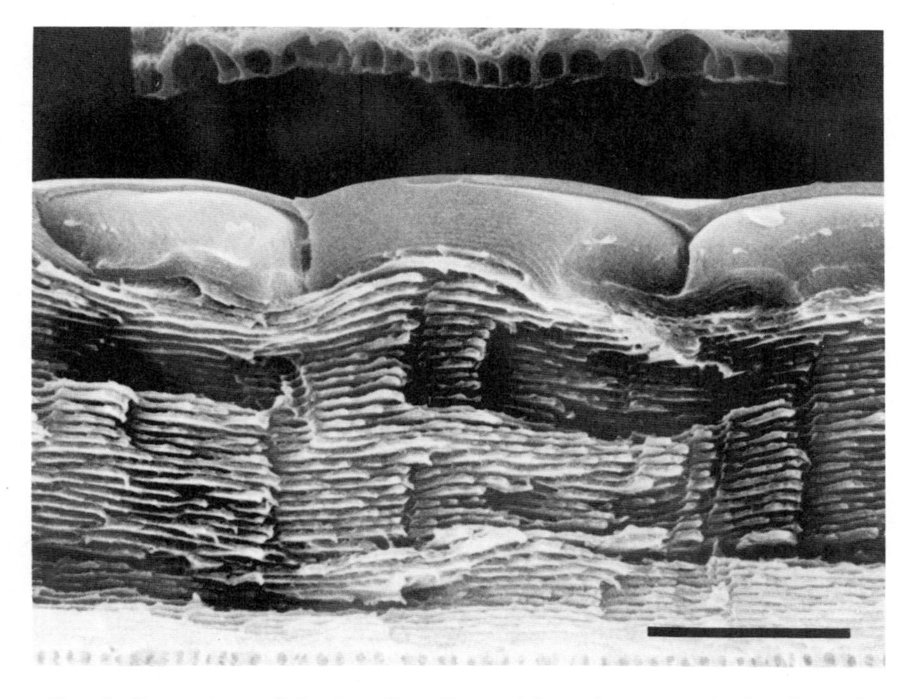

FIG. 1. Comparison of chorions from *Drosophila melanogaster* (top) and *Bombyx mori* (bottom). Scanning electron micrographs of transverse rips through mature eggshells are shown. Bar, 5 μm. In both micrographs the vitelline membrane encompassing the oocyte would be toward the bottom. Note that the fly chorion at most resembles the innermost (trabecular) layer of the moth chorion: both structures largely consist of air spaces separated by multiple vertical pillars. However, the bulk of the moth chorion is made up of horizontal lamellae of various types, which are not evident in the fly chorion. These fibrous lamellae consist of proteins of the α/β moth chorion superfamily, which are discussed in the text; the composition of the trabecular layer is as yet unknown. Courtesy of G. D. Mazur. From Kafatos *et al.* (1985).

attributes: they are exclusively expressed in follicular epithelial cells, and with a very narrow temporal specificity.

II. Sequences of Chorion Genes and Proteins

A. Silk Moths

1. *The α/β Superfamily*

Extensive sequence information is now available for the major chorion components in both moths and flies. These proteins have proved to be numerous, but belong to a small number of families. Early work involved partial protein sequencing, while the studies of the last 5 years have been almost exclusively at the DNA level.

In silk moths such as the cultivated *Bombyx mori* and the wild oak silk moth *Antheraea polyphemus*, more than 100 major chorion components of low molecular weight can be resolved by two-dimensional gel electrophoresis (Regier *et al.*, 1982; Bock *et al.*, 1986). The proteins are classified according to the distinctive properties of three regions: the central domain and the flanking NH_2- and COOH-terminal ends (left and right arms). Two fundamental groups of proteins are evident, α or β (Fig. 2; Lecanidou *et al.*, 1986). Each group shows identities ranging from less than 50 to more than 99% and is accordingly subdivided into multiple families, subfamilies, types, and copies. The proteins include the early proline-rich CA family; the A family, which includes members of various developmental specificities; and the late high-cysteine HcA family (present only in *Bombyx*). Symmetrically, the β-proteins include the CB, B, and HcB families.

The α- and β-protein groups are defined by the central domain sequences. The β-proteins are highly similar in the 48-residue central domain: even among families the sequence identities are 63–79% at the amino acid level, and there are no deletions or insertions. This high conservation most probably reflects a universal requirement for a compact secondary structure in this domain, apparently consisting of eight very short antiparallel β-sheet strands alternating with β-turns (Hamodrakas *et al.*, 1985). The central domain of the β-proteins (65–71 residues) is almost as well conserved, presumably for similar reasons. Distant but recognizable homology also exists between α and β central domains, supporting the concept of two branches of a single superfamily (Fig. 3).

The arms are much more variable in both sequence and length and help define the families and their subfamilies. Presumably these vari-

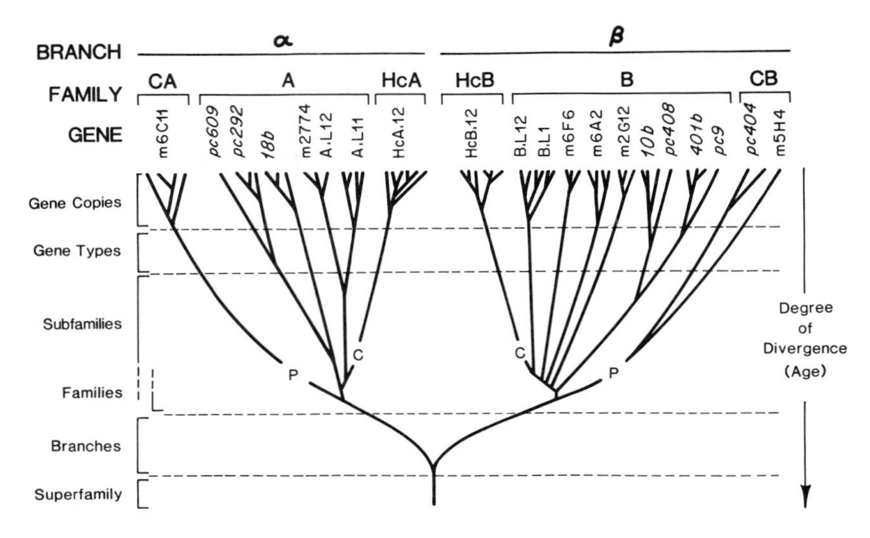

FIG. 2. Diagram of sequence homologies and nomenclature of genes in the α/β moth chorion superfamily. The diagram is largely based on the central domain and is meant as a reasonable representation of our present knowledge of the superfamily; it underestimates the number of gene copies for the most abundant gene types, such as *18b, A.L12, A.L11, HcA.12, HcB.12, B.L12,* and *401b*, and it does not show to scale the degrees of sequence divergence and presumed age of the genes. The superfamily consists of two distinct branches, α and β. Major subdivisions of each branch are gene families and subfamilies. The CA and HcA families are distinguished from the A family by arms that are proline-rich (P) or cysteine-rich (C), respectively; similar differences distinguish among the CB, HcB, and B families. Four A, six B, and two CB subfamilies are presently known and range from high to moderate divergence. Progressively more related genes are classified as different gene types (e.g., *pc609* versus *pc292*) or as nonidentical copies of the same gene type (e.g., *m2774* versus *A.L12*). The examples of genes listed are from *Bombyx mori* (plain codes) or from *Antheraea polyphemus* (italicized codes). From Lecanidou *et al.* (1986).

ations reflect a greater variety of functions (e.g., thickness, stacking properties, or cross-linking of the fibers, ultimately reflected in the variable layers of the moth chorion) (Fig. 1). A notable feature of the arms is the existence of tandemly repetitive peptides (Regier *et al.*, 1978; Regier and Kafatos, 1985). Many A and B proteins bear glycine- and tyrosine-rich repeats (GYGGL or variants thereof). Cysteine and glycine repeats (CG or CGG), which must be responsible for the disulfide cross-links of the mature chorion (Blau and Kafatos, 1979), are found near the ends of all components; such repeats are remarkably extensive in the high-cysteine proteins, where they occupy essentially the entire arm length (Rodakis and Kafatos, 1982; Burke and Eickbush, 1986). Expansion or contraction of the various repeat arrays is a prominent feature of chorion gene evolution.

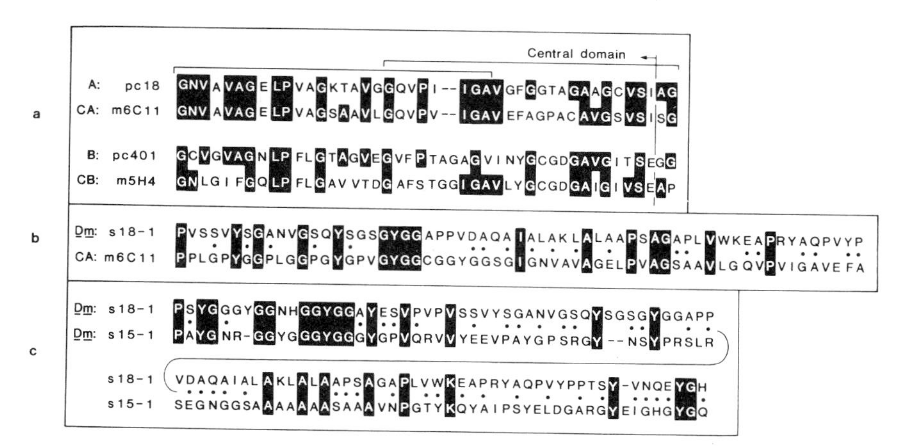

FIG. 3. Distant chorion protein homologies. (a) Homology between the α- and β-branches of the moth chorion superfamily. Typical α-sequences belonging to the A and CA families are compared with typical β-sequences belonging to the B and CB families. Amino acid residues present in one or both of the α- as well as β-sequences are shown in white against a black background. Additional sequence matches limited to one or the other branch are indicated by vertical tick marks. The regions compared encompass the -COOH-terminal part of the central domain and the first two residues of the right arm. Sequence identitites, neglecting conservative replacements, are 40% for A × B, 40% for A × CB, 40% for CA × B, and 33% for CA × CB. Good alternative alignments offset by 18 residues are also possible (horizontal bars), suggesting a subrepeat structure for the central domain (cf. Hamodrakas *et al.*, 1985). (b) Sequence homology of the *Drosophila melanogaster* chorion component, *s18-1* (approximately the middle third of the sequence), with the *B. mori* chorion component *m6C11* (CA family; left arm except for the first 10 residues and central domain except for the last 11 residues). (c) Sequence homology of the *D. melanogaster* chorion components *s18-1* (approximately the middle half of the sequence) and *s15-1* (all but the first 15 and the last 4 residues of the large exon). The matches of b and c were detected by the Lipman–Pearson (1985) FASTP computer program and also take into account conservative replacements (●). Respectively, they showed 70 initial and 73 optimized alignment scores (3.8 and 2.0 standard deviations above the mean of scores for 50 randomly scrambled sequences) and 54 initial and 74 optimized alignment scores (2.3 and 2.4 standard deviations above the mean for 50 randomly scrambled sequences). From Kafatos *et al.* (1986).

An intriguing observation is that similarities in the arms can be seen across the families and even across the two branches of the superfamily: (1) precise GYGGL repeats are found in the left arm of some A and B subfamilies (Tsitilou *et al.*, 1983); (2) the right arms of HcA and HcB proteins are highly similar, consisting of tandem GCGCGCX repeats [where X is G, R, or C (Burke and Eickbush, 1986)]; and (3) the HcA, HcB, A, and B families show marked patchy distributions of high sequence homology (Eickbush and Burke, 1985; Spoerel *et al.*, 1986). These capricious similarities apparently result from sequence transfers between chorion genes, through gene conversion-like events. The first strong evidence for such events came from the observation of a remarkable similarity in the 3′-untranslated sequences of HcA and HcB genes [not observed in other α and β components (Iatrou *et al.*, 1984)]. Subsequent careful comparisons between chorion gene sequences have revealed a wide occurrence of putative gene conversions, both within and among families. Elegant and compelling evidence has been assembled for the HcA and HcB genes (Eickbush and Burke, 1986), where the sequence transfers (gene conversions) are distributed in well-defined gradients, suggesting that they are promoted by the simple repeats of the right arm.

2. *The E Family*

The α/β superfamily encodes the major chorion structural proteins, but not all components involved in eggshell morphogenesis. In the silk moth chorion, air channels and their funnel-like outer extensions (aeropyles) are shaped by the localized secretion of special proteins, which are not part of the mature chorion structure (Regier *et al.*, 1982; Mazur *et al.*, 1982). These architectural, or "filler," proteins, named E1 and E2, are products of a distinct small family of genes, which in *A. polyphemus* are developmentally very late and predominantly expressed by a localized subpopulation of follicle cells (Regier *et al.*, 1984).

E1 and the amino-terminal domain of E2 show a periodicity of hydrophobic residues, corresponding to a distantly related repeat. At least in an 80-residue stretch of that domain, E1 and E2 components are homologous. However, they are not related to the α/β superfamily, nor to any other sequence in the protein data bases (Regier, 1986). The rest of the E2 sequence consists largely of an impressive tandem array of hydrophilic repeats, which frequently expands or contracts. One sequenced E2 component includes 74 such repeats, largely hexapeptides, conforming to the consensus $\mathrm{KKD}^{\mathrm{GVG}}_{\mathrm{DES}}$ or to the related consensus $\mathrm{RV}^{\mathrm{G}}_{\mathrm{S}}\mathrm{GVG}$. The hexapeptides tend to be embedded in higher order repeats (e.g., of 18 and 66 residues), in a manner reminiscent of other structural proteins (e.g., Pustell *et al.*, 1984).

B. *Drosophila*

Compared to moth chorion, the *Drosophila* chorion proteins are fewer in number (Waring and Mahowald, 1979; Margaritis *et al.*, 1980). The proteins encoded by the three genes that have been definitively sequenced (Wong *et al.*, 1985; Levine and Spradling, 1985) are only distantly related to each other, and their similarities to the moth chorion proteins are quite limited (Fig. 3). The fly sequences share an alternation of hydrophobic and hydrophilic segments (suggestive of similar three-dimensional structure), reasonably even spacing of the abundant tyrosines which are known to cross-link the final chorion structure (Petri *et al.*, 1976), at least one alanine-rich segment, and a few GYGG or related repeats (reminiscent of the GYGGL repeats of moth A and B proteins). They do not, however, show a highly conserved internal segment, akin to the central domain of moth chorion sequences. The shared features and the alignments shown in Fig. 3 suggest that the fly chorion genes are indeed homologous to each other and that they may belong to another branch of the same superfamily as the α and β moth chorion genes. However, these putative homologies are clearly very distant, and their statistical significance is only marginal.

III. Chromosomal Organization and Developmental Expression of Chorion Genes

In both flies and moths, chorion genes are highly clustered. However, the details of their organization differ in the two groups.

In the cultivated and inbred moth, *B. mori*, the chorion locus maps genetically on chromosome 2 ($n = 28$); it consists of two segments that may total approximately 1000 kb of DNA (Goldsmith and Basehoar, 1978; Goldsmith and Clermont-Rattner, 1979, 1980; Goldsmith and Kafatos, 1984). A chromosomal walk has yielded 270 kb from one of the segments, encompassing at least 64 A, B, HcA, and HcB genes of middle and late developmental specificity (Eickbush and Kafatos, 1982; T. Eickbush, N. Spoerel, and H. Nguyen, unpublished results). These genes are invariably organized as divergently transcribed α/β pairs (A/B or HcA/HcB). The early chorion genes appear to be clustered (but not necessarily arranged as pairs) in the other segment of the locus (T. Eickbush, personal communication). Although extensive polymorphisms hinder chromosomal walking in the wild moth, *A. polyphemus*, it is clear that the middle and late A and B genes of that species are also paired and clustered (Jones and Kafatos, 1981). The very late E genes are arranged in tandem, with E1 and E2 separated by approximately 7.5 kb (Hatzopoulos and Regier, 1986). Interestingly, all moth chorion genes have a single intron, located within the signal peptide-encoding sequence, irrespective of their family affiliation and chromosomal arrangement (Fig. 4).

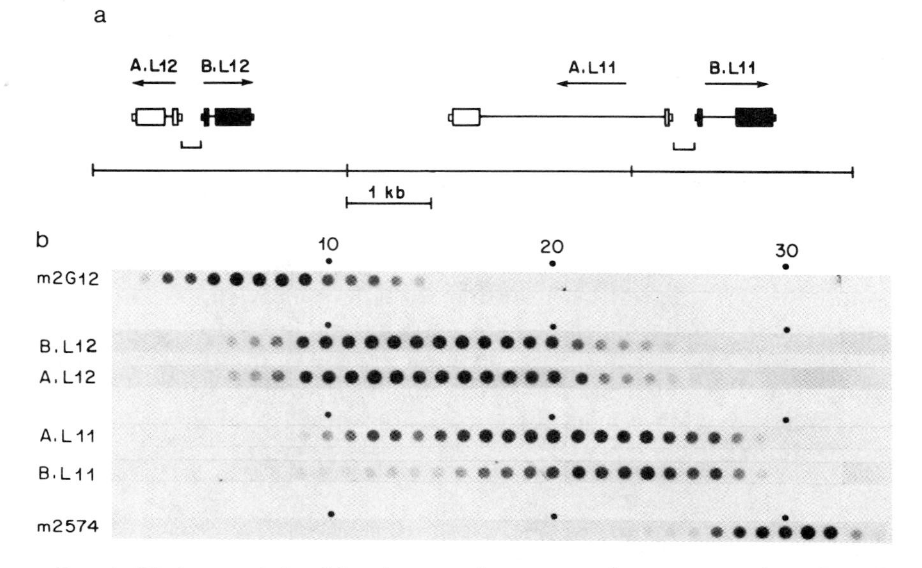

FIG. 4. Distinct periods of developmental expression for two types of coordinately expressed chorion gene pairs, *A/B.L12* and *A/B.L11* (early middle and late middle, respectively). (a) Diagrammatic representation of two linked *Bombyx mori* chorion gene pairs, *A/B.L11* and *A/B.L12*. Note that each pair consists of one α- (▭) and one β- (▬) gene. The paired genes are divergently transcribed. Each gene bears a single intron of of variable length; in contrast, the 5′-flanking intergenic DNAs (⊔) are of constant length (see Fig. 5). (b) Developmental RNA dot blots. Total RNA was isolated from sequential, individual *B. mori* follicles numbered from the start of choriogenesis (except for numbers 30–34, each of which represents a pool of four adjacent follicles). Aliquots of the RNAs were dot blotted and hybridized with short 5′-end-specific probes that discriminate between *A.L12, A.L11, B.L12,* and *B.L11* (36, 74, 30, and 31 hybridizing nucleotides, respectively, hybridized at approximately 18, 32, 16, and 24°C below T_m, respectively, and all washed at approximately 1°C below T_m). There are 8 *A/B.L12*-like and 12 *A/B.L11*-like gene pairs in the genome. For each probe, the exposure time was adjusted to yield approximately similar intensities during the peak of transcript accumulation. Note the coordinate appearance and disappearance of transcripts from the paired genes. For comparison, the temporal profiles of transcripts corresponding to an early B gene type (*m2G12*) and to the late HcB family (*m2574*) are also shown. From Spoerel *et al.* (1986).

The predominant paired arrangement of A, B, HcA, and HcB genes has some invariable features of special interest. As exemplified by pairs *A/B.L11* and *A/B.L12* (Fig. 4), the members of a pair are very tightly linked and expressed completely coordinately, i.e., with the same temporal specificity. The shared 5′-flanking DNA that separates their cap sites is only 300 ± 40 bp long (Jones and Kafatos, 1980; Iatrou and Tsitilou, 1983; Rodakis *et al.*, 1984; Spoerel *et al.*, 1986;

Burke and Eickbush, 1986). In contrast, the introns and the 3′-flanking DNA separating adjacent gene pairs are highly variable in length and sequence. The variations are frequently due to insertions or deletions and may involve transposable elements (Jones and Kafatos, 1980; Spoerel *et al.*, 1986; Burke and Eickbush, 1986). It is tempting to speculate that the highly constrained and short length of the 5′-flanking DNA in chorion gene pairs is functionally significant, i.e., that this DNA of 300 ± 40 bp contains regulatory elements which ensure the developmentally specific and coordinate expression of the flanking genes. Evidence is now available in support of this speculation (see below).

The major chorion genes of *Drosophila* are clustered, but always arranged in tandem. Two short, 5- to 10-kb clusters have been identified (Spradling, 1981; Griffin-Shea *et al.*, 1982; Parks *et al.*, 1986): one on the X chromosome (7F1–2) and one on the third chromosome (66D12–15), containing six and four chorion genes, respectively.

The clustered arrangement of chorion genes in both moths and flies is not unusual for members of multigene families. In the moths, clustering is probably related to the ability of these genes to evolve in a concerted manner, through sequence exchanges (see above). Developmental significance can probably only be ascribed to the paired organization, which correlates with coordinate expression, not to the global organization: although all 15 late HcA/HcB pairs are contiguous (Eickbush and Kafatos, 1982), A/B gene pairs of two different developmental specificities are intermixed in the chromosome (H. Nguyen, N. Spoerel, and G. Beltz, unpublished results). Similarly, in *Drosophila* the unit of expression appears to be the individual gene: virtually every gene within each of the two clusters is expressed with slightly different developmental kinetics (Griffin-Shea *et al.*, 1982; Parks *et al.*, 1986). These conclusions are supported by the normal expression of individual genes or gene pairs that have been introduced into the *Drosophila* germ line via P-element-mediated transformation (see below).

The clustered arrangement of *Drosophila* chorion genes is also related to developmentally regulated gene amplification (Spradling and Mahowald, 1980, 1981). Amplification does not occur in moths (Jones and Kafatos, 1981; Hatzopoulos and Regier, 1986) and is unnecessary since the gene multiplicity is high and choriogenesis lasts for 2 or more days (Paul and Kafatos, 1975; Nadel and Kafatos, 1980; Bock *et al.*, 1986). In contrast, the quantitative demands of choriogenesis are extremely high in *Drosophila melanogaster*: the genes are present in a single copy per genome, and choriogenesis is completed in 5 hours, with the period of synthesis of any one protein substantially shorter. In this case, the demands are met by specific amplification of both chorion clusters, beginning several hours before choriogenesis. The entire cluster in each locus amplifies as a unit (approximately 20-fold for the X and 60- to 80-fold for the third chromosome). The genes of the third chromosome tend to be expressed somewhat later, and their amplification continues for a longer time (Orr *et al.*, 1984). Multiple rounds of amplification apparently begin from a single origin and extend bidirectionally to indeterminate positions, generating amplification gradients over 80- to 100-kb chromosomal domains (Spradling, 1981; Osheim and Miller, 1983). cis-Acting elements which are essential for amplification, or modulate its level, have been mapped by transformation procedures (deCicco and Spradling, 1984; Kalfayan *et al.*, 1985; Kafatos *et al.*, 1985, 1986; Orr-Weaver and Spradling, 1986; C. Delidakis, unpublished results).

IV. The 5′-Flanking DNA Sequences of Chorion Genes

Detailed knowledge of the evolution of chorion structural genes, and of their chromosomal arrangement and developmental properties, permits a meaningful comparison of DNA sequence elements that might be involved in their cis regulation. We have focused our attention on the 5′-flanking DNA.

Interestingly, the intergenic (5′-flanking) DNA of the moth chorion gene pairs, while nearly constant in length, shows a stepwise pattern of sequence variation (Spoerel *et al.*, 1986). This pattern cannot be explained solely by gene conversion, since it is much more discontinuous and not congruent with the variability of the structural genes, and since the 5′-flanking DNA shows a minimum of conversion events (Eickbush and Burke, 1986). In *B. mori*, three distinct types of 5′-flanking DNA have been defined by hybridization and sequence analysis (Iatrou and Tsitilou, 1983; Spoerel *et al.*, 1986; Burke and Eickbush, 1986; H. T. Nguyen, unpublished results): one is characteristic of the 15 late HcA/HcB gene pairs, one is found in 8 early middle A/B pairs (typified by *A/B.L12*), and the third is found in 14 late middle A/-B pairs (typified by *A/B.L11*). A few additional A/B gene pairs have 5′-flanking sequences not belonging to these three types and are currently being studied, structurally and developmentally. A reasonable working model would relate any sequence similarities among the three predominant types of 5′-flanking DNA to their common developmental properties (sex-, tissue-, and choriogenesis-specific expression of the respective genes), while ascribing consistent sequence differences to the differences in temporal specificity. Recognition of important sequence similarities and differences should be facilitated by the availability of multiple nonidentical copies for each developmental class—differences among copies can be presumed to be functionally unimportant. A number of short internal repeats and common elements have thus been detected in the 5′-flanking sequences. For example, when four late and late middle 5′-flanking DNAs are compared (Fig. 5), short common elements become apparent, provided that minor deletions or insertions are allowed for alignment (Kafatos *et al.*, 1986). The early middle type of 5′-flanking DNA is more disparate (Spoerel *et al.*, 1986).

As might be expected from the distant homologies of their coding regions and from the differences in their temporal regulation, the three sequenced *Drosophila* chorion genes have distinct 5′-flanking sequences that do not cross-hybridize with each other. However, as in the case of the three types of *Bombyx* chorion 5′-flanking sequences, careful comparisons show the presence of short common elements (Wong *et al.*, 1985; Levine and Spradling, 1985). Among these, the most prominent is the hexanucleotide TCACGT, which is found upstream of all three *Drosophila* genes at almost identical positions (beginning at −65 to −58). The same sequence is also found upstream of three additional, partially sequenced *Drosophila* chorion genes (Wakimoto *et al.*, cited in Levine and Spradling, 1985; A. Georgi and D. King, unpublished observations), whereas it is absent from a control sample of 10 unrelated *Drosophila* 5′-flanking sequences totaling 6.5 kb in length (Wong *et al.*, 1985). Remarkably, the same element is also found in all three types of *Bombyx* 5′-flanking sequences (Spoerel *et al.*, 1986), at a very similar location upstream of the β-gene (Fig. 5). Furthermore, TCACGT or its reverse complement is also found in the 5′-flanking or intron sequences of *A. polyphemus* α/β gene pairs (N. Spoerel, unpublished observations) and even in the 5′-flanking DNA of E1 and E2 genes (Regier *et al.*, 1986). This common feature is particularly impressive, given the only very distant similarities of the coding regions in the major fly and moth chorion genes, and their clear lack of homology to the E family.

V. Evolutionary Conservation of Regulatory Elements among Flies and Moths

Despite the evolutionary distance and differences in organization and sequence, fly and moth chorion genes share important regulatory features. In both cases, they are only expressed in choriogenic follicle

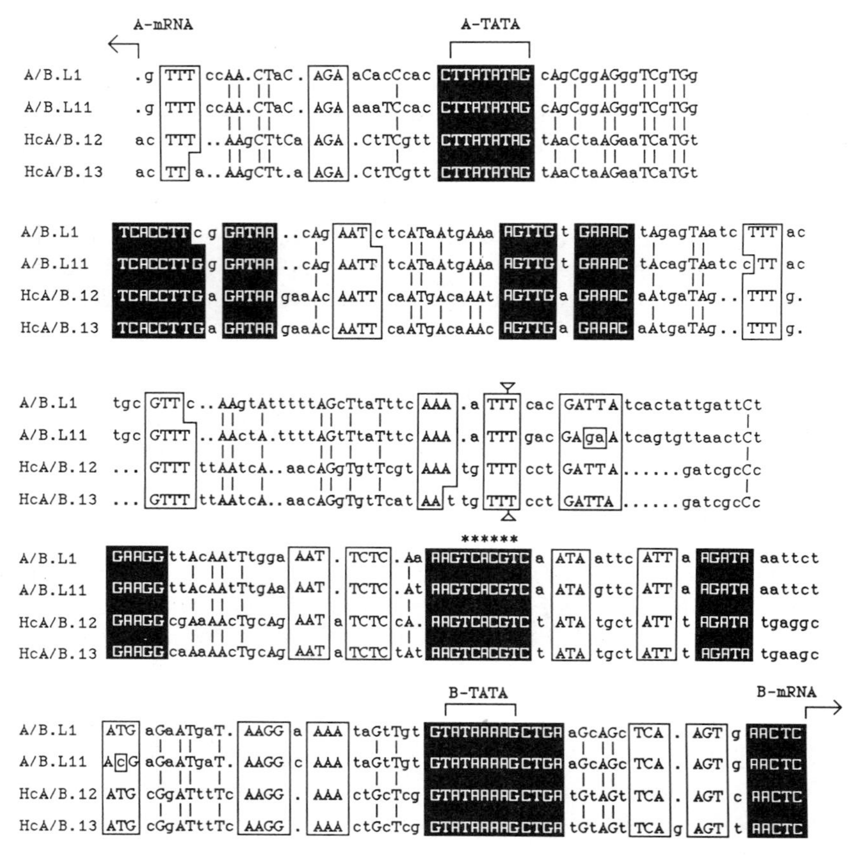

FIG. 5. Conservation of sequence elements in the intergenic regions of two late HcA/-HcB and two late middle A/B gene pairs. The region shown for *A/B.L11* corresponds to the bracket in Fig. 4a. Blocks of three or more nucleotides, found in comparable locations in at least one late and one late middle sequence, are boxed; the blocks are shown in white on a black background, if they are five or more nucleotides long and are present in all four sequences. Dots represent gaps introduced for alignment. All matching nucleotides are shown in capitals, and if present in all four sequences they are emphasized by vertical lines; mismatched nucleotides are shown in lowercase letters. The TATA boxes are indicated, and the TCACGT element characteristic of chorion genes is emphasized with asterisks. From Kafatos *et al.* (1986).

cells, toward the end of oogenesis. Reasoning that development tends to be a conservative "tinkerer" during evolution, we wondered whether moth chorion genes might be expressed normally if transferred into flies. We found that indeed they can be (Mitsialis and Kafatos, 1985).

Transgenic flies have been constructed using P-element transposons that encompass a single *B. mori* chorion gene pair and flanking DNA, plus the ry^+ gene as a visible marker. One construct includes the 3.1-kb *Eco*RI fragment bearing the *A/B.L12* gene pair (Fig. 4), and another includes a 3.8-kb fragment bearing a closely related pair, *A/-B.L9*. In all 10 independent transformant lines examined, the A and B moth genes are expressed exclusively in the ovary, as they are within the moth. Moreover, the temporal specificity of their expression is completely coordinate, as in the moths, and closely parallels the temporal specificity of the endogenous fly gene, *s15-1* (Fig. 6). Moths and flies are so different that temporal specificities cannot be easily cross-referenced. Nevertheless, these results clearly show that trans regulators (regulatory proteins?) of fly origin can effectively recognize cis-regulatory elements of moth chorion genes, resulting in developmentally appropriate gene expression.

We have been able to approximately localize the conserved regulatory elements using functional tests. As hypothesized earlier, such elements are located within the short A/B intergenic region. When a 272-bp moth DNA fragment consisting exclusively of the 5′-flanking region of *A/B.L12* is fused with the bacterial chloramphenicol acetyltransferase (CAT) gene, and is transferred to *Drosophila* using a P-element vector, CAT activity is detected exclusively in the ovary (Fig. 7). Indeed, the expression is limited to the follicles and shows late developmental specificity, approximating that of *s15-1* (S. A. Mitsialis, unpublished results).

Similarly, transformation has been used to study the regulatory elements of fly chorion genes (Kafatos *et al.*, 1985; Kalfayan *et al.*, 1985; Wakimoto *et al.*, 1986; Mariani *et al.*, 1987). Elements sufficient to support the normal sex, tissue, and temporal specificity of the *Drosophila s15-1* gene are localized within a 402-bp fragment, extending from −370 to +32 bp, relative to the cap site of the gene (C. Romano, unpublished results): when that fragment is fused to the alcohol dehydrogenase (*Adh*) gene and is transferred to Adh^- *Drosophila* using a P-element vector, mRNA from the chorion-*Adh* fusion construct is detected exclusively in late-stage follicles (C. Romano, unpublished results). Within that *s15-1* regulatory region, elements involved in conferring temporal specificity are being further defined by deletion and transformation analyses (Mariani *et al.*, 1987).

Once functionally equivalent regulatory elements of moths and flies were localized within 0.3 and 0.4 kb, respectively, their tentative identification through sequence comparisons was undertaken. Although equivalent elements need not share primary structure, the 272

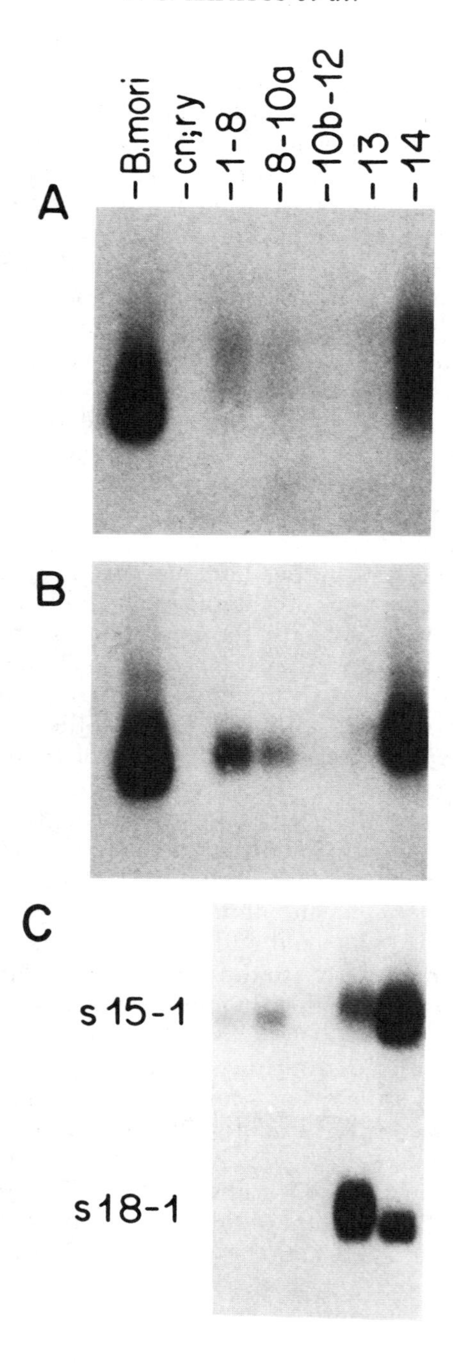
-B.mori
-cn;ry
-1-8
-8-10a
-10b-12
-13
-14
A
B
C
s15-1
s18-1

nucleotides of the *A/B.L12* intergenic region, upstream of the *B.L12* cap site, and the corresponding portion of the *s15-1* 5′-flanking region, do show some sequence similarities (Fig. 8). In the alignment shown, there are 50% matches, 28% mismatched, and 22% gaps in one or the other sequence. The significance of the matches is increased by their clustering (142/154 matches occur in blocks of two or more bases), and by the fact that the gaps tend to cancel each other, i.e., that the matches are located at approximately equal distances from the respective cap site (within 11 nucleotide positions from each other). Furthermore, with due allowance for a few deletions–insertions, substantial blocks of nucleotide identity can be detected: for example, with one single-nucleotide gap, 15 out of 16 nucleotides encompassing the B-TATA box are identical between *A/B.L12* and *s15-1*; with four single-nucleotide gaps (two of which cancel out), 17/18 nucleotides encompassing the chorion-specific TCACGT element are identical; the uninterrupted heptanucleotide GTAGAAT is present in both sequences, centered at position −195. With a somewhat greater tolerance for variable distance from the cap site, the nonanucleotide AGTGTATTC is detected at positions −235 and −274 of *A/B.L12* and *s15-1*, respectively; partial matches to the same sequence (TATT; AGT/GTXTTC) are also encountered at more equivalent positions (⊗ in Fig. 8).

The first step in verifying experimentally the significance of these shared sequence elements has already been taken: as Fig. 7 shows, when the sequence GTCACGTT, including the TCACGT element of the *A/B.L12–cat* fusion construct, is substituted by CAGATCTG in a linker–scanner mutation, the follicle-specific CAT expression is completely abolished. Clearly, detailed understanding of the developmentally important chorion cis-regulatory elements is obtainable, through a dialectical interplay of careful comparative analysis of the sequence patterns, plus experimental modification of the sequences

---

FIG. 6. Temporal specificity of silk moth chorion gene expression in transformed *Drosophila*. Total RNA (15 μg/lane) was isolated from follicles of an *A/B.L12*-transformant line, representing the various stages (1–14) of *Drosophila* oogenesis. Total ovarian RNA (15 μg) from the parental line (*cn;ry*) was used as a negative control, and total *Bombyx mori* ovarian RNA (0.05 μg) was used as a positive control. The RNAs were glyoxylated, fractionated by electrophoresis, blotted onto nylon membrane filters, and hybridized with A or B gene-specific single-stranded cRNA probes (A or B, respectively) or nick-translated probes specific for the *Drosophila s15-1* or *s18-1* gene (C). Note that the moth A and B genes are coordinately expressed, with a bimodal temporal pattern resembling that of the endogenous *s15-1* chorion gene, but distinct from that of gene *s18-1*. From Mitsialis and Kafatos (1985).

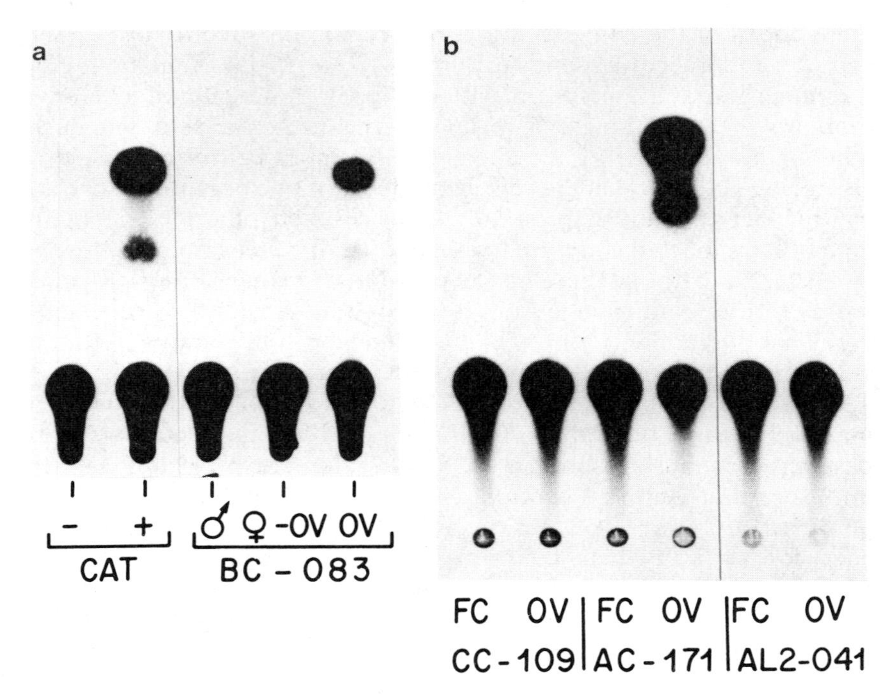

FIG. 7. Moth regulatory DNA necessary for chorion-specific gene expression in *Drosophila*. A 272-bp intergenic DNA fragment from the *A/B.L12* chorion gene pair (⊔ in Fig. 4a, cf. Fig. 8 for sequence) was fused to the bacterial chloramphenicol acetyltransferase (CAT) gene within a transposable P-element vector. Gene activity was assayed through the presence of CAT enzyme activity in extracts, causing the conversion of radioactive chloramphenicol (large bottom spot in all lanes) to acetoxy derivatives (top two spots in some lanes). An extract of the parental fly line is inactive, but addition of exogenous enzyme yields detectable activity (reconstruction standard, left two lanes of a). In the transformant line BC-083, in which the *cat* gene is driven by the B.L12 promoter, activity is absent in male and ovariectomized female extracts, but present in ovarian extracts (lanes 3, 4, and 5 of a). (b) Compares extracts of ovaries (OV) and ovariectomized female carcass (FC). Line CC-109 is a control transformant, with the vector alone. Line AC-171 is a transformant in which the *cat* gene is driven by the A.L12 promoter and shows ovary-specific activity. Line AL2-041 is a similar transformant in which the TCACGT element has been disrupted by a linker–scanner mutation; the ovarian-specific expression is eliminated by that mutation.

followed by testing *in vivo* function through transformation assays. In parallel, we will be attempting to identify and compare the trans-regulatory elements at the molecular level. Understanding cis and trans regulation of individual genes will only be an initial (but crucially important) step in understanding the overall developmental program of choriogenesis.

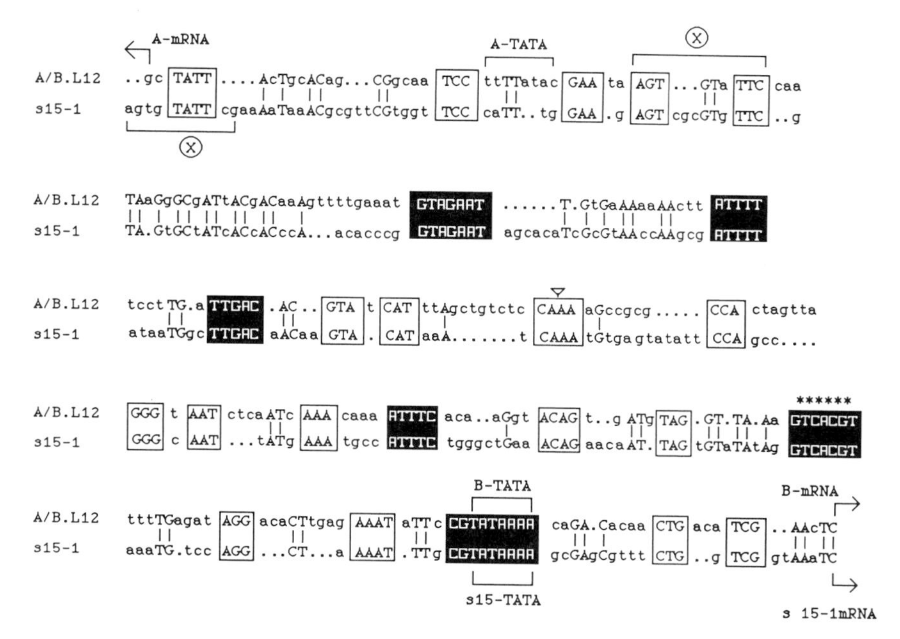

FIG. 8. Shared sequence elements in the 5′-flanking DNA of the *Drosophila melanogaster s15-1* gene and the *B. mori A/B.L12* chorion gene pair, which is expressed in transgenic flies with the same developmental specificity as *s15-1*. Conventions as in Fig. 5. The symbol ⊗ indicates an exact nonanucleotide match at slightly different locations in the two sequences. From Kafatos *et al.* (1986).

ACKNOWLEDGMENTS

Some of the material in this review is similarly summarized in Kafatos *et al.* (1986). We thank our colleagues for permission to quote unpublished results and E. Fenerjian for secretarial assistance. Work at Harvard was supported by NIH and NSF grants to F. C. Kafatos and at Athens by grants from the General Secretariat for Research and Technology to G. C. Rodakis, R. Lecanidou, and S. G. Tsitilou. H. T. Nguyen holds an NIH postdoctoral fellowship, N. Spoerel and S. A. Mitsialis are fellows of the Charles A. King Trust, and B. D. Mariani is a fellow of the Jane Coffin Childs Fund for Medical Research.

REFERENCES

Blau, H. M., and Kafatos, F. C. (1979). Morphogenesis of the silkmoth chorion: Patterns of distribution and insolubilization of the structural proteins. *Dev. Biol.* **72**, 211–225.

Bock, S. C., Tiemeier, D. C., Mester, K., and Goldsmith, M. R. (1983). Differential patterns in the temporal expression of *Bombyx mori* chorion genes. *Wilhelm Roux's Arch. Dev. Biol.* **192,** 222–227.

Bock, S. C., Campo, K., and Goldsmith, M. R. (1986). Specific protein synthesis in cellular differentiation. VI. Temporal expression of chorion gene families in *Bombyx mori* strain C108. *Dev. Biol.* **117,** 215–225.

Burke, W. D., and Eickbush, T. H. (1986). The silkmoth late chorion locus. I. Variation within two paired multigene families. *J. Mol. Biol.* **190,** 343–356.

deCicco, V. A., and Spradling, A. C. (1984). Localization of a *cis*-acting element responsible for the developmentally regulated amplification of *Drosophila* chorion genes. *Cell* **38,** 45–54.

Eickbush, T. H., and Burke, W. D. (1985). Silkmoth chorion gene families contain patchwork patterns of sequence homology. *Proc. Natl. Acad. Sci. U.S.A.* **82,** 2814–2818.

Eickbush, T. H., and Burke, W. D. (1986). The silkmoth late chorion locus. II. Gradients of gene conversion in two paired multigene families. *J. Mol. Biol.* **190,** 357–366.

Eickbush, T. H., and Kafatos, F. C. (1982). A walk in the chorion locus of *Bombyx mori. Cell* **19,** 633–643.

Goldsmith, M. R., and Basehoar, G. (1978). Organization of the chorion genes of *Bombyx mori,* a multigene family. I. Evidence for linkage to chromosome 2. *Genetics* **90,** 291–310.

Goldsmith, M. R., and Clermont-Rattner, E. (1979). Organization of the chorion genes of *Bombyx mori,* a multigene family. II. Partial localization of three gene clusters. *Genetics* **92,** 1173–1185.

Goldsmith, M. R., and Clermont-Rattner, E. (1980). Organization of the chorion genes of *Bombyx mori,* a multigene family. III. Detailed composition of three gene clusters. *Genetics* **96,** 201–212.

Goldsmith, M. R., and Kafatos, F. C. (1984). Developmentally regulated genes in silkmoths. *Annu. Rev. Genet.* **18,** 443–487.

Griffin-Shea, R., Thireos, G., and Kafatos, F. C. (1982). Organization of a cluster of four chorion genes in *Drosophila* and its relationship to developmental expression and amplification. *Dev. Biol.* **91,** 325–336.

Hamodrakas, S. J., Etmektzoglou, T., and Kafatos, F. C. (1985). Aminoacid periodicities and their structural implications for the evolutionarily conservative central domain of some silkmoth chorion proteins. *J. Mol. Biol.* **186,** 583–589.

Hatzopoulos, A. K., and Regier, J. C. (1986). Organization of regionally expressed silkmoth chorion genes. *Mol. Cell. Biol.* **6,** 3215–3220.

Hinton, H. E. (1981). "Biology of Insect Eggs." Pergamon, Oxford.

Iatrou, K., and Tsitilou, S. G. (1983). Coordinately-expressed chorion genes of *Bombyx mori*: Is developmental specificity determined by secondary structure recognition? *EMBO J.* **2,** 1431–1440.

Iatrou, K., Tsitilou, S. G., and Kafatos, F. C. (1984). DNA sequence transfer between two high-cysteine chorion gene families in *Bombyx mori. Proc. Natl. Acad. Sci. U.S.A.* **81,** 4452–4456.

Jones, C. W., and Kafatos, F. C. (1980). Structure, organization and evolution of developmentally-regulated chorion genes in a silkmoth. *Cell* **22,** 855–867.

Jones, C. W., and Kafatos, F. C. (1981). Linkage and evolutionary diversification of developmentally regulated multigene families: Tandem arrays of the 401/18 chorion gene pair in a silkmoth. *Mol. Cell. Biol.* **1,** 814–828.

Kafatos, F. C., Mitsialis, S. A., Spoerel, N., Mariani, B., Lingappa, J. R., and Delidakis, C. (1985). Studies on the developmentally regulated expression and amplification of insect chorion genes. *Cold Spring Harbor Symp. Quant. Biol.* **50,** 537–547.

Kafatos, F. C., Mitsialis, S. A., Nguyen, H. T., Spoerel, N., Tsitilou, S. G., and Mazur, G.

D. (1987). Evolution of structural genes and regulatory elements for the insect chorion. *In* "Development as a Evolutionary Process" (R. Raff and E. C. Raff, eds.), pp. 161–178. Liss, New York.

Kalfayan, L., Levine, J., Orr-Weaver, T., Parks, S., Wakimoto, B., deCicco, D., and Spradling, A. (1985). Localization of sequences regulating *Drosophila* chorion gene amplification and expression. *Cold Spring Harbor Symp. Quant. Biol.* **50**, 527–535.

Lecanidou, R., Rodakis, G. C., and Kafatos, F. C. (1986). Evolution of the silkmoth chorion gene superfamily: The CA and CB gene families. *Proc. Natl. Acad. Sci. U.S.A.* **83**, 6514–6518.

Levine, J., and Spradling, A. C. (1985). DNA sequence of a 3.8 kilobase pair region controlling *Drosophila* chorion gene amplification. *Chromosoma* **92**, 136–142.

Lipman, D. J., and Pearson, W. R. (1985). Rapid and sensitive protein similarity searches. *Science* **227**, 1435–1441.

Margaritis, L. H. (1985). Structure and physiology of the eggshell. *In* "Comprehensive Insect Physiology, Biochemistry and Pharmacology" (G. A. Kerkut and L. I. Gilbert, eds.), Vol. 1, pp. 153–230. Pergamon, Oxford.

Margaritis, L. H., Kafatos, F. C., and Petri, W. H. (1980). The eggshell of *Drosophila melanogaster.* I. Fine structure of the layers and regions of the wild-type eggshell. *J. Cell Sci.* **43**, 1–35.

Mariani, B. D., Lingappa, J. R., and Kafatos, F. C. (1987). A negative *cis*-regulator is partially responsible for temporal control of expression in a *Drosophila* chorion gene. *Proc. Natl. Acad. Sci. U.S.A.* (submitted).

Mazur, G. D., Regier, J. C., and Kafatos, F. C. (1982). Order and defects in the silkmoth chorion, a biological analogue of a cholesteric liquid crystal. *In* "Insect Ultrastructure" (H. Akai and R. C. King, eds.), Vol. 1, pp. 150–183. Plenum, New York.

Mitsialis, S. A., and Kafatos, F. C. (1985). Regulatory elements controlling chorion gene expression are conserved between flies and moths. *Nature (London)* **317**, 453–456.

Nadel, M. R., and Kafatos, F. C. (1980). Specific protein synthesis in cellular differentiation. IV. The chorion proteins of *Bombyx mori* and their programs of synthesis. *Dev. Biol.* **75**, 26–40.

Orr, W., Komitopoulou, K., and Kafatos, F. C. (1984). Mutants suppressing in *trans* chorion gene amplification in *Drosophila. Proc. Natl. Acad. Sci. U.S.A.* **81**, 3773–3777.

Orr-Weaver, T. L., and Spradling, A. C. (1986). *Drosophila* chorion gene amplification requires an upstream region regulating *s*18 transcription. *Mol. Cell. Biol.* **6**, 4624–4633.

Osheim, Y. N., and Miller, O. L. (1983). Novel amplification and transcriptional activity of chorion genes in *Drosophila melanogaster. Cell* **33**, 543–553.

Parks, S., Wakimoto, B., and Spradling, A. (1986). Replication and expression of an X-linked cluster of *Drosophila* chorion genes. *Dev. Biol.* **117**, 294–305.

Paul, M., and Kafatos, F. C. (1975). Specific protein synthesis in cellular differentiation. II. The program of protein synthetic changes during chorion formation by silkmoth follicles, and its implementation in organ culture. *Dev. Biol.* **42** 141–149.

Petri, W. H., Wyman, A. R., and Kafatos, F. C. (1976). Specific protein synthesis in cellular differentiation. III. The eggshell proteins of *Drosophila melanogaster* and their program of synthesis. *Dev. Biol.* **49**, 185–199.

Pustell, J., Kafatos, F. C., Wobus, U., and Bäumlein, H. (1984). Balbiani ring DNA: Sequence comparisons and evolutionary history of a family of hierarchically repetitive protein-coding genes. *J. Mol. Evol.* **20**, 281–295.

Regier, J. C. (1986). Evolution and higher-order structure of architectural proteins in silkmoth chorion. *EMBO J.* **5,** 1981–1986.

Regier, J. C., and Kafatos, F. C. (1985). Molecular aspects of chorion formation. *In* "Comprehensive Insect Physiology, Biochemistry and Pharmacology" (G. A. Kerkut and L. I. Gilbert, eds.), Vol. 1, pp. 113–151. Pergamon, Oxford.

Regier, J. C., Kafatos, F. C., Goodfliesh, R., and Hood, L. (1978). Silkmoth chorion proteins: Sequence analysis of the products of a multigene family. *Proc. Natl. Acad. Sci. U.S.A.* **75,** 390–394.

Regier, J. C., Mazur, G. D., Kafatos, F. C., and Paul, M. (1982). Morphogenesis of silkmoth chorion: Initial framework formation and its relation to synthesis of specific proteins. *Dev. Biol.* **92,** 159–174.

Regier, J. C., Hatzopoulos, A. K., and Durot, A. C. (1984). Molecular cloning of region-specific chorion-encoding RNA sequences. *Proc. Natl. Acad. Sci. U.S.A.* **81,** 2796–2800.

Regier, J. C., Hatzopoulos, A. K., and Durot, A. C. (1986). Patterns of region-specific chorion gene expression and identification of shared 5′ flanking genomic elements. *Dev. Biol.* (in press).

Rodakis, G. C., and Kafatos, F. C. (1982). The origin of evolutionary novelty in proteins: How a high-cysteine chorion protein has evolved. *Proc. Natl. Acad. Sci. U.S.A.* **79,** 3551–3555.

Rodakis, G. C., Lecanidou, R., and Eickbush, T. H. (1984). Diversity in a chorion multigene family created by tandem duplication and a putative gene-conversion event. *J. Mol. Evol.* **20,** 265–273.

Spoerel, N., Nguyen, H. T., and Kafatos, F. C. (1986). Gene regulation and evolution in the chorion locus of *B. mori*: Structural and developmental characterization of four eggshell genes and their flanking DNA regions. *J. Mol. Biol.* **190,** 23–35.

Spradling, A. C. (1981). The organization and amplification of two clusters of *Drosophila* chorion genes. *Cell* **27,** 193–202.

Spradling, A. C., and Mahowald, A. P. (1980). Amplification of genes for chorion proteins during oogenesis in *Drosophila melanogaster. Proc. Natl. Acad. Sci. U.S.A.* **77,** 1096–1100.

Spradling, A. C., and Mahowald, A. P. (1981). A chromosome inversion alters the pattern of specific DNA replication of *Drosophila* follicle cells. *Cell* **27,** 203–209.

Tsitilou, S. G., Rodakis, G. C., Alexopoulou, M., Kafatos, F. C., Ito, K., and Iatrou, K. (1983). Structural features of B family chorion sequences in the silkmoth, *Bombyx mori,* and their evolutionary implications. *EMBO J.* **2,** 1845–1852.

Wakimoto, B. T., Kalfayan, L. J., and Spradling, A. C. (1986). Developmentally regulated expression of *Drosophila* chorion genes introduced at diverse chromosomal positions. *J. Mol. Biol.* **187,** 33–45.

Waring, G. L., and Mahowald, A. P. (1979). Identification and time of synthesis of chorion proteins in *Drosophila melanogaster. Cell* **16,** 599–607.

Wong, Y.-C., Pustell, J., Spoerel, N., and Kafatos, F. C. (1985). Coding and potential regulatory sequences of a cluster of chorion genes in *Drosophila melanogaster. Chromosoma* **92,** 124–135.

THE SIGNIFICANCE OF SPLIT GENES TO DEVELOPMENTAL GENETICS

Antoine Danchin

Unité de Régulation de l'Expression Génétique, Institut Pasteur, F-75724 Paris Cedex 15, France

I. Introduction

Regulation of gene expression is a central question of biology. Models derived from the extensive study of prokaryotes, mainly *Escherichia coli,* have dominated the conceptual genetic framework of most scientists for more than 20 years. A major understanding, derived from the seminal work of Francois Jacob and Jacques Monod, brought evidence for coordinate regulation of gene expression at the level of transcription, formalized as the model of the operon. Unfortunately, the outstanding achievements of Jacob and Monod had the drawback that other ways of considering different modes of gene expression were overlooked. The operon model is based on a special route for the "information" flow specified by the hereditary material: DNA segments are transcribed into messenger RNA molecules, which are then translated into proteins. In this model, messenger RNA is an essentially *labile* molecular species. This ensures that control of gene expression is fast and efficient. The model holds true for bacteria only: messenger RNA molecules of eukaryotic cells are usually much more stable. Moreover, as has been experimentally recognized more recently, eukaryotic messenger RNA is not the direct product of DNA. Instead, nuclear transcripts must be *processed* before they can undergo translation in the cytoplasm. In addition, the processing of nuclear pre-mRNA molecules seems most unusual (as compared with processing of preribosomal RNA in prokaryotes, for instance). Indeed, there is *splicing* of distant RNA segments, which, although transcribed from the genes, do not appear as transcription products at the final stage of gene expression (cytoplasmic mRNA). Thus, it has been found that many eukaryotic genes are split into coding and noncoding segments, termed exons and introns, respectively, by Gilbert (1978). Because of the frequent confusion between the levels of RNA and DNA, several other terms have been used instead of "intron," the most frequent one being "intervening sequence" (IVS); I shall hereafter use intron when no confusion between DNA and RNA is to be expected, in particular, d-intron for the gene (DNA) sequence and r-intron for the pre-mRNA intervening sequence when there is ambiguity.

Since their discovery, split genes have been submitted to intense investigation. Many review articles have dealt with the subject and I shall not repeat here the content of these articles: this would mean citing more than a 1000 references. I shall mainly keep to the most recent work (since 1983) and ask the reader to go through former review articles for older references (Abelson, 1979; Cech and Bass, 1986; Breathnach and Chambon, 1981; Crick, 1979; Nevins, 1983; Padgett *et al.*, 1986).

The function of split genes in the nucleus is not yet understood, whereas significant knowledge has been accumulated in the case of organelles, mainly mitochondria (see, for instance, Dujon, 1981; Jacq *et al.*, 1982; Kotylak *et al.*, 1985; Tabak and Grivell, 1986). I shall therefore separate this article into four sections. The first one will deal with split genes in organelles (essentially *Saccharomyces cerevisiae* mitochondria), and will be centered on the *RNA-maturase* concept proposed by P. Slonimski and his group. In the second part I shall discuss the present state of knowledge in the case of nuclear split genes. But, rather than remaining at a descriptive level on the role of nuclear introns, I shall propose that this (usually unexpected) organization of eukaryotic genes is deeply related to specific features of their expression, also rooted in another structural characteristic of eukaryotes, their nuclear envelope. In a nutshell, split genes and the nuclear envelope, associated in a specific regulatory structure required for gene expression, will be described as missing links which permit reconciliation of preformationism and epigenesis. Finally I shall summarize data and speculations on the origin and fate of introns in an evolutionary perspective.

II. The Concept of Maturase

A. Split Gene Organization

Introns were first discovered in an animal virus, in an immunoglobulin gene, and in the ovalbumin gene. Since this discovery, thousands of introns have been identified, and it has been recognized that the vast majority of metazoan genes are split genes. Indeed, genes without introns are exceptions, apart from genes from unicellular organisms such as *S. cerevisiae*, in which only certain genes contain introns. Several classes of introns have been described.

1. Short or very short introns, found in many tRNA genes.
2. Much longer introns, probably folded into complex secondary structures, found in mitochondrial genes as well as in nuclear genes coding for structural RNA molecules. Two groups have been defined in this case; group-I introns, for instance, comprise the intervening sequences in nuclear and mitochondrial rRNA and many mitochondrial mRNAs, e.g., introns found in the *E. coli* bacteriophage T4 (Chu *et al.*, 1984, 1986; Sjöberg *et al.*, 1986) and in several chloroplast tRNA precursors (Davies *et al.*, 1982; Michel *et al.*, 1982; Cech *et al.*, 1983; Michel and Dujon, 1983; Waring and Davies, 1984). These introns share a set of highly conserved structural elements that dictate a specific folding pattern which results in direct apposition of the intron–exon border by alignment against an "internal guide sequence." The three-dimensional geometry of this structure seems to be of major importance, in that several of the precursors have been shown to undergo autocatalytic splicing *in vitro* in the complete absence of protein (Kruger *et al.*, 1982; Garriga and Lambowitz, 1984; Van der Horst and Tabak, 1985; Arnberg *et al.*, 1986).
3. Long introns (from a few hundred nucleotides to tens of kilobases long), for which very little secondary structure or few consensus sequences, namely, the so-called donor (GU) and acceptor (AG) universal sequences (Breathnach and Chambon, 1981; Mount, 1982), have been described, are found in genes specifying proteins.

Genes which direct synthesis of proteins that do not display highly repetitive structures (such as globin genes) usually possess few introns (e.g., two in globin genes), while genes coding for proteins displaying highly repetitive structures are generally interrupted by a large number of introns, which are usually rather short and display common primary sequence features (Upholt and Sandell, 1986).

As stated above, it appears that some introns are extremely large, and it is not surprising that open reading frames (ORFs) can be found in their sequence. It is usually not known whether this corresponds to expressed genes, except in cases of alternative splicing (see below), raising the question of the raison d'être of such large quantities of "junk" DNA. A quite exotic gene arrangement has been discovered in *Drosophila melanogaster,* in which Henikoff and co-workers (1986) have described a gene within a gene. In their work on the structure and organization of the *Gart* locus they found a sequence on the noncoding strand within one of the *Gart* introns which showed some of the properties of gene-coding sequences. The intronic gene has its own intron, so that a 71-bp region of the DNA is transcribed from both strands and is excised from both primary transcripts. The gene appears to code for a pupal cuticle protein. It has a similar organization to, and sequence homology with, previously described larval cuticle proteins. In general, however, it appears that exons found in long introns correspond to transcripts having the same orientation: this is the case, for instance, of the μ constant region of immunoglobulin genes, where appropriate (unknown) controls permit a shift to δ constant region expression with complete splicing out of the μ messenger (Rabbitts, 1978; Rabbitts *et al.*, 1980; Maki *et al.*, 1981).

Soon after the biochemical discovery of intervening sequences in eukaryotic genes (d-introns), genetic evidence suggested that yeast mitochondrial genes might reveal a split organization. It had indeed been known for some time that mutations of several mitochondrial genes fell into multiple complementation groups. It was discovered then confirmed and extended in *S. cerevisiae* mitochondria that two genes displayed a mosaic structure composed of exons and introns (see Jacq *et al.*, 1982, for review): the gene specifying synthesis of the amino acid sequence of cytochrome *b* (*cob-box*), as well as the *oxi3* gene specifying synthesis of subunit I of cytochrome oxidase. It appeared, therefore, that, owing to the powerful approach afforded by genetic techniques, mitochondrial genes might provide an excellent tool for investigating the structure and function of introns. As respiratory organelles, mitochondria are indeed dispensible devices of yeast cells. Multiple mutants can therefore be easily collected that are not sensitive to selective pressure; they are the objects of choice for the study of introns.

In lower organisms the study of the mosaic organization of mitochondrial genes is generally less advanced than in yeast, but concordant data indicate that lower organisms often harbor introns. Introns in various yeast mtDNAs are related both to each other and to introns present in other fungal and plant mtDNAs and even in certain chloroplast DNAs. A most characteristic feature of fungal (including yeast) mitochondrial introns is that many contain long ORFs, usually linked in phase with the preceding exon and covering most of the intron sequence. They can be grouped into two main families (groups I and II), each of whose members display extensive similarities in the amino acid sequence of the proteins they encode, in the possession of short, conserved sequence elements, and in potential RNA secondary structure. The two latter features are also characteristic of the introns present in the nuclear-encoded rRNA genes of *Tetrahymena* and *Physarum,* raising the possibility that both types of introns use essentially similar splicing mechanisms (e.g., see Waring and Davies, 1984). As a specific example, in lower eukaryotes, fungi (*Aspergillus nidulans* and *Neurospora crassa*), and plants, the mitochondrial cytochrome *b* gene is split by at least one intron harboring a long ORF, and another intron is present in the large ribosomal RNA gene in both fungal species, as well as in yeast strains exhibiting polarity of recombination. The intron of the large rRNA gene (intron *rI1*) has unique properties. First, it is the only intron involved in the phenomenon of polarity of recombination (Dujon *et al.*, 1976, and references therein), as in the following example: in crosses between two strains, only one of which contains the intron, there is a unidirectional transfer of the intron from the gene that contains it to the gene lacking it, so that the vast majority of the progeny receive the mosaic gene. The flanking regions are also transferred with an efficiency decreasing with their distance from the intron. Second, it contains an ORF, which is not necessary for correct splicing but which is required for polarity of recombination (preferential transfer of the d-intron sequence) (see below).

Analysis of RNA molecules present in mitochondria has been possible after cloning DNA fragments of introns and exons and using Northern blotting. This has permitted the monitoring of splicing in different physiological conditions and the investigation of the requirement of factors or macromolecular synthesis in the process. Mitochondrial protein synthesis is required for the splicing of most introns except *bI1, bI5, aI5,* and the intron located in ribosomal RNA, *rI1.* Several mitochondrial RNA r-introns are thus able to excise from a pre-mRNA molecule in the complete absence of protein synthesis. It was found in early work that excision of the corresponding sequences yielded RNA molecules that had apparently been circularized

(Halbreich *et al.*, 1980), but nothing was known about the mechanism of circularization. The corresponding introns were members of group I, originally defined in fungal mitochondrial introns on the basis of conservation of both primary sequence and potential secondary structure (see Michel and Dujon, 1983); group-I introns can be discriminated by specific features, namely, by conserved nucleotide sequences that may be recognized by appropriate enzymes, or by eventual self-splicing, as demonstrated by Kruger *et al.* (1982) in the case of a nuclear ribosomal RNA r-intron (see below). It is therefore not entirely unexpected that certain group-I introns in *Neurospora* and yeast mitochondria will also undergo self-splicing when their transcripts are incubated under similar conditions (Arnberg *et al.*, 1986, and references therein).

B. Intron-Encoded Proteins as Helpers in RNA Splicing

An initially surprising observation, that yeast mitochondrial mutants deficient in the synthesis of a functional cytochrome *b* were clustered in different complementation groups scattered through major portions of the mitochondrial genome, was resolved when electron microscopic mapping of mRNA–DNA heteroduplexes and Northern blotting analyses revealed the split character of the gene and allowed assignments of mutations to various exons and introns. Two categories of mutations, differing by their complementation phenotype, were found in d-introns: some were cis dominant, others were trans recessive. Whereas cis dominance was indeed expected, because an intron must be the *substrate* for some kind of splicing enzyme, trans recessiveness was somewhat unexpected, because it strongly suggested that, at least in specific instances, introns would specify synthesis of a *diffusible* product required for splicing activity. The unexpected was realized when sequencing data obtained by Jacq *et al.* (1980) supported the idea that the intron-encoded product was a protein. Indeed, the very fact that introns are spliced out of the final active RNA species would argue against their use as templates for quantitative (but not qualitative) synthesis of important proteins. The true extent of the mosaic nature of the *cob-box* gene, of the gene (*coxI*) coding for subunit I of cytochrome *c* oxidase, and of 23 S RNA gene was fully revealed by elucidation of the DNA sequence, which showed that several introns are in fact genes within genes, in that they contain long reading frames often in phase with the preceding exon. More than 20 different open reading frames located in introns of mitochondrial genes in *Saccharomyces, Schizosaccharomyces, Kluyveromyces, Aspergillus, Neurospora,* and *Podospora* have been discovered by DNA sequencing. The proteins that could be translated from these ORFs display various degrees of structural similarities among themselves and can be classified into two main structural groups. In the majority of cases the function of the proteins is still a matter of conjecture. In a now classical paper, Lazowska *et al.* (1980) proposed a model for the involvement of some of these intron-encoded proteins (called maturases) in splicing. In a few cases in which the analysis of mutants disclosed or suggested a function, a variety of biologically important roles was uncovered (see below).

It has now been shown for four yeast mitochondrial proteins encoded by introns 2 and 4 (*bI2* and *bI4*) of the cytochrome *b* gene and by introns 1 and 4 (*aI1* and *aI4*) of the *coxI* gene that they are required for splicing, being endowed with the "RNA-maturase" activity. The demonstration of this activity is based on several lines of evidence (for a review, see Kotylak *et al.*, 1985). (1) Mutations that induce translation termination or missense mutations that modify the structure of the protein are unable to excise the r-intron from the pre-mRNA. (2) These mutations are recessive and can be complemented in trans, even by another splicing-deficient mutant, on the condition, however, that it carries an intact intronic ORF. (3) Single amino acid replacement can activate an intron-encoded protein that normally does not have RNA-maturase activity but acquires it upon mutation (Dujardin *et al.*, 1982). (4) Maturases are trans acting, but, with two interesting exceptions, they seem to be capable only of mediating excision of their cognate intron ("matricidal" effect). The exceptions are the highly homologous fourth introns of the cytochrome *b* and *coxI* genes. Mutants with changes in the intronic reading frame of the cytochrome *b* gene are pleiotropic and fail to synthesize either cytochrome, due to an inability to splice the fourth intron of both genes. Conversely [item (3)], a single amino acid change in the *coxI* intronic reading frame leads to a functional maturase which can suppress the deleterious effects on splicing caused by mutations in the cytochrome *b* intron. (5) Physiological experiments are also consistent with the existence of maturases: nonsense and missense mutants are suppressed by antibiotics, allowing translation errors (Dujardin *et al.*, 1984). (6) Finally, the synthesis of maturases is regulated by negative feedback, due to their "matricidal" activity; the maturases accumulate and can be easily detected in mutants which do not excise the r-introns; the lengths of accumulated polypeptide chains are colinear with the positions of translation-termination mutations and the proteins react, as predicted, with antibodies raised against synthetic oligopeptides or translated from genetic constructs (Guiso *et al.*, 1984; Jacq *et al.*, 1984).

All these experiments show that RNA maturases are essential for RNA splicing of some r-introns but do not explain the mechanism of their activity. Based on the limited characterization of protein that accumulates in splicing-deficient mutants, it would appear that some maturases are active as fusion proteins, whereas others may undergo proteolytic cleavage, resulting in removal of domains encoded by upstream exons; some may also be directly synthesized from ORFs entirely contained within introns. Little is known about how maturases promote splicing, however, because no maturase has yet been isolated in an active form [although the bI4 protein has been partially purified from yeast mitochondria and from genetic constructs expressed in *E. coli*; the study of its biochemical properties *in vitro* has begun and direct proof of its activity has been obtained; its synthesis in the cytoplasm—i.e., with the changing of the mitochondrial code into cytoplasmic code after localized mutagenesis of appropriate codons by a protein targeted to enter mitochondria—resulted in the concept that removal of an intron could be obtained from a defective mutant (Banroques *et al.*, 1986)]. The ability of some precursor RNAs to self-splice (see below) makes a direct catalytic role (i.e., involving activation energy) less likely and tends to favor models in which these proteins stabilize the secondary or tertiary structure of RNA, thereby increasing efficiency and perhaps also specificity of the cutting and ligation steps.

Thus introns are required for yeast mitochondrial RNA splicing. In at least one case (*bI4*) they have a known physiological function, coupling the synthesis of cytochrome *b* and synthesis of coxI. Their actual evolutionary fitness has still, however, to be evaluated, because no clear advantage has yet been discovered in comparisons of the growth of strains with and without introns.

III. Ontogeny

A. Dynamics of Nuclear Intron Splicing

1. *Self-Splicing Introns*

The very existence of splicing prompts major biochemical questions. How is it performed? What are the biochemical elements involved? How accurately can specificity be achieved? What are the physiological functions involved? In 1981, Cech and his colleagues serendipitously made a major discovery in this respect: they found that an r-intron located in the large subunit of a nuclear rRNA was able to undergo spontaneous splicing (Kruger *et al.*, 1982). The large ribosomal RNA precursor of *Tetrahymena thermophila* can be self-spliced without proteins in the presence of Mg^{2+} and guanosine mononucleotide *in vitro* (Cech *et al.*, 1981; Kruger *et al.*, 1982). Several fungal mitochondrial intervening sequences have also been reported to be self-splicing (Garriga and Lambowitz, 1984; van der Horst and Tabak, 1985). These intervening sequences [group I, but also group II (Van der Veen *et al.*, 1986) introns] share a set of conserved sequences that are involved in determining their structures and mediating splicing and related reactions (Davies *et al.*, 1982; Michel *et al.*, 1982; Michel and Dujon, 1983; Waring *et al.*, 1982, 1984; Cech *et al.*, 1983; Arnberg *et al.*, 1986; Burke *et al.*, 1986).

Splicing of this class of RNA involves two distinct cleavage–ligation reactions, both occurring through transesterification mechanisms. Precursor RNA is initially cleaved by covalent addition of a guanosine molecule to the 5′ end of the r-intron, releasing the 5′ exon from covalent attachment to the r-intron. Subsequently, the 5′ exon adds to the 3′ exon in a ligation step. The detailed mechanism of the reaction has recently been reviewed by Cech and Bass (1986).

2. *Consensus Sequences and Lariat Formation*

Nucleotide sequence analysis has revealed that nuclear introns display consensus sequences at their extremities (Mount, 1982). However, the consensus that can be proposed in nearly all instances is rather poor: /GU·····AG/, with loose constraints in the 5′- and 3′-exon sequences. Biochemical analysis of the reaction has recently been performed by several groups (reviewed in Padgett *et al.*, 1986). The proposed mechanism is that splicing involves cleavage at the 5′-splice site of the intron, yielding a 5′-phosphorylated guanosine residue at the 5′ terminus of the intron, which then becomes joined to the 2′ hydroxyl group of the branch acceptor nucleotide within the intron. In the next step, the cleaved 3′ end of the first exon, likely to be bound to a splicing complex (see below), attacks the phosphate between the last (3′) nucleotide of the first intron and the first (5′) nucleotide of the second exon, releasing the intron as a lariat while joining the two exons. The branch acceptor nucleotide (usually an adenosine residue) does not seem to be embedded in a highly specific sequence, because, in the case of r-introns of higher eukaryotes, as shown by Wieringa *et al.* (1984), deletions of substitutions that remove all but 15 nucleotides adjoining the 3′-splice site of the large β-globin intron do not impair splicing. In the case of yeast nuclear gene introns, however, an internal sequence (UACUAAC; see Langford *et al.*, 1984) is essential for splicing; Pikielny *et al.* (1983) have mapped a reverse transcription stop in this region. Ruskin *et al.* (1984) have found that all β-globin introns contain a consensus sequence located in the region between 19 and 37 nucleotides upstream of the 3′-splice site. The consensus sequence is Y-N-Y-U-R-$\underline{A}$-Y (the underlined A represents the branchpoint). It loosely matches the yeast sequence. The sequence in which the branchpoint is embedded, identified by Padgett *et al.* (1984) for adenovirus, conforms to this consensus sequence except that the fifth residue is a pyrimidine rather than a purine. Some of the intron deletions examined by Wieringa *et al.* (1984) do not contain a consensus sequence but are efficiently spliced; its seems therefore that branch formation may occur elsewhere if a consensus sequence is lacking. This sometimes results in splicing of exons that are not adjacent in the wild-type mRNA (e.g., Brandt *et al.*, 1984; Speck and Strominger, 1985). Thus, *in vitro* and *in vivo* analyses of a number of splicing intermediates, as well as intron nucleotide sequence determinations, have suggested that, in higher eukaryotes, specificity is only poorly determined by the RNA sequence alone, whereas stringent sequence requirements are clear in the case of yeast.

3. *The Coptosome*

In spite of the existence of alternative splicing mechanisms, the maturation of pre-mRNA molecules usually does not contribute to diversity of gene expression. The question of splicing specificity and accuracy is, however, crucial to higher eukaryotic gene expression. Since self-splicing is certainly not a general phenomenon (and does not seem to occur in the case of pre-mRNA introns), this requires the existence of a catalytic architecture that must embody elements for both a "general" splicing machinery and molecules directing specificity. It must be emphasized that accuracy is a very stringent requirement: there can be no variation in the position of exon–intron borders, because shift of one nucleotide completely destroys the translation frame. In addition, it is observed that despite the existence of a large number of putative splice sites (because the apparent requirements in primary structure are quite loose), only the proper ones are usually used. Moreover, when splicing contributes to diversity in gene expression (e.g., Singer *et al.*, 1980; Nawa *et al.*, 1984; Rozek and Davidson, 1986, and references therein), the problem seems even more crucial because it then appears that this diversity, requiring accurate choice of alternative splicing sites, is performed despite a large background of putative sites (including the ones that are used in other environments (see Solnick, 1985a; Ruskin *et al.*, 1986, and references therein).

A large complex containing several small nuclear RNAs (U RNAs) has been described that permits general splicing (Flint, 1983, and Padgett *et al.*, 1986, for reviews; Brody and Abelson, 1985; Chabot *et al.*, 1985; Skoglund *et al.*, 1986; Vijayraghavan *et al.*, 1986). This complex has been termed "splicase" or "spliceosome." A better nomenclature would use only greek roots (as in other biological terms, such as chromosome), and I propose "coptosome" (from *κοπτω*, *I sew*) rather than such an oxymoron! If specificity is really a problem (and the extreme abundance of introns having unrelated structure suggests that it might be so), then it means that genes directing it must be present at a high level. We have seen how the specificity problem was solved in yeast mitochondria; it seems, therefore, worth asking whether a similar solution involving a heterodox translation of intron sequences might exist.

B. Nuclear Translation Revisited

Nuclear translation has been an open question for a very long time, and it was an object of debate until 1978 (Goidl and Allen, 1978). Nobody would argue against the fact that there remains a detectable amount of protein synthesis when the cytoplasmic translation machinery is blocked (for instance, by antibiotics such as cycloheximide); the question is, however, to ascribe this residual protein synthesis to its proper source. The general consensus is that if there is protein synthesis in the nucleus, its amount, as compared with protein synthesis in the cytoplasm, must be very low (probably less than 0.5%). However, it is also admitted that, if there is such nuclear protein synthesis, the corresponding proteins would be regulatory proteins of prime importance.

The molecular features of this hypothetical polypeptide synthesis show that the nuclear machinery must differ from the cytoplasmic machinery. In fact, the residual protein synthesis appears to be sensitive to inhibition by antibiotics (such as chloramphenicol) usually specific against prokaryotic protein synthesis machinery. This explains why it has been, at least in part, attributed to contaminating organelle protein synthesis. Authors claim that they have found a residual incorporation of amino acids, stimulated by aurintricarboxylic acid (Chatterjee *et al.*, 1977) and distinct from contaminating mitochondrial protein synthesis. In addition, a large body of data documents the observation that there is abundant *de novo* protein synthesis in the perinuclear space when new syntheses are induced (Leduc *et al.*, 1968; Avrameas, 1970), and polysomes have been visualized attached not only to the external face of the nuclear envelope but also inside the nucleus (cf. Franke and Scheer, 1974). These may, however, be preparation artifacts due to contaminating extranuclear polysomes.

Nuclear RNA is modified inside the nucleus before it is spliced and translocated to the cytoplasm. It has been shown, for instance, that capping and poly(A) addition (Salditt-Georgieff *et al.*, 1980) occur inside the nucleus before translocation. It appears therefore that nuclear RNA possesses the distinctive features required for initiation of protein synthesis with eukaryotic ribosomes (Kozak, 1978). This raises the question of the nature of the nuclear ribosomes that would be responsible for the hypothetical nuclear translation. From the above considerations one may conjecture that they possess features of eukaryotic ribosomes; on the other hand, their pattern of sensitivity to antibiotics would indicate that they differ from their cytoplasmic counterparts: thus, they might permit translation using a genetic code different from the cytoplasmic code. It is well established that ribosomal RNAs are synthesized in the nucleolus and are processed in the nucleus. They assemble into "immature" ribosomal particles after association with ribosomal proteins synthesized in the cytoplasm and shuttled into the nucleus by a specific, but as yet undefined, active process (Goldstein, 1974; Tsurugi and Ogata, 1984). These preribosomes, lacking a few ribosomal proteins, then translocate to the cytoplasm in an unknown way, but most likely through the nuclear pores (Schumm *et al.*, 1979). There the remaining ribosomal proteins attach to the immature particles to form mature 40 and 60 S ribosomal subunits. The nuclear preribosomes could very well perform a nuclear translation with some ambiguity in the coding process (see, for instance, ambiguity generated in some altered prokaryotic ribosomes) and be actively involved in the general splicing machinery: they would be translocated into the cytoplasm as the pre-mRNA is excised. If maturation of ribosomes were involved as a preliminary step for nuclear translation, one would expect a role of the nucleolus—where ribosomal RNA is transcribed and the first maturation steps of the ribosome occur—in the processing of hnRNA.

The second candidates as actors in the nuclear translation machinery are the granules found at the inner and outer faces of the nuclear pores. These structures consist of openings in the nuclear envelope;

they are delimited by two annuli, each made of eight granules somewhat larger than ribosomes and most likely consisting of an association of RNA fibers and proteins [see Franke and Scheer (1974) and other papers in the series "The Cell Nucleus" (H. Busch, ed.)]. One frequently sees that the granules of the outer annulus are close to and in continuity with polyribosomes bound to the outer nuclear membrane or extending to the cytoplasm. In addition, one observes that fibrillar threads are firmly attached with the granules on either annulus, especially the inner one. These fibers are connected to the nuclear matrix and are made of RNAs bound to proteins, and in some instances one may observe a ribosome-like particle bound to the ribonuclear protein (RNP) fiber in the middle of the nuclear pore lumen. Moreover, purification of pore complex material has revealed that a good deal of it is made of ribosomal RNA (Shaper *et al.*, 1979; Davis and Blobel, 1986, and references therein). This suggests that pore complexes are sites of final processing and assembly of ribosomes, at least at the inner side of the pore. Taken together, all these data are consistent with a translation process occurring at the nuclear envelope border, more precisely at the nuclear pores. This translation would be concomitant with maturation of both ribosomes and messenger RNA as nuclear RNA–preribosome complex crosses the pore lumen.

Thus, I think that the hypothesis of a nuclear translation (be it only translation at the nuclear border) deserves consideration. There does not exist, however, compelling evidence for it, although a large amount of data are consistent with this somewhat unorthodox hypothesis. The unexplained effect on some eukaryotic cell types of antibiotics inhibiting prokaryotic (and mitochondrial) protein synthesis might be ascribed to alteration of nuclear protein synthesis in particularly permeable cells [cf. toxicity of aminoglycosides and chloramphenicol action on lymphocytes (Yunis and Bloomberg (1964); Igaraoti *et al.* (1971); and see below]. When searching for open reading frames, many introns are found to harbor large coding regions, even if one keeps with the standard genetic code. Moreover it is difficult to define without ambiguity the nature of an open reading frame, especially if one considers the UGA codon: it has been found recently, for instance, that this codon might be translated by a specific tRNA molecule carrying a selenocysteine residue (Chambers *et al.*, 1986)! Finally, when one looks at the consensus found at the 5′ end of the intron, one frequently finds that the upstream exon reading frame is *not* interrupted, whereas a termination codon is present immediately out of frame (see, for instance, Garman *et al.*, 1986), which might imply that an incoming ribosome is part of the splicing machinery (Danchin, 1982). This is particularly relevant if one compares the self-splicing introns of bacteriophage T4, where the situation is precisely the opposite: a termination codon ends the first exon sequence (Chu *et al.*, 1986; Sjöberg *et al.* (1986). Another suspicion of an involvement of translation at the nuclear level comes from the study of the feedback that occurs from the cytoplasm into the nucleus; it has been found, for instance, that structures very similar to the prokaryotic attenuators exist in eukaryotes, and this corresponds in the former types of organisms to coupling between transcription and translation (see Yanofsky, 1983, for references and discussion)! A further demonstration that a coupling exists between transcription and translation has been obtained by Dabeva *et al.* (1986), who have found that there is autogenous regulation of splicing of the transcript of a yeast ribosomal protein gene. Whereas translation of the ribosomal protein is certainly occurring in the cytoplasm it is nevertheless clear that the corresponding protein intereferes with splicing, therefore occurring in the nucleus. An obvious refinement of such regulation would be that translation occurs at the nuclear envelope, and there interferes (positively or negatively) with splicing.

The reader who has followed up to now may already think I have indulged in too much speculation. I shall, however, add a few more speculative reflections to the ones already made. My main contention will be, in the following sections, that a major feature of eukaryotic gene expression is regulation by m-protein-dependent splicing of nuclear pre-mRNAs. Not only do I separate from the frequent view that introns are "selfish" DNA fragments, remaining without functions other than their own propagation since early terrestrial life, but I think that introns are in fact one of the most important and specific means evolved in eukaryotic cells to regulate gene expression, involving the nuclear envelope as a major instrument. I shall try here to show how this regulation might operate and what would be the consequence of such a regulatory process.

C. Membrane Heredity

Life is perfectly well expressed in prokaryotes, which are organisms extraordinarily well adapted to rapidly changing environments. (Imagine the life cycle of the plain *E. coli*: aerobic life in a poor environment, followed by extremely acidic pH, then alkaline pH, together with strong detergents and an almost complete absence of oxygen, competition with at least 40 different species for food supplied in very different conditions, in the presence of high concentrations of proteolytic enzymes, etc., and finally, within seconds, return to aerobic life . . . .) This remarkable adaptation shows that there exists no intrinsic reason for the requirement of a frontier separating the organized chromosomal DNA from the cytoplasm. Eukaryotic organisms, and especially metazoans, appear to have evolved, contrary to prokaryotes, by multiplying the frontiers (nuclear envelope, but also all sorts of cellular tissues that play the role of "skins" that separate organs or organisms from the changes in the environment of the external medium, so that variations are more and more buffered as species evolution proceeds).

For prokaryotes, rapidity of adaptation is derived from exquisite regulatory means allowing the coordinate expression of many genes. For this purpose, groups of genes are expressed as polycistronic operons, and appropriate metabolic controls of initiation and termination of transcription, coupled to regulation of translation rates as well as fast mRNA turnover, permit rapid and fine coordinate dosage of gene expression (Danchin and Ullmann, 1980). On the contrary, eukaryotic cells do not seem to possess—at least as a general control mechanism—polycistronic operons; controls of transcription initiation and termination do exist in eukaryotes, but the mRNA turnover is usually slow, and regulation at the level of initiation of translation seems important. All these observations suggest that eukaryotic gene expression differs to a large extent from prokaryotic gene expression. On the other hand, eukaryotes, especially metazoans, are differentiated organisms. This requires the presence of stable, hereditary diversity of given expression states of the genetic program for each class of cell type. Following Slonimski (1980), I shall propose that the function of the nuclear envelope is precisely to permit *heredity of expression of a given state of the genetic program.*

How can an envelope ensure, without alteration of the DNA structure as a prerequisite, a hereditary expression of the genome? An old elegant experiment of Cohn and Horibata (1959a,b) gives us an illustration of what may happen. Using the well-known lactose operon of *E. coli,* they have shown that memory of a past event may be kept for many generations when its molecular site is localized in the bacterial envelope: appropriate conditions of the external medium can maintain conditions for stable reproduction of the envelope state from generation to generation, while the envelope state can direct specific expression of a set of genes.

Thus, I wish to emphasize the separation between genetic heredity (i.e., hereditary transmission through the DNA sequence) and epigenetic heredity [i.e., transmitted from generation to generation but not resulting from alteration of a DNA sequence (Simionovitch, 1976)]. This distinction is somewhat related to the Darwinian separation between atavism (i.e., a trait which may be observed in some individual, but which may reappear after several generations) and heredity. It is also reminiscent of the well-known separation in physical sciences between *type* and *state* of a material system.

D. The Model

Bearing in mind these considerations, it is possible to contemplate the ultimate consequences of extending to the nucleus the mitochondrial maturase hypothesis (Slonimski, 1980); this can be visualized in the following model:

1. The eukaryotic cell nuclear envelope harbors control elements of gene expression responsible for epigenetic heredity.
2. Control elements are encoded by specialized introns of split genes. They are m-proteins translated at the nuclear envelope from nuclear pre-mRNA sequences.
3. The nuclear envelope possesses a limited number of binding sites for active m-proteins at loci where mRNA molecules are spliced from pre-mRNA and translocated into the cytoplasm (coptosome).
4. m-Proteins bound at the nuclear envelope constitute a network, i.e., they are associated by quaternary interactions, involving specific domains (introtypes), and are embedded in a specific structure of the envelope.
5. When a cell divides, m-proteins of the parental nuclear envelope are distributed among daughter nuclei. New sites for m-protein binding are available at this stage. When all things are kept equal, the network of introtypes ensures a stable reproduction of the envelope state (i.e., free sites are occupied by newly synthesized m-proteins reproducing exactly the situation present in the parent). Under appropriate conditions new m-protein species may colonize the available sites, thus forming a new introtype network.

I shall now detail the consequences of the model and see how this would fit with the available experimental data.

1. The Nuclear Envelope

Before justifying postulates and investigating some of their consequences I shall briefly outline the main features of the nuclear organization that seem relevant. The cell nucleus is a highly organized cell compartment (see Cook and Laskey, 1984). Apart from chromatin, composed of DNA, histones, regulatory proteins, and transcription machinery, ribonuclear protein fibers constitute the network, the nuclear matrix which connects to the nuclear envelope, itself composed of a lamina densa and nuclear pores (Davis and Blobel, 1986). The nuclear matrix consists of acidic proteins and RNA and is connected to the nuclear envelope. This envelope consists of a double-membrane structure internally coated by the lamina densa, made of "structural" hydrophobic proteins associated with a large amount of other ill-characterized proteins. The nuclear pores provide hydrophilic channels connecting the nucleoplasm with the cytoplasm and are thought to be the sites of nucleocytoplasmic exchange of macromolecules. Of particular interest is the possible role of the nuclear envelope in the processing of heterogeneous nuclear RNA and its translocation as messenger RNA to the cytoplasm. It has been shown that rapidly labeled nuclear RNA is preferentially attached to the nuclear matrix and Clawson *et al.* (1980) have proposed that mRNA can be translocated from the RNP to the inner nuclear membrane connected to the nuclear matrix. Herlan *et al.* (1979) have suggested that processing might occur as the nuclear RNA moves along the nuclear matrix toward the nuclear pores. Clawson and Smuckler (1980) have further substantiated this point. These authors have indeed shown that the complexity of RNA associated with the nuclear envelope is very high, thus showing that a predominant fraction of DNA transcripts is ultimately bound at such sites. It is clear from their results that the corresponding RNA sequences are not all present in cytoplasmic RNA and that the population is as complex as total nuclear RNA. Since the only nucleocytoplasmic connecting passages are the nuclear pores, these observations would suggest that splicing occurs at the nuclear pores themselves.

We have seen that likely localizations for nuclear translation are the inner and outer annuli delimiting the nuclear pores. We therefore make the tentative hypothesis that nuclear pores harbor the general splicing machinery (coptosome). This machinery is able to recognize the consensus /GU·····AG/ exon–intron sequences together with other secondary and tertiary RNA sequences required for splicing. It would be expected to yield a large amount of inaccurate splicing if there did not exist messenger proteins bound at appropriate sites around the nuclear pores. In this respect it seems interesting to note that lack of capping of pre-mRNA, although it does not prevent splicing, results in inaccurate splicing (Green *et al.*, 1983; Krainer *et al.*, 1984). We see, therefore, the nuclear matrix and lamina densa as structures where pre-mRNA is en route to the cytoplasm and as the reservoir of m-proteins. We assume that protein domains coded by introns (introtypes) allow m-protein to be located at sites where splicing is brought about. A typical example of an m-protein might be the small T antigen of SV40.

A central question for the theory—as well as for any theory involving membrane heredity—is therefore the problem of continuity of m-protein organization as the cell divides. Heredity will require that not only DNA replicates identical to itself. I certainly do not claim to have a definite answer to this question, but I emphasize again the fact that any non-DNA regulatory mechanism presents the same question. The organization of the cell nucleus seems entirely designed for preserving the continuity of the states of the controlling elements as nuclei divide. On the one hand, DNA is organized into chromosomes, which maximize the possibilities of continuity. DNA replication follows stringent and accurate replication rules and, in many cases, diploidy allows maintenance of continuity in DNA sequences. On the other hand, transcription continuity does exist and nuclear transcripts are folded into RNP fibers which organize the nucleoplasm in such a way that RNA processing is strictly coupled to the state of the nuclear envelope. This complex organization is yet poorly understood, and it does certainly involve splicing processes different from the hypothetical m-protein-dependent splicing. Finally, there is continuity of the two main components of the nuclear envelope, the nuclear matrix (and lamina densa) and the nuclear pores. When the cell divides, the organization of the nuclear envelope dissolves (Jost and Johnson, 1981), but its proteins are not degraded (Ely *et al.*, 1978) and are reassembled in the two daughter nuclei (Gerace and Blobel, 1980). Moreover, it appears that nuclear pores are not destroyed during division and that they segregate in the two daughter nuclei linked to the chromosome (Maul, 1977).

I admittedly cannot propose a detailed molecular mechanism for this fundamental continuity, especially because it may appear paradoxical as the lamina is spread throughout the cytoplasm during cell division, when there is complete nuclear envelope breakdown. However, this apparent destruction is certainly not random: at telophase, lamina reconstitutes in a way which shows that, although spread into the cytoplasm, it must be linked to an appropriate structural network (a role of cytoskeletal structures has been shown). Moreover, Jost *et al.* (1979) have shown that lamina is certainly involved in control of gene expression as reactivation of chick erythrocytes in heterokaryons is preceded by a considerable influx of laminar proteins into the erythrocyte nucleus. At present it appears likely that nuclear envelope fragmentation results from mechanical disruption of the membrane by microtubules, as proposed by Bajer and Mole-Bajer (1969). Reassembly would be triggered by structures bound to the DNA (including nuclear pores?) thanks to a cooperative action of the cytoskeleton. Thus the apparent disruption of the nuclear envelope would look more like an expansion followed by contraction of an organized structure.

2. *Regulation of Gene Expression by m-Protein-Dependent Splicing: Hypothetical Quantitation*

The progeny of a given cell must change its original protein synthesis from the initial predifferentiated state (protocyte) to the final differentiated state. This requires a change in pre-mRNA processing, corresponding to a complete replacement of the initial set of m-proteins (specifying the set of proteins that label the protocyte) by a new set (specifying the new markers), and this is done usually within 10 generations. This corresponds to an m-protein set of roughly $2^{10} = 1000$ molecules per nuclear envelope. Such a figure may be correlated to other data known, for instance, for immunoglobulin synthesis, and for the nucleus of lymphocytes as they differentiate. The average nucleus surface is about 100 μm^2; when it is occupied by contiguous pores, this corresponds to 1000–2000 pores. A secreting lymphocyte, for instance, has a nucleus covered with pores and the figure found is therefore of the same order of magnitude as that of m-proteins required for splicing immunoglobulin pre-mRNA. The turnover rate of these pre-mRNA molecules in the nucleus has been evaluated to be of the order of 1–2 minutes and this corresponds to the fastest RNA processing known in eukaryotic cells. The turnover of the pre-mRNAs corresponding to the doubling time of the cell would be about 1000 times slower. This matches with the following turnover: one m-protein molecule situated at one pore allows the doubling of the pre-mRNA processing of one gene into mature mRNA in one generation.

A given pore must permit the processing and translocation of several different pre-mRNA molecules. Since (in the model) processing requires the existence of a translational step, the time course of the process will be at least several seconds. As a consequence, within one generation time, a given nuclear pore will permit the maturation of at most 10^6–10^7 individual pre-mRNA molecules (assuming only one m-protein-dependent splicing event per gene). How does one correlate this figure with the expected number of m-proteins found at the nuclear envelope? The mean surface occupied by an m-protein would be of the order of 10 nm^2, and taking into account the fact that 90% of the nuclear surface must be occupied by pores and structural proteins, this leaves at most 10^6 m-proteins per nucleus. Once again this figure is of the order of magnitude of the pre-mRNA-translocating events occurring in one generation. This means that a given m-protein has the occasion to direct splicing from 1 to 10 times in one generation in the worst case, because of competition with the other m-proteins present around the nuclear pores. As a consequence, the number of identical m-proteins is multiplied at the nuclear envelope and may reach concentrations as high as one-tenth of the mRNA concentration. But if this is so, taking an average figure of pre-mRNA turnover of 10 minutes (corresponding to 100 m-proteins), then the nucleus cannot harbor more than 10^4 different m-protein species.

This line of reasoning is certainly an oversimplification, because many other arrangements would permit much variation of the estimates, but it is only used as an illustrative example. Here we reach a crucial point of our argument: it appears that the number of available sites for m-proteins in a nucleus is at least one order of magnitude smaller than the number of genes expected to be present in the mammalian genome. This strongly suggests, even for the most primitive eukaryotic cells, that there must exist a strong competition among m-proteins for their binding site at the nuclear envelope. The result of this competition will be selective stabilization of one expression of the genome, among several possible.

The selective stabilization process (see, for instance, Changeux and Danchin, 1976) requires that the number of m-proteins translated after a new nucleus has been assembled be larger than the number of sites available at its envelope. Selection will correspond to proper binding of some of these m-proteins in a network where interaction between introtypes will be fundamental. In other words, we see the regulatory state of a nuclear envelope as derived from a network of introtypes and anti-introtypes, each specifying its complement, so that a "replication" of the envelope structure can occur when the cell divides. If there is no gross alteration in the transcription pattern, then the state of the nuclear envelope of daughter nuclei will be completely determined by the state of the parental one.

The model is founded on the hypothesis that a given state of the nuclear envelope can direct a given expression of the genome. More precisely we assume, as did, for instance Dulbecco (1979), that translation of nuclear RNA into messenger RNA is the raison d'être of the nuclear envelope (see also Danchin and Slonimski, 1981). In addition, our scheme for m-protein-dependent splicing has an important consequence: expression of a given gene is not only regulated by its transcription into pre-mRNA but also by its maturation into mRNA. And, since maturation depends on the presence of m-proteins at the nuclear envelope, expression of a given gene will be finally dependent on the presence of a given m-protein at a specific binding site. The state of the nuclear envelope will therefore govern gene expression and any hereditary conservation of the state of a nuclear envelope will result in epigenetic heredity of the expression of a set of genes.

E. Implications of the Model: From Egg to Old Age

Because competition between m-proteins is stabilized in a selective fashion by the interactions between introtypes, only organized sets of m-proteins can be simultaneously present in a given nucleus. This means that there exist only domains of stability where only limited variation can occur: this constitutes the cell types.

1. Egg Determination

As a consequence, the genomic expression in a cell will reflect the organization of its nuclear envelope; in other words, the future development of the progeny of a cell is determined by its nuclear state, or, else, the phenotype is the direct image of the nuclear envelope state. The fundamental question of egg determination, for instance, might be asked back to the egg nuclear envelope organization. Fragmentation of this envelope as the initial cell divided would provide a general development scheme: as the nuclear envelope is a mosaic of porelike structures, the epigenetic heredity deriving from this organization will also have a mosaic-like character. One could well see the first stages of embryogenesis as direct consequences of the preorganization of the nuclear cell envelope. Another consequence of the model is that one might obtain epigenetic variants of a cell population rather than mutants (Simionovitch, 1976). This has been claimed by many authors but disputed by others (see Gorczynski and Steele, 1981; Brent *et al.*, 1981).

Each differentiated cell corresponds to an organized set of m-protein pathways which have led from the original egg to the differentiated state. The stability derives from the continuity rule based on the nuclear organization and on the specific restrictions imposed by introtype interactions, as stated above. This has, however, specific drawbacks. Competition for a limited number of sites, all things being kept equal, results necessarily in invasion, in a sudden cooperative fashion, of the majority of available sites by a minority of m-protein types [this is a consequence of the fluctuation in finite large numbers; see Azencott and Ruget (1977)]. Therefore, after a finite number of divisions the cell type must alter. Thus the preorganization of the egg nuclear envelope might result in a finite number of cell types and, eventually, a finite number of cells. This might be the specific cause of the well-defined small number of cells of certain primitive metazoans.

To escape the constraints imposed by competition of m-proteins for their binding sites at the nuclear envelope, the cell may have derived a strategy which extends the available surface of the nucleus. This may correspond to the lamellae found in many oocytes and interpreted as superpositions of several pore structures; this may also explain the formation of macronuclei in protists when differentiated cells have already undergone a large number of divisions. Finally, the major importance of the web of m-proteins for gene expression, especially in the egg, might be the underlying constraint of the allometry rules which imposes a limit on the dimensions of mononucleate cells.

2. *Cell Differentiation*

As we have seen, introtypes determine the possible arrangements of m-proteins [which should be present at the nuclear envelope: the nuclear lamina has been shown to be the site of expression of cell-type-specific proteins (Benavente *et al.*, 1985; Stick and Hausen, 1985)]. This gives a first, fundamental function to introns. Another aspect of intron function might also be exploitation of the organization of exons into RNA segments coding for individual, well-structured domains (see also below).

Appropriate splicing might couple a given domain to a series of different other domains, according to the presence of one or another m-protein at the cell envelope [it should be emphasized here that autogenous regulation of splicing (see Dabeva *et al.*, 1986) by protein domains expressed through alternative splicing might play a role similar to that attributed to m-proteins]. For instance, a given catalytic subunit might be associated with different regulatory structural subunits such that regulation or location of the catalytic activity in the cell is altered. It is most likely that such combinatorial splicing is fundamental in differentiation of tissues (Kemp *et al.*, 1980), not only by allowing varied locations of proteins at the cytoplasmic membrane [where the interaction between cells occurs; see, for instance, Singer *et al.* (1980)], but also by permitting redistribution of various catalytic activities inside the cell (King and Piatigorsky, 1983; Medford *et al.*,

1984; Breitbart *et al.*, 1985; Tunacliffe *et al.*, 1986; de Ferra *et al.*, 1985, and references therein). Receptors of various hormones, for instance, might be located at different places according to the presence of specific domains corresponding to combinatorial splicing of various exons. The fact that several oncogenes are analogs of receptors might be relevant in this respect (notice that viral oncogenes are often devoid of introns) (Ben-Neriah *et al.*, 1986). Modulation of catalytic activities could also follow modulation of binding domains: this might be illustrated by the case of acetylcholinesterase, in which the globular catalytic subunit is linked to fibrillar domains which appear to vary according to the tissue considered (Massoulié and Bon, 1982). Some of the variations of ubiquitous proteins such as actin or myosin, which appear to vary from cell type to cell type, might also be accounted for by a similar mechanism.

As in the case of the immunoglobulin-producing cells, many stem cells will be pluripotential (Antoine *et al.*, 1979) and the interaction with proper environmental effectors will select a specific differentiation pathway. The model predicts that, during specification, certain cells will be able to express *together* several of their potentialities, whereas they will end as expressing only one. The differentiation of neural stem cells in the sympathetic ganglia of rats might be an example of such a mechanism. Indeed, one observes that such cells can differentiate into neurones producing either catecholamines or acetylcholine as neurotransmitters, according to environmental parameters, and that there exists a step where they produce both transmitters (Weber, 1979; Le Douarin, 1980).

In summary, introns would therefore allow combinatorial regulation through differential splicing of various introns mediated by various m-proteins [this would correspond to the trans-acting proteins that are postulated to explain developmentally regulated splicing; see, e.g., Rozek and Davidson (1986)] and through cytoplasmic translation of various combinations of exons. This would ensure an extremely rich actualization of the eukaryotic genetic program, permitting maintenance of an optimum homeostasis of the cell. In counterpart, however, the model implies that the pathways leading to differentiated cells might end in many pathological alterations, expected either as deriving from spontaneous deviations of the normal scheme or as experimental alterations of the scheme.

Before considering pathological alterations of the splicing regulation process one should consider how it may normally be modulated by outside effectors. An effector (hormone, growth factor, etc.) would bind to a cytoplasmic membrane receptor and there be internalized by receptor-mediated endocytosis (see Yamamoto, 1985, for the case of corticoid receptors). Then it would go to the nucleus and interact with m-protein domains corresponding to the receptor-binding domain; this in turn would affect splicing, binding of the m-protein to the nuclear envelope site, or continuity of this binding as the cell divides. As a result the m-protein network will shift from one state to another state and the differentiated state of the cell will therefore be changed. A model of the specification of the immune response has been developed in detail elsewhere (Danchin and Slonimski, 1981) and can be seen as an illustration of what may happen, in addition or in parallel to the direct effects of transcription of various effectors.

Many experimental pathogenic deviations of this scheme may be thought of. The less specific is the temperature or solvent effect on differentiation. Heat shock has been found to interfere with differentiation, and it has been shown to affect splicing specifically (Yost and Lindquist, 1986). It also is known that dimethyl sulfoxide alters cell differentiation (especially of Friend leukemia cells) or that progeny of alcoholic mothers are strongly affected in their developmental pattern; this may be accounted for by many models of differentiation but fits perfectly well with the crucial role we give to quaternary interactions in splicing. More specific would be alterations of the nuclear translation machinery by appropriate antibiotics. The literature has given a list of pathological side effects of many antibiotics especially active against prokaryotic organisms. The best-documented effects are brought about by aminoglycosides and chloramphenicol. In the former case one observes essentially ototoxity due to specific cell degeneracy in the ciliated cells responsible for the first transformations of auditory signals (Wersall *et al.*, 1969; Igaraoti *et al.*, 1971); whether this is related to the process of protein synthesis is uncertain. The case of chloramphenicol is different: this antibiotic inhibits prokaryotic protein synthesis (Weisberger and Wolfe, 1964) as well as protein synthesis in eukaryotic organelles; it is also assumed to interfere with the hypothetical nuclear protein synthesis. The most clear-cut effects of chloramphenicol occurs in the blood system. Early studies have endeavored to discover its molecular basis without success. The only clear-cut evidence shows that the antibiotic prevents *de novo* commitment of immunoglobulin synthesis in lymphocytes (Ambrose and Coons, 1963). The effect of chloramphenicol occurs at very low concentrations (less than 1/20 of the concentration required to block mitochondrial protein synthesis) and, in the first days of the specification of the immune response, already committed cells are much less sensitive to this antibiotic (Couderc *et al.*, 1983). Other effects have also been demonstrated in different tissues, such as nerve cells (Ramirez, 1973; Bertolini and Poggioli, 1981). These results appear to fit well with the model; they should be considered, however, with caution because they are very indirect. Finally, it was found that splicing was affected in a differential manner in adenovirus in the presence of translation inhibitors, thus showing a direct effect of translation on splicing (Persson *et al.*, 1981).

3. *Aging and Cancer*

As we have seen, competition among m-proteins for a limited number of binding sites at the nuclear envelope will result in sudden invasion of the envelope by a small number of different m-protein species. As a consequence, if it has kept its differentiated form, the cell will lose some of the components required for continuing division: a differentiated cell, therefore, will often die after a finite number of divisions. This prediction corresponds to a very controversial discussion about *in vitro* culture of normal differentiated tissues (Hayflick, 1979); it appears, however, that, at least in some cases, differentiated cells do indeed cease dividing after a precisely defined number of generations. Our model will therefore tend to explain aging in a way similar to the "error catastrophe" theory (Orgel, 1963), and not as due to errors in the replication, transcription, or translation machineries, but in the accuracy required for precise continuity of reproduction of the nuclear envelope state as cells divide. The alternative to this aging constraint imposed on each type of differentiated cell will be that the network of m-protein which invades the available sites can direct expression of genes necessary for rapid cell multiplication at the expense of cell differentiation. The alternative to cell progeny death is therefore uncontrolled cell multiplication. Aging will result in lack of ability to multiply or in tumor-forming activity. This sheds a new light on the way cells might turn out to be cancerous, and I think the consequences of the unorthodox view presented above should therefore be investigated in detail.

Other perturbations may also result in stable alterations leading to aging and cancer: overproduction of complex protein subunits containing alterations in the amino acid sequences joining two well-defined domains [this is analogous to heavy chain disease; see Seligmann *et al.* (1979)] in the case of immunoglobulins, synthesis of aberrant proteins containing one domain coupled to another unusual domain [this might be found in some cases of amyloidosis; see Rosenthal and Franklin (1977)], absence of protein synthesis while nuclear RNA sequences containing the mRNA sequence are detectable (defects in RNA processing), mutations found in one gene (especially located in introns) affecting the expression of other genes, etc. In addition, one should find many pathological defects leading to alterations of specific differentiation steps. If, in particular, a splicing defect has prevented the formation of membrane-bound receptors of specific effectors, or has interfered with receptor-mediated endocytosis, differentiation relying upon interaction of the effector at the nuclear envelope will be defective.

Apart from these pathological spontaneous alterations, corresponding most often to genetic defects, one will find alterations induced by environmental factors. Analogs of hormones or growth factors will interfere with the normal differentiation steps and eventually cause epigenetic hereditary modifications (this is general and could be expected for other models of differentiation; our prediction, however, is more precise, because we predict that the modifying events must occur at the nuclear envelope). This may be caused in two ways, either by interaction of molecules with the exon types of m-proteins, thus perturbing splicing, or interference with the organization of the introtype network; in particular, proteins organized as are the m-proteins (i.e., composed of a hydrophilic domain coupled to a hydrophobic one, similar to those binding to the nuclear envelope) might underly the separation between the two types of action. Inducers will act on the DNA and trigger transcription of a set of pre-mRNA which, after maturing to mRNA, allows uncontrolled cell multiplication; promoters will permit maturation of these pre-mRNAs by allowing the proper positioning of the corresponding m-proteins at the nucleus.

The fact that frameshift mutagens are usually powerful carcinogens might also be related to their potential action on translation of the nuclear envelope of specific intron sequences (change in the reading frame), whereas many viruses might be oncogenic not only because of such action (insertion in the host DNA) but because they produce m-protein-like products [see, for instance, small T and middle T antigens in SV40 or polyoma virus; see Tooze (1980), for review] or because they interfere with proper formation of the nuclear envelope, especially during mitosis [this might be the function of certain oncogenes, which perturb protein phosphorylation, since phosphorylation of some lamina proteins appears to be involved in formation–disruption of the nuclear envelope; see Gerace and Blobel (1980)]. Alternatively, the phosphorylation pattern of proteins of the nucleolus (Daskal *et al.*, 1980) might be modified with immediate consequences to the pre-mRNA maturation process.

IV. Phylogeny

As our knowledge of molecular biology progresses, we are in a position to gain more insight into the origin of life and the evolution of the first living organisms. Whereas the most popular thinking has long been to perceive prokaryotes as primitive organisms, this is no longer the case, and many data substantiate the viewpoint that life originated from single cells that bear a resemblance to eukaryotic organisms. It seems also clear that before evolution of present-day organisms, which have elaborate means to synchronize duplication of their genome with membrane division, a more random division process was at work, in which fusions and divisions occurred frequently. In addition, recent data on the catalytic function of RNA molecules (see Cech and Bass, 1986, for review) have given RNA a central position in the very first steps of the origin of present-day organisms: primitive cells could be seen as having split genomes composed of a collection of RNA molecules, permitting synthesis of proteins necessary for the completion of "statistical" organisms, defined by their capacity to reproduce a significant number of similar organisms. Evolution has then selected for two main forms, based on accuracy of reproduction: this required, first, a more stable genome, and DNA could fill a gap that the more reactive RNA could not, and, second, "buffering" of the environmental pertubations. Thus cells with a genome as compact as possible were selected, cells that could adjust to their metabolism as quickly as possible, in a variable environment—these are Eubacteria and Archaebacteria; a second way to adapt was to multiply the physical frontiers with the environment, even though it might slow the pace of reproduction, and after quite a few trials and errors, including fusions and divisions, this constituted the eukaryotic cell, with its many membranes (and genomes), followed by evolution toward the metazoa.

A major question to be asked about the existence of split genes is that of the origin of introns. Two divergent explanations prevail: (1) introns are recent "discoveries" of RNA structures and are able to spread into the genomes of mitochondria or higher cell nuclei, behaving as "selfish" nucleic acids [see Orgel and Crick (1981), and Doolittle and Sapienza (1981), for the behavior of selfish DNA] or (2) introns are very old remnants of original living cells and are continuously evolving from an original structure that displayed the special catalytic properties of RNA molecules, as discovered by Cech and his co-workers. A definite and adequate answer can be proposed after the study of mitochondrial introns of lower eukaryotes, and I shall present it first. More speculative proposals stem from the study of higher eukaryotic pre-mRNA structures, and will be presented at the end.

A. The Maturase Descent

Evidence presented above, mainly derived from experiments performed in P. Slonimski's laboratory, has strongly substantiated the existence and function of RNA maturases in splicing. A surprising observation is the apparent redundancy in the high number of maturases required for completing a functional mRNA. Several different types of functions have been recognized in the intron-encoded ORFs in addition to the function of maturase. Some seem to exhibit cryptic function, suggesting that they are strongly related phylogenetically: the maturase coded by intron *bI4* is necessary for the splicing of intron *aI4*, but a single mutation in the reading frame of intron *aI4* permits synthesis of a maturase that could restore correct splicing (Dujardin *et al.*, 1982). Whereas redundancy might address the question of splicing specificity, it has been strongly suggested that other functions are encoded by introns.

The "polarity of recombination" phenomenon (see Dujon *et al.*, 1976), for instance (which results from the presence of two allelic forms, ω^+ and ω^-, of the gene coding for the large ribosomal RNA, differing by the presence in ω^+ and absence in ω^- of an intron of ~1 kb inserted into the rRNA-coding sequence), is explained by the observation that the intron behaves like an infectious or transposable element (see Jacquier and Dujon, 1983). Analysis of mutants defective in the polarity of recombination showed that the intron-encoded protein is essential for the transposition. It induces *in vivo* a double-stranded break in ω^- DNA by recognizing a specific sequence exactly at the place where the intron is inserted; the genetically engineered constructs expressed in *E. coli* confirm the endonuclease activity of the protein (Colleaux *et al.*, 1986, and references therein). Interestingly, the recognition site resembles the site recognized by the mating-type switching endonuclease of the yeast nucleus (Kostriken and Heffron, 1984).

A further analysis has then shown that ORFs present in group-II introns are related to each other and to retroviral reverse transcriptase (Michel and Lang, 1985). This corresponds to a third intron-encoded function, which could also be important in the spread of introns.

A fourth function of maturases has recently been discovered (Kotylak *et al.*, 1985). The experiments that led to this unexpected discovery are based on genetic crosses between two sets of yeast strains, *S. cerevisiae* and *Saccharomyces douglasii*, followed by molecular analyses of mtDNA and mtRNA. Intron elimination, yielding a functional *coxI* gene from a couple of defective mutants, is achieved by a quite unexpected process which reshuffles completely, at the DNA level, the *coxI* gene. In these experiments, recombinants have acquired three introns, *aI1, aI2,* and *aI3,* from the *S. cerevisiae* mtDNA, have eliminated the intron *aI4-α* (present in *S. douglasii*), making the exon *A4* continuous, and have kept all the downstream introns, such as *aI4-β*, etc. (and its exon–intron boundary with *A4*), as in *S. douglasii*. The *coxI* gene is now a chimera: the first half of it has the *S. cerevisiae* structure and the other half has the *S. douglasii* structure. It was thus demonstrated that a recombination event (crossing-over or gene conversion) had taken place. The first position of this event could be localized in the homologous parts of exons *A4* and *A4-β* while the second position must be located in (or upstream of) the exon *A1*. What was extraordinary was that rearrangement of the *coxI* gene between heterologous DNA genes was dependent upon the presence of the active maturase coded by intron *bI4*. Thus, the potential for creating a new genomic organization is much greater than in the case of the ω^+ protein, which acts in selfish way by spreading the intron *rI1* that encodes it, not creating any new genes but simply increasing the frequency of the preexisting one. It should be noted here that this "recombinase" potential of maturases is precisely of the type that is assumed to exist (and not yet identified) in the case of immunoglobulin genes (see Gough, 1981, for a review).

Maturases are thus endowed with several activities on nucleic acids, mainly related to shuffling of segments of RNA or DNA. It seems clear that RNA molecules must have played an important role in the first evolutionary steps in the origin of life (and this hypothesis is more prominent now that RNA molecules have been shown to possess catalytic activities; see below). The question is therefore asked whether maturases are remnants of a past situation, when isolated RNA molecules were combined to generate new types of catalytic functions. Comparison of the central consensus in maturases has revealed a DGDG sequence that seems to be highly preserved, and it has been found (P. Slonimski, personal communication) that the same sequence, together with other conserved amino acids, was also present in the reading frame of an intron found in a gene from the archaebacterium *Desulfurococcus mobilis* (Kjems and Garrett, 1985). This adds support to the idea that maturases are very "old" proteins, perhaps involved in the generation of evolved genomes. In that respect it seems clear that eukaryotes have undergone many less generations than fast-growing bacteria. Thus the absence of introns in eubacteria (one should note, however, the presence of introns in bacteriophage T4) and the group-I intron present inside the gene for thymidylate synthase, harboring an ORF (Chu *et al.*, 1986; Sjöberg *et al.*, 1986), which is usually perceived as suggesting a recent invasion, cannot be taken as an indication of the absence of introns in the original protobiont. It appears that two divergent processes might be at work: loss of introns present from very early times, and spreading of transposon-like introns.

It is likely that, evolutionarily, protein folding was mainly a means for lowering the entropy of activation in chemical reactions, rather than the energy of activation, the latter being accomplished mainly by specific chemicals that serve as present-day coenzymes. Thus maturases are better seen as providing acceleration of naturally occurring processes, such as self-splicing, and as permitting splicing accuracy. Coevolution of the RNA molecule and the protein structure would have led to correct splicing of RNA no longer able to perform the reaction at a reasonable rate in the absence of the maturase. The fact that the maturase is coded by the RNA it is acting upon certainly permits such coevolution. It can also be interpreted as another indication of the presence of maturase-like proteins in the early steps of evolution.

This "egg and hen" problem (who was first: intron without ORF or intron with ORF?) is better understood as follows: coevolution would have led to the divergence of ORF functions and intron structure, leading to group-I and -II RNAs. In turn, each class would have led to the evolution of self-splicing forms that could dispense with the ORF product, thus permitting evolution either to loss of function and disappearance, or to new functions (see Scheme 1).

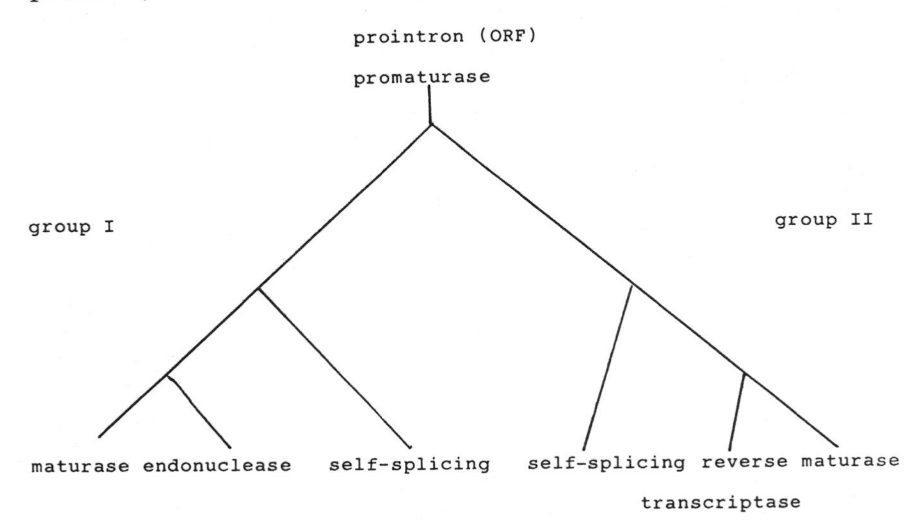

SCHEME 1

In this process introns are perceived as coding rather than noncoding entities.

B. Pre-mRNA Introns

Investigation of the present-day structure of exons and introns as well as of splicing tells much about the constitution of genes and genomes. Self-excision means that the converse (i.e., integration) is also possible: a population of introns might evolve to synthesize a complex genome, and, indeed, Zaug and Cech (1985) have shown that introns are able to oligomerize, even in the absence of proteins. Also, the fact that mitochondrial introns could encode proteins related to endonuclease, reverse transcriptase, and recombinase suggests that they might be involved in the constitution of a very primitive DNA genome. But what is known about today's proteins? Speculation on the phylogenetic origin of introns led to the idea that intron localization might be related to structural features of proteins. Gilbert (1978) proposed, for instance, that introns might delimit domains of the protein that had been associated through exon shuffling during phylogenesis. And indeed, this is consistent with the localization of introns in genes, such as the immunoglobulin gene (or related genes such as the T-cell receptors or the histocompatibility genes). This is, however, certainly different in other cases and Gō (for a review, see Gō, 1985) has proposed that introns might border DNA sequences coding for structural elements smaller than domains, which she called "modules." While this hypothesis is consistent in the case of the globin genes (and corresponds to the study of Gō, who has been able to predict that an intron has been lost during evolution in the central globin module of vertebrate genes), it probably cannot be generalized to all introns, and it seems likely that classification of introns will require collection of more data before it will be understood. In any case it seems likely that introns that are relics of the past are placed at locations that are related to the structure of the gene product. Finally, there are cases where introns do not seem to be present in certain sites. A thorough structural analysis of exon length, suggesting that proteins are composed of building blocks that were limited by the statistical presence of translation-termination codons is, however, somewhat challenging to the structural hypothesis (Senapathy, 1986; see also Rogers, 1986). A consequence of this latter work, coupled to the hypothesis of the coevolution of primordial "catalytic" introns coding for nucleic acid-"wielding" proteins (following P. Slonimski's terminology), would be that evolution would have provided a means to synthesize RNA adducts to existing introns, the adducts having more random structures and some coding capacity. By trial and error these new sequences would have led to present-day exons (see Naora and Deacon, 1982, and references therein), while introns were losing their own coding capacity (Moore, 1983); thus, by genetic takeover (Cairns-Smith, 1982), exons would have replaced introns in their coding potential by providing a much larger set of potential activities. In any event, the presence of introns having somewhat related structures would be extremely important in phenomena like gene conversion and would have a major role in evolution (Dover, 1986).

Evolution of higher eukaryotic introns, which may be derived from group-II introns, is visible in that they are no longer able to undergo self-splicing. A complex machinery has replaced a restricted mechanism to give it a broad range of specificities. In this process, as we have seen, there is formation of a lariat, followed by transesterification with the upstream exon: this opens the possibility for trans splicing, with another incoming exon (Inoue *et al.,* 1985; Solnick, 1985b). A general trans-splicing mechanism may even be at work in exotic organisms such as protozoa (Borst, 1985). This would increase genome fluidity, if reverse transcriptase was involved. Reverse transcription is certainly still playing an important function in genome evolution, and it is probably a major source of intron loss: this is easily seen in certain viral sequences of oncogenes that lack the intron sequences present in their cellular counterparts. This is also visible in the fact that pseudogenes often correspond to sequences without introns (see Chen *et al.,* 1982, for instance). A selective pressure for the presence of introns is, however, perceptible in the fact that proper translocation of mRNA into the cytoplasm has been shown in several instances to require the presence of introns (Gruss and Khoury, 1980). Indeed, if they play a major role in differentiation, as I have proposed in the preceding section, there must exist a very strong pressure for their conservation in crucial positions. In this respect it seems worth noticing that yeasts have only a few introns that seem to be located in the 5′ regions of the genes (thus suggesting 3′–5′ orientation of their elimination).

A final function might have played a major role in intron evolution. Gene expression is organized in such a way that there is a large amplification of the response from the gene to the final protein. Typically, in prokaryotes, one mRNA molecule is translated 20 times. In eukaryotes the amplification factor is extremely large because mRNAs are very stable entities, and it is typical to obtain 1000 proteins from a single mRNA molecule. It becomes therefore obvious that the problem of accuracy at the transcription level is quite compelling, because any error at this level will be reproduced in the final product 1000-fold. Although there certainly exist appropriate rewriting rules enacted during transcription, it is likely that other correction processes can operate on the transcript. And I propose that m-protein-dependent splicing can perform such a function. Indeed, mRNA can only exist after splicing of its nuclear precursor. How is it possible to protect against error in the exons by intron splicing? One mechanism would be that the entire pre-mRNA structure is folded into a tertiary conformation involving both introns and exons so that proper folding would be prevented when errors are present in the exons. Alternatively, it could be that a product (a polypeptide, etc.) specified by the exon is required for splicing. Whereas no sequence data have yet been analyzed to assess the former hypothesis, the second one fits very well within the m-protein model, since the exon domain of the maturase is supposed to act as a docking adaptor that accurately fits the pre-mRNA molecule to the catalytic site of a splicing enzyme. Thus, besides permitting regulation of gene expression, maturases would contribute to accuracy of mRNA formation.

V. General Conclusion

The fate of highly speculative hypotheses is often that they rapidly fall into oblivion because it is soon found that they are inadequate in some aspect. This was the case, for instance, of the ingenious theory of protein synthesis and replication proposed in 1952 by Dounce. This theory, which was well ahead of many further hypotheses, was obscured by the discovery of the DNA double helix in 1953, although it must have been a most important constructive step on the pathway to the double helix. The theory presented above might even be less accurate than the theory of Dounce and, indeed, it has a drawback in common with Dounce's theory: Dounce postulated that protein synthesis occurred in the nucleus! We do not make such an impossible claim, but assume that there might exist some specific protein-synthesizing machinery at the inner border of the nuclear envelope. . . .

Despite the difficult judgment of history I have dared to propose a quite unorthodox series of hypotheses, but I hope that, be it only for the sake of stimulating reflection about gene organization and genetic and epigenetic heredity, the theory will have some constructive consequences. The recent major discovery of Cech *et al.* about intron structure and catalytic properties might open a new way for looking at introns and even create a new understanding of the functions of nucleic acids, but this increases the need for a largely open series of reflections about intron structure and function; the present attempt is merely meant to suggest some interesting ways for considering this problem.

Acknowledgments

The theory presented herein owes much to P. Slonimski, and his contribution is clearly visible in many aspects of the material discussed. It is only owing to circumstantial events that he did not present the model himself in 1986. An account of his ideas can be found in his 1980 Comptes Rendus article. I thank him for all the exciting discussions we have had in the past 8 years, and I wish to dedicate this paper to him.

References

Abelson, J. (1979). RNA processing and the intervening sequence problem. *Annu. Rev. Biochem.* **48,** 1035–1069.

Ambrose, C. T., and Coons, A. H. (1963). Studies on antibody production. VIII. The inhibitory effect of chloramphenicol on the synthesis of antibody in tissue culture. *J. Exp. Med.* **117,** 1075–1088.

Antoine, J. C., Bleux, C., Avrameas, S., and Liacopoulos, P. (1979). Specific antibody-secreting cells generated from cell producing immunoglobulins without detected antibody function. *Nature (London)* **277,** 218–219.

Arnberg, A. C., Van der Horst, G., and Tabak, H. F. (1986). Formation of lariats and circles in self-splicing of the precursor to the large ribosomal RNA of yeast mitochondria. *Cell* **44,** 235–242.

Avrameas, S. (1970). Immunoenzyme techniques: Enzymes as markers for the localization of antigens and antibodies. *Int. Rev. Cytol.* **27,** 349–385.

Azencott, R., and Ruget, G. (1977). Mélanges d'équation différentielles et grands écarts à la loi des grands nombres. *Z. Wahrscheinlichkeitstheor. Verw. Gebiete* **38,** 1–54.

Bajer, A., and Mole-Bajer, J. (1969). Formation of spindle fibers, kinetochore orientation and behaviour of nuclear envelope during mitosis in endosperm. *Chromosoma* **27,** 448–484.

Banroques, J., Delahodde, A., and Jacq, C. (1986). Mitochondrial RNA maturase gene transferred to the yeast nucleus can control the mitochondrial mRNA splicing. *Cell* **46,** 837–844.

Benavente, R., Krohne, G., and Franke, W. W. (1985). Cell type specific expression of nuclear lamina proteins during development of *Xenopus laevis*. *Cell* **41,** 177–190.

Ben-Neriah, Y., Bernards, A., Paskind, M., Daley, G. Q., and Baltimore, D. (1986). Alternative 5′ exons in c-abl mRNA. *Cell* **44,** 577–586.

Bertolini, A., and Poggioli, R. (1981). Chloramphenicol administration during brain development: Impairment of avoidance learning in adulthood. *Science* **213,** 238–239.

Borst, P. (1985). Discontinuous transcription and antigenic variation in trypanosomes. *Annu. Rev. Biochem.* **55,** 701–732.

Brandt, C. R., Morrison, S. L., Birshtein, B. K., and Milcarek, C. (1984). Loss of a consensus splice signal in a mutant immunoglobulin gene eliminates the CH_1 domain exon from the mRNA. *Mol. Cell. Biol.* **4,** 1270–1277.

Breathnach, R., and Chambon, P. (1981). Organization and expression of eucaryotic split genes coding for proteins. *Annu. Rev. Biochem.* **50,** 349–383.

Breitbart, R. E., Nguyen, H. T., Medford, R. M., Destree, A. T., Mahdavi, V., and Nadal-Ginard, B. (1985). Intricate combinatorial patterns of exon splicing generate multiple regulated troponin isoforms from a single gene. *Cell* **41,** 67–82.

Brent, L., Rayfields, L. S., Chandler, P., Fierz, W., Medawar, P. B., and Simpson, E. (1981). Supposed lamarckian inheritance of immunological tolerance. *Nature (London)* **290,** 508–512.

Brody, E., and Abelson, J. (1985). The "Spliceosome": Yeast premessenger RNA associates with a 40S complex in a splicing-dependent reaction. *Science* **228,** 963–967.

Burke, J. M., Irvine, K. D., Kaneko, K. J., Kerker, B. J., Oettgen, A. B., Tierney, W. M., Williamson, C. L., Zaug, A. J., and Cech, T. R. (1986). Role of elements 9L and 2 in self-splicing. *Cell* **45,** 167–176.

Cairns-Smith, A. G. (1982). "Genetic Takeover." Cambridge Univ. Press, London and New York.

Cech, T. R., and Bass, B. L. (1986). Biological catalysis by RNA. *Annu. Rev. Biochem.* **55,** 599–629.

Cech, T. R., Zaug, A. J., and Grabowski, P. J. (1981). *In vitro* splicing of the ribosomal RNA precursor of *Tetrahymena*: Involvement of a guanosine nucleotide in the excision of the intervening sequence. *Cell* **27,** 487–496.

Cech, T. R., Kyle-Tanner, N., Tinoco, I., Jr., Weil, B. R., Zuker, M., and Perlman, P. S. (1983). Secondary structure of the *Tetrahymena* ribosomal RNA intervening sequence: Structural homolog with fungal mitochondrial intervening sequences. *Proc. Natl. Acad. Sci. U.S.A.* **80,** 3903–3907.

Chabot, B., Black, D. L., LeMaster, D. M., and Steitz, J. A. (1985). The 3′ splice site of pre-messenger RNA is recognized by small nuclear ribonucleoprotein. *Science* **230,** 1344–1349.

Chambers, I., Frampton, J., Goldfarb, P., Affara, N., McBain, W., and Harrison, P. R. (1986). The structure of the mouse glutathione peroxidase gene: The selenocysteine in the active site is encoded by the "termination" codon, TGA. *EMBO J.* **5,** 1221–1227.

Changeux, J. P., and Danchin, A. (1976). Selective stabilization of developing synapses as a mechanism for the specification of neuronal networks. *Nature (London)* **264,** 705–712.

Chatterjee, N. K., Dickermann, H. W., and Beach, T. A. (1977). Isolation of a distinct pool of polyribosomes from nuclei of adenovirus infected HeLa cells. *Arch. Biochem. Biophys.* **183,** 228–241.

Chen, M. J., Shimada, T., Davis-Moulton, A., Harrison, M., and Nienhuis, A. W. (1982). Intronless human dihydrofolate reductase genes are derived from processed RNA molecules. *Proc. Natl. Acad. Sci. U.S.A.* **79,** 7435–7439.

Chu, F. K., Maley, G. F., Maley, F., and Belfort, M. (1984). Intervening sequence in the thymidylate synthase gene of bacteriophage T4. *Proc. Natl. Acad. Sci. U.S.A.* **81,** 3049–3053.

Chu, F. K., Maley, G. F., West, D. K., Belfort, M., and Maley, F. (1986). Characterization of the intron in the phage T4 thymidylate synthase gene and evidence for its self-excision from the primary transcript. *Cell* **45,** 157–166.

Clawson, G. A., and Smuckler, E. A. (1980). On the nature of nuclear envelope associated RNA. *Biochem. Biophys. Res. Commun.* **96,** 812–816.

Clawson, G. A., James, J., Woo, C., Friend, D., Moody, D., and Smuckler, E. A. (1980).

Pertinence of nuclear envelope nucleoside triphosphatase activity to ribonucleic acid transport. *Biochemistry* **19**, 2748–2756.

Cohn, M., and Horibata, K. (1959a). Inhibition by glucose of the induced synthesis of the β-galactoside enzyme system of *E. coli*. Analysis of maintenance. *J. Bacteriol.* **78**, 601–612.

Cohn, M., and Horibata, K. (1959b). Analysis of the differentiation and of the heterogeneity within a population of *E. coli* undergoing induced β-galactosidase synthesis. *J. Bacteriol.* **78**, 613–623.

Colleaux, L., D'Auriol, L., Betermier, L., Cottarel, M., Jacquier, G., Galibert, A., and Dujon, B. (1986). Universal code equivalent of a yeast mitochondrial intron reading frame is expressed into *E. coli* as a specific double strand endonuclease. *Cell* **44**, 521–533.

Cook, P. R., and Laskey, R. A. (1984). Higher order structure in the nucleus. *J. Cell Sci. Suppl.* **1.**

Couderc, J., Perrodon, Y., Ventura, M., Liacopoulos, P., and Danchin, A. (1983). Specification of the immune response: Its suppression induced by chloramphenicol *in vitro*. *Biosci. Rep.* **3**, 19–29.

Crabtree, G. R., and Kant, J. A. (1982). Organization of the rat α-fibrinogen gene: Alternative mRNA splice patterns produce the A and B (′) chains of fibrinogen. *Cell* **31**, 159–166.

Crick, F. (1979). Split genes and RNA splicing. *Science* **204**, 264–271.

Dabeva, M. D., Post-Beittenmiller, M. A., and Warner, J. R. (1986). Autogenous regulation of splicing of the transcript of a yeast ribosomal protein gene. *Proc. Natl. Acad. Sci. U.S.A.* **83**, 5854–5857.

Danchin, A. (1982). From mitochondria to nuclei: Regulation of split genes expression. *In* "Cell Function and Differentiation" (G. Akayunoglou, A. E. Evangelopoulos, J. Georgatsos, G. Palaiologos, A. Trakellis, and C. P. Tsiganos, eds.), Part B, pp. 375–396. Liss, New York.

Danchin, A., and Slonimski, P. P. (1981). Generation of immune specificity: A working hypothesis. *Biosystems* **13**, 259–266.

Danchin, A., and Ullmann, A. (1980). The coordinate expression of polycistronic operons in bacteria. *Trends Biochem. Sci.* **5**, 51–52.

Daskal, Y., Smetana, K., and Busch, H. (1980). Evidence from studies on segregated nucleoli that nucleolar silver staining proteins C23 and B23 are in the fibrillar component. *Exp. Cell Res.* **127**, 285–291.

Davies, R. W., Waring, R. B., Ray, J. A., Brown, T. A., and Scazzocchio, C. (1982). Making ends meet: A model for RNA splicing in fungal mitochondria. *Nature (London)* **300**, 719–724.

Davis, L. I., and Blobel, G. (1986). Identification and characterization of a nuclear pore complex protein. *Cell* **45**, 699–709.

Do Choi, Y., Grabowski, P. J., Sharp, P. A., and Dreyfuss, G. (1986). Heterogeneous nuclear ribonucleoproteins: Role in RNA splicing. *Science* **231**, 1534–1539.

Doolittle, W. F., and Sapienza, C. (1980). Selfish genes, the phenotype paradigm and genome evolution. *Nature (London)* **284**, 601–603.

Dounce, A. L. (1952). Duplicating mechanism for peptide chain and nucleic acid synthesis. *Enzymologia* **15**, 251–273.

Dover, G. A. (1986). Molecular drive in multigene families: How biological novelties arise, spread and are assimilated. *Trends Genet.* **2**, 159–165.

Dujardin, G., Jacq, C., and Slonimski, P. P. (1982). Single base substitution in an intron of oxidase gene compensates splicing defects of the cytochrome *b* gene. *Nature (London)* **298**, 628–632.

Dujardin, G., Lund, P., and Slonimski, P. P. (1984). The effect of paromomycin and [psi] on the suppression of mitochondrial mutation in *Saccharomyces cerevisiae*. *Curr. Genet.* **9**, 21–30.

Dujon, B. (1981). *In* "Molecular Biology of the Yeast *Saccharomyces*: Life Cycle and Inheritance" (J. N. Strathern, E. W. Jones, and J. R. Broach, eds.), pp. 505–535. Cold Spring Harbor Laboratory, Cold Spring Harbor, New York.

Dujon, B., Bolotin-Fukuhara, M., Coen, D., Deutsch, J., Netter, P., Slonimski, P. P., and Weill, L. (1976). Mitochondrial genetics. XI. Mutations at the mitochondrial locus *omega* affecting the recombination of mitochondrial genes in *Saccharomyces cerevisiae*. *Mol. Gen. Genet.* **143**, 131–165.

Dulbecco, R. (1979). Contribution of microbiology to eukaryotic gene biology: New directions for microbiology. *Microbiol. Rev.* **43**, 443–452.

Ely, S., D'Arcy, E., and Jost, E. (1978). Interaction of antibodies against nuclear envelope associated proteins from rat liver nuclei with rodent and human cells. *Exp. Cell. Res.* **116**, 325–331.

Ferra, de, F., Engh, H., Hudson, L., Kamholz, J., Puckett, C., Molineaux, S., and Lazzarini, R. A. (1985). Alternative splicing accounts for the four forms of myelin basic protein. *Cell* **43**, 721–727.

Fischer, P. A. (1986). Nuclear structure: In search of an organizing principle. Higher order structure in the nucleus. *J. Cell Sci. Suppl.* **1**.

Flint, S. J. (1983). Processing of an mRNA precursors in eukaryotic cells. *In* "Processing of RNA" (D. Apirion, ed.), p. 151. CRC Press, Boca Raton, Florida.

Franke, W. W., and Scheer, U. (1974). Structures and functions of the nuclear envelope. *In* "The Cell Nucleus" (H. Busch, ed.), Vol. I, pp. 219–347. Academic Press, New York.

Garman, R. D., Doherty, P. J., and Raulet, D. H. (1986). Diversity, rearrangement, and expression of murine T cell gamma genes. *Cell* **45**, 733–742.

Garriga, G., and Lambowitz, A. M. (1984). RNA splicing in *Neurospora* mitochondria: Self-splicing of a mitochondrial intron *in vitro*. *Cell* **38**, 631–641.

Gerace, L., and Blobel, G. (1980). The nuclear envelope laminin is reversibly depolymerized during mitosis. *Cell* **19**, 277–287.

Gilbert, W. (1978). Why genes in pieces? *Nature (London)* **271**, 501.

Gō, M. (1985). Protein structures and split genes. *Adv. Biophys.* **19**, 91–131.

Goidl, J. A., and Allen, W. R. (1978). Does protein synthesis occur within the nucleus? *Trends Biochem. Sci.* **3**, N225–N228.

Goldstein, L. (1974). Movement of molecules between nucleus and cytoplasm. *In* "The Cell Nucleus" (H. Busch, ed.), Vol. 1, pp. 387–438. Academic Press, New York.

Gorczynski, R. M., and Steele, E. J. (1981). Simultaneous yet independent inheritance of somatically acquired tolerance to two distinct H-2 antigenic haplotype determinants in mice. *Nature (London)* **289**, 678–681.

Gough, N. (1981). Gene rearrangement can extinguish as well as activate and diversify immunoglobulin genes. *Trends Biochem. Sci.* **6**, 300–302.

Green, M. R., Maniatis, T., and Melton, D. A. (1983). Human globin pre-mRNA synthesized *in vitro* is accurately spliced in *Xenopus* oocyte nuclei. *Cell* **32**, 681–694.

Gruss, P., and Khoury, G. (1980). Rescue of a splicing defective mutant by insertion of an heterologous intron. *Nature (London)* **286**, 634–637.

Guiso, N., Dreyfus, M., Siffert, O., Danchin, A., Spyridakis, A., Gargouri, A., Claisse, M., and Slonimski, P. P. (1984). Antibodies against synthetic oligopeptides allow identification of the mRNA maturase encoded by the second intron of the yeast *cob-box* gene. *EMBO J.* **3**, 1769–1772.

Halbreich, A., Pajot, P., Foucher, M., Grandchamp, C., and Slonimski, P. P. (1980). A pathway of specific splicing steps in cytochrome b mRNA processing revealed in yeast mitochondria by mutational blocks within the intron and characterization of circular RNA derived from a complementable intron. *Cell* **19,** 321–329.

Hayflick, L. (1979). Cell biology of aging. *Fed. Proc., Fed. Am. Soc. Exp. Biol.* **38,** 1847–1850.

Henikoff, S., Keene, M. A., Fechtel, K., and Fristrom, J. W. (1986). Gene within a gene: Nested *Drosophila* genes encode unrelated proteins on opposite DNA strands. *Cell* **44,** 33–42.

Herlan, G., Eckert, W., Kaffenberger, W., and Wunderlich, F. (1979). Isolation and characterization of an RNA-containing nucleal matrix from *Tetrahymena* macronuclei. *Biochemistry* **18,** 1782–1788.

Igaraoti, M., Lundquist, P. G., Alford, B. R., and Migata, H. (1971). Experimental ototoxicity of gentamycin in squirrel monkeys. *J. Infect. Dis.* **124** (Suppl.), 114–119.

Inoue, T., Sullivan, F. X., and Cech, T. R. (1985). Intermolecular exon ligation of the rRNA precursor of *Tetrahymena*: Oligonucleotides can function as 5′ exons. *Cell* **43,** 431–437.

Jacq, C., Lazowska, J., and Slonimski, P. P. (1980). Sur un nouveau mécanisme de la régulation de l'expression génétique. *C. R. Acad. Sci. Paris. Ser. D* **290,** 89.

Jacq, C., Pajot, P., Lazowska, J., Dujardin, G., Claisse, M., Groudinsky, O., De la Salle, H., Grandchamp, C., Labouesse, M., Gargouri, A., Guiard, B., Spyridakis, A., Dreyfus, M., and Slonimski, P. P. (1982). In Slonimski, P. P., Borst, P. and Attardi, G. (eds.). Mitochondrial genes. *Cold Spring Harbor Monogr. Ser.* **12,** 155–183.

Jacq, C., Banroques, J., Becam, A. M., Slonimski, P. P., Guiso, N., and Danchin, A. (1984). Antibodies against a fused "LacZ-yeast mitochondrial intron" gene product allow identification of the mRNA-maturase encoded by the fourth intron of the yeast *cob-box* gene. *EMBO J.* **3,** 1567–1572.

Jacquier, A., and Dujon, B. (1983). The intron of the mitochondrial 21S rRNA gene: Distribution in different yeast species and sequence comparison between *Kluyveromyces thermotolerans* and *Saccharomyces cerevisiae. Mol. Gen. Genet.* **192,** 487–499.

Jost, E., and Johnson, R. T. (1981). Nuclear lamina assembly, synthesis and disaggregation during the cell cycle in synchronized HeLa cells. *J. Cell Sci.* **47,** 25–53.

Jost, E., D'Arcy, A., and Ely, S. (1979). Transfer of mouse nuclear envelope specific proteins to nuclei of chick erythrocytes during reactivation in heterokaryons with mouse A9 cells. *J. Cell Sci.* **37,** 97–109.

Kemp, D. J., Harris, A. W., and Adams, J. M. (1980). Transcripts of the immunoglobulin C gene vary in structure and splicing during lymphoid development. *Proc. Natl. Acad. Sci. U.S.A.* **77,** 7400–7404.

King, C. R., and Piatigorsky, J. (1983). Alternative RNA splicing of the murine A-crystallin gene: Protein-coding information within an intron. *Cell* **32,** 707–712.

Kjems, J., and Garrett, R. A. (1985). An intron in the 23S ribosomal RNA gene of the archaebacterium *Desulfurococcus mobilis. Nature (London)* **318,** 675–677.

Kostriken, R. G., and Heffron, F. (1984). The product of the HO gene is a nuclease: Purification and characterization of the enzyme. *Cold Spring Harbor Symp. Quant. Biol.* **49,** 89–96.

Kotylak, Z., Lazowska, J., Hawthorne, D. C., and Slonimski, P. P. (1985). Intron encoded proteins of mitochondria: Key elements of gene expression and genomic evolution. *In* "Achievements and Perspectives of Mitochondrial Research" (E. Quagliariello and E. Palmieri, eds.), Vol. II, pp. 1–20. Elsevier, Amsterdam.

Kozak, M. (1978). How do eukaryotic ribosomes select initiation regions in messenger RNA? *Cell* **15,** 1109–1123.

Krainer, A. R., Maniatis, T., Ruskin, B., and Green, M. R. (1984). Normal and mutant human β-globin pre-mRNAs are faithfully and efficiently spliced in vitro. *Cell* **36**, 993–1005.

Kruger, K., Grabowski, P. J., Zaug, A. J., Sands, J., Gottschling, D. E., and Cech, T. R. (1982). Self-splicing RNA: Autoexcision and autocyclization of the ribosomal RNA intervening sequence of *Tetrahymena*. *Cell* **31**, 147–157.

Langford, C. J., Klinz, F. J., Donath, C., and Gallwitz, D. (1984). Point mutations identify the conserved, intron-contained TACTAAC box as an essential splicing signal sequence in yeast. *Cell* **36**, 645–653.

Lazowska, J., Jacq, C., and Slonimski, P. P. (1980). Sequence of introns and flanking exons in wild type and *box3* mutants of cytochrome b reveals an interlaced splicing protein coded by an intron. *Cell* **22**, 333–348.

Le Douarin, N. M. (1980). The ontogeny of the neural crest in avian embryo chimaeras. *Nature (London)* **286**, 663–669.

Leduc, E. H., Avrameas, S., and Bouteille, M. (1968). Ultrastructural localization of antibody in differentiating plasma cells. *J. Exp. Med.* **127**, 109–118.

Maki, R., Roeder, W., Traunecker, A., Sidman, C., Wabl, M., Rashke, W., and Tonegawa, S. (1981). The role of DNA rearrangement and alternative RNA processing in the expression of immunoglobulin delta genes. *Cell* **24**, 353–365.

Massoulié, J., and Bon, S. (1982). The molecular forms of cholinesterase and acetylcholinesterase in vertebrates. *Annu. Rev. Neurosci.* **5**, 57–106.

Maul, G. G. (1977). Nuclear pore complexes. Elimination and reconstruction during mitosis. *J. Cell Biol.* **74**, 492–500.

Medford, R. M., Nguyen, H. T., Destree, A. T., Summers, E., and Nadal-Ginard, B. (1984). A novel mechanism of alternative RNA splicing for the developmentally regulated generation of troponin T isoforms from a single gene. *Cell* **38**, 409–421.

Michel, F., and Dujon, B. (1983). Conservation of RNA secondary structure in two intron families including mitochondrial-, chloroplast- and nuclear-encoded members. *EMBO J.* **2**, 33–38.

Michel, F., and Lang, B. F. (1985). Mitochondrial class II introns encode proteins related to the reverse transcriptases of retroviruses. *Nature (London)* **316**, 641–643.

Michel, F., Jacquier, A., and Dujon, B. (1982). Comparison of fungal mitochondrial introns reveals extensive homologies in RNA secondary structures. *Biochimie* **64**, 867–881.

Moore, G. P. (1983). Slipped-mispairing and the evolution of introns. *Trends Biochem. Sci.* **8**, 411–414.

Mount, S. M. (1982). A catalogue of splice junctions. *Nucleic Acids Res.* **10**, 459–472.

Naora, H., and Deacon, N. J. (1982). Relationship between the total size of exons and introns in protein-coding genes of higher eukaryotes. *Proc. Natl. Acad. Sci. U.S.A.* **79**, 6196–6200.

Nawa, H., Kotani, H., and Nakanishi, S. (1984). Tissue-specific generation of two preprotachykinin mRNAs from one gene by alternative RNA splicing. *Nature (London)* **312**, 729–734.

Nevins, J. R. (1983). The pathway of eukaryotic mRNA formation. *Annu. Rev. Biochem.* **52**, 441–466.

Nevins, J. R., and Wilson, M. C. (1981). Regulation of adenovirus-2 gene expression at the level of transcriptional termination and RNA processing. *Nature (London)* **290**, 113–118.

Orgel, L. E. (1963). The maintenance of the accuracy of protein synthesis and its relevance to ageing. *Proc. Natl. Acad. Sci. U.S.A.* **49**, 517–521.

Orgel, L. E., and Crick, F. H. C. (1980). Selfish DNA: The ultimate parasite. *Nature (London)* **284,** 604–607.

Padgett, R. A., Konarska, M. M., Grabowski, P. J., Hardy, S. F., and Sharp, P. A. (1984). Lariat RNA's as intermediates and product in the splicing of messenger RNA precursors. *Science* **225,** 898–903.

Padgett, R. A., Grabowski, P. J., Konarska, M. M., Seiler, S., and Sharp, P. A. (1986). Splicing of messenger RNA precursors. *Annu. Rev. Biochem.* **55,** 1119–1150.

Persson, H., Monstein, H. J., Akusjärvi, G., and Philipson, L. (1981). Adenovirus early gene products may control viral mRNA accumulation and translation *in vivo. Cell* **23,** 485–496.

Pikielny, C. W., and Rosbash, M. (1985). mRNA splicing efficiency in yeast and the contribution of nonconserved sequences. *Cell* **41,** 119–126.

Pikielny, C. W., Teem, J. L., and Rosbash, M. (1983). Evidence for the biochemical role of an internal sequence in yeast nuclear mRNA introns: Implications for U1 RNA and metazoan mRNA splicing. *Cell* **34,** 395–403.

Rabbitts, T. H. (1978). Evidence for splicing of interrupted immunoglobulin variable and constant region sequences in nuclear RNA. *Nature (London)* **275,** 291–296.

Rabbitts, T. H., Forester, A., Dunnick, W., and Bentley, D. L. (1980). The role of gene deletion in the immunoglobulin heavy chain switch. *Nature (London)* **283,** 351–356.

Ramirez, G. (1973). Synaptic plasma membrane protein synthesis: Selective inhibition by chloramphenicol *in vivo. Biochem. Biophys. Res. Commun.* **50,** 452–458.

Reed, R., and Maniatis, T. (1985). Intron sequences involved in lariat formation during pre-mRNA splicing. *Cell* **41,** 95–105.

Rogers, J. (1986). Introns between protein domains: Selective insertion or frameshifting? *Trends Genet.* **2,** 223.

Rosenthal, C. J., and Franklin, E. C. (1977). Amyloidosis and amyloid protein. *Recent Adv. Clin. Immunol.* **1,** 41–76.

Rozek, C. E., and Davidson, N. (1986). Differential processing of RNA transcribed from the single-copy *Drosophila* myosin heavy chain gene produces four mRNAs that encode two polypeptides. *Proc. Natl. Acad. Sci. U.S.A.* **83,** 2128–2132.

Ruskin, B., Krainer, A. R., Maniatis, T., and Green, M. R. (1984). Excision of an intact intron as a novel lariat structure during pre-mRNA splicing in vitro. *Cell* **38,** 317–331.

Ruskin, B., Pikielny, C. W., Rosbash, M., and Green, M. R. (1986). Alternative branch points are selected during splicing of a yeast pre-mRNA in mammalian and yeast extracts. *Proc. Natl. Acad. Sci. U.S.A.* **83,** 2022–2026.

Salditt-Georgieff, M., Harpold, M., Chen-Kiang, S., and Darnell, J. E., Jr. (1980). The addition of 5′ cap structures occurs clearly in hnRNA synthesis and prematurely terminated molecules are capped. *Cell* **19,** 69–78.

Schumm, D. E., Niemann, M. A., Palayaor, T., and Webb, T. E. (1979). *In vivo* equivalence of a cell-free system from rat liver for ribosomal RNA processing and transport. *J. Biol. Chem.* **254,** 12126–12130.

Seligmann, M., Mihaesco, E., Preud'homme, J. L., Danon, F., and Brouet, J. C. (1979). Heavy chain diseases: Current findings and concepts. *Immunol. Rev.* **48,** 145–167.

Senapathy, P. (1986). Origin of eukaryotic introns: A hypothesis, based on codon distribution statistics in genes, and its implications. *Proc. Natl. Acad. Sci. U.S.A.* **83,** 2133–2137.

Shaper, J. H., Pardoll, D. M., Kaufmann, S. H., Barrack, E. R., Vogelstein, B., and Coffey, D. S. (1979). The relationship of the nuclear matrix to cellular structure and function. *Adv. Enzyme Regul.* **17,** 213–248.

Simionovitch, L. (1976). On the nature of hereditable variation in cultured somatic cells. *Cell* **7**, 1–11.

Singer, P. A., Singer, H. H., and Williamson, A. R. (1980). Different species of messenger RNA encode receptor and secretory IgM chains differing at their carboxy termini. *Nature (London)* **285**, 293–300.

Sjöberg, B. M., Hahne, S., Mathews, C. Z., Mathews, C. K., Rand, K. N., and Gait, M. J. (1986). The bacteriophage T4 gene for the small subunit of ribonucleotide reductase contains an intron. *EMBO J.* **5**, 2031–2036.

Skoglund, U., Andersson, K., Strandberg, B., and Daneholt, B. (1986). Three-dimensional structure of a specific pre-messenger RNP particle established by electron microscope tomography. *Nature (London)* **319**, 560–564.

Slonimski, P. P. (1980). Eléments hypothétiques de l'expression des gènes morcelés: Protéines messagères de la membrane nucléaire. *C. R. Acad. Sci. (Paris)* **290 D**, 331–334.

Solnick, D. (1985a). Alternative splicing caused by RNA secondary structure. *Cell* **43**, 667–676.

Solnick, D. (1985b). Trans splicing of mRNA precursors. *Cell* **42**, 157–164.

Speck, S. H., and Strominger, J. L. (1985). Analysis of the transcript encoding the latent Epstein-Barr virus nuclear antigen. I: A potentially polycistronic message generated by long-range splicing of several exons. *Proc. Natl. Acad. Sci. U.S.A.* **82**, 8305–8309.

Stick, R., and Hausen, P. (1985). Changes in the nuclear lamina composition during early development of *Xenopus laevis*. *Cell* **41**, 191–200.

Tabak, H. F., and Grivell, L. A. (1986). RNA catalysis in the excision of yeast mitochondrial introns. *Trends Genet.* **2**, 51–55.

Tooze, J. (1980). "Molecular Biology of Tumor Viruses." Cold Spring Harbor Laboratory, Cold Spring Harbor, New York.

Tsurugi, K., and Ogata, K. (1984). Effects of cell sap, ATP, and RNA synthesis on the transfer of ribosomal proteins into nuclei and nucleoli in a rat liver cell-free system. *Eur. J. Biochem.* **145**, 83–89.

Tunnacliffe, A., Sims, J. E., and Rabbitts, T. H. (1986). T3 pre-mRNA is transcribed from a non-TATA promoter and is alternatively spliced in human T cells. *EMBO J.* **5**, 1245–1252.

Upholt, W. B., and Sandell, L. J. (1986). Exon/intron organization of the chicken type II procollagen gene: Intron size distribution suggests a minimal intron size. *Proc. Natl. Acad. Sci. U.S.A.* **83**, 2325–2329.

Van der Horst, G., and Tabak, H. F. (1985). Self-splicing of yeast mitochondrial ribosomal and messenger RNA precursors. *Cell* **40**, 759–766.

Van der Veen, R., Arnberg, A. C., Van der Horst, G., Bonen, L., Tabak, H. F., and Grivell, L. A. (1986). Excised group II introns in yeast mitochondria are lariats and can be formed by self-splicing *in vitro*. *Cell* **44**, 225–234.

Vijayraghavan, U., Parker, R., Tamm, J., Iimura, Y., Rossi, J., Abelson, J., and Guthrie, C. (1986). Mutations in conserved intron sequences affect multiple steps in the yeast splicing pathway, particularly assembly of the spliceosome. *EMBO J.* **5**, 1683–1695.

Waring, R. B., and Davies, R. W. (1984). Assessment of a model for intron RNA secondary structure relevant to RNA self-splicing—A review. *Gene* **28**, 277–291.

Waring, R. B., Davies, R. W., Schazzocchio, C., and Brown, T. A. (1982). Internal structure of a mitochondrial intron of *Aspergillus nidulans*. *Proc. Natl. Acad. Sci. U.S.A.* **79**, 6332–6336.

Waring, R. B., Brown, T. A., Ray, J. A., Scazzocchio, C., and Davies, R. W. (1984). Three variant introns of the same general class in the mitochondrial gene for cytochrome oxidase subunit 1 in *Aspergillus nidulans. EMBO J.* **3,** 2121–2128.

Weber, M. (1979). Cellular migration and axonal outgrowth in the peripheral nervous system. *In* "The Role of Intercellular Signals, Navigation Encounter Outcome" (J. G. Nicholls, ed.), pp. 97–118. Verlag Chemie, Weinheim.

Weisberger, A. S., and Wolfe, S. (1964). Effect of chloramphenicol on protein synthesis. *Fed. Proc., Fed. Am. Soc. Exp. Biol.* **23,** 976–983.

Wersall, J., Lundquist, P. G., and Bjorkroth, B. (1969). Ototoxicity of gentamycin. *J. Infect. Dis.* **119,** 410–423.

Wieringa, B., Hofer, E., and Weissmann, C. (1984). A minimal intron length but no specific internal sequence is required for splicing the large rabbit β-globin intron. *Cell* **37,** 915–925.

Yamamoto, K. R. (1985). Steroid receptor regulated transcription of specific genes and gene networks. *Annu. Rev. Genet.* **19,** 209–252.

Yanofsky, C. (1983). Prokaryotic mechanisms in eukaryotes? *Nature* (*London*) **302,** 751–752.

Yost, H. J., and Lindquist, S. (1986). RNA splicing is interrupted by heat shock and is rescued by heat shock protein synthesis. *Cell* **45,** 185–193.

Yunis, A. A., and Bloomberg, G. R. (1964). Chloramphenicol toxicity: Clinical features and pathogenesis. *Prog. Hematol.* **4,** 138–159.

Zaug, A. J., and Cech, T. R. (1985). Oligomerization of intervening sequence RNA molecules in the absence of proteins. *Science* **229,** 1060–1064.

GENE TRANSFER INTO MICE

George Scangos* and Charles Bieberich

Department of Biology, The Johns Hopkins University, Baltimore, Maryland 21218

I. Preface

The first transgenic mice were made just over 6 years ago. Many types of questions have been addressed with the technology, and the field has matured considerably. Transgenic mice provide a powerful assay system to characterize temporal and spatial patterns of gene expression, to analyze the role of specific molecules in complex biological phenomena, and to alter the physiology of whole animals in defined ways. Several reviews that cover part or all of the material discussed here have been published (Alt *et al.*, 1985; Palmiter and Brinster, 1985, 1986; Gordon and Ruddle, 1985). In this review, we have divided the published work into broad areas to which transgenic mice have made or are likely to make a significant contribution—tissue- and developmental stage-specific gene expression, oncology, immunology, development, and understanding of genetic diseases.

II. Introduction

Only since 1977 has it been possible to introduce exogenous DNA into cultured mammalian cells (Wigler *et al.*, 1977; Maitland and McDougall, 1977; Bachetti and Graham, 1977). In the ensuing years, gene transfer techniques have become widely used by diverse researchers to address a variety of genetic and cell biological problems. In some respects, the development of gene transfer technologies has

* Present address: Molecular Therapeutics, Inc., West Haven, Connecticut 06516.

complemented the vast power of molecular genetics, in that cloned genes can be characterized, modified, and then returned to cells to test function. Gene transfer techniques also have been utilized to clone genes (Kuhn *et al.*, 1984; Lin *et al.*, 1985) and will be very important in structure–function analyses. Despite the knowledge gained as a result of these experiments and the obvious power of the techniques, there are limitations inherent in cell culture that make it difficult or impossible to address many types of questions. Regulation of genes expressed in specific types of differentiated cells is best studied *in vivo*, and dissection of complex phenomena that require the interactions of different cell types (for example, neurobiology, immunology, and development) will require *in vivo* manipulations. To these ends, in the late 1970s, a number of workers developed microinjection techniques for the introduction of exogenous DNA into the genomes of intact animals. In 1980, Jon Gordon, Frank Ruddle, and their colleagues were the first to report the successful introduction of foreign DNA into the genomes of mice by microinjection of embryos (Gordon *et al.*, 1980), followed a few months later by several other groups (Costantini and Lacy, 1981; Brinster *et al.*, 1981; E. Wagner *et al.*, 1981; T. Wagner *et al.*, 1981). It soon became apparent that the foreign DNA could be expressed (Brinster *et al.*, 1981; E. Wagner *et al.*, 1981; T. Wagner *et al.*, 1981) and could be integrated into the genomes of germ-line cells and passed to offspring in Mendelian fashion (Costantini and Lacy, 1981; Gordon and Ruddle, 1981; Stewart *et al.*, 1982). Mice that contain exogenous DNA integrated into their genomes have been termed transgenic (Gordon and Ruddle, 1981), and the introduced DNA is called a transgene (see Table 1).

In the microinjection protocol (Fig. 1), DNA is loaded into a micropipet and expelled into one of the pronuclei of a fertilized mouse egg. The injected embryos are implanted into the oviducts of pseudopregnant foster mothers, and some develop to term. Thorough reviews of the technical aspects of the procedure have been written (Gordon and Ruddle, 1983; Hogan *et al.*, 1986). Many factors can affect the frequency of success of the experiments, including the strains of mice used, the concentration and form of the DNA (Brinster *et al.*, 1985), and obviously the skill of the experimenter, but in most laboratories where the technique is performed routinely, 20–30% of mice born from injected embryos are transgenic. Thus, although the technique is technically demanding and requires skill, an experienced experimenter can generate more than one transgenic mouse per day of injections.

Microinjection of DNA directly into the pronucleus of fertilized

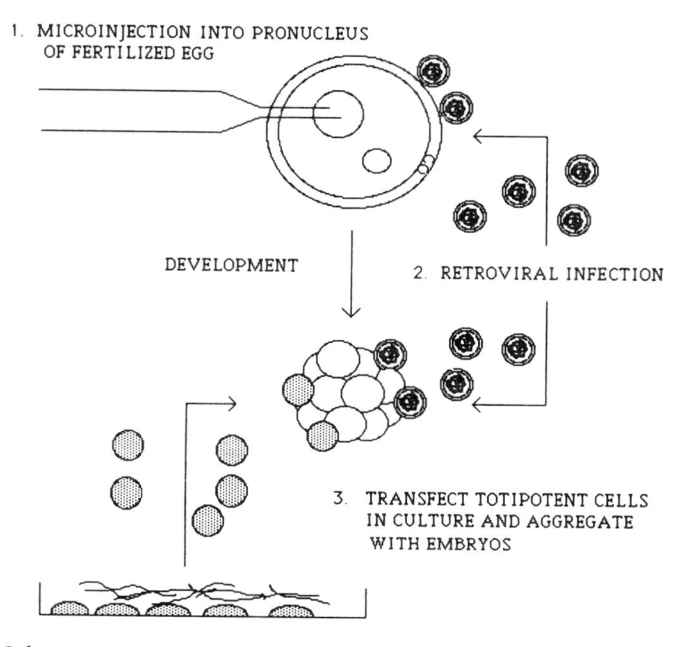

FIG. 1. Schematic representation of gene transfer into mouse embryos. Genes can be introduced into single-cell fertilized eggs by direct microcapillary injection of DNA into one of the two pronuclei (1) or by retroviral infection (2). Genes can be introduced into later stage embryos by retroviral infection or by introducing into the embryos totipotent cells which previously had been transfected with DNA in culture (3).

eggs is only one method of introducing DNA into the genomes of mice (Fig. 1). The first successful deliberate introduction of exogenous genetic material into a mouse germ line involved viral vectors (Jaenisch and Mintz, 1974; Jaenisch, 1976). In early experiments, in which naturally occurring viruses were introduced, several aspects of the retroviral life cycle seemed to point to some significant advantages over the microinjection protocol. These included relative ease of infection, single-copy integrations flanked by retroviral long terminal repeats (LTRs), and high efficiency. These properties combined to generate significant interest in the use of retroviruses as vectors, and considerable effort has been expended to utilize retroviruses as vectors for the introduction of cloned DNA. Van der Putten *et al.* (1985) and Jahner *et al.* (1985) infected preimplantation mouse embryos with recombinant retroviruses and were able to demonstrate that viral genomes were present in the mice and were passed on through the germ line. These results demonstrate that retroviral vectors hold considerable promise,

TABLE 1
Genes Transferred to Mice[a]

Transgene	Species	Reference
α-Actin/globin	Rat/human	Shani (1986)
αa(I) collagen/CAT	Mouse/bacterial	Khillan *et al.* (1986)
αA-crystallin/CAT	Mouse/bacterial	Overbeek *et al.* (1985)
BKV early region	BKV	Small *et al.* (1986b)
E_α MHC class II	Mouse	Pinkert *et al.* (1985), Yamamura *et al.* (1985), Le Meur *et al.* (1985)
Elastase I	Rat	Swift *et al.* (1984)
Elastase I/growth hormone	Rat/human	Ornitz *et al.* (1985a,b)
Elastase I/SV40 early region	Rat/SV40	Ornitz (1985b)
α-Fetoprotein	Mouse	Krumlauf *et al.* (1985)
β-Globin	Rabbit	T. Wagner *et al.* (1981), Lacy *et al.* (1983)
β-Globin	Human	E. Wagner *et al.* (1981), Stewart *et al.* (1982), Townes *et al.* (1985a), Costantini *et al.* (1986)
δ–β-Globin	Mouse/human	Humphries *et al.* (1985)
Gonadotropin-releasing hormone	Mouse	Mason *et al.* (1986)
Growth hormone	Human	Hammer *et al.* (1984), Wagner *et al.* (1985)
Growth hormone	Rat	Hammer *et al.* (1984)
HBV surface antigen	HBV	Chisari *et al.* (1985), Babinet *et al.* (1985)
Immunoglobulin κ	Mouse	Storb *et al.* (1984, 1986)
Immunoglobulin μ	Mouse	Weaver *et al.* (1985, 1986), Storb *et al.* (1986)
Immunoglobulin $\mu + \kappa$	Mouse	Rusconi and Kohler (1985), Storb *et al.* (1986)
Immunoglobulin μ	Human	Yamamura *et al.* (1986)
Insulin	Human	Burki and Ullrich (1982), Bucchini *et al.* (1986)
Insulin/SV40 early region	Rat/SV40	Hanahan (1985)
JCV early region	JCV	Small *et al.* (1986a,b)

MHC class I	Pig	Frels *et al.* (1985)
MHC class I	Mouse	Bieberich *et al.* (1986)
MMTV LTR/thymidine kinase	MMTV/HSV	Ross and Solter (1985)
MT-I/growth hormone	Mouse/rat	Palmiter *et al.* (1982a)
MT-I/growth hormone	Mouse/human	Palmiter *et al.* (1983b)
MT-I/growth hormone	Mouse/bovine	Hammer *et al.* (1985)
MT-I/growth hormone releasing factor	Mouse/human	Hammer *et al.* (1985)
MT-I/growth hormone/SV40 early region	Mouse/human/SV40	Messing *et al.* (1985), Palmiter *et al.* (1985b)
MT-I/HPRT	Mouse/human	Stout (1985)
MT-I/HBV surface antigen	Mouse/HBV	Chisari *et al.* (1985)
MT-I/placental lactogen	Mouse/human	Palmiter (1985a)
MT-I/somatostatin	Mouse/rat	Low *et al.* (1985)
MT-I/thymidine kinase	Mouse/HSV	Palmiter *et al.* (1982b)
MT-IIA/growth hormone	Human/human	Hammer (1985b)
MMTV/*c-myc*	MMTV/mouse	Stewart *et al.* (1984)
Myosin light chain	Rat	Shani (1985)
RSV LTR/CAT	RSV/bacterial	Overbeek *et al.* (1986)
SV40 early region	SV40	Brinster *et al.* (1984), Palmiter *et al.* (1985b), Small *et al.* (1985)
SV40 large T antigen	SV40	Palmiter *et al.* (1985b)
Thymidine kinase	HSV	T. Wagner *et al.* (1981)
Transferrin	Chicken	McKnight *et al.* (1983)

[a] This is a table of published genes which have been transferred to mice. Hybrid transgenes are denoted in a 5′–3′ orientation. Abbreviations: BKV, BK virus; CAT, chloramphenicol acetyltransferase; HBV, hepatitis B virus; HPRT, hypoxanthine phosphoribosyltransferase; HSV, herpes simplex virus; JCV, JC virus; LTR, long terminal repeat; MHC, major histocompatibility complex; MMTV, mouse mammary tumor virus; MT, metallothionein; RSV, Rous sarcoma virus; SV40, Simian virus 40.

and almost certainly will be useful tools for the introduction of genes into germ-line and somatic tissues. At present, there remain technical problems to overcome, including limits on the size of the insert, potential interference with tissue- and developmental stage-specific expression by the retroviral LTRs, and intereference between transcriptional and processing signals on the insert and viral DNAs. Additionally, one must weigh the time and difficulty of constructing the appropriate retroviral construct versus the time to produce mice with the microinjection protocol. Thus, although it is clear that retroviral vectors will be very useful for some purposes, it is unclear whether they will provide a method of choice for a wide variety of applications.

Another potential method for the introduction of exogenous DNA into animals involves the introduction of DNA into totipotent stem cell lines in culture. The cells then can be screened or selected to detect those that contain and/or express the foreign DNA. Cells with the desired properties then can be injected into the blastocoel cavity or aggregated with eight-cell embryos, where they can differentiate and become the progenitors of differentiated cells in tissues of the adult animal. Theoretically, this is an attractive approach, since genes can be introduced *in vitro* and the cells can be manipulated to obtain those with the desired characteristics. In many cases, it has been difficult to maintain stem cell lines in a diploid state so that they retain their totipotent potential, and the resulting animals are always mosaic. If the cells contributed to the formation of germ cells, then the animals can be bred to generate progeny, and some fraction of the offspring will be transgenic. Since all of the animals are mosaic, and only some of them have germ cells derived from the stem cell lines, this procedure is difficult in practice, and has not been pursued as widely as microinjection and retroviral infection. Several recent reports, however, have demonstrated success in introducing foreign genes into mice by this methodology (Lovell-Badge *et al.*, 1985; Stewart *et al.*, 1985; Wagner *et al.*, 1985). A recent report by Robertson *et al.* (1986) demonstrated successful introduction of recombinant retroviruses into a totipotent stem cell line in culture. The infected cells were introduced into blastocyst-stage embryos and contributed to the makeup of adult animals. The stem cells also contributed to the germ line, and offspring of the chimeras contained retroviral insertions. These data demonstrate that this approach also is a feasible and useful technique.

Pronuclear microinjection remains the most widely used method of introducing DNA into the germ line of mouse embryos. The relatively high efficiency and relative ease with which the technique can be performed by experienced investigators indicate that the technique will be useful for many years to come. Almost all of the work discussed in the remainder of this review is based on introduction of genes by pronuclear microinjection.

III. Integration

Injected DNA integrates into the embryonic genome, although the biochemistry of integration has not been elucidated. Typically, integration occurs at one site in the genome, but a significant fraction of transgenic mice contain multiple integration sites (Gordon and Ruddle, 1985; Small *et al.*, 1986a). Lo (1983) reported that iontophoretic injection of DNA more often leads to multiple integration sites. An integration site may contain anywhere from 1 to more than 100 copies of the injected DNA. When multiple copies are present, they usually are arranged in tandem head-to-tail arrays, although rearrangements and deletions are common. Additionally, cases in which copies of the injected DNA are separated by host DNA have been reported (Covarrubias *et al.*, 1986), and the origin of the intervening DNA can be far from the integration site. Covarrubias *et al.* (1986) also reported a large deletion accompanying an integration site, while two other characterized insertion sites had a 1-kb deletion and 15-kb duplication (Palmiter and Brinster, 1986). An integration site characterized in our own laboratory also involved a large deletion or rearrangement. Although the precise nature of the rearrangement has not been elucidated, it is clear that the sequences that flank the transgene were not contiguous in the mouse genome prior to integration. Thus, it is clear that integration of transgenes into the mouse genome often is complex, and that there can be large rearrangements of both the donor and target loci during or subsequent to integration.

To explain the frequent occurrence of multiple copies of the injected DNA at a single integration site, Brinster *et al.* (1985) proposed that the limiting step in integration was the frequency of random chromosomal breaks. Since linear DNA seems to integrate more efficiently than circular DNA, they proposed that the ends of the injected DNA interacted with those breaks to initiate integration. To explain the presence of multiple copies, they proposed that additional molecules integrated into the first by homologous recombination. Thus, they proposed that homologous recombination among injected molecules was a frequent event, but that homologous recombination among injected and host DNA was an infrequent event. There is precedent for this hypothesis since homologous recombination among introduced molecules in cultured mammalian cells occurs at high frequency (Small and Scangos, 1983; DeSaint Vincent and Wahl, 1983; Shapira *et al.*, 1983), although integration via homologous recombination is quite rare (Smithies *et al.*, 1985; Thomas *et al.*, 1986; Lin *et al.*, 1985). Palmiter *et al.* (1985a) demonstrated that homologous recombination among exogenous molecules also could occur in transgenic mice. It is not known whether the recombination occurs prior or subsequent to integration of the first molecule, although in cultured cells there are data that indicate recombination occurs prior to integration (Folger *et al.*, 1982). Additionally, it has been difficult to distinguish classical homologous recombination from gene conversion. There are data that suggest that both processes occur (Shapira *et al.*, 1983), although the relative frequencies of each have been difficult to ascertain.

There is no discernible pattern to the integration of exogenous DNA into the genomes of transgenic mice. Integrations into many different autosomes (Lacy *et al.*, 1983) as well as the X (Krumlauf *et al.*, 1985a) chromosome have been documented, and integration into sites of homology has not been reported. It seems likely that integration does not occur truly randomly and that certain chromatin structures, perhaps open chromatin compatible with gene expression or DNA replication, are more accessible to integration than heterochromatic regions. This hypothesis would explain why a large proportion of transgenic mice express transgenes at significant levels while a large part of the genome is genetically inactive. The inability to direct the site of integration remains a significant limitation of the transgenic mouse technology.

Although integration sites often have deletions and/or rearrangements, it is not clear at this point whether the transgenes serve as markers of spontaneously occurring rearrangements, or whether the transgenes somehow play a role in generating the rearrangements. Since insertion sites that induce mutations in the homozygous state are the most interesting, they have been preferentially cloned and characterized. Since these insertional mutations are a result of interruption of host sequences, and since large rearrangements are more likely to interrupt essential host functions, it is likely that mice with extensive rearrangements at the sites of integration are over-represented in this group. It is therefore difficult to determine the frequency with which integration of microinjected DNA is accompanied by large rearrangements of the host genome.

As opposed to integration of microinjected DNA, integration of retroviral vectors seems to occur without extensive rearrangement of the host target sequences, and mimics naturally occurring retroviral integrations (Jahner *et al.*, 1985; van der Putten *et al.*, 1985). There is only one copy of the retroviral genome integrated at each site, and host DNA at the integration site is not extensively rearranged.

IV. Translocations

If, in fact, integration occurs preferentially at sites of chromosomal breaks, then it would not be surprising if chromosomal translocations frequently accompanied integration. Furthermore, a simple model predicts that translocations might occur at the site of integration so that the host sequences that flank the transgene might originate from two different chromosomes. The available data are consistent with the first of these hypotheses: at least three laboratories have identified transgenic mice in which there are chromosomal translocations. Overbeek *et al.* (1986) described a line of mice that contained the long terminal repeat of Rous sarcoma virus (RSV) ligated to a bacterial chloramphenicol acetyltransferase (CAT) gene. When the founder (the mouse derived directly from an injected embryo) or its offspring were mated with wild-type mice, approximately half of the embryos died before day 8 of gestation. Of the surviving half, 50% carried the RSV–CAT gene and also produced small litters. This curious genetic pattern differed from previously reported insertional mutants in that the effect was dominant, and that some positive offspring survived. They hypothesized that the effect was due to a translocation involving the chromosome on which the integration occurred. The translocation chromosome containing the integration site would be unbalanced unless the translocation partner was present. Thus the 50% of transgenic offspring that inherited the translocation partner survived, while the 50% without the translocation partner did not. The negative offspring by definition did not receive the chromosome with the insertion, but 50% did inherit the translocation partner. Thus, half of the positive and half of the negative offspring would have unbalanced genotypes and thus fail to survive. The presence of translocations was confirmed cytogenetically.

A reciprocal translocation that led to male sterility in a line of transgenic mice has been identified (J. Gordon and F. Ruddle, personal communication). In this line, in which the females also were subfertile, there was abnormal tetrad formation in primary spermatocytes. The chromosomes involved in the reciprocal translocation were 2 and 12, and the integration site of the transgene was in one of the translocation partners.

In our own lab, we have identified a transgenic mouse line with a translocation. This mouse line contains a construct consisting of the *Drosophila* heat-shock HSP70 promoter ligated to the human growth hormone gene. The mouse, termed H30, had two integration sites, which we called A and B. Of over 250 mice of subsequent generations analyzed to date, no mice with only the A integration site have been identified. Approximately 50% of the offspring have the B integration site, and of those, approximately 50% contain the A integration site as well, but no mice with the A integration site alone have been identified. This inheritance pattern was consistent with a reciprocal translocation between the chromosomes containing the A and B translocation sites. One need only postulate that the imbalance provided by the A translocation chromosome alone was lethal, while the imbalance provided by the B translocation chromsome alone was not lethal. The presence of both the A and B chromosomes would provide a balanced genotype. This hypothesis predicts that in crosses between two AB heterozygotes, mice of the genotypes A, AA, and AAB should be inviable and thus not be found among living progeny, while mice of genotypes B, AB, and ABB should be present. This prediction was confirmed. Chromosomal analyses demonstrated a translocation or deletion involving chromosome 19. The A integration site has been mapped to the abnormal 19, while identification of the B integration site is in progress. We have molecularly cloned the A integration site as well as unique-sequence flanking DNA from both sides of the integration. The unique-sequence flanking regions are being used as probes of a hybrid panel to assign them to chromosomes. If a translocation occurred at the site of integration, then the two flanking sequences might map to different chromosomes in a wild-type mouse.

Other sterile or semisterile mice either homozygous (G. Scangos and C. Bieberich, unpublished observations) or heterozygous (Pinkert *et al.*, 1985) for transgenic loci have been identified. Since reduced fertility or sterility is a common consequence of radiation-induced translocations (Searle, 1981), it may be that these lines also contain translocations and that translocations occur more commonly than previously suspected in transgenic mice. It is likely that translocations represent only the largest form of DNA rearrangements that occur at sites of integration, and that a significant fraction of transgenic mice contain rearrangements of some sort at the site of integration. This question will be resolved when a number of integration sites have been molecularly cloned and characterized.

V. Insertional Mutagenesis

Since integration of injected DNA can occur throughout the genome, it follows that some of the integrations will interrupt essential regions of the recipient mouse genome. In a diploid organism many of the mutations will be recessive and apparent only when present in the homozygous state. This phenomenon provides a potentially powerful tool to generate and characterize developmentally important mutations, since the introduced sequence can be cloned along with the flanking region of the mouse genome. Thus, one can generate insertional mutations, characterize the phenotype, and clone the region in which the insertion has induced the mutation.

To date, nine insertional mutations have been reported in the literature (Schnieke *et al.*, 1983; Wagner *et al.*, 1983; Palmiter *et al.*, 1983a; Mark *et al.*, 1985; Woychik *et al.*, 1985; Overbeek *et al.*, 1986; Shani, 1986; Gerlinger *et al.*, 1986), and several others have been identified but not yet reported. In five of the reported cases, the insertions induced prenatal lethalities, although in only one case has the interrupted gene been identified (Schnieke *et al.*, 1983). In that case, a retroviral insertion occurred in the first intron of the $\alpha1(I)$ collagen gene. Mice homozygous for the insertion died at day 12 of embryogenesis due to death of erythropoietic and mesenchymal cells and hemorrhageing from major blood vessels. The mechanisms by which the insertion into the intron interrupted gene function are not clear. Interestingly, the gene carrying the insertion was not transcribed, even though the promoter was intact (Hartung *et al.*, 1986). This result suggests that integration of exogenous DNA can modulate expression of nearby transcription units, and that mutations need not be the result of direct interruption of transcribed regions.

The inserted gene and flanking DNA have been isolated from one additional prenatal homozygous-lethal mutation (Covarrubias *et al.*, 1986). That mouse, designated HUGH/3, contained an insert of the human growth hormone gene. Mice homozygous for the insertion died on days 4–5 of gestation, just after implantation. The integration site contained approximately five copies of the growth hormone sequences distributed in three regions separated by mouse DNA. There was additionally a deletion of at least 10 kb of mouse genomic sequences. Thus this integration site was complex and involved rearrangement of host sequences. The complexity of the integration site makes it difficult to identify the mechanisms by which the integration led to embryonic lethality, and to identify a meaningful region of the corresponding uninterrupted site.

Of the remaining characterized prenatal lethalities, one occurs early in gestation, prior to implantation (Mark *et al.*, 1985), while the remainder are postimplantation. In our laboratory we have identified a homozygous-lethal mutation which seems to occur at approximately the time of birth. This line of mice, termed D19, contains a transgene encoding a heterologous class I major histocompatibility antigen. Normal numbers of embryos are present throughout gestation, but many of the animals die within a day of birth. No homozygous adults have been obtained in this line, and many of the dead neonates are homozygous. The gross pathology of the dead homozygotes is normal.

Three mutations that do not result in embryonic lethality have been reported. In the first case, a line of mice containing the mouse mammary tumor virus long terminal repeat fused to a *myc* gene developed a limb deformity (Woychik *et al.*, 1985). In homozygous mice the bones of both forelimbs and hindlimbs were fused, and many of the bones in the foot were fused or missing. The inserted DNA and flanking region were cloned, and initial characterization indicated that about 1 kb of mouse DNA had been deleted during integration. Use of unique-sequence flanking DNA as a probe of hybrid panels demonstrated that the integration was on chromosome 2.

The abnormal limb phenotype was very similar to spontaneous mouse mutation termed *ld* (for limb deformity), which previously had been mapped to mouse chromosome 2. To test for complementation, heterozygous transgenic mice were mated with heterozygous *ld* mice. Since both alleles are recessive, animals expressing the limb deformity phenotype would be expected only if the insertion and the spontaneous mutation were allelic. Several animals exhibiting the phenotype were obtained from these crosses, suggesting that the insertion and mutation were in fact allelic. This point was confirmed by identifying restriction-fragment-length polymorphisms (RFLPs) that cosegregated with the *ld* allele using flanking DNA from the integration site as a probe.

This example points out the potential power of insertional mutagenesis, since the insertion provided a molecular handle to identify a gene involved in limb development. Although the spontaneous mutation had existed for over 20 years, little progress had been made because there was no obvious way to identify the responsible gene or gene product. The DNA flanking the integration now can be used as a probe to examine the location and temporal pattern of transcription, and sequence information might be used as a clue to the localization of the corresponding protein and to generate antibodies. Ultimately, a characterization of gene expression and protein localization and function might generate significant insights into mammalian morphogenetic development.

A second insertional mutation also generated limb deformities (Overbeek *et al.*, 1986), although in that case the deformity was limited to fused toes. No further characterization of this phenotype has been reported.

The third insertional mutation affecting adult mice was reported in a strain of mice containing a metallothionein promoter fused to a thymidine kinase gene (Palmiter *et al.*, 1983a). The salient observation was that the transgene was not transmitted by males, although it was passed on to progeny at normal frequency by females. The males were fertile and produced normal litters, but none of their offspring contained the transgene. It was postulated that the insertion must affect a postmeiotic function of sperm, so that sperm carrying the insertion were defective.

Almost all of the insertional mutations described are recessive. For embryonic lethal insertions, most dominant mutations would be lost since the founder mice, derived from injected embryos, probably would fail to develop. If they were mosaic, so that they escaped the mutation and did reach maturity, any of their offspring that inherited the sequence would fail to develop. These mice would be included in the class of mosaics that failed to pass on the sequences, and would be difficult to distinguish from those that did not contain the integration in the germ line. Among mutations that are not lethal, the only dominant one seems to affect sperm, and probably functions postmeiotically when the cells are haploid. Other mutations, which seem to act dominantly, probably are a result of translocations that create an unbalanced genome. Indeed, for a diploid mammalian organism, most dominant insertional mutations probably will turn out to be translocations creating an imbalance in a chromosomal region rather than disruptions of a single gene.

It is difficult to ascertain the frequency with which integrations of exogenous DNA lead to insertional mutations. A review of the small body of literature in which attempts have been reported (Mark *et al.*, 1985; Shani, 1986; Overbeek *et al.*, 1986; Wagner *et al.*, 1983; Palmiter *et al.*, 1983a; Woychik *et al.*, 1985) indicates a frequency of between 15 and 20%, although this number is likely to be an overestimate, since reports of failures to identify insertional mutations probably are under-represented in the literature. In our laboratory, we have identified one neonatal lethal mutation out of 13 transgenic lines bred to homozygosity. Additionally, we have identified one translocation, four cases of male sterility, and several cases of semisterility (all of which may represent additional translocations).

It is clear from these data that insertional mutagenesis is a fairly frequent event in transenic mice and that the technique will be useful to identify developmentally important genes. What also is clear is that the use of insertional mutants must be approached with caution. The prevalence of rearrangements, complex integration patterns, and translocations will make it difficult or impossible in many cases to identify the segment of cellular DNA whose disruption is responsible for the observed phenotype. In that respect, retroviral insertions, which occur cleanly, with a duplication of only a few base pairs of cellular DNA, may be preferable for generating and studying insertional mutations. Even with retroviral insertions, however, it is clear that the insertion can have effects on the transcription of sequences not directly adjacent to the site of insertion (Hartung *et al.*, 1986). Thus, although this is a promising and potentially powerful technique with which to identify developmentally interesting genes, it is important to proceed cautiously.

VI. Gene Expression

Insertional mutations essentially are a by-product of transgenic experiments. Most of the reported mutations occurred in mice that had been generated for other purposes. One of the areas in which transgenic mice have made a significant contribution is in the understanding of tissue-specific and developmental stage-specific gene expression. It is now feasible to begin to define, on a molecular level, the factors which govern developmental gene expression in both a temporal and spatial fashion in animals.

Several strategies have been used successfully to approach the problem of tissue-specific and developmental regulation in transgenics. One technique is to add *in vitro* modified copies of a native gene to the mouse genome. The modifications allow the investigator to distinguish transcription from transgenic and endogenous loci on the basis of the molecular weight of resulting mRNAs or by hybridization to specific probes. This is a particularly useful approach because it allows direct comparison of transcription from endogenous and transgenic loci. However, it should be pointed out that *in vitro* modifications have the potential to introduce artifacts in expression by affecting mRNA stability or tissue-specific patterns of expression in unpredictable ways. Another common strategy is to place a reporter gene under the control of potential cis-acting elements from a gene of interest. The prokaryotic gene for the enzyme chloramphenicol acetyltransferase and the SV40 DNA tumor virus large T-antigen gene have successfully been used for this purpose.

One rather surprising conclusion that can be drawn is that species barriers do not seem to play a role in tissue- or stage-specific expression, since many genes from humans (Townes *et al.*, 1985a; Bucchini *et al.*, 1986; Chada *et al.*, 1986; Yamamura *et al.*, 1986), rats (Swift *et al.*, 1984; Shani, 1985, 1986), pigs (Frels *et al.*, 1985), and rabbits (Wagner *et al.*, 1981) are expressed appropriately in transgenic mice. Additionally, many genes that have been introduced into mice are expressed in appropriate tissue- and developmental stage-specific patterns, even though integration occurs at many sites throughout the genome. These results demonstrate that the primary cis-acting determinants of correct expression are evolutionarily conserved and reside close to the structural gene, so that appropriate expression is not dependent upon integration into the proper chromosomal locus. If trans-acting factors are involved in expression of these genes, then they must recognize their binding sites regardless of chromosomal position.

Chada *et al.* (1986) demonstrated that when the human fetal γ-globin gene was introduced into transgenic mice, it was expressed along with the mouse embryonic gene, implying that the trans-acting factors that regulate the mouse gene were also active on the human gene, normally expressed at a later stage of development. Other genes, including β-globin (Chada *et al.*, 1985, 1986; Townes *et al.*, 1985a; Magram *et al.*, 1985; Kollias *et al.*, 1986), actin (Shani, 1986), α-fetoprotein (Krumlauf *et al.*, 1985), αA-crystallin (Overbeek *et al.*, 1985), and αA(I) collagen (Khillan *et al.*, 1985), also were expressed in appropriate developmental patterns in transgenic mice. Taken together, these data demonstrate a remarkable evolutionary conservation of both cis- and trans-acting factors involved in tissue- and developmental stage-specific gene expression, and make it clear that while chromosomal location probably is important to fine tune the level of gene expression, position effects usually are not the primary determinants of the pattern of gene expression. Since it is unlikely that globin genes integrate in early embryos only into chromosomal domains that will be "open" in erythroid cells, while elastase genes integrate into domains that will be open in pancreatic cells, these data imply that the more open conformation of the chromatin of active genes is a result, rather than a cause, of expression. In contrast to other mammalian genes, the chicken transferrin gene was expressed at significant levels in kidney of transgenic mice, a tissue in which the mouse gene is poorly expressed (McKnight *et al.*, 1983). This result may indicate that the signals that have been conserved evolutionarily among mammalian species have not been so tightly conserved among birds and mammals.

Although the pattern of expression of many genes is appropriate, the level of expression can vary. In some lines of mice there is no detectable expression, while in others, expression can vary from a few percent of the wild-type allele to wild-type levels. There seem to be gene-specific differences in the level of expression when introduced into mice. For example, elastase is expressed at levels that approach or exceed the endogenous gene (Swift *et al.*, 1984), while globins, although expressed in erythroid cells, often are expressed at levels far below their endogenous counterparts (Chada *et al.*, 1985, 1986; Townes *et al.*, 1985a). These results imply that chromosomal position is more important for maximal expression of some genes than for others.

For the metallothionein gene, which is inducible in response to heavy metals, both the basal level of expression and the degree of inducibility differ from strain to strain, although for a given construct, the tissue distribution of expression is more or less consistent (Palmiter *et al.*, 1982a,b, 1983b; Hammer *et al.*, 1985; Low *et al.*, 1985; Chisari *et al.*, 1985; Swanson *et al.*, 1985). Thus, chromosomal position does not seem to affect the pattern of expression, but does seem to exert secondary effects that regulate the level of expression and inducibility. It seems likely that in lines of mice that do not express transgenes that are expressed well in other lines, the integration has occurred within heterochromatic regions incompatible with gene expression in particular differentiated tissues.

Occasionally transgenic mice express transgenes in inappropriate tissues, almost always at a low level (Lacy *et al.*, 1983; Shani, 1986; Swift *et al.*, 1984; Pinkert *et al.*, 1985). In early experiments in which plasmid vector sequences were not removed from DNA to be injected, globin gene expression was detected at a very low level in muscle and testis, in animals in which no expression was detectable in erythroid cells (Lacy *et al.*, 1983). It was determined subsequently that expression of globin as well as many other genes was adversely affected by the bacterial plasmid or phage sequences: removal of the sequences was necessary for proper expression (Townes *et al.*, 1985a; Chada *et al.*, 1986). Other genes for which this effect has been documented include actin (Shani, 1986) and α-fetoprotein (Krumlauf *et al.*, 1985). Although the mechanisms that are responsible for this phenomenon are unknown, the data make it clear that given the time and effort required to generate transgenic mice, it is advisable to remove vector sequences prior to injection of the DNA.

For many genes, important transcriptional regulatory sequences have been defined by deletion or linker-scanning analysis of 5′ regions. For tissue-specific genes, the most reliable assay system is *in vivo,* since the full range of different cell types in all stages of differentiation cannot be duplicated in culture. The elastase I gene, expressed primarily in exocrine cells of the pancreas, has been carefully characterized in transgenic mice. A fragment containing the entire rat structural gene and 7.2 kb of 5′-flanking sequences was appropriately expressed in transgenic mice at levels approximately equal to the endogenous gene (Swift *et al.*, 1984). Expression in other cells was reduced by at least a factor of 10^3–10^4, demonstrating that tissue-specific regulation of this gene was quite tight. Deletion analysis of both 5′ and 3′-flanking DNA demonstrated that 205 bp of 5′-flanking DNA was sufficient to direct expression to the pancreas (Ornitz *et al.*, 1985a,b).

The elastase gene provides an elegant example of the elucidation of important 5′-controlling sequences. The demonstration of negative transcriptional modulators (silencers), which may be the negative regulatory equivalent of enhancers in some genes (Brand *et al.*, 1985), makes analysis of gene expression somewhat more complicated, especially for hybrid genes, in which regions of different genes are joined. Since important regulatory regions may be within or 3′ to structural genes, one cannot assume that the pattern of expression of hybrid genes will be directed solely by the promoter–enhancer region. A good example of this is provided by the metallothionein promoter. Although most genes fused to this promoter are expressed in preditable patterns, some metallothionein (MT) fusion genes are expressed in aberrant tissues. An example of aberrant expression is provided by a MT–growth hormone gene, which had novel developmental specificity in the nervous system of transgenic mice (Swanson *et al.*, 1985). The fusion gene was expressed in a subset of neurons, although neither the metallothionein gene nor the growth hormone gene is expressed in neuronal tissues. Another example of complex regulatory interactions is provided by a construct that contained MT–hGH gene adjacent to an SV40 molecule from which the enhancer had been removed. Mice containing this construct developed hepatocellular carcinomas, which could be explained since the MT 5′ region directs expression to the liver. However, they also developed insulinomas (tumors of the pancreas) and peripheral neuropathy due to expression in Schwann cells (Messing *et al.*, 1985). Additional examples of unexpected patterns of expression have been seen following joining of diverse DNA segments (Low *et al.*, 1986; Townes *et al.*, 1985b). All of these results can be explained by postulating that many genes have intragenic or 3′-regulatory sequences that are important in tissue-specific expression. They may act independently or in concert with 5′-regulatory sequences to direct appropriate expression.

VII. Developmental Stage-Specific Gene Expression

The mechanisms which govern cell-specific gene regulation could be involved in the timing of expression as well. Transgenic mice have been successfully used to begin to define the DNA sequences required for proper developmental gene expression. Several examples are discussed below.

The pattern of globin gene expression in normal human and mouse development has been well characterized, thereby making it particularly amenable to study in transgenic mice. Initial experiments with a human adult β-globin transgene showed that human globin expression could be restricted to erythroid cells in adult transgenic mice (Chada *et al.*, 1985; Townes *et al.*, 1985a). In subsequent experiments, a hybrid mouse–human adult β-globin gene was introduced and was shown to be expressed in erythroid cells and regulated appropriately during development (Magram *et al.*, 1985). The chimeric construct contained the 5′ portion of a mouse adult β-globin gene along with 1.2 kb of flanking DNA fused to the 3′ portion of a human adult β-globin gene with 2 kb of 3′-flanking human DNA. The hybrid gene behaved qualitatively as though it was an endogenous mouse β-globin gene, becoming transcriptionally active around day 12 of gestation.

Recently an elegant experiment has been reported which employed the globin gene system to shed light on the evolution of developmentally regulated expression. During the course of vertebrate evolution, several patterns of developmental expression have arisen. The majority of mammals, including mice, express two types of globin, embryonic and adult. Humans and some other primates express an additional type of globin termed fetal. Fetal globins are thought to be descendants of a proto-γ gene whose expression pattern was embryonic, and which has subsequently been recruited as a fetal gene. Chada *et al.* (1986) demonstrated that when the human fetal γ-globin gene was introduced into mice, it was coordinately regulated with the mouse embryonic gene. This implies that the regulatory sequences and trans-acting factors which regulated the ancestral proto-γ gene have been well conserved. Furthermore, the authors concluded that the recruitment of the proto-γ gene most likely involved a switch in the timing of expression of a trans-acting factor rather than a mutation in cis-acting DNA sequences.

Kollias *et al.* (1986) have independently confirmed and extended these findings. They generated transgenic lines carrying a human fetal γ-globin gene, a human adult β-globin gene, or a hybrid construct consisting of the 5′ region of a human fetal γ-globin gene joined to the 3′ portion of a human adult β-globin gene. Their findings with human fetal and adult globin expression in transgenics were consistent with those discussed above. Interestingly, the hybrid γ–β gene was turned on appropriately during development and continued to be expressed in adult blood. Apparently, the 3′ β-globin sequences were able to stimulate transcription from the normally silent γ-globin promoter in adult mice. These results demonstrate the potential importance of intragenic and 3′ sequences in regulation of tissue-specific expression and point out the need for caution when interpreting expression patterns of hybrid constructs.

Another developmental gene which has been characterized in transgenic mice is the α-fetoprotein (AFP) gene. The AFP gene, evolutionarily related to serum albumin genes, is restricted in its expression to the visceral endoderm of the yolk sac, fetal gut, and fetal liver. Expression in liver is regulated temporally as well, declining 10,000-fold in transcription rate soon after birth. Krumlauf *et al.* (1985) constructed AFP minigenes and produced transgenic mice. Analysis of expression on the RNA level clearly demonstrated that 7 kb of DNA 5′ to the AFP gene was sufficient to direct correct tissue- and stage-specific regulation. Although initially the level of expression of the AFP minigenes was significantly lower than that of the endogenous AFP gene, removal of plasmid sequences prior to injection resulted in a high level of expression in all animals tested.

The developmental patterns of expression of several genes have been assayed by linking their 5′ regions to the prokaryotic reporter gene, chloramphenicol acetyltransferase. Measurement of CAT activity then gave an indication of the developmental and tissue-specific activity of the regulatory region. Developmental regulation of an αA-crystallin–CAT fusion gene was demonstrated by Overbeek *et al.*

(1985). The CAT construct and the endogenous αA-crystallin gene became active at approximately day 12.5 of embryonic development, suggesting coordinate regulation of the two genes. Similarly, Khillan *et al.* (1985) showed that in mice carrying an $\alpha 2$ type I collagen–CAT gene, CAT activity was detected beginning at day 8.5 of development, which coincides with the expected time of appearance of type I collagen in normal mice.

Shani (1986) observed developmental regulation of a hybrid construct containing the 5′ two-thirds of the rat skeletal actin gene linked to the 3′ end of a human embryonic globin gene. The chimeric gene behaved in a manner similar to the endogenous mouse skeletal actin gene: relative to the level of expression in skeletal muscle, the transgene was expressed at a high level in cardiac muscle of neonatal mice and at a lower level in adult mice.

Several general conclusions can be drawn regarding developmental regulation of transgenes:

1. When 5′-flanking sequences are included and vector sequences are removed, expression is generally qualitatively correct; transgenes are activated at the appropriate stage of development in specific cell types. The only exceptions are some hybrid constructs, which can develop new developmental specificities when diverse controlling regions are joined.
2. Quantitatively, expression varies from a few percent of the endogenous level of several orders of magnitude above the endogenous level. One explanation for this observation is that the level of expression is affected by chromosomal position. Alternatively, sequences farther upstream or downstream could be required for the fine control of transcription.
3. For the transgenes from organisms other than mice which have been analyzed to date, factors which regulate stage-specific expression have been remarkably well conserved in vertebrate evolution.

VIII. Oncology

Transgenic mouse technology has provided an effective complement to the genetic and biochemical characterizations of *onc* genes that have advanced so rapidly in recent years. Although many *onc* genes have been isolated from viruses and transformed cells (Bishop and Varmus, 1985), it has been difficult to demonstrate that particular cellular genes are directly involved in the etiology of specific types of tumors. One way to address this problem is to return the *onc* genes, either in their native form or after specific manipulation, to animals, and to assess simply if the animals develop tumors in predictable patterns. Appearance of tumors may demonstrate that the *onc* gene plays a direct role in tumor formation. The frequency and timing of tumor development, as well as the tissues affected, can provide information about the activity of the regulatory region as well as the protein encoded by the *onc* gene, and can suggest whether or not the particular *onc* gene is sufficient for tumor induction or if other types of changes have to occur. Several *onc* genes have been introduced into transgenic mice, and some have generated tumors.

The early region of SV40 was the first potentially tumor-inducing sequence to be introduced into transgenic mice (Brinster *et al.*, 1984; Palmiter *et al.*, 1985b; Small *et al.*, 1985). Several of the mice carrying the SV40 early region developed choroid plexus papillomas, and in later experiments the SV40 early region, and more specifically the SV40 enhancer and large T antigen, were shown to be responsible for the development of tumors (Palmiter *et al.*, 1985b). Several lines of mice were produced in which positive offspring developed the choroid plexus tumors at reproducible times—every mouse in a given line developed tumors at the same age, although the age varied among different lines. Expression of T antigen was at a high level in tumor tissue, often was seen in thymus and kidney, and occasionally was detected at low levels in other tissues. In addition to the choroid plexus papillomas, thymic hypertrophy and kidney pathology were observed in some mice. Since expression of SV40 T antigen was low or undetectable in most unaffected tissues and in affected tissues prior to the appearance of pathology, these experiments strongly suggested that SV40 T antigen was directly involved in the development of the tumors.

The above experiments suggested that the specificity of the tumors for the choroid plexus was a result of high-frequency expression in that tissue. They did not rule out the possibility that the choroid plexus and/or other tissues are more susceptible to the action of T antigen than others. Thus the development of tumors in a specific tissue may be a result of the tissue tropism of the enhancer driving expression of the *onc* gene, as well as the susceptibility of a given tissue to the action of that gene product. Since SV40 T antigen seems to have a wide spectrum of transforming potential, it seemed likely that the targeting to the choroid plexus was a result of the pattern of expression. This prediction has been borne out by experiments in which the SV40 T-antigen genes have been ligated to regulatory regions of other genes (Hanahan, 1985; Ornitz *et al.*, 1985b). When the

SV40 enhancer was deleted from the construct that also carried the MT–growth hormone fusion genes, totally different pathology resulted (Messing *et al.*, 1985; Palmiter *et al.*, 1985b): mice developed peripheral demyelination, hepatocellular carcinomas, and islet cell adenomas.

Hanahan (1985) introduced a construct containing the large T-antigen gene litigated to the 5′ region of the rat insulin II gene and detected T-antigen expression exclusively in the β-cells of the endocrine pancreas. Mice harboring this construct developed tumors consisting entirely of β-cells, which differed from naturally occurring insulinomas, which normally contain both β- and δ-cells. In contrast to the case with the intact SV40 early region, which apparently was activated in the choroid plexus tumors, there was detectable expression of the SV40 T antigen in islet of Langerhans cells prior to tumor development. The T antigen was expressed in all islets examined. Additionally, approximately 4 or 5 of the 100 or so islets in a pancreas developed into solid β-cell tumors, although most islets showed hyperplasia and all had increased β-cell density. These results demonstrate that although T antigen was expressed in all islets, only a small percentage developed into tumors, and thus suggest that expression of T antigen was not sufficient for tumor production. All islets had evidence of abnormal cell proliferation, which probably was a result of T-antigen production. Other changes might be necessary for the development of tumors. The tumors in these mice also were heritable. These experiments are interesting not only because they characterize the spectrum of action of T antigen, but because tumor induction by T antigen is a sensitive and specific marker for the tissues in which it is expressed. In this case, expression was limited to cells in the islets of Langerhans, and probably β-cells, demonstrating that the 5′ end of the insulin gene was sufficient to direct expression to those cells. Similar experiments in which the T antigen was litigated to the elastase promoter resulted in tumors of the exocrine pancreas (Ornitz *et al.*, 1985b), again demonstrating that cell specificity was imparted by the 5′ region of the gene.

Early regions of two human papovaviruses (BK virus and JC virus) closely related to SV40 also have been introduced into transgenic mice. Antibodies to each of the viruses are detectable in 70–80% of adults. JC has been strongly implicated in the fatal demyelinating syndrome, progressive multifocal leukoencephalopathy (PML), which occurs predominantly in immunocompromised individuals (Johnson, 1983). Multiple glial tumors also have been observed in PML patients

(GiaRusso and Koeppen, 1978). BK virus (BKV) is found predominantly in the kidney, where it induces a subclinical infection (McCance, 1983). In immunosuppressed patients, BKV can be excreted in the urine, although infections usually are asymptomatic (Tooze, 1981). The nucleotide sequence and amino acid homology of the T-antigen and viral capsid protein genes of SV40, BKV, and JC virus (JCV) are between 70 and 80%. The T antigens are most conserved at the amino-terminal end (Frisque *et al.*, 1984). The most divergent region of the viral genomes encompasses the origin of replication and enhancer sequences. JCV and BKV early regions were introduced into transgenic mice, where they induced distinct and reproducible types of pathology. JCV induced adrenal medullary neuroblastomas and dysmyelination in the central nervous system (Small *et al.*, 1986a,b), thus mimicking its behavior in humans. BKV induced primary renal, hepatic, and lung carcinomas (Small *et al.*, 1986b). JC T antigen was expressed at high levels in the tumor tissue, and at lower but detectable levels in histologically normal tissue. Expression of JC T antigen also was high in oligodendrocytes of animals that developed dysmyelination (G. Scangos, unpublished data). BK was expressed in many tissues of the animal and in high amounts in tumor tissue. Thus, each of the closely related viruses induced distinct and tissue-specific pathologies. Subsequent experiments in which hybrid viruses are introduced will determine to what extent differences in the T antigens and regulatory regions are responsible for the different pathologies.

Lacey *et al.* (1986) introduced bovine papilloma virus (BPV) sequences into transgenic mice, which developed fibropapillomas of the skin. Thus, in this case as well, the pathology induced by the microinjected viral DNA was analogous to that induced in the natural host. In the case of BPV, the viral sequences were integrated in normal tissue, while extrachromosomal BPV DNA was detected in all tumors.

Transgenic mice containing the *myc* oncogene also have been produced. The *myc* oncogene has been implicated in the etiology of certain lymphomas and leukemias because of the reproducible occurrence of chromosomal translocations close to the *myc* gene in the malignant cells (Leder *et al.*, 1983). To elucidate the role in tumor etiology, Stewart *et al.* (1984) introduced into mice constructs in which the protooncogene c-*myc* was placed under the control of the hormonally inducible regulatory sequences from mouse mammary tumor virus (MMTV). Thirteen lines of transgenic mice were generated, and individuals of two lines developed mammary adenocarcinomas. Eleven of the twelve lines showed expression of *myc* in salivary glands, and expression was detected in intestine and mammary glands of two of three lines tested. Tumors appeared in multiparous females, and in the two tumor-prone lines, tumor formation segregated with the *myc* gene in offspring, indicating that the *myc* gene was involved in tumor formation. In the tumor-bearing mice, tumors arose in a subset of the mammary glands, suggesting that *myc* was not sufficient for tumor development. In addition to *myc*, an additional genetic or epigenetic change, as well as a change in hormonal environment brought about by pregnancy, seemed to be necessary for tumor development.

A subsequent report from the same group (Leder *et al.*, 1986) described a line of mice with a *myc*–MMTV construct which was expressed in many tissues. The line was subject to a variety of tumors of testicular, breast, B and T lymphocyte, and mast cell origin. Similar findings were reported after introduction of a c-*myc* gene whose expression was controlled by immunoglobulin enhancers. This construct induced lymphoid malignancies in transgenic mice (Adams *et al.*, 1985). Despite its participation in a wide variety of tumors, the *myc* gene did not interfere with normal development and many tissues that expressed significant amounts of the transgene were unaffected. These studies demonstrated that *myc* was capable of contributing to malignacy in a wide variety of, but probably not all, tissues, and thus are the beginning of attempts to define the spectrum of action of the myc protein. Since not all cells of affected organs became neoplastic, these studies again demonstrated that *myc* expression was not sufficient to induce neoplasia.

IX. Immunology

The production of transgenic mice will be particularly useful in dissecting complex biological phenomena that require interaction of several different cell types—a characteristic that limits cell-culture studies. An example of such an area is the development of immunocompetence in mammals. Molecular cloning and characterization of genes involved in immunological phenomena, particularly those encoding antibodies, histocompatibility antigens, T-cell receptors, and immune response genes, have generated considerable information about the organization and expression of those genes. However, returning the genes to animals is perhaps the only way to assess their function accurately.

A. Immunoglobulin Genes

Several groups have introduced functionally rearranged murine (Brinster *et al.*, 1983; Grosschedl *et al.*, 1984; Rusconi and Kohler, 1985; Storb *et al.*, 1986) or human (Yamamura *et al.*, 1986) immunoglobulin (Ig) genes into the mouse germ line to study control of Ig gene rearrangement. Immunoglobulin heavy chains are assembled by correctly joining three gene segments V_H, D_H, and J_H in B lymphocytes (reviewed in Tonegawa, 1983). Similarly, Ig light chains are assembled from V_L and J_L segments. Only one heavy chain gene and one light chain gene are functionally rearranged and expressed in a normal B cell, a process called allelic exclusion. Several models have been proposed to explain this phenomenon (Altenburger *et al.*, 1980; Alt *et al.*, 1984; Wabl and Steinberg, 1982). One model postulates that functional rearrangements occur infrequently so that the probability of two functional rearrangements occurring in the same cell is very low. A second model suggests that further rearrangement is prevented through a negative-feedback mechanism once a functional rearrangement occurs. A third model hypothesizes that B cells with more than one functional rearrangement are eliminated. The production of transgenic mice carrying germ-line-rearranged heavy or light chain genes or both has generated considerable insight into the mechanisms of Ig gene regulation and allelic exclusion.

The first Ig gene to be introduced into the mouse germ line was a κ light chain gene of the myeloma MOPC-21 (Brinster *et al.*, 1983). Initial experiments demonstrated that the κ transgene was expressed in spleen but not in liver. A more detailed analysis of expression showed that the κ transgene was restricted in its expression to lymphoid tissues (Storb *et al.*, 1984). Furthermore, in lymphoid cell subpopulations there was a positive correlation between the level of transgene mRNA and the relative number of B cells. These data suggested that the microinjected κ gene was being specifically expressed in B lymphocytes despite its presence in rearranged form in all tissues, including all of the lymphoid tissues.

To examine the effect of the κ light chain transgene on allelic exclusion of endogenous κ chain genes, Ritchie *et al.* (1984) produced hybridomas from spleen cells of transgenic mice. They concluded that rearrangement of endogenous Ig light chains was turned off only when a heavy chain was expressed along with the transgenic κ light chain. In other words, the presence of a complete Ig molecule correlated with a shutdown of light chain rearrangement. The expression of the transgenic κ light chain in the absence of heavy chain was not sufficient to establish allelic exclusion of endogenous light chain genes.

Groschedel *et al.* (1984) produced transgenic mice carrying a rearranged Ig heavy chain. High levels of expression were observed in B and, additionally, T cells from several mouse lines. A low level of expression was also detected in cardiac tissue. The T cells normally do not rearrange their Ig heavy or light chain genes. However, the observed high level of expression of a "pre-rearranged" Ig heavy chain gene in transgenic mice strongly suggests that T cells may contain the transcriptional factors requisite for Ig gene expression.

Subsequent experiments by Weaver *et al.* (1985) examined the effect of the Ig μ transgene on rearrangement of endogenous Ig alleles: 40% of Abelson murine leukemia virus (AMuLV)-transformed pre-B cells from the Ig μ-transgenic mice contained J_H genes in their germ-line configuration, whereas all AMuLV transformants from a normal mouse had rearranged J_H alleles. Apparently, the presence of the Ig μ transgene inhibited rearrangement of endogenous alleles early in B-cell ontogeny. To assess heavy chain DNA rearrangement late in B-cell development, hybridomas were made from spleen cells of transgenics. Germ-line configuration J_H alleles were present in 10% of transgenic hybridomas, suggesting that inhibition of rearrangement by the Ig transgene was maintained to later stages of B-cell differentiation. The presence of the Ig heavy chain transgene had no effect on light chain rearrangement. Although the presence of a functional Ig heavy chain obviously had an inhibitory effect on endogenous allele rearrangement, the inhibition was far from complete. This suggested that the timing and level of transgene expression was an important determinant in control of endogenous allele rearrangement.

Weaver *et al.* (1986) have carefully examined the repertoire of Ig genes expressed in their μ heavy-chain-transgenic mice. Although most of the hybridomas from transgenic spleens did not express the transgene, a high percentage did make antibodies similar, on a functional level, to that encoded by the transgene. It appeared that a rearranged heavy chain gene, when introduced into the germ line of mice, was able to activate cellular regulatory mechanisms which led to expression of endogenous genes coding for antigen-binding regions which mimicked that of the transgene.

Rearranged Ig heavy and light chains have been introduced together into the mouse germ line by Rusconi and Kohler (1985). An examination of hybridomas from lymph node cells of a single transgenic line showed that most clones which expressed the transferred heavy and light chains also expressed endogenous light chains. Several also co-expressed endogenous heavy chains. Apparently, low levels of expression of the transgenic heavy and light chains together did not prevent rearrangement of endogenous alleles. However, higher levels of expression did appear to establish allelic exculsion.

Storb *et al.* (1986) have recently characterized transgenic mice carrying κ light (167-κ) and μ (167-μ) heavy chain genes which encode antiphosphorylcholine antibodies. The injected μ genes were either complete or truncated such that the membrane terminus was deleted, leaving only the secreted terminus (167-μ Δmem). The constructs were introduced in the following combinations: 167-κ alone, 167-μ alone, 167-κ with 167-μ, 167-μ Δmem, and 167-κ with 167-μ Δmem. All of the transgenes were expressed in lymphoid tissues. Transgenic μ mRNA but not transgenic κ mRNA was found in T cells, in agreement with the results of Groschedl *et al.* (1984) for a different μ transgene. Cytoplasmic μ protein was found in 60% of transgenic thymocytes, but none was detected on the surface of T cells. Mice carrying a complete 167-μ transgene seemed to express increased levels of endogenous 167-κ mRNA. Conversely, 167-κ transgenics produced more 167-μ mRNA. The authors speculated that B cells from 167-μ and 167-κ transgenics, which fortuitously express endogenous 167-κ or 167-μ alleles, are selected for and expanded faster than other B cells. Although the 167-μ Δmem transgenics expressed high levels of transgene mRNA, they did not show increased levels of endogenous 167-κ mRNA nor did there appear to be any inhibition of rearrangement of endogenous heavy or light chain rearrangement in these mice.

A human γ_1 heavy chain gene was introduced into mice by Yamamura *et al.* (1986). A low level of expression was detected in spleen only and was inducible by lipopolysaccharide but not by concanavalin A, suggesting B-cell-specific expression.

Several conclusions can be drawn from the Ig transgenic experiments:

1. With regard to the process of allelic exclusion, it now seems most likely that shutoff of rearrangement is mediated by the expression of a sufficient level of complete immunoglobulin molecules.
2. In general, Ig transgenes are expressed at detectable levels only in lymphoid tissues; however, T cells, which do not normally rearrange Ig genes or express functional Ig mRNA, can efficiently transcribe mouse but not human heavy chain genes.
3. The presence of a rearranged heavy chain gene in all tissues of a

mouse has profound effects on the Ig repertoire of those mice. It will be of great interest to determine the mechanisms which govern transgenic versus endogenous gene expression.

B. MHC Antigen Genes

Germ-line transformation of mice will also be valuable in examining the functions of the major histocompatibility complex (MHC) antigens, which are encoded by a highly conserved family of genes that are part of the growing "immunoglobulin supergene family." Genes for two major classes of integral membrane proteins, designated class I and II, reside within the MHC. Both are known to be critical components of the immune recognition process: foreign antigens must be presented to T cells in the context of a class I or class II antigen to properly trigger the response which leads to the elimination of the foreign antigen. MHC antigens are thought to play a role in several other processes (Ivanyi, 1978) and it is clear that the full range of their functions has yet to be elucidated.

The power of mouse genetics has contributed greatly to sorting out the biology of MHC antigens. Hundreds of inbred, hybrid, congenic, and spontaneous MHC mutant lines have been carefully developed, characterized, and maintained. Transgenic mouse production can clearly augment the above list of mouse reagents. Lines which differ in a single MHC antigen can easily be derived, a goal which is difficult, if not impossible, to achieve through standard breeding–selection schemes. Furthermore, precisely defined mutations can be introduced into antigen genes *in vitro* and their effects assayed *in vivo*. This approach provides a powerful new tool with which to dissect the functions of MHC antigens.

To date, murine class I (Bieberich *et al.*, 1986) and class II (Yamamura *et al.*, 1985; Le Meur *et al.*, 1985; Pinkert *et al.*, 1985) and porcine class I (Frels *et al.*, 1985) genes have been introduced into the mouse germ line by microinjection. Several groups have generated transgenics carrying the class II gene E_{α} with varying amounts of 5'-flanking sequences. All of the E_{α} transgenics made with constructs having 2 kb or more of 5'-flanking DNA expressed the transgene in an appropriate tissue-restricted fashion, with one exception (Pinkert *et al.*, 1985). The E_{α} gene product was expressed on the surface of B cells in association with an endogenous E_{β} antigen to form an I–E molecule that appeared to be immunologically functional both *in vivo* and *in vitro*. Expression of the class II transgene was inducible by γ-interferon in macrophages, as are endogenous class II antigens. Interestingly, deletion of 5'-flanking DNA to 1.4 kb appeared to eliminate the elements required for expression in B cells but not in macrophages or dendritic cells (Flavell *et al.*, 1986).

We have introduced the murine class I gene D^d into an inbred background by microinjection. The D^d transgene was expressed in all tissues examined although the steady-state level of D-specific mRNA varied in a manner that paralleled the expression of the endogenous K class I antigen (Bieberich *et al.*, 1986). Expression of the class I transgene was shown to be inducible by interferons and repressible by human adenovirus 12 transformation, as are normal class I antigens (C. Bieberich *et al.*, unpublished observations), and it appears to be immunologically functional both *in vitro* and *in vivo*. More recently, we have generated transgenics carrying the D^d gene under the control of the mouse metallothionein I promoter. The level of *D* transgene expression is inducible by heavy metals primarily in liver, intestine, and kidney of these lines.

X. Correction of Genetic Deficiencies

The finding that foreign genes usually are expressed appropriately in transgenic mice raised the possibility of correcting genetic defects present in inbred strains, and several groups have attempted such experiments. Hammer *et al.* (1984) introduced a metallothionein–growth hormone construct into *little* mice, which are an inbred line of dwarf mice. Several transgenic lines that expressed the heterologous growth homrone gene in the liver were produced, and they grew approximately three times larger than their nontransgenic littermates. Additionally, the fertility of *little* males was elevated.

Three groups have corrected an immunodeficiency present in lines of inbred mice. The deficiency is of a gene encoding the class II molecule E_α. Mice with the defect are unable to mount an immune response against poly(L-glutamic acid-L-lysine-L-phenylalanine) (Yamamura *et al.*, 1985). Mice with a transgenic E_α gene were capable of mounting a normal immune response against the antigen, demonstrating that the introduced E_α gene was immunologically functional (Le Meur *et al.*, 1985; Pinkert *et al.*, 1985; Yamamura *et al.*, 1985). The genes in these mice were expressed and associated with E_β on the cell surface. The genes also were inducible in response to interferon.

Costantini *et al.* (1986) introduced mouse and human β-globin genes into β-thallasemic mice that had a deletion of the β-globin gene. Both transgenes produced functional β-globin chains, leading to a reduction or elimination of the anemia and abnormalities of the red blood cells. The mouse β-globin transgene was expressed at a low level, and the anemia was reduced in intensity. The human β-globin gene was expressed at significant levels, and the human β-chains associated with the mouse α-chains to produce functional hemoglobin. The thallasemic mice with the human transgene were cured of the thallasemia.

Recently, Mason *et al.* have introduced a gene encoding gonadotropin-releasing hormone (GnRH) into hypogonadal mice (Mason *et al.*, 1986). The gene was expressed in hypothalamic neurons of transgenic mice which were reproductively completely normal. Pituitary and serum concentrations of leuteinizing hormone, follicle-stimulating hormone, and prolactin were restored.

Isola and Gordon (1986) introduced into mice a gene encoding dihydrofolate reductase (DHFR) that was approximately 270-fold more resistant to methotrexate than the normal DHFR protein. All of the eight lines tested for methotrexate resistance were significantly more resistant than controls. Although these experiments did not correct a genetic defect, they, along with the above experiments, demonstrate that foreign genes can have significant effects on the physiology of transgenic animals. Genes encoding proteins deficient in the host animals are expressed appropriately, and the proteins can correct the genetic effects. Extra genes can confer novel properties on the organism which carries them.

It is important to note that in all of these experiments, the transgene integrates at random sites and does not replace the defective gene. Thus, transgenic mice carry both defective and functional genes, which can segregate in subsequent generations. The mutation leading to the defect is not repaired, but is complemented in the 50% of the offspring that inherit the transgene. For this reason, and for obvious ethical reasons, this approach is not viable for the correction of human genetic deficiencies. Rather, correction of genetic defects in mice is likely to generate insight into disease mechanisms and to produce models of human diseases (Costantini *et al.*, 1986).

XI. Transgenic Mice as Models of Human Disease

The above experiments demonstrate that introduced genes can significantly alter the phenotype of transgenic animals. These observations raised the possibility that it might be possible to alter the genetic composition of mice so that they would mimic human diseases and thus provide animal models for important diseases. Babinet *et al.* (1985) and Chisari *et al.* (1985) introduced constructs expressing the surface antigen of the hepatitis virus into transgenic mice and demonstrated expression. The hope is that these mice will mimic the chronic carrier state of hepatitis virus infection. Small *et al.* (1986a,b) introduced the early region of the JC virus, which has been implicated as a causative agent in the demyelinating disease progressive multifocal leukoencephalopathy. Transgenic mice harboring the JC-early region developed a dysmyelination in the central nervous system. JCV T antigen could be detected in oligodendrocytes and some astrocytes of the central nervous system, indicating that the virus has the same tissue tropism in mice that it exhibits in humans. These data demonstrate that the concept of introducing novel genetic information into mice so that they will mimic human disease is a workable one.

XII. Areas of Future Research

While it is clear that transgenic mice have contributed significant insights into a wide range of biological problems, some aspects of the technology limit its utility. It is important to address these limitations so that a wider range of questions can be addressed in the future. Areas of research that will enhance the technology are briefly presented below.

1. *Improvement of constructs.* A significant limitation is the absence of tightly regulatable promoters that can be turned on and off during development and in the adult mouse. These promoters would allow the introduction and controlled expression of genes that encode potentially toxic molecules. Additionally, measurement of phenotypic differences with genes on and off may define physiological effects of expression of specific proteins. For retroviruses, the development of vectors that allowed normal or near-normal levels of tissue- and stage-specific expression is an important goal.
2. *Improvement of technology.* Identification of conditions that allow targeting of genes to specific chromosomal locations, perhaps via homologous recombination, will permit many new experiments to be performed. The ability to target genes would mean that functional genes can be replaced with mutant ones, or conversely, mutant genes with functional ones. The ability to generate single or low-copy-number integration sites by microinjection without significant chromosomal rearrangement also will be important. Finally, as more transgenic lines are made, it becomes increasingly expensive and cumbersome to maintain them. The development and refinement of techniques for cryogenic preservation of embryos, in a manner analogous to cell lines, will become a necessity.

Transgenic mice will continue to provide important information about gene regulation and insertional mutations probably will fortuitously lead to the identification of important developmental genes. As more genes encoding proteins involved in complex biological systems, such as the immune and nervous systems, are cloned, transgenic technology will provide powerful assay systems with which to define the role of those proteins. Continued progress of molecular biology and biochemistry in these areas, coupled with increasingly sophisticated gene transfer technology, indicates that an important and dramatic new understanding of the function of mammalian organisms will occur in the next decade.

References

Adams, J. M., Harris, A. W., Pinkert, C. A., Corcoran, L. M., Alexander, W. S., Cory, S., Palmiter, R. D., and Brinster, R. L. (1985). The c-myc oncogene driven by immunoglobulin enhancers induces lymphoid malignancy in transgenic mice. *Nature (London)* **318,** 533–538.

Alt, W. F., Yancopoulos, G. D., Blackwell, K. T., Wood, C., Thomas, E., Boss, M., Coffman, R., Rosenberg, N., Tonegawa, S., and Baltimore, D. (1984). Ordered rearranagement of immunoglobulin heavy chain variable region segments. *EMBO J.* **3,** 1209–1219.

Alt, F., Blackwell, T. K., and Yancopoulos, G. D. (1985). Immunoglobulin genes in transgenic mice. *Trends Genet.* **1,** 231–236.

Altenburger, W., Steinmetz, M., and Zachau, J. (1980). Functional and non-functional joining in immunoglobulin light chain genes of a mouse myeloma. *Nature (London)* **287,** 603–607.

Babinet, C., Farza, H., Morello, D., Hadchouel, M., and Pourcel, C. (1985). Specific expression of hepatitis B surface antigen (HBsAg) in transgenic mice. *Science* **230,** 1160–1163.

Bachetti, S., and Graham, F. L. (1977). Transfer of the gene for thymidine kinase to thymidine kinase-deficient human cells by purified herpes simplex viral DNA. *Proc. Natl. Acad. Sci. U.S.A.* **74,** 1590–1594.

Bieberich, C., Scangos, G., Tanaka, K. and Jay, G. (1986). Regulated expression of a murine class I gene in transgenic mice. *Mol. Cell. Biol.* **6,** 1339–1342.

Bishop, J. M., and Varmus, H. (1985). *In* "RNA Tumor Viruses" (R. Weiss, N. Teich, H. Varmus, and J. Coffin, eds.). Cold Spring Harbor Laboratory, Cold Spring Harbor, New York.

Brand, A. H., Breeden, L., Abraham, J., Sternglanz, R., and Nasmyth, K. (1985). Characterization of a silencer in yeast: A DNA sequence with properties opposite to those of a transcriptional enhancer. *Cell* **41,** 41–48.

Brinster, R. L., Chen, H. Y., Trumbauer, M. E., Senear, A. W., Warren, R., and Palmiter, R. D. (1981). Somatic expression of herpes thymidine kinase in mice following injection of a fusion gene into eggs. *Cell* **27,** 223–231.

Brinster, R. L., Ritchie, K. A., Hammer, R. E., O'Brien, R. L., Arp B., and Storb, U. (1983). Expression of a microinjected immunoglobulin gene in the spleen of transgenic mice. *Nature (London)* **306,** 332–336.

Brinster, R. L., Chen, H. Y., Messing, A., van Dyke, T., Levine, A. J., and Palmiter, R. D. (1984). Transgenic mice harboring SV40 T-antigen genes develop characteristic brain tumors. *Cell* **37,** 367–379.

Brinster, R. L., Chen, H. Y., Trumbauer, M. E., Yagle, M. K., and Palmiter, R. D. (1985).

Factors affecting the efficiency of introducing foreign DNA into mice by microinjecting eggs. *Proc. Natl. Acad. Sci. U.S.A.* **82**, 4438–4442.

Bucchini, D., Ripoche, M.-A., Stinnakre, M.-G., Desbois, P., Lores, P., Monthioux, E., Absil, J., Lepesant, J.-A., Pictet, R., and Jami, J. (1986). Pancreatic expression of human insulin gene in transgenic mice. *Proc. Natl. Acad. Sci. U.S.A.* **83**, 2511–2515.

Burki, K., and Ullrich, A. (1982). Transplantation of the human insulin gene into fertilized mouse eggs. *EMBO J.* **1**, 127–131.

Chada, K., Magram, J., Raphael, K., Radice, G., Lacy, E., and Costantini, F. (1985). Specific expression of a foreign β-globin gene in erythroid cells of transgenic mice. *Nature (London)* **314**, 377–380.

Chada, K., Magram, J., and Costantini, F. (1986). An embryonic pattern of expression of a human fetal globin gene in transgenic mice. *Nature (London)* **319**, 685–689.

Chisari, F. V., Pinkert, C. A., Milich, D. R., Filippi, P., McLachlan, A., Palmiter, R. D., and Brinster, R. L. (1985). A transgenic mouse model of the chronic hepatitis B surface antigen carrier state. *Science* **230**, 1157–1160.

Costantini, F., and Lacy, E. (1981). Introduction of a rabbit β-globin gene into the mouse germ line. *Nature (London)* **294**, 92–94.

Costantini, F., Chada, K., and Magram, J. (1986). Correction of murine β-thalassemia by gene transfer into the germ line. *Science* **233**, 1192–1194.

Covarrubias, L., Nishida, Y., and Mintz, B. (1986). Early postimplantation embryo lethality due to DNA rearrangements in a transgenic mouse strain. *Proc. Natl. Acad. Sci. U.S.A.* **83**, 6020–6024.

DeSaint Vincent, B. R., and Wahl, G. M. (1983). Homologous recombination in mammalian cells mediates formation of a functional gene from two overlapping gene fragments. *Proc. Natl. Acad. Sci. U.S.A.* **80**, 2002–2006.

Flavell, R. A., Allen, H., Burkly, L. C., Sherman, D. V., Waneck, G. L., and Widera, G. (1986). Molecular biology of the H-2 histocompatibility complex. *Science* **233**, 437–443.

Folger, K. R., Wong, E. A., Wahl, G., and Capecchi, M. (1982). Patterns of integration of DNA microinjected into cultured mammalian cells: Evidence for homologous recombination between injected plasmid DNA molecules. *Mol. Cell. Biol.* **2**, 1372–1387.

Frels, W. J., Bluestone, J. A., Hodes, R. J., Capecchi, M. R., and Singer, D. S. (1985). Expression of a microinjected porcine class I major histocompatibility complex gene in transgenic mice. *Science* **228**, 577–580.

Frisque, R. J., Bream, G. L., and Cannella, M. T. (1984). Human polyomavirus JC virus genome. *J. Virol.* **51**, 458–469.

Gerlinger, P., Le Meur, M., Irrmann, C., Renard, P., Wasylyk, C., and Wasylyk, B. (1986). B-Lymphocyte targeting of gene expression in transgenic mice with the immunoglobulin heavy-chain enhancer. *Nucleic Acids Res.* **14**, 6565–6577.

GiaRusso, M. H., and Koeppen, A. H. (1978). Atypical progressive multifocal leukoencephalopathy and primary cerebral malignant lymphoma. *J. Neurol. Sci.* **35**, 391–398.

Gordon, J. W., and Ruddle, F. H. (1981). Integration and stable germ line transmission of genes injected into mouse pronuclei. *Science* **214**, 1244–1246.

Gordon, J. W., and Ruddle, F. H. (1983). Gene transfer into mouse embryos: Production of transgenic mice by pronuclear injection. *In* "Methods in Enzymology" (R. Wu, L. Grossman, and K. Moldave, eds.), Vol. 101, pp. 411–433. Academic Press, New York.

Gordon, J. W., and Ruddle, F. H. (1985). DNA-mediated genetic transformation of mouse embryos and bone marrow—A review. *Gene* **33**, 121–136.

Gordon, J. W., Scangos, G. A., Plotkin, D. J., Barbosa, J. A., and Ruddle, F. H. (1980). Genetic transformation of mouse embryos by microinjection of purified DNA. *Proc. Natl. Acad. Sci. U.S.A.* **77**, 7380–7384.

Grosschedl, R., Weaver, D., Baltimore, D., and Costantini, F. (1984). Introduction of a mu immunoglobulin gene into the mouse germ line: Specific expression in lymphoid cells and synthesis of functional antibody. *Cell* **38**, 647–658.

Hammer, R. E., Palmiter, R. D., and Brinster, R. L. (1984). Partial correction of murine hereditary disorder by germ-line incorporation of a new gene. *Nature (London)* **311**, 65–67.

Hammer, R. E., Brinster, R. L., Rosenfeld, M. G., Evans, R. E., and Mayo, K. E. (1985). Expression of human growth hormone-releasing factor in transgenic mice results in increased somatic growth. *Nature (London)* **315**, 413–416.

Hanahan, D. (1985). Heritable formation of pancreatic B-cell tumours in transgenic mice expressing recombinant insulin/simian virus 40 oncogenes. *Nature (London)* **315**, 115–122.

Hartung, S., Jaenisch, R., and Breindl, M. (1986). Retrovirus insertion inactivates mouse α1(I) collagen gene by blocking initiation of transcription. *Nature (London)* **320**, 365–367.

Hogan, B. L. M., Costantini, F., and Lacy, E. (1986). "Manipulation of the Mouse Embryo: A Laboratory Manual." Cold Spring Harbor Lab., Cold Spring Harbor, New York.

Humphries, R. K., Berg, P., DiPietro, J., Bernstein, S., Baur, A., Nienhaus, A. W., and Anderson, W. F. (1985). Transfer of human and murine globin-gene sequences into transgenic mice. *Am. J. Hum. Genet.* **37**, 295–310.

Isola, L. M., and Gordon, J. W. (1986). Systemic resistance to methotrexate in transgenic mice carrying a mutant dihydrofolate reductase gene. *Proc. Natl. Acad. Sci. U.S.A.* **83**, 9621–9625.

Ivanyi, P. (1978). Some aspects of the H-2 system, the major histocompatibility system in the mouse. *Proc. R. Soc. London* **202**, 117–123.

Jaenisch, R. (1976). Germ line integration and Mendelian transmission of the exogenous Moloney leukemia virus. *Proc. Natl. Acad. Sci. U.S.A.* **73**, 1260–1264.

Jaenisch, R., and Mintz, B. (1974). Simian virus 40 sequences in DNA of healthy adult mice derived from preimplantation blastocysts injected with viral DNA. *Proc. Natl. Acad. Sci. U.S.A.* **71**, 1250–1254.

Jahner, D., Haase, K., Mulligan, R., and Jaenisch, R. (1985). Insertion of the bacterial gpt gene into the germ line of mice by retroviral infection. *Proc. Natl. Acad. Sci. U.S.A.* **82**, 6927–6931.

Johnson, R. T. (1983). Evidence for polyomaviruses in human neurological diseases. *In* "Polyomaviruses and Human Neurological Diseases," (J. Sever and D. Madden, eds.), pp. 489–520. Plenum, New York.

Khillan, J. S., Schmidt, A., Overbeek, P. A., de Crombrugghe, B., and Westphal, H. (1986). Developmental and tissue-specific expression by the α_2 type I collagen promoter in transgenic mice. *Proc. Natl. Acad. Sci. U.S.A.* **83**, 725–729.

Kollias, G., Wrighton, N., Hurst, J., and Grosveld, F. (1986). Regulated expression of human $^{A}\gamma$-, β-, and hybrid γ–β-globin genes in transgenic mice: Manipulation of the developmental expression patterns. *Cell* **46**, 89–94.

Krumlauf, R., Hammer, R. E., Tilghman, S. M., and Brinster, R. L. (1985a). Developmental regulation of α-fetoprotein genes in transgenic mice. *Mol. Cell. Biol.* **5**, 1639–1648.

Krumlauf, R., Chapman, V. M., Hammer, R. E., Brinster, R., and Tilghman, S. M. (1985b). Differential expression of α-fetoprotein genes on the inactive X chromosome in extraembryonic and somatic tissues of transgenic mice. *Nature (London)* **319**, 224–226.

Krumlauf, R., Hammer, R. E., Brinster, R., Chapman, V. M., and Tilghman, S. M. (1985c). Regulated expression of α-fetoprotein genes in transgenic mice. *Cold Spring Harbor Symp. Quant. Biol.* **50**, 371–378.

Kuhn, L. C., McClelland, A., and Ruddle, F. H. (1984). Gene transfer, expression, and molecular cloning of the human transferrin receptor gene. *Cell* **37**, 95–103.

Lacy, E., Roberts, S., Evans, E. P., Burtenshaw, M. D., and Costantini, F. (1983). A foreign β-globin gene in transgenic mice: Integration at abnormal chromosomal positions and expression in inappropriate tissues. *Cell* **34**, 343–358.

Lacey, M., Alpert, S., and Hanahan, D. (1986). Bovine papillomavirus genome elicits skin tumours in transgenic mice. *Nature (London)* **322**, 609–612.

Leder, P., Battey, J., Lenoir, G., Moulding, C., Murphy, W., Potter, H., Stewart, T., and Taub, R. (1983). Translocations among antibody genes in human cancer. *Science* **222**, 765–771.

Leder, A., Pattengale, P. K., Kuo, A., Stewart, T. A., and Leder, P. (1986). Consequences of widespread deregulation of the c-myc gene in transgenic mice: Multiple neoplasms and normal development. *Cell* **45**, 485–495.

Le Meur, M., Gerlinger, P., Benoist, C., and Mathis, D. (1985). Correcting an immune-response deficiency by creating E_α gene transgenic mice. *Nature (London)* **316**, 38–42.

Lin, F.-L., Sperle, K., and Sternberg, N. (1985). Recombination in mouse L cells between DNA introduced into cells and homologous chromosomal sequences. *Proc. Natl. Acad. Sci. U.S.A.* **82**, 1291–1395.

Lin, P.-F., Lieberman, H. B., Yeh, D.-B., Xu, T., Zhao, S.-Y., and Ruddle, F. H. (1985). Molecular cloning and structural analysis of murine thymidine kinase genomic and cDNA sequences. *Mol. Cell. Biol.* **5**, 3149–3156.

Lo, C. (1983). Transformation by iontophoretic microinjection of DNA: Multiple integrations without tandem insertions. *Mol. Cell. Biol.* **3**, 1803–1814.

Lovell-Badge, R. H., Bygrave, A. E., Bradley, A., Robertson, E., Evans, M. J., and Cheah, K. S. E. (1985). Transformation of embryonic stem cells with the human type-II collagen gene and its expression in chimeric mice. *Cold Spring Harbor Symp. Quant. Biol.* **50**, 707–711.

Low, M. J., Hammer, R. E., Goodman, R. H., Habener, J. F., Palmiter, R. D., and Brinster, R. L. (1985). Tissue-specific post-translational processing of pre-somatostatin encoded by a metallothionein-somatostatin fusion gene in transgenic mice. *Cell* **41**, 211–219.

Low, M. J., Lechan, R. M., Hammer, R. E., Brinster, R. L., Habener, J. F., Mandel, G., and Goodman, R. H. (1986). Gonadotroph-specific expression of metallothionein fusion genes in pituitaries of transgenic mice. *Science* **231**, 1002–1004.

McCance, D. J. (1983). *In* "Polyomaviruses and Human Neurological Diseases," (J. L. Sever and D. L. Madden, eds.), pp. 343–357. Liss, New York.

McKnight, G. S., Hammer, R. E., Kuenzel, E. A., and Brinster, R. L. (1983). Expression of the chicken transferrin gene in transgenic mice. *Cell* **34**, 335–341.

Magram, J., Chada, K., and Costantini, F. (1985). Developmental regulation of a cloned adult β-globin gene in transgenic mice. *Nature (London)* **315**, 338–340.

Maitland, H., and McDougall, J. (1977). Biochemical transformation of mouse cells by fragments of herpes simplex virus DNA. *Cell* **11**, 233–241.

Mark, W. H., Signorelli, K., and Lacy, E. (1985). An insertional mutation in a transgenic mouse line results in developmental arrest at day 5 of gestation. *Cold Spring Harbor Symp. Quant. Biol.* **50**, 453–463.

Mason, A. J., Pitts, S. L., Nikolics, K., Szonyi, E., Wilcox, J. N., Seeburg, P. H., and Stewart, T. A. (1986). The hypogonadal mouse: Reproductive functions restored by gene therapy. *Science* **234,** 1372–1378.

Messing, A., Chen, H. Y., Palmitaer, R. D., and Brinster, R. L. (1985). Peripheral neuropathies, hepatocellular carcinomas and islet cell adenomas in transgenic mice. *Nature (London)* **316,** 461–463.

Ornitz, D. M., Palmiter, R. D., Hammer, R. E., Brinster, R. L., Swift, G. H., and MacDonald, R. J. (1985a). Specific expression of an elastase-human growth hormone fusion gene in pancreatic acinar cells of transgenic mice. *Nature (London)* **313,** 600–603.

Ornitz, D. M., Palmiter, R. D., Messing, A., Hammer, R. E., Pinkert, C. A., and Brinster, R. L. (1985b). Elastase 1 promoter directs expression of human growth hormone and SV40 T-antigen genes to pancreatic acinar cells in transgenic mice. *Cold Spring Harbor Symp. Quant. Biol.* **50,** 399–409.

Overbeek, P. A., Chepelinsky, A. B., Khillan, J. S., Piatigorsky, J., and Westphal, H. (1985). Lens-specific expression and developmental regulation of the bacterial chloramphenicol acetyltransferase gene driven by the murine αA-crystallin promoter in transgenic mice. *Proc. Natl. Acad. Sci. U.S.A.* **82,** 7815–7819.

Overbeek, P. A., Lai, S.-P., Van Quill, K. R., and Westphal, H. (1986). Tissue-specific expression in transgenic mice of a fused gene containing RSV terminal sequences. *Science* **231,** 1574–1577.

Palmiter, R. D., and Brinster, R. L. (1985). Transgenic mice. *Cell* **41,** 343–345.

Palmiter, R. D., and Brinster, R. L. (1986). Germline transformation of mice. *Annu. Rev. Genet.* **20,** 465–499.

Palmiter, R. D., Brinster, R. L., Hammer, R. E., Trumbauer, M. E., Rosenfeld, M. G., Birnberg, N. C., and Evans, R. M. (1982a). Dramatic growth of mice that develop from eggs microinjected with metallothionein-growth hormone fusion genes. *Nature (London)* **300,** 611–615.

Palmiter, R. D., Chen, H. Y., and Brinster, R. L. (1982b). Differential regulation of metallothionein-thymidine kinase fusion genes in transgenic mice and their offspring. *Cell* **29,** 701–710.

Palmiter, R. D., Wilkie, T. M., Chen, H. Y., and Brinster, R. L. (1983a). Transmission distortion and mosaicism in an unusual transgenic mouse pedigree. *Cell* **36,** 869–877.

Palmiter, R. D., Norstedt, G., Gelinas, R. E., Hammer, R. E., and Brinster, R. L. (1983b). Metallothionein-human GH fusion genes stimulate growth of mice. *Science* **222,** 809–814.

Palmiter, R. D., Hammer, R. E., and Brinster, R. L. (1985a). Expression of growth hormone genes in transgenic mice. *In* "Banbury Report 20: Genetic Manipulation of the Early Embryo" (F. Costantini and R. Jaenisch, eds.), pp. 123–132. Cold Spring Harbor Laboratory, Cold Spring Harbor, New York.

Palmiter, R. D., Chen, H. Y., Messing, A., and Brinster, R. L. (1985b). SV40 enhancer and large T-antigen are instrumental in development of choroid plexus tumors in transgenic mice. *Nature (London)* **316,** 457–460.

Pinkert, C. A., Widera, G., Cowing, C., Heber-Katz, E., Palmiter, R. D., Flavell, R. A., and Brinster, R. L. (1985). Tissue-specific, inducible and functional expression of the $E_\alpha{}^d$ MHC class II gene in transgenic mice. *EMBO J.* **4,** 2225–2230.

Ritchie, K. A., Brinster, R. L., and Storb, U. (1984). Allelic exclusion and control of endogenous immunoglobulin gene rearrangement in κ transgenic mice. *Nature (London)* **312,** 517–20.

Robertson, E., Bradley, A., Kuehn, M., and Evans, M. (1986). Germ-line transmission of genes introduced into cultured pluripotential cells by retroviral vector. *Nature (London)* **323,** 445–448.

Ross, S. R., and Solter, D. (1985). Glucocorticoid regulation of mouse mammary tumor virus sequences in transgenic mice. *Proc. Natl. Acad. Sci. U.S.A.* **82,** 5880–5884.

Rusconi, S., and Kohler, G. (1985). Transmission and expression of a specific pair of μ and κ genes in a transgenic mouse line. *Nature (London)* **314,** 330–334.

Schnieke, A., Harbers, K., and Jaenisch, R. (1983). Embryonic lethal mutation in mice induced by retrovirus insertion into the α1(I) collagen gene. *Nature (London)* **304,** 315–320.

Searle, A. G. (1981). *In* Genetic Variants in Strains of the Laboratory Mouse." (M. C. Green, ed.). Fischer, Stuttgart.

Shani, M. (1985). Tissue-specific expression of rat myosin light-chain 2 gene in transgenic mice. *Nature (London)* **314,** 283–286.

Shani, M. (1986). Tissue-specific and developmentally regulated expression of a chimeric actin/globin gene in transgenic mice. *Mol. Cell. Biol.* **6,** 2624–2631.

Shapira, G., Stachelek, J. L., Letsou, A., Soodak, L. K., and Liskay, R. M. (1983). Novel use of synthetic oligonucleotide insertion mutants for the study of homologous recombination in mammalian cells. *Proc. Natl. Acad. Sci. U.S.A.* **80,** 4827–4831.

Small, J., and Scangos, G. A. (1983). Recombination during transfer into mouse cells can restore the function of deleted genes. *Science* **219,** 174–176.

Small, J. A., Blair, D. G., Showalter, S. D., and Scangos, G. A. (1985). Analysis of a transgenic mouse containing simian virus 40 and v-myc sequences. *Mol. Cell. Biol.* **5,** 642–648.

Small, J. A., Scangos, G. A., Cork, L., Jay, G., and Khoury, G. (1986a). The early region of human papovavirus JC induces dysmyelination in transgenic mice. *Cell* **46,** 13–18.

Small, J. A., Khoury, G., Jay, G., Howley, P. M., and Scangos, G. A. (1986b). Early regions of JC virus and BK virus induce distinct and tissue-specific tumors in transgenic mice. *Proc. Natl. Acad. Sci. U.S.A.* **83,** 8288–8292.

Smithies, O., Grett, R. G., Boggs, S. S., Koralewski, M. A., and Kucherlapati, R. S. (1985). Insertion of DNA sequences into the human chromosomal β-globin locus by homologous recombination. *Nature (London)* **317,** 230–234.

Stewart, C. L., Vanek, M., and Wagner, E. F. (1985). Expression of foreign genes from retroviral vectors in mouse teratocarcinoma chimaeras. *EMBO J.* **4,** 3701–3709.

Stewart, T. A., Wagner, E. F., and Mintz, B. (1982). Human β-globin gene sequences injected into mouse eggs, retained in adults, and transmitted to progeny. *Science* **217,** 1046–1048.

Stewart, T. A., Pattengale, P. K., and Leder, P. (1984). Spontaneous mammary adenocarcinomas in transgenic mice that carry and express MTV/myc fusion genes. *Cell* **38,** 627–637.

Storb, U., O'Brien, R. L., McMullen, M. D., Gollahon, K. A., and Brinster, R. L. (1984). High expression of cloned immunoglobulin kappa gene in transgenic mice is restricted to B lymphocytes. *Nature (London)* **310,** 238–241.

Storb, U., Pinkert, C., Arp, B., Engler, P., Gollahon, K., Manz, J., Brady, W., and Brinster, R. L. (1986). Transgenic mice with μ and κ genes encoding antiphosphorylcholine antibodies. *J. Exp. Med.* **164,** 627–641.

Stout, J. T., Chen, H. Y., Brennand, J., Caskey, C. T., and Brinster, R. L. (1985). Expression of human HPRT in the central nervous system of transgenic mice. *Nature (London)* **317,** 250–252.

Swanson, L. W., Simmons, D. M., Arriza, J., Hammer, R., Brinster, R., Rosenfeld, G. M., and Evans, R. M. (1985). Novel developmental specificity in the nervous system of transgenic animals expressing growth hormone fusion genes. *Nature (London)* **317,** 363–366.

Swift, G. H., Hammer, R. E., MacDonald, R. J., and Brinster, R. L. (1984). Tissue-specific expression of the rat pancreatic elastase 1 gene in transgenic mice. *Cell* **38,** 639–646.

Thomas, K. R., Folger, K. R., and Capecchi, M. R. (1986). High frequency targeting of genes to specific sites in the mammalian genome. *Cell* **44,** 419–428.

Tonegawa, S. (1983). Somatic generation of antibody diversity. *Nature (London)* **302,** 575–581.

Tooze, J. (1981). "The Molecular Biology of Tumor Viruses." Cold Spring Harbor Laboratory, Cold Spring Harbor, New York.

Townes, T. M., Lingrel, J. B., Chen, H. Y., Brinster, R. L., and Palmiter, R. D. (1985a). Erythroid-specific expression of human β-globin genes in transgenic mice. *EMBO J.* **4,** 1715–1723.

Townes, T. M., Chen, H. Y., Lingrel, J. B., Palmiter, R. D., and Brinster, R. L. (1985b). Expression of human β-globin genes in transgenic mice: Effects of a flanking metallothionein-human growth hormone fusion gene. *Mol. Cell. Biol.* **5,** 1977–1983.

van der Putten, H., Bottari, F. M., Miller, A. D., Rosenfeld, M. G., Fan, H., Evans, R. M., and Verma, I. M. (1985). Efficient insertion of genes into the mouse germ line via retroviral vectors. *Proc. Natl. Acad. Sci. U.S.A.* **82,** 6148–6152.

Wabl, M., and Steinberg, C. (1982). A theory of allelic and isotypic exclusion for immunoglobulin genes. *Proc. Natl. Acad. Sci. U.S.A.* **79,** 6976–6978.

Wagner, E. F., Stewart, T. A., and Mintz, B. (1981). The human β-globin gene and a functional thymidine kinase gene in developing mice. *Proc. Natl. Acad. Sci. U.S.A.* **78,** 5016–5020.

Wagner, E. F., Covarrubias, L., Stewart, T. A., and Mintz, B. (1983). Prenatal lethalities in mice homozygous for human growth hormone gene sequences integrated in the germ line. *Cell* **35,** 647–655.

Wagner, E. F., Keller, G., Gilboa, E., Ruther, U., and Stewart, C. L. (1985). Gene transfer into murine stem cells and mice using retroviral vectors. *Cold Spring Harbor Symp. Quant. Biol.* **50,** 691–700.

Wagner, T. E., Hoppe, P. C., Jollick, J. D., Scholl, D. R., Hodinka, R., and Gault, J. B. (1981). Microinjection of a rabbit β-globin gene in zygotes and its subsequent expression in adult mice and their offspring. *Proc. Natl. Acad. Sci. U.S.A.* **78,** 6376–6380.

Weaver, D., Costantini, F., Imanishi-Kari, T., and Baltimore, D. (1985). A transgenic immunoglobulin mu gene prevents rearrangement of endogenous genes. *Cell* **42,** 117–127.

Weaver, D., Reis, M. H., Albanese, C., Costantini, F., Baltimore, D., and Imanishi-Kari, T. (1986). Altered repertoire of endogenous immunoglobulin gene expression in transgenic mice containing a rearranged mu heavy chain gene. *Cell* **45,** 247–259.

Wigler, M., Silverstein, S., Lee, L. S., Pellicer, A., Cheng, Y. C., and Axel, R. (1977). Transfer of purified herpes virus thymidine kinase gene to cultured mouse cells. *Cell* **11,** 223–232.

Woychik, R. P., Stewart, T. A., Davis, L. G., D'Eustachio, P., and Leder, P. (1985). An inherited limb deformity created by insertional mutagenesis in a transgenic mouse. *Nature (London)* **318,** 36–40.

Yamamura, K., Kikutani, H., Folsom, V., Clayton, L. K., Kimoto, M., Akira, S., Kashiwamura, S., Tonegawa, S., and Kishimoto, T. (1985). Functional expression of a microinjected E_{α}^{d} gene in C57BL/6 transgenic mice. *Nature (London)* **316,** 67–69.

Yamamura, K., Kudo, A., Ebihara, T., Kamino, K., Araki, K., Kumahara, Y., and Watanabe, T. (1986). Cell-type specific and regulated expression of a human γ1 immunoglobulin gene in transgenic mice. *Proc. Natl. Acad. Sci. U.S.A.* **83,** 2152–2156.

THE MOLECULAR BASIS OF THE EVOLUTION OF SEX

H. Bernstein,* F. A. Hopf,† and R. E. Michod‡

*Department of Microbiology and Immunology, College of Medicine,
The University of Arizona, Tucson, Arizona 85724

†Optical Sciences Center and ‡Department of Ecology and Evolutionary Biology,
The University of Arizona, Tucson, Arizona 85721

I. The Benefit of Sex Is a Major Unsolved Problem

Historical Comments on the Variation Hypothesis

The benefit of sexual reproduction (sex) is a major unsolved problem in evolutionary biology (Williams, 1975; Maynard Smith, 1978; Bell, 1982). Traditionally, it has been thought that the benefit of sex is the genetic variation (new combination of alleles) it produces. In this review, we will present evidence and arguments for an alternative view, that the primary benefit of sex is repair of DNA and masking of mutations (Bernstein *et al.*, 1981, 1984, 1985a–c). For convenience, we refer to the classical idea as the "variation hypothesis," and our alternative view as the "repair hypothesis." As early as 1889, Weisman alluded to precursors of both ideas. He suggested that the advantage of sex was the production of varied progeny (variation hypothesis), but also cited the suggestion by contemporaries that the advantage was rejuvenation at each generation, an idea related to the repair hypothesis (see Section IX). The variation hypothesis has dominated thinking in the intervening period. For most of this period the hypothesized advantages of variation were ones that acted at the level of the group or species and not at the level of the individual (see

ADVANCES IN GENETICS, Vol. 24

below). It was not until the early 1960s that group selection came under criticism and today its use in evolutionary explanation requires careful analysis of factors relating to population structure. Since the early 1960s, the variation hypothesis has provided a basis for numerous population genetic models which seek to explain the benefit of sex by selection at the level of the individual organism (for reviews, see Williams, 1975; Maynard Smith, 1978; Bell, 1982).

1. Group Selection and Short-Term Costs

In sharp contrast to his general view that characters evolve as a result of their benefit (or lack thereof) to individuals, Fisher (1958, pp. 49–50, 135–162) singled out sexual reproduction as "the possible exception . . . which could be interpreted as evolved for the specific rather than for the individual advantage." Weisman (1889), Fisher (1930, 1958), and Muller (1932) all suggested reasons why the genetic variation generated by sex might benefit species in the long term by increasing their rate of evolution. The view that sex is beneficial at the level of the species is not without its problems (for review, discussion, and criticism, see Bell, 1982). However, more important, are the large short-term costs imposed on individuals each generation by sex. These short-term costs of sex were largely ignored by early writers but have been stressed in recent work (see Section VII,C). On theoretical grounds it is difficult for selection at the level of the species to overcome large short-term costs to individuals. Additional evidence that short-term costs are important comes from data on the geographic distribution of parthenogens. These data indicate clearly that the occurrence of parthenogenesis is very often correlated with situations in which the short-term costs of sex are large (Bell, 1982). Thus the question of whether sex provides long-term benefits through group selection has been superseded, rather than resolved, by the more pressing issue of short-term benefits.

2. The Current Dilemma

Numerous theoretical investigations have approached the problem of the short-term benefit of sex by assuming that selection of sex is synonomous with selection for increased rates of exchange of alleles at linked loci. In modeling this idea, a gene product that promotes recombination is assumed to be a neutral modifier, having no direct effect on fitness itself, but causing exchange of alleles at other loci which are themselves assumed to have direct effects on fitness (see, for example, Kimura, 1956; Nei, 1967, 1969; Turner, 1967; Eshel and Feldman, 1970; Feldman and Balkau, 1972, 1973; Teague, 1976; Strobeck *et al.*, 1976; Charlesworth *et al.*, 1979; Feldman *et al.*, 1980;

Holsinger and Feldman, 1983; Kondrashov, 1984). There are several problems with using a neutral modifier approach to modeling the evolution of recombination.

First, it is well known that recombination enzymes are not neutral, as is assumed in the neutral modifier theory, but have direct effects on fitness since they are employed in a major form of DNA repair (see Section V).

Second, studies of meiotic tetrads in yeast and helf-tetrads in *Drosophila* show that, on average, greater than 60% of breakage and reunion events between homologous chromosomes (physical recombinations) fail to give rise to exchange of alleles flanking the short region of physical exchange (see Whitehouse, 1982, Tables 19 and 38, and p. 321; Chovnick *et al.*, 1970). Thus, if the advantage of sex is allelic exchange, most physical recombination events should be of little or no benefit. On the repair hypothesis, the benefit of physical recombination is realized whether or not flanking alleles are exchanged. To avoid confusion in our further discussion we will use the term physical recombination when we refer to recombination in the sense of physical breakage and reunion of two homologous chromosomes or DNA molecules (Lewin, 1985, Chap. 34). When we refer to recombination as a synonym for exchange of alleles (e.g., King and Stansfield, 1984) we will use the term allelic recombination. The data show conclusively that allelic recombination is an occasional, but not a necessary, consequence of physical recombination.

Third, to select for increased rates of allelic recombination, neutral modifier models depend on conditions that are not broadly applicable. The generality of sex in the biota suggests that the benefits of sex are equally general, yet the modifier models depend on restrictive conditions to work. It is beyond the scope of this review to summarize the particular conditions used in each of the models [see Maynard Smith (1978) and Bell (1982) for detailed discussions of these issues]. However, it is clear from the reviews of Maynard Smith and Bell that despite considerable theoretical work, no model based on allelic recombination is both general in scope and consistent with available evidence.

II. Informational Noise: Damage, Mutation, and Recombination

A. Accurate Transfer of Information Is Basic to Life

Living systems are highly ordered structures. This order depends on information encoded in the sequence of nucleotide base pairs in

DNA. A fundamental problem faced by all living organisms is the accurate transmission of this genetic information from one generation to the next. We use the term informational noise to refer to changes which are introduced at random into such an information sequence. In genetic systems, it is common to distinguish two forms of informational noise: allelic recombination and mutation. Processes which produce informational noise in a genetic system should generally reduce fitness unless they also produce a benefit which offsets this deleterious effect.

As discussed in more detail below, in genetic systems, informational noise can be produced during the repair of DNA damages and during the replication of DNA. In general, the benefits of repair and replication are self-evident, although their relevance to the evolution of sex requires some development. In Sections IV–VII we develop the idea that the two basic aspects of sex, physical recombination and outcrossing, are fundamental adaptive responses to DNA damage and mutation. We argue that recombination repairs DNA damage and outcrossing masks mutations in reproductive systems which have recombination. We first review the properties of DNA damage and mutation that bear on the evolution of sex.

B. Contrast between Damage and Mutation

DNA is the genetic material of all organisms, except for certain simple viruses which use RNA instead (see Section VII,A,3). We discuss damage and mutation in the context of DNA with the understanding that RNA is basically similar. DNA damages are structural irregularities such as breaks, depurinations, cross-links, thymine dimers, and modified bases. Damages usually interfere with replication (Rupp and Howard-Flanders, 1968; Cleaver, 1969; Burck *et al.*, 1979) and transcription (Zieve, 1973; Hackett and Sauerbier, 1975; Nocentini, 1976; Leffler *et al.*, 1977). They are not replicated. By contrast, mutations are alternations in the base pair sequence of DNA. They can result from base pair substitutions, additions, or deletions, or from rearrangement of the standard base pairs. Mutations do not generally alter the physical regularity of the DNA. Mutations can be replicated and thus can be inherited.

The characteristic of damage which we think is crucial for the evolution of sex is that damage can be recognized by enzymes and be repaired, provided an intact template is available to provide a source for the information lost because of the damage. Since a mispaired base is an intermediate in the formation of a mutation, enzymes that recognize and remove a mispaired base may prevent a mutation from arising. However, if replication resolves the mispairing before it is corrected, there is no enzymatic mechanism for determining which daughter DNA is correct (in the sense of carrying the original information). Mutations can only be removed by natural selection acting on their expressed phenotype.

In summary, damages can be repaired but not replicated whereas mutations can be replicated but not repaired.

C. General View: Natural Selection Acts to Reduce Informational Noise

1. Damage Generates Mutational and Recombinational Noise

DNA damages differ from mutations in that they cannot be replicated. Thus DNA damages are not noise in the informational sense since they are not composed of the four characters in the genetic "alphabet." However, damages can cause informational noise.

When a DNA damage occurs, one of three basic consequences is possible: (1) The damage may be repaired by a process which does not produce mutations. (2) The damage may be repaired inaccurately, leading to mutation (see Walker, 1984, for review). (3) The damage may be left unrepaired. Within these basic categories there are several alternatives which we discuss from the point of view of whether they produce informational noise. Repair without mutation may either (1a) result in allelic recombination or (1b) not result in allelic recombination. Although inaccurate repair (2) has the same alternatives, little is known about the biochemical mechanism and it will not be discussed further. Unrepaired damages may (3a) lead to death of the cell or (3b) be passed on intact to a daughter cell. Category (3b) applies to subtle lesions which are structurally similar to regular bases and consequently do not block replication. Such damages are diluted out during subsequent cell divisions. However, they may cause replication errors leading to mutation.

Under the first option, repair without mutation, the repair may involve replacing damaged information in one strand of the DNA molecule by copying the information from the other intact strand (excision repair). Alternatively, the source of intact information may be another homologous DNA molecule (recombinational repair). In the case of recombinational repair, if the two parental molecules contain different alleles, the resulting recombinant molecule may differ genetically from both parents [category (1a)]. Thus damage that is recombinationally repaired can generate allelic recombination, which is a form of inheritable change. As mentioned above, damages can also generate mutations by inaccurate repair [category (2)] or by causing replication errors [category (3b)].

In summary, DNA damage generates two types of inheritable change, mutation and allelic recombination. Since both of these alter the genetic information randomly they are forms of informational noise and natural selection should act to reduce them.

2. *Reduction of Mutational Variation*

Mutations result from replication errors occurring either during chromosome duplication or during repair of damage. Error rates are impressively low, varying from about 10^{-8} to 10^{-11} per replicated nucleotide, depending on the organism (Drake, 1974). This implies that during evolution there has been strong selection for accuracy-determining mechanisms, since, in the absence of a DNA polymerase, complementary base pairing is by itself relatively inaccurate, having an error rate of approximately 10^{-2} (Loeb *et al.*, 1978). Hartman (1980) has reviewed evidence that in bacteria the high level of accuracy results from the combined effects of (1) base selection by complementary base pairing, (2) DNA polymerase discrimination, (3) a 3′- to 5′-exonucleolytic proofreading function of DNA polymerase that removes newly inserted mismatched bases, and (4) postreplicative repair of mismatched bases. In addition, the machinery that carries out replication is complex, and this complexity is thought to reflect adaptations for promoting accuracy (Alberts *et al.*, 1980).

3. *Recombinational Variation Should Also Be Reduced*

Although mutation provides the essential raw material for evolution, the first-order effect of the replicative machinery is to reduce mutational variation. Recombinational variation, like mutational variation, also provides raw material for evolution. Moreover, we think that, just as mutational variation is a by-product of DNA replication, recombinational variation (allelic recombination) is a by-product of recombinational repair of DNA (i.e., physical recombination; Section V). Therefore, just as the replicative machinery has been selected for accuracy, i.e., to reduce mutational variation, so, we think has the recombinational repair machinery been selected to reduce recombinational variation (Section VIII,A,2; Bernstein *et al.*, 1986). In both cases, we believe, the main selective force is the immediate reduction in fitness caused by informational noise (random changes in the genetic information). This view is incompatible with all theories which assume that sex is maintained by an advantage of recombinational variation. The assumption that sex is maintained by an advantage of recombinational variation is, in our view, analogous to the assumption that DNA replication is maintained by the advantage of the beneficial mutational variants it sometimes produces.

III. Overview of Repair Hypothesis

There are two fundamental aspects of sex: (1) physical recombination, in the sense of breakage and reunion of two different DNA molecules, and (2) outcrossing, in the sense that the two DNA molecules involved in recombination come from different individuals. Recombination is clearly a more basic aspect of sex than outcrossing as evidenced by the various reproductive systems that have abandoned outcrossing but have retained recombination (Section VII,C). In contrast, there are relatively few reproductive systems, such as the diploid hybridogenetic forms of *Poeciliopsis* fishes (Schultz, 1961, 1966, 1969; Cimino, 1972), in which outcrossing is maintained, in the sense that two genomes from different individuals come together in the same nucleus, yet recombination does not occur.

In Section II we argued that there are two intrinsic sources of informational noise in the genetic material: DNA damage and errors produced during DNA replication. The repair hypothesis has two parts corresponding to these two problems. These are (1) DNA damage decreases fitness and this selects for recombination, the first aspect of sex, and (2) mutations are most often deleterious and recessive, and when recombination is present they select for outcrossing, the second aspect of sex.

Dougherty (1955) first proposed that sex originated as a genetic repair process. Bernstein (1977, 1979, 1983), Martin (1977), Walker (1978), and Maynard Smith (1978, p. 7) also discuss the evolution of recombination as a repair process. The combination of (1) and (2), which is needed for a complete explanation, was presented by Bernstein *et al.* (1981, 1984, 1985a–c).

IV. DNA Damage in Nature

A. Single- and Double-Strand Damages Present Different Problems

DNA damages are of two types, those that affect only one strand of DNA, and those that affect both strands. Examples of the latter are double-strand breaks, cross-links, and the structures formed when replication proceeds past a single-strand damage, leaving a gap in the new strand opposite the damage.

Single-strand damages can be removed by excision repair enzymes. In such cases the undamaged DNA strand acts as a template for replacing the excised damaged DNA, and recombination does not occur. Double-strand damages cannot be repaired by such a mechanism since there is no intact strand within the DNA duplex to serve as a template. Double-strand damages are critical for our hypothesis since they can only be repaired by physical recombination with another homologous DNA. We next present evidence that such damages are prevalent.

B. Double-Strand Damages

The highly reactive superoxide radical (O_2^-) and hydrogen peroxide (H_2O_2) are thought to occur frequently in cells as by-products of cellular respiration (Harman, 1981). The relative rates at which various types of DNA damage are produced by H_2O_2 have been estimated by Massie *et al.* (1972). Modified bases (single-strand damages) and double-strand damages are formed in an approximate ratio of 60 : 1. The two types of double-strand damage, cross-links and double-strand breaks, occur with roughly equal frequency. The oxidatively produced single-strand damages can be removed from DNA by excision repair enzymes that are specific to the types of damages. For example, repair glycosylases which are specific to oxidatively modified bases have been identified which remove (1) thymine glycol in calves and humans (Breimer, 1983) and in *Escherichia coli* (Breimer and Lindahl, 1984), (2) hydroxymethyluracil in mice (Hollstein *et al.*, 1984), and (3) formamidopyrimidine in rodents (Margison and Pegg, 1981) and *E. coli* (Chetsanga and Lindahl, 1979). The occurrence of these enzymes, which are specific for oxidative damage in such widely different organisms, indicates that oxidative damages are prevalent.

Recently, Cathcart *et al.* (1984) have estimated that about 320 thymine glycol residues are removed by repair from the average human cell per day, and that an additional comparable number of hydroxymethyl residues are also removed. If at least 600 oxidatively modified bases occur per human cell per day [Ames *et al.* (1985) estimated several thousand], then about 10 additional double-strand damages should also occur (since oxidatively modified bases and double-strand damages are produced in a ratio of about 60 : 1). Since about 90% of oxidative reactions occurs in the cells' mitochondria, about 90% of the

DNA damage is expected to be in the mitochondrial DNA and 10% in the DNA of the cell nucleus. If we make the conservative assumption that only 1% of oxidative damages occurs in nuclear DNA, the number of double-strand damages per cellular genome per day would still be 0.1. Since even a single unrepaired double-strand damage can block replication and cause cell death (e.g., Cole, 1971), a frequency of 0.1 per day should substantially reduce fitness. Although we have emphasized oxidative damages, there might be other important natural sources of double-strand damages as well.

In summary, we argued in this section that double-strand damages can only be dealt with by recombinational repair. Furthermore, we reviewed evidence that double-strand damages are an important problem in nature. In the next section, we review evidence that recombinational repair is widespread and is efficient in overcoming a variety of damages, particularly double-strand damages.

V. Recombinational Repair Is Efficient, Versatile, and Prevalent

The following four observations indicate the presence of recombinational repair in an organism. First, survival after treatment with a DNA-damaging agent is substantially enhanced when two or more chromosomes are present in a cell, rather than one, and when normal recombination functions are present. Second, mutations in genes responsible for recombination increase the sensitivity of organisms to DNA-damaging agents. Third, DNA-damaging agents increase levels of recombination. Fourth, mutants blocked in other repair pathways have increased levels of damage-induced recombination. All four lines of evidence are found in viruses, bacteria, and yeast. In *Drosophila* and other multicellular organisms, such as mammals and plants, only some of these kinds of evidence have been obtained.

A. Viruses

1. Multiplicity Reactivation

Recombinational repair was first described by Luria (1947) in studies of bacteriophages T2, T4, and T6. When he irradiated these phages with ultraviolet light, he found that the ability to produce progeny phage was much higher when cells were infected by more than one phage than when they are infected by only one. This higher survival rate was interpreted as being due to recombination (Luria and

Dulbecco, 1949) and was designated *multiplicity reactivation.* It has since been shown that multiplicity reactivation is effective against widely different DNA-damaging agents, including ultraviolet light, mitomycin C, nitrous acid, X rays, and ^{32}P (see Bernstein, 1981, and Bernstein and Wallace, 1983, for reviews). Doses of ultraviolet light, mitomycin C, and HNO_2 which allow only 1% survival of progeny-forming ability in single infections allow, respectively, about 70%, 65%, and 44% survival in multiple infections (Epstein, 1958; Nonn and Bernstein, 1977; Holmes *et al.*, 1980). This indicates that multiplicity reactivation in phage T4 is efficient in overcoming lethal effects of DNA damage.

The conclusion that multiplicity reaction is a form of recombinational repair is based on three lines of evidence: (1) by definition, it depends on at least two homologous phage chromosomes being present in the same cell; (2) it depends on phage gene functions that are also required for recombination of phage genetic markers (allelic recombination); and (3) under conditions where multiplicity reactivation occurs, the frequency of allelic recombination increases. Since point (1) is the defining characteristic of multiplicity reactivation and has been described above, we will now discuss points (2) and (3).

With respect to point (2), multiplicity reactivation of UV-irradiated phage T4 depends on seven gene products (gps), which are required for allelic recombination. These are gp uvsx, which promotes homologous pairing and strand exchange of DNA molecules (Griffith and Formosa, 1985); gp 32, which is a helix-destabilizing protein that enhances the pairing reaction; gps 46 and 47, which have an exonuclease activity that produces single-strand gaps from nicks and may promote formation of joint DNA molecules; and gps 59, uvsw, and y, whose functions are not yet understood (for reviews, see Bernstein, 1981; Bernstein and Wallace, 1983). Similar requirements for recombination functions have been shown for multiplicity reactivation of phage damaged by nitrous acid (Nonn and Bernstein, 1977), mitomycin C (Holmes *et al.*, 1980), and H_2O_2 (D. Chen, personal communication).

Phage λ also undergoes multiplicity reactivation after UV treatment (Kellenberger and Weigle, 1958; Baker and Haynes, 1967; Huskey, 1969). This depends either on the host recombination function *recA* or on a phage recombination function *red*, so that the absence of both results in loss of multiplicity reactivation (Huskey, 1969).

With respect to point (3) above, allelic recombination in phage T4 is increased by ultraviolet light (Epstein, 1958), nitrous acid (Fry, 1979), mitomycin C (Holmes *et al.*, 1980), psoralen plus near-UV light (Miskimins *et al.*, 1982), X rays (Harm, 1958), and ^{32}P (Symonds and Ritchie, 1961).

Chan (1975) showed that recombination is stimulated by UV light and that this stimulation is substantially greater in the excision repair-defective mutant *den* V^-. This result implies that UV-induced DNA lesions which are ordinarily repaired by excision repair are handled by recombinational repair when excision repair is defective.

Multiplicity reactivation of UV-irradiated phage T4 is apparently error free since it occurs without introduction of new mutations (Yarosh *et al.*, 1980). Similarly, multiplicity reactivation of phage ϕX174 treated with proflavin plus light is also not mutagenic (Piette *et al.*, 1978).

Phages other than T4 and λ which have been shown to undergo multiplicity reactivation are T1 (Tessman and Ozaki, 1957), T2, T5, and T6 (Luria, 1947; Luria and Dulbecco, 1949), and ϕX174 (Piette *et al.*, 1978), which infect *E. coli*, and the *Salmonella* phage Vi (A. Bernstein, 1957). In addition, a number of animal viruses undergo multiplicity reactivation. These include herpes virus (Hall and Scherer, 1981; Hall *et al.*, 1980), adenovirus type 12 and simian virus 40 (Yamamoto and Shimojo, 1971), influenza virus (Barry, 1961), reovirus (McClain and Spendlove, 1966), and poxvirus (Abel, 1962).

We conclude that, in general, multiplicity reactivation (hence, recombinational repair) is a common form of repair among viruses and it is effective against a variety of genome-damaging agents.

2. *Postreplication Recombinational Repair*

In addition to multiplicity reactivation, phage T4 also undergoes a form of recombinational repair known as postreplication recombinational repair (for review, see Bernstein and Wallace, 1983). The mechanism, as originally proposed by Harm (1964), involves interaction of the two sister DNA molecules formed upon replication. It employes many of the same gene functions that are used in multiplicity reactivation.

3. *Prophage Reactivation*

Some phages are also able to undergo another type of recombinational repair called prophage reactivation (for reviews, see Bernstein, 1981; Devoret *et al.*, 1975). In this case recombination is between a damaged infecting phage chromosome and a homologous phage genome integrated into the bacterial DNA and existing in a prophage state. Prophage reactivation of UV-irradiated phage depends on the *E. coli recA* gene product.

Overall, the evidence indicates that recombinational repair is a major pathway for overcoming genome damages among viruses.

B. Bacteria

In this section we present evidence that different specific types of double-strand damages, as well as other types of damage, are overcome by recombinational repair, and in some cases that this has been shown to be very efficient.

1. Cross-Links

When *E. coli* cells are treated with psoralen plus near-UV light, interstrand cross-links are produced in their DNA. Cole and collaborators have extensively studied the repair of these cross-links (e.g., Cole *et al.*, 1976, 1978). They found that cross-links are repaired by a process in which there is first an uncoupling of the cross-link from one strand, leaving a gap in that strand. This is followed by the restoration of the sequence by recombinational exchange with another homologous chromosome. The recombination step requires the recA protein which promotes pairing of homologous DNA molecules and strand exchange (Flory *et al.*, 1984). A wild-type cell can recover from 65 psoralen cross-links (Cole, 1971) even though a single unrepaired cross-link is sufficient to inactivate a cell. These results indicate that recombinational repair is very efficient in overcoming this type of lesion.

2. Double-Strand Breaks

When *E. coli* cells are irradiated with X rays, double-strand breaks are induced in their DNA. Experiments by Krasin and Hutchinson (1977) indicated that these breaks are repaired by recombinational repair. They showed that when four or five chromosomes are present in the cell, 50–70% of the double-strand breaks are removed during a period of postirradiation incubation. However, no repair of double-strand breaks was detected when an average of only 1.3 genomes were present per cell, even though under these conditions repair of single-strand breaks was efficient. Double-strand break repair proved to depend on *recA* function as well as other gene functions inducible by DNA damage (Krasin and Hutchinson, 1981). Experiments with a *recA*$^-$ mutant suggested that a single, unrepaired double-strand break in a chromosome is lethal to the cell.

3. Gaps Opposite Thymine Dimers

When *E. coli* cells are treated with UV light, pyrimidine dimers are produced in the DNA. Upon replication, gaps are formed in the newly synthesized daughter strands opposite the dimers (Rupp and Howard-

Flanders, 1968). These gaps opposite dimers can be regarded as double-strand damages since information is lost in both strands of the DNA in the same region. Rupp *et al.* (1971) showed that one single-strand exchange occurs for every one or two unremoved dimers in the replicated DNA. They proposed that the postreplicative gaps are filled by recombinational transfer of undamaged single-strand segments from the homologous sister chromosome. This permits the dimer to be bypassed during replication. It takes about 50 pyrimidine dimers to kill each cell when postreplication recombinational repair is the only significant repair process present (Howard-Flanders *et al.*, 1968). This implies that postreplication recombinational repair permits efficient bypass of pyrimidine dimers. Postreplication recombinational repair requires the recA protein (Smith and Meun, 1970).

4. *Recombinational Repair during Conjugation*

Escherichia coli donor males can be UV irradiated to induce thymine dimers in their DNA prior to conjugation with an F^- female cell. As the donor DNA replicates during its conjugal transfer, gaps are formed in the new strands opposite the pyrimidine dimers. The gaps are then filled by *recA*-dependent recombination with the female F^- DNA (Howard-Flanders *et al.*, 1968). The frequency of allelic recombination is also increased by UV irradiation of the donor.

5. *Recombinational Repair Is Effective against Many Types of Damage*

DeFlora *et al.* (1984) obtained evidence that recombinational repair is efficient and versatile in overcoming lethal damages to DNA. They reported the results of testing a wide variety of compounds for genotoxicity in three *E. coli* strains. These were (1) wild type, (2) an excision repair-defective strain (*uvrA*$^-$ *polA*$^-$), and (3) a strain defective both in excision repair and recombinational repair (*uvrA*$^-$ *recA*$^-$ *lexA*$^-$). After testing 135 compounds, the "*uvrA*$^-$ *recA*$^-$ *lex*$^-$ strain was found to be considerably superior to the *uvrA*$^-$ *polA*$^-$ strain . . . in detecting genotoxic agents." These results and those discussed in the previous sections indicate that in *E. coli* recombinational repair is effective against a wide variety of DNA damages.

6. *Recombination Is Increased by Mutations in Mismatch Repair Genes*

The *E. coli* mismatch repair system recognizes noncomplementary base pairs in DNA and replaces mispaired bases (Radman and Wagner, 1984). This process apparently involves localized excision and resynthesis. DNA regions in which GATC sequences are fully adenine methylated appear to be resistant to mismatch repair. However, in the DNA regions immediately behind the replicative fork, GATC sequences are transiently undermethylated, and this allows mismatch repair to operate only on newly synthesized strands and thus to remove replication errors. This repair system depends on the following gene functions: *mutH*, *mutL*, *mutS*, *mutU*, and *dam*, where the latter codes for adenine methylase. Mutations in each of these genes were found to have abnormally high levels of recombination in bacterial conjugation (Feinstein and Low, 1986) or between chromosomal duplications (Konrad, 1977). This suggests that when mismatch repair is faulty damages arise which are handled by recombinational repair.

7. *Recombinational Repair in Bacteria Other Than E. coli*

As discussed in Sections IV,B,1–4, the recA protein is required for recombinational repair in *E. coli*. The presence in other bacterial species of a gene similar to *recA* implies a capacity for recombinational repair in these oganisms. Genes similar to *recA* have been identified in the following organisms by their ability to functionally substitute for the *recA* gene and/or by the immunological cross-reaction of their product with the *recA* protein: *Proteus vulgaris*, *Erwinia carotovora*, and *Shigella flexneri* (Keener *et al.*, 1984), *Proteus mirabilis* (Eitner *et al.*, 1982; West *et al.*, 1983), *Salmonella typhimurium* (Pierre and Paoletti, 1983), *Bacillus subtilis* (de Vos *et al.*, 1984; Lovett and Roberts, 1985), and *Vibrio cholerae* (Goldberg and McKalanos, 1986). This suggests that recombinational repair is common in bacteria.

In the case of *B. subtilis*, Dodson and Hadden (1980) have presented evidence for a postreplication recombinational repair processes which is similar to the more well-studied example in *E. coli*. This process depends on the *B. subtilis recE* gene, which is similar to the *E. coli recA* gene (deVos *et al.* 1984; Lovett and Roberts, 1985). In the photosynthetic cyanobacterium *Synechococcus* R2, double-strand gaps in DNA were shown to be repaired by a recombinational process (Kolowsky and Szalay, 1986).

C. Recombinational Repair in Eukaryotes

In Sections V,A and V,B, we presented evidence that recombinational repair is common in viruses and bacteria and that it efficiently overcomes a variety of damages, particularly double-strand damages.

In this section we extend these conclusions to eukaryotes. First, we consider simple fungi, then we discuss multicellular eukaryotes.

1. Fungi

Most of the experiments of recombinational repair in fungi were done with the yeast, *Saccharomyces cerevisiae.*

a. Cross-Link Repair. Experiments on the recombinational repair of cross-links in *S. cerevisiae* (Magana-Schwencke *et al.*, 1982) indicate that the mechanism is similar to that in *E. coli*. As in the *E. coli* studies, cross-links were introduced by psoralen plus near-UV light. The initial step in the repair process is an incision, presumably leading to the unlinking of the cross-link from one DNA strand to form a gap. The rejoining of the strands across the gap depends on the *rad-51* gene product, which is involved in recombination. When this repair process is deficient only one cross-link is lethal, whereas in repair-proficient cells it takes about 122 psoralen cross-links per genome to induce one lethal hit. This implies that recombinational repair of cross-links is very efficient.

b. Double-Strand Break Repair. In *S. cerevisiae*, ionizing radiation has been used to induce double-strand breaks. Repair of these damages requires the presence of two chromosomes, either sister chromatids or homologous chromosomes (Resnick and Martin, 1976). Gene product 52, which is needed for spontaneous recombination (see Kunz and Haynes, 1981, for review), is also required for repair of double-strand breaks (Ho, 1975; Resnick and Martin, 1976). These observations indicate that repair of double-strand breaks in *S. cerevisiae* is a recombinational process. Resnick (1976) has proposed a specific recombinational repair model for double-strand breaks. Repair of double-strand breaks is also efficient in *S. cerevisiae*. According to Resnick and Martin (1976), it took an average of 35 double-strand breaks in wild type to produce one lethal hit, whereas, in a *rad-52* mutant, lacking recombinational repair, it took only 2.2 double-strand breaks to produce a lethal hit.

Double-strand breaks induced by methyl methanesulfonate also appear to be repaired by recombinational repair, since two homologous chromosomes are required (Chlebowicz and Jachymczyk, 1979). However, in this case, genes required for recombination were not tested.

c. Repair of Other Types of Damage. When 101 heterogeneous chemicals including carcinogens and procarcinogens were assayed for stimulation of mitotic recombination in *S. cerevisiae*, 44 proved to be recombinogenic (Simmon, 1979). This suggests that recombinational repair may be used against a variety of different kinds of DNA damage.

d. Excision Repair Mutants Have Increased UV-Induced Recombination. Mutants of the *rad-3* group, which are defective in excision repair of UV-induced thymine dimers, have an increased frequency of UV-induced recombination at low UV doses (Kunz and Haynes, 1981). This suggests that thymine dimers that are not removed by excision repair are handled by recombinational repair.

e. Recombinational Repair in the Fission Yeast Schizosaccharomyces pombe. In a recent review of the literature on recovery, repair, and mutagenesis in the fission yeast *S. pombe,* Phipps *et al.* (1985) characterized recombinational repair as a major mechanism of DNA repair in this organism. Gentner *et al.* (1978) described recombinational repair of UV-induced damage as depending on (1) the presence of two DNA copies, either sister chromatids or homologous chromosomes, and (2) on the *rad-1* gene product, which is also necessary for allelic recombination.

f. Recombinational Repair in the Smut Fungus Ustilago maydis. Kmiec and Holloman (1982, 1984) have obtained evidence that the *rec-1* gene product of the smut fungus *U. maydis* has properties similar to the recA protein of *E. coli* in that it promotes pairing of homologous chromosomes. Since a *rec-1* mutant undergoes abnormal meiosis and has increased sensitivity to UV, ionizing radiation, and nitrosoguanidine, the *rec-1* gene is presumed to be involved in recombinational repair (Holliday *et al.,* 1976).

2. *Drosophila*

The existence of recombinational repair in *Drosophila* is suggested by the finding that mutants which are defective in allelic recombination are also sensitive to several DNA-damaging agents. Mutants defective in genes *mei-9* and *mei-41* have decreased spontaneous meiotic recombination and increased sensitivity to X rays, UV light, methyl methanesulfonate, nitrogen mustard, and 2-acetylaminofluorene (Baker *et al.,* 1976). Mutants defective in another gene, *mus-101,* have a similar phenotype (Boyd, 1978). The presence of recombinational repair is also suggested by the observation that recombination is stimulated by a variety of DNA-damaging agents, including UV light (Prudhommen and Proust, 1973; Martensen and Green, 1976), X rays (Abbadessa and Burdick, 1963), and mitomycin C (Schewe *et al.,* 1971). These findings indicate that recombinational repair occurs in multicellular eukaryotes as well as in the simpler eukaryotes, prokaryotes, and viruses.

3. *Mammalian Cells*

Repair of double-strand breaks introduced by ionizing radiation occurs in hamster ovary cells, and X-ray-sensitive mutant cells are defective in double-strand rejoining (Kemp *et al.*, 1984). Since double-strand breaks in *E. coli* and yeast are repaired by recombinational repair, a similar pathway may be used in hamster ovary cells. However, experiments by Resnick and Moore (1979) suggest that such a mechanism is relatively inefficient in these cells. On the other hand, recent work by Brenner *et al.* (1986) showed that repair of double-strand breaks in mouse L cells by recombination between homologous plasmids is very efficient.

Hollstein and McCann (1979) compiled references to 18 chemical carcinogens and mutagens (DNA-damaging agents) that stimulate sister chromatid exchange in mammalian cells *in vitro*. Such stimulation is regarded as an indication of recombinogenic activity. Since stimulation of recombination by DNA-damaging agents generally accompanies recombinational repair in lower organisms, it may also reflect such a process in mammalian cells.

4. *Plants*

In plants, small supernumerary chromosomes, called B chromosomes, are common. These are not homologous to the usual autosomal chromosomes, and may not behave regularly during nuclear division. The B chromosomes increase chiasma frequency in many plant species, including maize (see Bell, 1982, p. 411, and Staub, 1984, for references). Staub (1984) has shown in maize that B chromosomes increase resistance to γ-irradiation-induced DNA damage. In this study, cells were irradiated prior to meiosis and resistance was assayed by pollen viability. These results suggest that B chromosomes are maintained in natural populations, at least in part, because they enhance recombinational repair.

Westerman (1967) summarized the results of nine studies of the effect of DNA-damaging treatments, given at meiosis, on recombination frequency. These studies were with plants (*Lilium*, *Tradescantia*, and *Chlamydomonas*), as well as with insects (*Drosophila* and *Schistocerca*). Treatments given during the period between premeiotic DNA replication and chiasma formation, i.e., during zygotene or early pachytene, generally increased recombination. This suggests that the response to increased DNA damage is increased recombinational repair.

5. *Conclusions*

In Section IV we argued that only recombinational repair can overcome double-strand damages, and that such damages are common in nature. In this section we reviewed evidence that recombinational repair is prevalent among viruses, bacteria, and fungi, and that it occurs in *Drosophila,* mammals, and plants. Recombinational repair is effective against a wide variety of different DNA-damaging agents and in particular is very efficient in overcoming double-strand damages both in *E. coli* and *S. cerevisiae.* These findings provide the basis for the first part of the repair hypothesis, that recombination is an adaptation for dealing with DNA damage.

Many of the systems for studying recombinational repair reviewed here were chosen for experimental convenience and thus involve asexually, rather than sexually, reproducing cells (e.g., mitotic recombination in yeast and sister chromosome recombination in *E. coli* and mammalian cells). If organisms can carry out efficient recombinational repair in vegetative cells, we presume that recombinational repair in the germ line should be at least as efficient, since the rates of recombination in meiosis are several orders of magnitude higher than in mitosis. In addition, as we argue in the next section, meiosis appears designed to promote recombinational repair of DNA.

VI. Recombinational Repair: A Basic Feature of Sex

In this section we focus on meiosis as the key stage of the sexual cycle in eukaryotes. We first briefly indicate the current state of understanding of the mechanism of meiotic recombination. Next we argue that this mechanism is an adaptation for DNA repair.

A. Meiotic Recombination: The Double-Strand Break Repair Model

During most of the past decade the model of Meselson and Radding (1975) has dominated thinking about the mechanism of general recombination. More recently, however, Szostak *et al.* (1983) proposed a model which they refer to as the double-strand break repair model. This model, like its earlier counterpart, was developed to explain the extensive data obtained from fine structure genetic analyses in fungi (as reviewed by Orr-Weaver and Szostak, 1985), as well as knowledge of the physical properties of DNA and the enzymatic reactions involved in its processing. The double-strand break repair model, however, accommodates recent experimental evidence not readily explained by the older Meselson–Radding model, and it is currently the most authoritative one for general recombination.

B. Meiosis Is Designed to Promote Repair of DNA

The initial steps in the double-strand break repair model are the formation of a double-strand break and then the conversion of the break into a double-strand gap. These features of the model are based on several lines of experimental evidence (Orr-Weaver and Szostak, 1985). One of the most significant experiments was the demonstration that when information is removed from yeast DNA to form a double-strand gap, this information can be precisely restored by recombination with a homologous chromosome. These gaps stimulate recombination by as much as 1000-fold, and the process requires the gene product 52, which is also essential for spontaneous recombination (Orr-Weaver *et al.*, 1981).

In principle, any double-strand damage can be removed enzymatically with the resulting formation of a double-strand gap. The double-strand break model provides a mechanism for coupling this removal with accurate restoration of the lost information from another homologous DNA molecule. Thus, although this model was designed to explain spontaneous recombination, it also suggests a general mechanism for recombinational repair. Although this model is consistent with the hypothesis that repair is the primary function of recombination, we stress that the correctness of the repair hypothesis does not necessarily depend on the correctness of the double-strand break model.

The process of meiosis includes recombination as a basic feature. In Section VII,D,1 we discuss the other basic feature of meiosis, which is the reduction in ploidy which occurs as a result of the second meiotic division. Many of the unique characteristics of meiosis, which distinguish it from mitosis, are adaptations to promote recombination. If recombination is designed for DNA repair, as suggested above, so must meiosis be designed for repair as well. In the next section we develop an evolutionary progression for the origin and evolution of sexual reproduction based on the repair hypothesis.

VII. The Origin and Evolution of Sex Explained as a Continuum by the Repair Hypothesis

A. Origin and Early Evolution of Sex in Haploid Systems

Based on the broad similarity of recombination mechanisms at the level of DNA in prokaryotes and eukaryotes, Dougherty (1955) proposed that sex in all organisms has a common ancestry. We think that the considerable information accumulated on the molecular processes of recombination since Dougherty's proposal (for reviews, see Stahl, 1979, and Whitehouse, 1982) reinforces this view.

Because the important early evolutionary stages of sex probably occurred at the level of the genetic material, it is unlikely that a fossil record of these events is preserved. Therefore, we approach a description of these early events from general considerations of information transfer (Bernstein *et al.*, 1984).

1. An Informational Approach

The two most fundamental properties of life may be regarded as the capacities to replicate and to encode information. In accord with the extensive experimental and theoretical work of Eigen and collaborators (Eigen, 1971; Eigen and Schuster, 1979; Eigen *et al.*, 1981), we assume that life arose as a self-replicating heteropolymer similar to RNA. The first resources for self-replication were probably ribonucleotides present in the aqueous environment. Natural selection may have arisen among RNA replicators as a result of competition for nucleotides.

Folded configurations of RNA were probably the first functional adaptations (Kuhn, 1972; Bernstein *et al.*, 1983; Darnell and Doolittle, 1986). These configurations were the first phenotypes, and they were determined by RNA base sequences, the first genotypes (Michod, 1983). The RNA replicators would then acquire adaptive configurations that promoted (1) increased rate and accuracy of replication, (2) protection of replicators from damage by (a) preventing the occurrence of damage and (b) coping with damage after it occurs, and (3) increased ability to incorporate nucleotides from the environment (Bernstein *et al.*, 1983). As the replicators evolved the capacity to specify enzymes, these three kinds of adaptations became associated with increasingly complex structures. Here we focus on ways the replicator can cope with damage.

2. *Sex as a Solution to the Problem of Lethal Genome Damage or Loss*

Damage to an RNA replicator might often block its replication or expression of its encoded information. In a population of independent replicators lacking repair, damaged ones simply die. It has been proposed that, although the RNA replicators were at first independent of each other, they evolved mutual dependencies based on joint use of encoded products (e.g., primitive enzymes) (Eigen, 1971; Eigen and Schuster, 1979; Eigen *et al.*, 1981). Eigen and co-workers further suggested an evolutionary progression from free RNA replicators to "hypercycles" (sets of mutually dependent replicators) and then to encapsulated hypercycles (see Bernstein *et al.*, 1984, for further discussion).

The problem of damage to the set of RNA replicators comprising a hypercycle becomes more acute when it becomes encapsulated within a simple membrane, such as a lipid bilayer. The advantage of encapsulation to form a protocell is that it allows the RNAs to localize their encoded products for more efficient use. If each RNA molecule within a protocell produces one or more distinct products that promote the survival and duplication of the protocell, then each RNA segment is informationally equivalent to a chromosome. Woese (1983) has proposed a universal cellular ancestor which had an RNA genome that was "physically disaggregated, comprising a collection of gene size pieces." A haploid organism with such a genome would be very vulnerable to damage, since a single lesion in any segment is potentially lethal to the protocell. These protocells are also vulnerable to loss of segments, since when the set of segments replicates, at least one copy of each segment must be passed to each daughter protocell. Failure to segregate properly leads to inviability. Encapsulation thus promotes efficient use of gene products, but at the cost of making the entire set of replicator segments vulnerable when only one is damaged or lost. This vulnerability could be reduced by maintaining more than one copy of each RNA segment in each protocell. This redundancy would allow a damaged or lost segment to be replaced by an additional replication of its homologue. However, to maintain extra copies of each segment is costly in terms of resources invested and lengthened generation time. The fitness (per capita rate of increase) of the protocell is reduced by these costs of redundancy. Thus, a basic problem for early protocells was to cope with damaged or lost genetic information while minimizing the costs of redundancy.

Bernstein *et al.* (1984, 1985a) have carried out a cost–benefit analysis in which the costs of maintaining redundancy were balanced against the costs of damage or loss. Under a wide range of circumstances the selected strategy is for each protocell to be haploid, but to periodically fuse with another haploid protocell to form a transient diploid. Such periodic fusions allow mutual reactivation of lethally damaged protocells. As long as one undamaged copy of each gene is present in the transient diploid, viable progeny can be formed. Formation of two viable haploid progeny requires an extra replication of the intact chromosomes homologous to any damaged or lost ones, and the splitting of the diploid protocell. We consider this cycle of haploid reproduction, involving fusion to a transient diploid state with subsequent splitting to the haploid state, to be the sexual cycle in its most primitive form. Haploid progeny would often be recombinant in the sense that they would have the genes of their two parents in new combinations. Although recovery from genome damage or loss is the adaptive function of this simple sexual cycle, allelic recombination would arise as a by-product. Without this sexual cycle, damaged protocells would simply die.

3. *Extant Segmented Single-Stranded RNA Viruses Provide a Model for the Earliest Form of Sex*

Our model for the primitive sexual cycle is not merely hypothetical. It corresponds to the established sexual behavior of some of the simplest extant organisms known, the segmented RNA viruses. One example is influenza virus, whose genome is composed of single-stranded RNA divided into eight physically separate segments (Lamb and Choppin, 1983). Six segments encode one polypeptide each, and are thus equivalent to individual genes. The other two segments each encode two or three polypeptides. When two influenza viruses infect a cell, the genomes of the two viruses enter the cell, where they replicate. Most progeny viruses contain a genome composed of genes from both parents. When influenza viruses are UV irradiated they are able to undergo multiplicity reactivation (Barry, 1961). In DNA viruses, multiplicity reactivation results from enzyme-mediated exchange of DNA segments, leading to progeny that are free of lethal damages (see Section V,A). However, in segmented RNA viruses, multiplicity reactivation likely results from replication of undamaged RNA segments of the infecting virus, and the reassortment of these replicas to form complete, undamaged progeny viruses. Another segmented RNA virus, reovirus, also undergoes multiplicity reactivation (McClain and Spendlove, 1966), although in this case, the RNA segments are double stranded rather than single stranded.

In conclusion, the plausibility of our proposal for the origin of sexual reproduction in primitive protocells is supported by the occurrence of a similar process, i.e., multiplicity reactivation, in some of the simplest organisms known, the segmented RNA viruses.

4. Recombinational Repair on DNA-Containing Organisms

The genes of most DNA viruses are connected end to end to form one continuous DNA molecule. This arrangement probably evolved to promote reliable segregation of a complete set of genes to each progeny virus, which is an improvement over the unreliable assortment in the simple, segmented RNA viruses. In primitive RNA protocells, as in segmented RNA viruses, we think repair occurred simply by nonenzymatic reassortment of replicas of undamaged segments. This, we believe, evolved into recombinational repair involving enzyme-mediated breakage and exchange between DNA molecules from separate individuals, as presently occurs in DNA viruses and higher organisms (Section V). Although the linking of genes in DNA promotes reliable segregation of gene sets, it complicates repair because now double-strand damages can only be repaired by enzyme-mediated breakage and exchange between DNA molecules—in other words, by physical recombination.

B. Emergence of Diploidy

1. Coping with Mutation by Improving Replication Accuracy

As primitive haploid organisms evolved, their genes increased in number. This made them more vulnerable to lethal mutations, since mutation frequency per genome is proportional to the number of genes. The adaptive response of these haploid organisms, at first, was probably to improve the fidelity of replication so that the mutation rate per genome per replication remained low. This is suggested by the analysis of mutation rates in five haploid organisms by Drake (1974). These were the phages T4 and λ, the bacteria *E. coli* and *S. typhimurium,* and the fungus *Neurospora crassa.* Drake showed that with increasing genome size over a 1000-fold range, mutation rate per base pair per replication decreases in such a way that the mutation rate per genome per replication remains roughly constant at about 0.001–0.003 for all five organisms.

2. Coping with Mutation by Genetic Complementation and Diploidy

The diploid stage of the sexual cycle probably was initially transient. However, as the genome expanded, it became cost-effective to take advantage of the masking effect of diploidy, which protects against expression of deleterious recessive mutations through complementation (Maynard Smith, 1958, 1978; Bernstein *et al.*, 1981). Thus the diploid stage became the dominant stage of the sexual cycle in some lines of descent. The transition to diploidy may have occurred when the costs of maintaining an extra genome became small relative to the costs of maintaining the total cellular metabolism. Once this strategy is adopted, deleterious mutations can no longer be efficiently weeded out of the population by natural selection. Hence, they accumulate. This accumulation will continue until a balance is reached between the occurrence of new mutations and the selective loss of deleterious homozygous recessive alleles (Muller, 1932; Crow and Kimura, 1970). When this new balance point is reached there are many more deleterious recessive mutations in each individual. Should a diploid organism at this stage revert to a predominantly haploid life cycle, all of the accumulated deleterious recessive alleles are expressed immediately. Because of the accumulation of deleterious recessives, the advantage of shifting to diploidy is temporary. However, the transition back to haploidy is not a viable option and so diploidy is maintained. We next argue that the outcrossing aspects of sex are maintained by similar factors.

C. Outcrossing

1. *Outbreeding Has Large Costs Implying Correspondingly Large Benefits*

There are several kinds of short-term costs of outcrossing sex. These include the cost of mating (Bernstein *et al.*, 1985d; Hopf and Hopf, 1985), the cost of males (Williams, 1975; Maynard Smith, 1978), high recombinational load (Shields, 1982), and lower genetic relatedness between parent and offspring (Williams, 1980; Uyenoyama, 1984, 1985). The following quotation by Bell (1982) on the high costs of sex is typical of current thinking.

> Sex is no minor readjustment having an inconsiderable effect on fitness. Sex tears apart every genome in every generation, and builds them up anew, every one different; and in doing so it does not merely reduce fitness, but halves it. If a reduction in fitness of a fraction of one percent can cripple a genotype, what will be the consequence of a reduction of 50 percent? There can be only one answer: sex is powerfully selected against and rapidly eliminated wherever it appears. And yet this has not happened [pp. 77 and 78].

The high costs of sex are related to outcrossing, the second aspect of sex. Therefore, any fundamental explanation of outcrossing, to be plausible, must provide a benefit that is large enough to account for the costs. We will now review evidence that the benefit of masking deleterious mutations accounts for outcrossing (see Bernstein *et al.*, 1985b, for a more complete discussion).

2. *Different Mating Systems Vary in the Number of Recessive Mutations Masked at Equilibrium*

So far, we have reviewed evidence for the central role of recombinational repair in the evolution of sex. Recombinational repair requires the presence of two chromosomes in a common cytoplasm. There are two potential strategies for obtaining these two chromosomes: (1) a closed-system strategy where they both come from the same parent; (2) an open-system strategy where they come from different parents. If recombinational repair were the only selective advantage of sexual reproduction, the most effective strategy would be a closed system such as self-fertilization or automixis (uniparental production of eggs through ordinary meiosis followed by some internal process for restoring diploidy). This would avoid the major costs of sex enumerated above. Thus, we need to explain why the most common strategy is outcrosing sex, an open system.

We have argued that outcrossing arose in primitive protocells because of the cost of maintaining more than one set of chromosomes within a protocell (Section VII,A; see also Bernstein *et al.*, 1984, 1985a). Later, diploidy likely emerged as the predominant stage of the life cycle because of the advantage of masking deleterious recessive alleles through complementation (Section VII,B; see also Bernstein *et al.*, 1981). We now argue that the advantage of complementation also explains the maintenance of outcrossing in reproductive systems that have recombination (Bernstein *et al.*, 1985b; Lande and Schemske, 1985).

As discussed in Section II,C,2, mutations result from errors of replication. Improvements in replication accuracy have costs, which include increased energy use (Hopfield, 1974), additional gene products (Alberts *et al.*, 1980), or slower replication (Gillen and Nossal, 1976). Thus there are probably cost-effectiveness barriers to indefinite improvement in accuracy, and a finite spontaneous mutation rate is an intrinsic property of genome replication. Felsenstein (1974) has noted that in the short run, deleterious mutations affect fitness much more than beneficial ones because of their much higher rate of occurrence. Haldane (1937) argued that in a population at equilibrium, deleterious mutations are removed by selection at the same rate that they arise by mutation. He demonstrated that the average survivorship at equilib-

TABLE 1
Classification of Diploid Reproductive Systems

Reproductive system	Masking ability at equilibrium	Recombinational repair	Source of homologous chromosome
Automixis	Low ($\sim 2u$)	Yes	Self
Selfing	Low ($\sim 2u$)	Yes	Self
Outcrossing	Intermediate (~ 1 to $\sqrt{Nu}$)[a]	Yes	Another individual
Endomitosis	High ($\sim N$)	Limited[b]	Self
Apoximis	High ($\sim N$)	No	Not applicable
Vegetative	High ($\sim N$)	No	Not applicable

[a] N denotes the number of functional genes per genome, which, in higher organisms, is approximately 40,000.
[b] See text.

rium is solely a function of u, the rate of deleterious mutation per haploid genome, and is not affected by how individually harmful the mutations are.

Table 1 lists reproductive systems of diploid organisms. We have found by a generalization of Haldane's argument that in all of these diploid reproductive systems survivorship due to expression of deleterious recessive alleles is e^{-u} at equilibrium (Hopf *et al.*, 1987). For haploid organisms (not listed in Table 1), survivorship is also given by e^{-u}, but this case will be discussed below [point (5)]. Since the effect of u on survivorship is the same for all reproductive systems, u should not differ among these systems. Consequently, we assume u to be fixed, and consider below [point (4)] the effect on survivorship of switching from one reproductive system to another among those listed in Table 1.

First, we estimate a value for u. Data obtained by Mukai *et al.* (1972) indicate that in *Drosophila* $u = 0.3$ for lethal and nonlethal mutations combined. This gives a survivorship $e^{-u} \simeq 0.7$, which is neither so low as to be devastating or so close to 1.0 as to indicate a negligible effect of mutational load. If the mutational load were negligible we think it would lead to an increase in the number of genes, N. Since u should be proportional to N, selection for increased N would also increase u. Therefore, we consider 0.7 to be an approximate general value for e^{-u}, and 0.3 a general value of u.

The second column of Table 1 lists the number of lethal mutations accumulated at equilibrium in the different reproductive systems.

The differences among these systems in the number of accumulated recessive mutations are substantial. Those reproductive systems effective at masking recessive mutations accumulate many. Those which are ineffective at masking accumulate few mutations.

3. *The Advantage of Changing from a Mating System That Masks Few Deleterious Mutations to One That Masks Many*

The effect of mutational load on survivorship makes all systems equally competitive at equilibrium. However, the mutational load also creates a transient selective advantage to moving downward in Table 1, i.e., toward greater masking, and a transient disadvantage to moving upward toward diminished masking. We have already discussed the similar transient advantage of the shift from haploidy to diploidy and the disadvantage of the reverse shift (Section VII, B). To illustrate the transient advantages and disadvantages of shifting between reproductive systems, we first consider a population fixed for selfing. Although individuals in this population will have accumulated few mutations at equilibrium (Table 1), new ones occur each generation at a frequency u, and a new mutant outcrossing individual will mask these mutations in its offspring. There is nearly complete complementation in an outcross because delterious alleles in the two partners are statistically unrelated to each other. Thus, the fitness of the outcrosser is not reduced at all by mutational load initially, and it has a survivorship of unity. However, the new outcrossers must pay the costs of outcrossing sex described in Section VII,C,1. Let C be the factor by which fitness is reduced due to these costs. Thus C is the ratio of the fitness of the selfer, taken as unity, to the fitness of the new outcrosser (<1), so that $C > 1$. We can now compare the cost C of the shift to outcrossing with the benefit of the shift. The benefit is given by $1/e^{-u}$ (or e^{u}), the factor by which fitness is increased due to the masking of deleterious recessive alleles when the shift occurs. If the cost is less than the benefit ($C < e^{u}$), then a gene for outcrossing should expand in the population. Taking u to have a general value of 0.3 as discussed above, the benefit e^{u} has a value of 1.4. Thus the shift to outcrossing should occur even if the costs are fairly large.

As outcrossing becomes fixed in a population, deleterious recessive alleles increase. Eventually the outcrosser has an equilibrium fitness that is reduced both by the costs of outcrossing and the mutational load. That is, the long-term effect of the transition to outcrossing is a net reduction of individual fitness. A successful shift back to selfing is inhibited, however, by the drastic consequences of unmasking the many accumulated deleterious recessive mutations. The shift from outcrossing to selfing will only succeed when the costs of outcrossing become very large. Among the first four reproductive systems listed in Table 1, which are those with full recombinational repair, outcrossing is favored. An intermediate level of outcrossing may be preferred over panmixia because of the need to preserve coadapted gene complexes (Shields, 1982).

4. *Costs of Switching from Outbreeding to Parthenogenesis or Vegetative Reproduction*

The transient advantage of complementation favors asexual systems in which the diploid maternal genome is passed down intact from mother to daughter, since this gives maximal masking of deleterious recessive alleles. However, the absence of recombinational repair in these asexual systems, as listed in the third column of Table 1, also needs to be considered in evaluating the success of a shift from outcrossing to asexual reproduction. Apomixis, a parthenogenetic system, is characterized by suppression of meiosis and its replacement by a single mitotic maturation division. In vegetative reproduction, meiosis is also absent. In a shift from outcrossing to apomixis or vegetative reproduction, repair of double-strand damages is probably largely abandoned, and thus these strategies are costly. However, the effect of DNA damage may be overcome in these cases by the cellular selection resulting from the death of damaged cells and the replication of undamaged cells. Cellular selection is effective when damages are infrequent and/or growth is rapid, but not otherwise.

Endomitosis, another parthenogenetic system, is characterized by two sequential premeiotic chromosome replications followed by an apparently normal meiosis to produce diploid eggs. Also, at the four-chromatid stage, pairing is between chromatids derived from only one initial chromosome (Cuellar, 1971; White, 1973; Cole, 1984). This reproductive system is used, for example, by whiptail lizards common in the deserts of the southwestern United States. Because there is no recombination between the non-sister homologues, the maternal genome should be passed on intact to daughters. If recombination between non-sister homologues were allowed, the transient advantage would be lost because of the immediate expression of accumulated recessives. Endomitosis might seem to be an ideal strategy since it reaps the benefit of meiotic repair while avoiding the expression of deleterious recessive alleles. However, double-strand damages occurring before the first premeiotic replication cannot be repaired by endomitosis, as all chromatid pairing partners are derived from the same chromosome and there is no intact template corresponding to the damaged site. This problem does not arise in conventional meiosis, because recombination is between non-sister chromatids. Hence, if double-strand damages are common before premeiotic replication, endomitosis is an unsatisfactory option.

In conclusion, shifting from outcrossing to any of the other reproductive systems in Table 1 results in an immediate reduction in fitness. This reduction is transient for selfing and automixis, since the few progeny that survive the unmasking of recessive mutations will be those with statistically fewer mutations. In contrast, the reduction in fitness upon shifting from outcrossing to either apomixis, endomitosis, or vegetative reproduction is permanent since these latter systems have reduced capacity for recombinational repair.

Lynch (1984) has summarized the literature indicating that parthenogens most often have lower reproductive rates than their sexual relatives, frequently less than 50%. As he discussed, these costs of parthenogenesis stem from poor hatching of eggs, resulting from developmental abnormalities, and in some cases reduced egg production. Newly arisen parthenogens seem to experience a greater reduction in fecundity than established ones. The costs of parthenogenesis predicted by theory and supported by the observed reduction in fecundity imply that parthenogens would only be competitive in situations where the costs of outcrossing are larger than the costs of parthenogenesis. One common situation of this type is in newly created natural habitats (such as produced by floods and fires), where finding a mate is difficult. Evidence reviewed by Bernstein *et al.* (1985d) supports the generalization that parthenogens are favored where the costs of finding a mate are high.

5. *The Special Problem of Multicellular Haploids and Haplodiploids*

We have argued that outcrossing is maintained by the transient advantage of masking mutations. However, this argument does not apply to mutations in genes that are expressed in the haploid stage of the life cycle since recessive mutations in such genes are not masked and can be eliminated by selection. In multicellular haploid organisms most genes may be of this type. If such genes were the only ones of significance, the benefit of outcrossing would be small and there should be little barrier to shifting to selfing or automixis. A similar problem exists in explaining outcrossing in haplodiploid insects. Recessive mutations are removed by selection in the haploid males and are largely masked in the diploid females independently of outcrossing. At first glance, these cases seem to argue against our hypothesis that the primary function of outcrossing is complementation. Unicellular haploids, as opposed to multicellular haploids, do not present a problem, since they need to outcross simply to bring homologous chromosomes together in the same cell for recombinational repair.

To counter the above argument, we pointed out (Bernstein *et al.*, 1985b) that in multicellular haploid organisms there are key genes that are only expressed in the diploid stage, such as the genes controlling meiosis. In fungi, for example, meiosis occurring in the diploid stage is one of their most complex functions. The multicellular haploid stage, although dominant in size, characteristically has a modular construction with little differentiation. In the multicellular haploid fungus *N. crassa,* Leslie and Raju (1985) found that recessive mutations affecting the diploid stage of the life cycle were frequent in natural populations. Based on this frequency they estimated the number of genes affecting the diploid stage to be minimally 435, and more likely about 1000. These mutations, when homozygous in the diploid stage, produced barren fruiting bodies with few sexual spores, or spores with maturation defects. The maintenance of outcrossing can readily be explained by the masking of such mutations through complementation. In multicellular haploid mushrooms and bracket fungi the conspicuous structures are often dikaryons containing nuclei from different parents (Fincham *et al.,* 1979). The dikaryons are functionally diploid since they permit masking of deleterious recessive alleles, and this makes outcrossing selectively advantageous.

Ferns, like fungi, have a multicellular haploid stage. However, unlike fungi, ferns have a large, complex diploid sporophyte stage with considerable tissue diversity. In at least one genus, *Osmunda,* numerous recessive lethal mutations are present in natural populations (Klekowski, 1973). These prevent normal development of the embryonic diploid sporophyte when homozygous. Thus, although these hermaphroditic plants are capable of self-fertilization, outcrossing is thought to be preferred at higher population densities because of the advantage of masking recessive lethal mutations (Klekowski, 1973).

In haplodiploid insects some genes are expressed only in the diploid phase (females) and not in the haploid phase (males). Data on the fraction of genes whose expression is limited to females were summarized by Kerr (1976) for various haplodiploid species. These estimates range from 14 to 46% depending on the species and kind of phenotype considered. He also calculated that 14% of the deleterius alleles in bee populations is limited to females. These results indicate that recessive mutations limited to females are sufficiently common to select for some degree of outcrossing in haplodiploids.

The evidence reviewed in this section suggests that there is a sufficient advantage to masking deleterious recessive alleles in the diploid stage of fungi, ferns, and haplodiploid insects to explain the occurrence of outcrossing in these organisms. However, we also expect the levels of inbreeding to be higher in these species, since there are fewer deleterious recessives to be expressed. This expectation appears to be confirmed in the case of haplodiploids (Hamilton, 1967).

In Section VII,C we presented evidence bearing on the second part of the repair hypothesis, that outcrossing is maintained by the advantage of masking deleterious recessive alleles. Reproductive forms which at first seemed difficult to explain on this basis, such as parthenogens and outcrossing multicellular haploids, prove on further analysis to have straightforward explanations in terms of the repair hypothesis.

D. Meiosis and the Germ Line

In Section VI we discussed meiotic recombination and argued that it is designed to promote repair of DNA. In this section we consider a second fundamental feature of meiosis, the reduction in ploidy which occurs, usually from diploid to haploid. This reduction prepares the daughter cells (gametes) for fusion, which restores diploidy.

1. Alternation of Diploid and Haploid Phases of Outcrossing Organisms

In Section VII,C,2 we pointed out that meiosis may be associated with either an open-system strategy such as outcrossing, in which fusion is between haploid cells from different individuals, or a closed-system strategy such as automixis, in which fusion is between cells from the same individual. We then presented an explanation for the commonness of outcrossing in nature based on the masking of deleterious mutations.

In the germ line of outcrossing multicellular organisms, there are typically many mitotic divisions. The fact that meiosis occurs only at one point in the sequence of cell divisions of the germ line raises several questions. How are DNA damages handled in the germ line by the mitotically dividing diploid cells? What is the special characteristic of the stage of the life cycle in which meiosis does occur that explains the need for it at that point? How does the repair hypothesis explain the lack of recombination in the male germ line of *Drosophila*, or more generally in the heterogametic sex of various species?

a. DNA Damage in the Mitotically Dividing Cells of the Germ Line. We assume that mitotically dividing germ-line cells in multicellular organisms are capable of the same kinds of repair processes as occur in somatic cells. The sequestering of germ cells early in development may be a strategy for reducing DNA damage by reducing the exposure of germ-line DNA to metabolic activity.

In addition, as can occur in the somatic line, cellular selection in the germ line can cope with lethal double-strand damages. For example, in the case of the frequently dividing diploid spermatagonia in animals, cells with unrepaired lethal DNA damages may simply die so that undamaged ones selectively replicate.

In conclusion, there is no obvious reason why DNA repair processes in mitotically dividing germ-line cells should be substantially different than in somatic cells, so long as the levels of damage are similar.

b. Meiosis Occurs at a Stage When Unrepaired Damages Are Exceptionally Costly. We think that there is a unique cost to unrepaired DNA damages in gametes that justifies the investment in producing them meiotically. A parent contributes a single gamete to each progeny of the next generation and this cell must be free of DNA damage. No other cell type in a multicellular organism so directly determines fitness as the gamete. In general, we think that this simple fact explains the timing of meiosis, a uniquely powerful but costly repair process.

Another factor which may determine the timing of meiosis is the common strategy among multicellular organisms of provisioning their eggs with metabolic products to be used by the developing embryo. This provisioning requires considerable metabolic activity, which produces reactive forms of oxygen as by-products. These in turn damage DNA. Thus the strategy of provisioning the egg may increase the need for meiotic DNA repair. Recombinational repair during meiosis, as well as other repair and protective processes, may barely be coping with germ-line damage, since in humans about half of all fertilized eggs fail to produce surviving embryos, as is typical of mammals generally (Austin, 1972), and in plants the failure of fertilized ovules to mature as seeds is common (Bawa and Webb, 1984). Although the causes of these failures have not been determined, it is likely that DNA damage is an important factor.

During the mitotic divisions of the germ line, the ability to replicate certifies that a cell is free of unrepaired double-strand damages. However, replication also leaves gaps opposite unrepaired single-strand damages, thus introducing new double-strand damages in daughter cells. One such damage is sufficient to kill a daughter cell. So, although replication ensures that the parental cell was free of double-strand damages, it does not ensure that either of its daughter cells is free of them. As discussed in Section VIII,A,1, there are reasons to expect that excision repair of single-strand damages is not very efficient. The only way of guaranteeing freedom from lethal damages is to produce daughter cells by a process which repairs all damages. This process, we believe, is meiotic recombination.

2. *Variation in Allelic Recombination between Males and Females*

Allelic recombination generally occurs in both males and females of a species. However, the rates often differ, the males having lower rates of recombination than females (Bell, 1982, Chap. 5). This bias is most extreme in *Drosophila* males, which do not undergo any allelic recombination or gene conversion, whereas females do (Chovnick *et al.*, 1970). The general trend of males having lower levels of recombination can be explained simply by the metabolic activity that is lower during spermatogenesis than during oogenesis, and hence lower levels of DNA-damaging oxidative compounds are produced during spermatogenesis than during oogenesis. However, certain lepidopteran females, such as silkworm moths (Sturtevant 1915; Tazima, 1964) and wax moths (White, 1945, p. 193), do not appear to undergo recombination, whereas males from these species do. These exceptions to the general trend are puzzling to us in light of our above explanation for the higher expected rate of damage during oogenesis than in spermatogenesis for most organisms. It should be noted that in Lepidoptera the females are heterogametic rather than the males, as in the case of *Drosophila*. Thus in these cases, lack of recombination is associated with heterogamy rather than a lower expected level of damage.

This completes our discussion of the origin and evolution of the major features of sexual reproduction which we interpreted as a continuum in terms of the repair hypothesis. In the next section, we consider ways of choosing, based on evidence, between the repair hypothesis and other, alternative explanations for the advantage of sex.

VIII. Alternative Theories

A. The Variation Hypothesis

In Section I, we discussed the historical development of explanations for the advantage of sex. These explanations were based almost exclusively on the advantage of variation, at first at the level of the group or species and more recently at the level of the individual organism. We also noted that most current studies of the variation hypothesis examine theoretically the evolution of a gene which promotes allelic recombination, but is otherwise neutral, having no direct effect on fitness itself. We then pointed out three problems with this neutral modifier approach to recombination. In this section, we suggest that an appropriate way of choosing between the variation hypothesis and the repair hypothesis is to examine the mechanisms of meiotic recombination and to ask whether it is designed to promote repair, variation, or both. Therefore, we next discuss two fundamental features of meiosis which, we believe, allow us to make a choice. These are premeiotic replication and the frequent occurrence of physical recombination without allelic recombination.

1. Function of Premeiotic Replication

Premeiotic replication is a general feature of meiosis. A definite function can be attributed to it on the repair hypothesis (Bernstein *et al.*, 1986). As discussed in Section V,B,3, studies using *E. coli* show that replication of DNA with single-strand damages leads to gaps in the new strand opposite the damages (Rupp and Howard-Flanders, 1968). These gaps promote recombinational repair (West *et al.*, 1982). Such gaps may serve as a universal initiator of recombinational repair since they have a molecular structure that is independent of the structure of the original damage. Excision of the damage opposite the single-strand gap by an enzyme, such as the nuclease controlled by *rad-52* in yeast (Resnick *et al.*, 1984), would result in a double-strand break. As discussed in Section VI,A, the current accepted model of meiotic recombination starts with a double-strand break (Szostak *et al.*, 1983).

Naturally occurring damages are probably a mixture of types, with many occurring at a low frequency. DNA likely undergoes transient structural variation during the course of normal gene expression and these variations are also probably of numerous types. As the DNA unwinds during replication such functional variations are likely released and damages should be readily distinguished by the DNA polymerase. Single-strand damages might be marked by the polymerase leaving a gap opposite the damage in the new strand, which would then direct recombinational repair as described above. This approach would avoid the difficulty of evolving excision repair enzymes to distinguish the variety of DNA damages that might occur from normal variations in DNA structure.

The evidence reviewed in Section V indicates that recombinational repair is able to overcome a wide variety of DNA damages, as expected by this view. Support for the assumption that excision repair enzymes may have difficulty distinguishing damages from normal structural variations comes from studies of phages. In one of the most well-studied excision repair systems, pyrimidine dimer excision repair in phage T4, the process is only 45–60% efficient (Pawl *et al.*, 1976).

In the variation hypothesis there is no apparent purpose to premeiotic replication. Instead, premeiotic replication is wasteful from the point of view of the variation hypothesis, because sister chromatid exchanges (SCEs) become possible and these do not bring about allelic recombination. If it could be shown that allelic recombination is promoted by suppressing SCEs while favoring non-SCEs, then it would imply that variation is being selected for in spite of an apparently pointless premeiotic replication. In males, the level of SCE for the X chromosome can be compared to the level of SCE in the autosomes. This allows one to determine if there is a preference for non-SCEs when both types of recombination are possible, since autosomes have both sister and non-sister homologues, whereas the X chromosome has only a sister homologue. According to Peacock (1970) there is no such preference, since SCEs occur "in the X chromosome, where they are as frequent per unit length, as in the autosomes." Thus the evidence indicates that SCE is not suppressed when a non-sister homologue is present and could serve as a partner for allelic recombination.

Our main point in this section is that a very general feature of meiosis, premeiotic replication, is unlikely to be an adaptation for recombinational variation but is explained by the hypothesis that meiotic recombination is an adaptation for DNA repair.

2. *Physical Recombination without Allelic Recombination*

Many studies of fungi have been performed in which meiotic recombination was detected by aberrant segregation at a genetically marked site, and allelic markers at flanking loci were scored as to whether or not they were exchanged. As mentioned in Section I, greater than 60% of physical recombination events between homologous chromosomes fail to give rise to exchange of flanking alleles outside the short region of physical exchanges (Whitehouse, 1982, p. 321). In 10 separate studies involving five different species the ratio of nonexchanged to exchanged flanking alleles on average was 66 : 34 [see Whitehouse (1982), Tables 19 and 38, for summaries of these data]. The average frequency for a species of nonexchange of flanking alleles varied from 80% in *Sordaria brevicollis* to about 50% in *S. cerevisiae*. Within the genome of a species, the percentage can vary from region to region. In addition to the fungal data, analysis of half-tetrads using compound autosomes in *Drosophila melanogaster* shows that only one-third of recombination events at the *rosy* cistron results in allelic recombination of flanking loci (Chovnick *et al.*, 1970).

The fact that in most cases physical recombination does not produce allelic recombination suggests that the common approach of modeling recombination in terms of the modification of the linkage relationship of genes is, at best, not a complete description of the phenomena to be explained. At worst this approach is misleading with respect to the evolutionary forces operating on recombination. The fact that physical recombination does not result in allelic recombination 66% of the time is inconsistent with the variation hypothesis unless it is a necessary consequence of the mechanisms of recombination. We have examined this possibility using the double-strand break repair model of Szostak *et al.* (1983) as the most authoritative model of general recombination currently available. We concluded (Bernstein *et al.*, 1986) that the low proportion of allelic recombination is not a necessary consequence of the recombination mechanism, and, in fact, that the mechanism is more complex than need be in order to reduce the level of allelic recombination.

B. The Parasitic DNA Hypothesis of Sex

Hickey (1982) proposed that "Sex itself, and especially outbreeding, is a product of parasitic genes." Dawkins (1982, p. 160) similarly suggested that there are replicating "engineers" of meiosis which "achieve their own replication success as a byproduct of forcing meiosis upon the organism." A similar proposal has also been made by Rose and Redfield (1986). The idea that sexual reproduction has arisen and been maintained as an unavoidable parasitic disease rather than an adaptation strikes us as implausible primarily because it attributes no advantage to sex to balance the costs (Section VII,C,1), which are usually substantial. It seems likely that some effective method would have evolved to eliminate parasitic genes, if they imposed these large costs at each generation without compensating benefit to the organism. For example, bacteria can deal with foreign parasitic DNA through restriction enzymes that recognize and degrade this DNA (Kornberg, 1980, pp. 333–340). Other methods also occur in microorganisms for inhibiting gene expression of competing DNA. It seems likely that higher organisms would have evolved similar ways of avoiding the high costs of sex if this process were maintained for the sole benefit of parasitic genes.

In the previous section, where we compared the variation hypothesis to the repair hypothesis, our approach was to ask whether general features of meiosis were adaptations for repairing DNA or producing genetic variation, these being presumed benefits to the organism. This approach cannot be used for a comparison of the parasitic DNA hypothesis and the repair hypothesis, since the former assumes that sex is not adaptive to the organism. This intrinsic feature of the parasitic DNA hypothesis makes it difficult to test. Perhaps if it could be shown that the genes which code for sex generally have the characteristics of transposable elements, or some similar entity, the hypothesis could be validated.

IX. Immortality of the Germ Line versus Mortality of the Somatic Line

A fundamental feature of sexual reproduction in multicellular organisms is the potential immortality of the germ line and the apparently inevitable aging of the individual organism. On the repair hypothesis, recombinational repair of DNA damage in the germ line can explain the potential immortality of this line. In contrast, the somatic line, which probably has much less capacity for recombinational repair, should accumulate lethal damages and be mortal (for further discussion, see Gensler and Bernstein, 1981, and Medvedev, 1981). A specific possibility, in line with our previous discussion, is that aging in humans may be due to the accumulation of double-strand damages produced by oxygen radicals in nondividing cells such as neurons. Double-strand damages might be irreversible in somatic cells but repairable by meiotic recombination in germ cells.

One of the most favorable organisms for studying aging is the unicellular eukaryotic protozoan *Paramecium tetraurelia*. Germ-line DNA is contained in a micronucleus, and the apparent equivalent of somatic cell DNA is present in multiple genome copies in the macronucleus. Successive acts of asexual duplication alternate with less frequent sexual reproduction (conjugation) or self-fertilization (autogamy). Aging, characterized by a decline in vitality, occurs during the asexual phase, and rejuvenation occurs following conjugation or autogamy. Smith-Sonneborn (1979) presented evidence that longevity during the asexual phase can be increased by induction of a DNA repair process. Holmes and Holmes (1986) reported the accumulation of single-strand breaks and/or apurinic sites occurring during the course of the asexual phase. These results support the earlier proposal of Martin (1977) that aging in these organisms is due to accumulation of DNA damage during the asexual cycle, and rejuvenation involves efficient DNA repair in the sexual phase.

X. Summary and Conclusions

Traditionally, sexual reproduction has been explained as an adaptation for producing genetic variation through allelic recombination. Serious difficulties with this explanation have led many workers to conclude that the benefit of sex is a major unsolved problem in evolutionary biology. A recent informational approach to this problem has led to the view that the two fundamental aspects of sex, recombination and outcrossing, are adaptive responses to the two major sources of noise in transmitting genetic information, DNA damage and replication errors. We refer to this view as the repair hypothesis, to distinguish it from the traditional variation hypothesis. On the repair hypothesis, recombination is a process for repairing damaged DNA. In dealing with damage, recombination produces a form of informational noise, allelic recombination, as a by-product. Recombinational repair is the only repair process known which can overcome double-strand damages in DNA, and such damages are common in nature. Recombinational repair is prevalent from the simplest to the most complex organisms. It is effective against many different types of DNA-damaging agents, and, in particular, is highly efficient in overcoming double-strand damages. Current understanding of the mechanism of recombination during meiosis suggests that meiosis is designed for repairing DNA. These considerations form the basis for the first part of the repair hypothesis, that recombination is an adaptation for dealing with DNA damage. The evolution of sex can be viewed as a continuum on the repair hypothesis. Sex is presumed to have arisen in primitive RNA-containing protocells whose sexual process was similar to that of recombinational repair in extent segmented, single-stranded RNA viruses, which are among the simplest known organisms. Although this early form of repair occurred by nonenzymatic reassortment of replicas of undamaged RNA segments, it evolved into enzyme-mediated breakage and exchange between long DNA molecules. As some lines of descent became more complex, their genome information increased, leading to increased vulnerability to mutation. The diploid stage of the sexual cycle, which was at first transient, became the predominant stage in some lines of descent because it allowed complementation, the masking of deleterious recessive mutations. Outcrossing, the second fundamental aspect of sex, is also maintained by the advantage of masking mutations. However, outcrossing can be abandoned in favor of parthenogenesis or selfing under conditions in which the costs of mating are very high. In multicellular organisms the potential immortality of the germ line may be due to recombinational repair during meiosis, whereas the apparent inevitable aging of the organism may be due to accumulated unrepaired DNA damages in the somatic line.

ACKNOWLEDGMENTS

This work was supported by NIH Grants GM27219 (H. Bernstein) and HD19949, HD00583, and GM36410 (R. E. Michod).

REFERENCES

Abbadessa, R., and Burdick, A. N. (1963). The effect of X-irradiation on somatic crossing over in *Drosophila melanogaster*. *Genetics* **48**, 1345–1356.

Abel, P. (1962). Multiplicity reactivation and maker rescue with vaccinia virus. *Virology* **17**, 511–519.

Alberts, B. M., Barry, J., Bedinger, P., Burke, R. L., Hibner, U., Liu, C.-C., and Sheridan, R. (1980). Studies of replication mechanisms with the T4 bacteriophage *in vitro* system. *In* "Mechanistic Studies of DNA Replication and Genetic Recombination, ICN–UCLA Symposia on Molecular and Cellular Biology," pp. 449–474. Academic Press, New York.

Ames, B. N., Saul, R. L., Schwiers, E., Adelman, R., and Cathcart, R. (1985). Oxidative DNA damage as related to cancer and aging: Assay of thymine glycol, thymidine glycol, and hydroxymethyluracil in human and rat urine. *In* "Molecular Biology of Aging: Gene Stability and Gene Expression" (R. S. Sohal, L. S. Birnbaum, and R. G. Cutter, eds.), pp. 137–144. Raven, New York.

Angulo, J. F., Schwencke, J., Moreau, P. L., Moustacchi, P., and Devoret, R. (1985). A yeast protein analogous to *Escherichia coli recA* protein whose cellular level is enhanced after UV irradiation. *Mol. Gen. Genet.* **201**, 20–24.

Austin, C. R. (1972). Pregnancy losses and birth defects. *In* "Reproduction in Mammals" (C. R. Austin and R. V. Short, eds.), Ch. 5. Cambridge Univ. Press, London.

Baker, B. S., Boyd, J. B., Carpenter, A. T. C., Green, M. M., Nguyen, T. D., Ripoll, P., and Smith, P. D. (1976). Genetic controls of meiotic recombination and somatic DNA metabolism in *Drosphila melanogaster*. *Proc. Natl. Acad. Sci. U.S.A.* **73**, 4140–4144.

Baker, R. M., and Haynes, R. H. (1967). UV-Induced enhancement of recombination among lambda bacteriophages in UV-sensitive host bacteria. *Mol. Gen. Genet.* **100**, 166–167.

Barry, R. D. (1961). The multiplication of influenza virus II. Multiplicity reactivation of ultraviolet irradiated virus. *Virology* **14**, 398–405.

Bawa, K. S., and Webb, C. J. (1984). Flower, fruit and seed abortion in tropical forest trees: Implications for the evolution of paternal and maternal reproductive patterns. *Am J. Bot.* **71,** 736–751.

Bell, G. (1982). "The Masterpiece of Nature: The Evolution and Genetics of Sexuality." Univ. of California Press, Berkeley.

Bernstein, A. (1957). Multiplicity reactivation of ultraviolet-irradiated Vi-phage II of *Salmonella typhi. Virology* **3,** 286–298.

Bernstein, C. (1979). Why are babies young? Meiosis may prevent aging of the germ line. *Perspect. Biol. Med.* **22,** 539–544.

Bernstein, C. (1981). Deoxyribonucleic acid repair in bacteriophage. *Microbiol. Rev.* **45,** 72–98.

Bernstein, C., and Wallace, S. S. (1983). DNA repair. *In* "Bacteriophage T4" (C. K. Mathews *et al.*, eds.), pp. 138–151. American Society for Microbiology, Washington, D. C.

Bernstein, H. (1977). Germ line recombination may be primarily a manifestation of DNA repair processes. *J. Theor. Biol.* **69,** 371–380.

Bernstein, H. (1983). Recombinational repair may be an important function of sexual reproduction. *BioScience* **33,** 326–331.

Bernstein, H., Byers, G. S., and Michod, R. E. (1981). Evolution of sexual reproduction: Importance of DNA repair, complementation and variation. *Am. Natl.* **117,** 537–549.

Bernstein, H., Byerly, H. C., Hopf, F. A., Michod, R. E., and Vemulapalli, G. K. (1983). The Darwinian dynamic. *Q. Rev. Biol.* **58,** 185–207.

Bernstein, H., Byerly, H. C., Hopf, F. A., and Michod, R. E. (1984). Origin of sex. *J. Theor. Biol.* **110,** 323–351.

Bernstein, H., Byerly, H. C., Hopf, F. A., and Michod, R. E. (1985a). The evolutionary role of recombinational repair and sex. *Int. Rev. Cytol.* **96,** 1–28.

Bernstein, H., Byerly, H. C., Hopf, F. A., and Michod, R. E. (1985b). Genetic damage, mutation and the evolution of sex. *Science* **229,** 1277–1281.

Bernstein, H., Byerly, H. C., Hopf, F. A., and Michod, R. E. (1985c). DNA repair and complementation: The major factors in the origin and maintenance of sex. *In* "Origin and Evolution of Sex" (H. O. Halvorsen, ed.), pp. 29–45. Liss, New York.

Bernstein, H., Byerly, H. C., Hopf, F. A., and Michod, R. E. (1985d). Sex and the emergence of species. *J. Theor. Biol.* **117,** 665–690.

Bernstein, H., Hopf, F. A., and Michod, R. E. (1986). Is meiotic recombination an adaptation for repairing DNA, producing genetic variation, or both? *In* "The Evolution of Sex: An Examination of Current Ideas" (B. Levin and R. Michod, eds.). Sinauer, New York, in press.

Boyd, J. B. (1978). DNA repair in *Drosophila. In* "DNA Repair Mechanisms" (P. C. Hanawalt, E. C. Friedberg, and C. F. Fox, eds.), pp. 449–452. Academic Press, New York.

Breimer, L. H. (1983). Urea-DNA glycosylase in mammalian cells. *Biochemistry* **2,** 4192–4203.

Breimer, L. H., and Lindahl, T. (1984). DNA glycosylase activities for thymine residues damaged by ring saturation, fragmentation, or ring contraction are functions of endonuclease III in *Escherichia coli. J. Biol. Chem.* **259,** 5543–5548.

Brenner, D. A., Smigocki, A. C., and Camerini-Otero, R. D. (1986). Double-strand gap repair results in homologous recombination in mouse L cells. *Proc. Natl. Acad. Sci. U.S.A.* **83,** 1762–1766.

Burck, K. B., Scraba, D. G., and Miller, R. C., Jr. (1979). Electron microscopic analysis of partially replicated bacteriophage T7 DNA. *J. Virol.* **32,** 606–613.

Cathcart, R., Schwiers, E., Saul, R. L., and Ames, B. N. (1984). Thymine glycol and thymidine glycol in human and rat urine: A possible assay for oxidative DNA damage. *Proc. Natl. Acad. Sci. U.S.A.* **81**, 5633–5637.

Chan, V. L. (1975). On the role of V gene ultraviolet-induced enhancement of recombination among T4 phages. *Virology* **65**, 266–267.

Charlesworth, D., Charlesworth, B., and Strobeck, C. (1979). Selection for recombination in partially self-fertilizing populations. *Genetics* **93**, 237–244.

Chetsanga, C., and Lindahl, T. (1979). Release of 7-methylguanine residues whose imidazole rings have been opened from damaged DNA by a DNA glycosylase from *E. coli. Nucleic Acids Res.* **6**, 3673–3683.

Chlebowicz, E., and Jachymczyk, W. J. (1979). Repair of MMS-induced DNA double-strand breaks in haploid cells of *Saccharomyces cerevisiae,* which requires the presence of a duplicate genome. *Mol. Gen. Genet.* **167**, 279–286.

Chovnick, A., Ballantyne, G. H., Baillie, D. L., and Holm, D. G. (1970). Gene conversion in higher organisms: Half-tetrad analysis of recombination within the rosy cistron of *Drosophila melanogaster. Genetics* **66**, 315–329.

Cimino, M. C. (1972). Egg-production, polyploidization and evolution in a diploid all-female fish of the genus *Poeciliopsis. Evolution* **26**, 294–306.

Cleaver, J. E. (1969). DNA repair in Chinese hamster cells of different sensitivities to ultraviolet light. *Int. J. Radiat. Biol.* **16**, 277–285.

Cole, C. J. (1984). Unisexual lizards. *Sci. Am.* **250**, 94–100.

Cole, R. S. (1971). Inactivation of *Escherichia coli*, F′ episomes at transfer and bacteriophage lambda by psoralen plus 360-nm light: Significance of deoxyribonucleic acid crosslinks. *J. Bacteriol.* **107**, 846–852.

Cole, R. S., Levitan, D., and Sinden, R. R. (1976). Removal of psoralen interstrand crosslinks from DNA of *Escherichia coli:* Mechanism and genetic control. *J. Mol. Biol.* **103**, 39–59.

Cole, R. S., Sinden, R. R., Yoakum, G. H., and Broyles, S. (1978). On the mechanism for repair of crosslinked DNA in *E. coli* treated with psoralen and light. *In* "DNA Repair Mechanisms, ICN–UCLA Symposia on Molecular and Cellular Biology" (P. C. Hanawalt, E. D. Friedberg, and F. C. Fox, eds.), pp. 287–290. Academic Press, New York.

Crow, J. F., and Kimura, M. (1970). "An Introduction to Population Genetics Theory." Harper, New York.

Cuellar, O. (1971). Reproduction and the mechanism of meiotic restitution in the parthenogenetic lizard *Cnemidophorus uniparens. J. Morphol.* **133**, 139–165.

Darnell, J. E., and Doolittle, W. F. (1986). Speculation on the early course of evolution. *Proc. Natl. Acad. Sci. U.S.A.* **83**, 1271–1275.

Dawkins, R. (1982). "The Extended Phenotype." Freeman, San Francisco.

DeFlora, S., Zanacchi, P., Camoirano, A., Bennicelli, C., and Badolati, B. S. (1984). Genotoxic activity and potency of 135 compounds in the Ames reversion test and in a bacterial DNA repair test. *Mutat. Res.* **133**, 161–198.

Devoret, R., Blanco, M., George, J., and Radman, M. (1975). Recovery of phage λ from ultraviolet damage. *In* "Molecular Mechanisms for Repair of DNA, Part A" (P. C. Hanawalt and R. B. Setlow, eds.), pp. 155–171. Plenum, New York.

de Vos, W. M., de Vries, S., and Venema, G. (1983). Cloning and expression of the *E. coli recA* gene in *Bacillus subtilis. Gene* **25**, 301–308.

Dodson, L. A., and Hadden, C. T. (1980). Capacity for postreplication repair correlated with transducibility in *rec*⁻ mutants of *Bacillus subtilis. J. Bacteriol.* **144**, 608–615.

Dougherty, E. C. (1955). Comparative evolution and the origin of sexuality. *Syst. Zool.* **4,** 145–190.
Drake, J. W. (1974). The role of mutation in bacterial evolution. *Symp. Soc. Gen. Microbiol.* **24,** 41–58.
Eigen, M. (1971). Self-organization of matter and the evolution of biological macromolecules. *Naturwissenschaften* **58,** 465–523.
Eigen, M., and Schuster, P. (1979). "The Hypercycle, a Principle of Natural Self-Organization." Springer-Verlag, Berlin.
Eigen, M., Gardiner, W., Schuster, P., and Oswatitsch, P. (1981). The origin of genetic information. *Sci. Am.* **244,** 88–118.
Eitner, G., Adler, B., Lanzov, V. A., and Hofemeister, J. (1982). Interspecies *recA* protein substitution in *Escherichia coli* and *Proteus mirabilis*. *Mol. Gen. Genet.* **185,** 481–486.
Epstein, R. H. (1958). A study of multiplicity reactivation in bacteriophage T4-K12 (λ) complexes. *Virology* **6,** 382–404.
Eshel, I., and Feldman, M. W. (1970). On the evolutionary effect of recombination. *Theor. Pop. Biol.* **1,** 88–100.
Feinstein, S. I., and Low, K. B. (1986). Hyper-recombining recipient strains in bacterial conjugation. *Genetics* **113,** 13–33.
Feldman, M. W., and Balkau, B. (1972). Some results on the theory of three gene loci. *In* "Population Dynamics" (T. N. E. Greville, ed.), pp. 357–383. Academic Press, New York.
Feldman, M. W., and Balkau, B. (1973). Selection for linkage modification. II. A recombination balance for neutral modifiers. *Genetics* **74,** 713–726.
Feldman, M. W., Christiansen, F. B., and Brooks, L. D. (1980). Evolution of recombination in a constant environment. *Proc. Natl. Acad. Sci. U.S.A.* **77,** 4838–4841.
Felsenstein, J. (1974). The evolutionary advantage of recombination. *Genetics* **78,** 737–756.
Fincham, J. R. S., Day, P. R., and Radford, A. (1979). "Fungal Genetics." Blackwell, Oxford.
Fisher, R. A. (1930). "The Genetical Theory of Natural Selection." Oxford Univ. Press, New York.
Fisher, R. A. (1958). "The Genetical Theory of Natural Selection." Dover, New York.
Flory, S. S., Tsang, J., Muniyappa, K., Bianchi, M., Gonda, D., Kahn, R., Azhderian, E., Egner, C., Shaner, S., and Radding, C. M. (1984). Intermediates in homologous pairing promoted by *recA* protein and correlations of recombination *in vitro* and *in vivo*. *Cold Spring Harbor Symp. Quant. Biol.* **49,** 513–523.
Fry, S. E. (1979). Stimulation of recombination in phage T4 by nitrous acid-induced lesions. *J. Gen. Virol.* **43,** 719–722.
Gensler, H. L., and Bernstein, H. (1981). DNA damage as the primary cause of aging. *Q. Rev. Biol.* **56,** 279–303.
Gentner, N. E., Werner, M. M., Hannan, M. A., and Nasim, A. (1978). Contribution of a caffeine-sensitive recombinational repair pathway to survival and mutagenesis in UV-irradiated *Schizosaccharomyces pombe*. *Mol. Gen. Genet.* **167,** 43–49.
Gillen, F. D., and Nossal, N. G. (1976). Control of mutation frequency by bacteriophage T4 DNA polymerase I. The tsCB120 antimutator DNA polymerase is defective in strand displacement. *J. Biol. Chem.* **251,** 5219–5224.
Goldberg, I., and McKalanos, J. J. (1986). Cloning of the *Vibrio cholerae recA* gene and construction of a *Vibrio cholerae recA* mutant. *J. Bacteriol.* **165,** 715–722.
Griffith, J., and Formosa, T. (1985). The *uvsX* protein of bacteriophage T4 arranges

single-stranded and double-stranded DNA into similar helical nucleoprotein filaments. *J. Biol. Chem.* **260**, 4484–4491.

Hackett, P. B., and Sauerbier, W. (1975). The transcriptional organization of the ribosomal RNA genes in mouse L cells. *J. Mol. Biol.* **91**, 235–256.

Haldane, J. B. S. (1937). The effect of variation on fitness. *Am. Nat.* **71**, 337–349.

Hall, J. D., and Scherer, K. (1981). Repair of psoralen-treated DNA by genetic recombination in human cells infected with herpes simplex virus. *Cancer Res.* **441**, 5033–5038.

Hall, J. D., Featherston, J. D., and Almy, R. E. (1980). Evidence for repair of ultraviolet light-damaged herpes virus in human fibroblasts by a recombination mechanism. *Virology* **105**, 490–500.

Hamilton, W. D. (1967). Extraordinary sex ratios. *Science* **156**, 477–488.

Harm, W. (1958). Multiplicity reactivation, marker rescue and genetic recombination in phage T4 following X-ray inactivation. *Virology* **5**, 337–361.

Harman, D. (1981). The aging process. *Proc. Natl. Acad. Sci. U.S.A.* **78**, 7124–7128.

Hartman, P. E. (1980). Bacterial mutagenesis: Review of new insights. *Environ. Mutagen.* **2**, 3–16.

Hickey, D. A. (1982). Selfish DNA: A sexually transmitted nuclear parasite. *Genetics* **101**, 519–531.

Ho, K. S. Y. (1975). Induction of DNA double-strand breaks by X-rays in a radiosensitive strain of the yeast *Saccharomyces cerevisiae*. *Mutat. Res.* **30**, 327–334.

Holliday, R., Halliwell, R. E., Evans, M. W., and Rowell, V. (1976). Genetic characterization of *rec-1*, a mutant of *Ustilago maydis* defective in repair and recombination. *Genet. Res.* **27**, 413–453.

Hollstein, M., and McCann, J. (1979). Short-term tests for carcinogens and mutagens. *Mutat. Res.* **65**, 133–226.

Hollstein, M. C., Brooks, P., Linn, S., and Ames, B. N. (1984). Hydroxymethyluracil DNA glycosylase in mammalian cells. *Proc. Natl. Acad. Sci. U.S.A.* **81**, 4003–4007.

Holmes, G. E., and Holmes, N. R. (1986). Accumulation of DNA damages in aging *Paramecium tetraurelia*. *Mol. Gen. Genet.* **204**, 108–114.

Holmes, G. E., Schneider, S., Bernstein, C., and Benstein, H. (1980). Recombinational repair of mitomycin C lesions in phage T4. *Virology* **103**, 299–310.

Holsinger, K. E., and Feldman, M. W. (1983). Linkage modification with mixed random mating and selfing: A numerical study. *Genetics* **103**, 323–333.

Hopf, F. A., and Hopf, F. W. (1985). The role of the Allee effect on species packing. *Theor. Pop. Biol.* **27**, 27–50.

Hopf, F. A., Michod, R. E., and Sanderson, M. (1987). Mutation load under different reproductive systems. In preparation.

Hopfield, J. J. (1974). Kinetic proofreading: A new mechanism for reducing errors in biosynthetic processes requiring high specificity. *Proc. Natl. Acad. Sci. U.S.A.* **71**, 4135–4139.

Howard-Flanders, P., Rupp, W. D., Wilkins, B., and Cole, R. S. (1968). DNA replication and recombination after ultraviolet-irradiation. *Cold Spring Harbor Symp. Quant. Biol.* **33**, 195–207.

Huskey, R. J. (1969). Multiplicity reactivation as a test for recombination function. *Science* **164**, 319–320.

Keener, S. L., McNamee, K. P., and McEntee, K. (1984). Cloning and characterization of *recA* genes from *Proteus vulgaris*, *Erwinia carotovora*, *Shigella flexneri*, and *Escherichia coli B/r*. *J. Bacteriol.* **160**, 153–160.

Kellenberger, G., and Weigle, J. (1958). Etude au moyen des rayons ultraviolets de l'interaction entre bacteriophage tempère et bacterie hôte. *Biochem. Biophys. Acta* **30**, 112–124.

Kemp, L. M., Sedgwick, S. G., and Jeppo, P. A. (1984). X-Ray sensitive mutants of Chinese hamster ovary cells defective in double-strand break rejoining. *Mut. Res. DNA Repair Rep.* **132**, 186–196.

Kerr, W. E. (1976). Population genetic studies in bees. 2. Sex-limited genes. *Evolution* **30**, 94–99.

Kimura, M. (1956). A model of a genetic system which leads to closer linkage by natural selection. *Evolution* **10**, 278–287.

King, R. C., and Stansfield, W. D. (1984). "A Dictionary of Genetics," 3rd Ed. Oxford Univ. Press, New York.

Klekowski, E. J. (1973). Genetic load in *Osmunda regalis* populations. *Am. J. Bot.* **60**, 146–154.

Kmiec, E. B., and Holloman, W. K. (1982). Homologous pairing of DNA molecules promoted by a protein from *Ustilago*. *Cell* **29**, 367–374.

Kmiec, E. B., and Holloman, W. K. (1984). Synapsis promoted by *Ustilago rec1* protein. *Cell* **36**, 593–598.

Kolowsky, K. S., and Szalay, A. A. (1986). Double-stranded gap repair in the photosynthetic procaryote *Synechococcus* R2. *Proc. Natl. Acad. Sci. U.S.A.* **83**, 5578–5582.

Kondrashov, A. S. (1984). Deleterious mutations as an evolutionary factor. I. The advantage of recombination. *Genet. Res.* **44**, 199–217.

Konrad, E. B. (1977). Method for isolation of *Escherichia coli* mutants with enhanced recombination between chromosomal duplications. *J. Bacteriol.* **130**, 167–172.

Kornberg, A. (1980). "DNA Replication." Freeman, San Francisco.

Krasin, F., and Hutchinson, F. (1977). Repair of DNA double-strand breaks in *Escherichia coli* which requires *recA* function and the presence of a duplicate genome. *J. Mol. Biol.* **116**, 81–89.

Krasin, F., and Hutchinson, F. (1981). Repair of DNA double-strand breaks in *Escherichia coli* cells requires synthesis of proteins that can be induced by UV light. *Proc. Natl. Acad. Sci. U.S.A.* **78**, 3450–3453.

Kuhn, H. (1972). Self-organization of molecular systems and evolution of the genetic apparatus. *Angew. Chem. Int. Ed.* **11**, 798–820.

Kunz, B. A., and Haynes, R. H. (1981). Phenomenology and genetic control of mitotic recombination in yeast. *Annu. Rev. Genet.* **15**, 57–89.

Lamb, R. A., and Choppin, P. W. (1983). The gene structure and replication of influenza virus. *Annu. Rev. Biochem.* **52**, 467–506.

Lande, R., and Schemske, D. W. (1985). The evolution of self-fertilization and inbreeding depression in plants. I. Genetic models. *Evolution* **39**, 24–40.

Leffler, S., Pulkrabak, P., Grunberger, D., and Weinstein, I. B. (1977). Template activity of calf thymus DNA modified by a dihydrodiol epoxide derivative of benzo(a)pyrene. *Biochemistry* **16**, 3133–3136.

Leslie, J. F., and Raju, N. B. (1985). Recessive mutations from natural populations of *Neurospora crassa* that are expressed in the sexual diplophase. *Genetics* **111**, 757–777.

Lewin, B. (1985). "Genes II." Wiley, New York.

Loeb, L. A., Weymouth, L. A., Kunkel, T. A., Gopinathan, K. P., Beckman, R. A., and Dube, D. K. (1978). On the fidelity of DNA replication. *Cold Spring Harbor Symp. Quant. Biol.* **43**, 921–927.

Lovett, C. M., Jr., and Roberts, J. (1985). Purification of a *recA* protein analogue from *Bacillus subtilis*. *J. Biol. Chem.* **260**, 3305–3313.

Luria, S. E. (1947). Reactivation of irradiated bacteriophage by transfer of self-reproducing units. *Proc. Natl. Acad. Sci. U.S.A.* **33**, 253–264.

Luria, S. E., and Dulbecco, R. (1949). Genetic recombinations leading to production of active bacteriophage from ultraviolet inactivated bacteriophage particles. *Genetics* **34**, 93–125.

Lynch, M. (1984). Destabilizing hybridization, general purpose genotypes and geographic parthenogenesis. *Q. Rev. Biol.* **59**, 257–290.

McClain, M. E., and Spendlove, R. S. (1966). Multiplicity reactivation of reovirus particles after exposure to ultraviolet light. *J. Bacteriol.* **92**, 1422–1429.

Magana-Schwencke, N., Henriques, J.-A. P., Chanet, R., and Moustacchi, E. (1982). The fate of 8-methoxypsoralen photoinduced crosslinks in nuclear and mitochondrial yeast DNA: Comparison of wild-type and repair deficient strains. *Proc. Natl. Acad. Sci. U.S.A.* **79**, 1722–1726.

Margison, G. P., and Pegg, A. E. (1981). Enzymatic release of 7-methylguanine from methylated DNA by rodent liver extracts. *Proc. Natl. Acad. Sci. U.S.A.* **78**, 861–865.

Martensen, D. V., and Green, M. M. (1976). UV-Induced mitotic recombination in somatic cells of *Drosophila melanogaster. Mutat. Res.* **36**, 391–396.

Martin, R. (1977). A possible genetic mechanism of aging, rejuvenation, and recombination in germinal cells. *ICN–UCLA Symp. Mol. Cell. Biol.* **7**, 355–373.

Massie, H. R., Samis, H. V., and Baird, M. B. (1972). The kinetics of degradation of DNA and RNA by H_2O_2. *Biochim. Biophys. Acta* **272**, 539–548.

Maynard Smith, J. (1958). "The Theory of Evolution." Penguin, Harmondsworth.

Maynard Smith, J. (1978). "The Evolution of Sex." Cambridge Univ. Press, London.

Medvedev, Z. A. (1981). On the immortality of the germ line: Genetic and biochemical mechanisms. A review. *Mech. Aging Dev.* **17**, 331–359.

Meselson, M. S., and Radding, C. M. (1975). A general model for genetic recombination. *Proc. Natl. Acad. Sci. U.S.A.* **72**, 358–361.

Michod, R. E. (1983). Population biology of the first replicators. *Am. Zool.* **23**, 5–14.

Miskimins, R., Schneider, S., Johns, V., and Bernstein, H. (1982). Topoisomerase involvement in multiplicity reactivation of phage T4. *Genetics* **101**, 157–177.

Mukai, T., Chigusa, S. I., Mettler, L. E., and Crow, J. F. (1972). Mutation rate and dominance of genes affecting viability in *Drosophila melanogaster. Genetics* **72**, 335–355.

Muller, H. J. (1932). Some genetic aspects of sex. *Am. Nat.* **66**, 118–138.

Nei, M. (1967). Modification of linkage intensity by natural selection. *Genetics* **57**, 625–641.

Nei, M. (1969). Linkage modification and sex difference in recombination. *Genetics* **63**, 681–699.

Nocentini, S. (1976). Inhibition and recovery of ribosomal RNA synthesis in ultraviolet irradiated mammalian cells. *Biochim. Biophys. Acta* **454**, 114–128.

Nonn, E., and Bernstein, C. (1977). Multiplicity reactivation and repair of nitrous acid-induced lesions in bacteriophage T4. *J. Mol. Biol.* **116**, 31–47.

Orr-Weaver, T. L., and Szostak, J. W. (1985). Fungal recombination. *Microbiol. Rev.* **49**, 33–58.

Orr-Weaver, T., Szostak, J., and Rothstein, R. (1981). Yeast transformation: A model system for the study of recombination. *Proc. Natl. Acad. Sci. U.S.A.* **78**, 6354–6358.

Pawl, G., Taylor, R., Minton, K., and Friedberg, E. C. (1976). Enzymes involved in thymine dimer excision repair in bacteriophage T4-infected *Escherichia coli. J. Mol. Biol.* **108**, 99–109.

Peacock, W. J. (1970). Replication, recombination and chiasmata in *Goniaea australasiae* (Orthoptera: Acrididae). *Genetics* **65,** 593–617.
Phipps, J., Nasim, A., and Miller, D. R. (1985). Recovery, repair and mutagenesis in *Schizosaccharomyces pombe. Adv. Genet.* **23,** 1–72.
Pierre, A., and Paoletti, C. (1983). Purification and characterization of *recA protein* from *Salmonella typhimurium. J. Biol. Chem.* **258,** 2870–2874.
Piette, J., Calberg-Bacq, C. M., and Van deVorst, A. (1978). Photodynamic effect of proflavine on ϕX174 bacteriophage, its DNA replicative form and its isolated single-stranded DNA: Inactivation, mutagenesis and repair. *Mol. Gen. Genet.* **167,** 95–103.
Prudhodmeau, C., and Proust, J. (1973). UV irradiation of polar cells of *Drosophila melanogaster* embryos. V. A study of the meiotic recombination in females with chromosomes of a different structure. *Mutat. Res.* **23,** 63–66.
Radman, R., and Wagner, R. (1984). Effects of DNA methylation on mismatch repair, mutagenesis, and recombination in *Escherichia coli. Curr. Top. Microbiol. Immunol.* **108,** 23–28.
Resnick, M. A. (1976). The repair of double-strand breaks in DNA: A model involving recombination. *J. Theor. Biol.* **59,** 97–106.
Resnick, M. A., and Martin, P. (1976). The repair of double-strand breaks in the nuclear DNA of *Saccharomyces cerevisiae* and its genetic control. *Mol. Gen. Genet.* **143,** 119–129.
Resnick, M. A., and Moore, P. D. (1979). Molecular recombination and the repair of DNA double-strand breaks in CHO cells. *Nucleic Acids Res.* **6,** 3145–3160.
Resnick, M. A., Chow, T., Nitiss, J., and Game, J. (1984). Changes in the chromosomal DNA of yeast during meiosis in repair mutants and the possible role of deoxyribonuclease. *Cold Spring Harbor Symp. Quant. Biol.* **49,** 639–649.
Rose, M. R., and Redfield, R. J. (1986). Is sex an adaptation? *Evol. Theor.* (in press).
Rupp, W. D., and Howard-Flanders, P. (1968). Discontinuities in the DNA synthesized in an excision-defective strain of *Escherichia coli* following ultraviolet irradiation. *J. Mol. Biol.* **31,** 291–304.
Rupp, W. D., Wilde, C. E., Reno, D. L., and Howard-Flanders, P. (1971). Exchanges between DNA strands in ultraviolet-irradiated *Escherichia coli. J. Mol. Biol.* **61,** 25–44.
Schewe, M. J., Suzuki, D. T., and Erasmus, U. (1971). The genetic effects of mitomycin C in *Drosophila melanogaster.* II. Induced meiotic recombination. *Mutat. Res.* **12,** 269–279.
Schultz, R. J. (1961). Reproductive mechanism of unisexual and bisexual strains of the viviparous fish *Poeciliopsis. Evolution* **15,** 302–325.
Schultz, R. J. (1966). Hybridization experiments with an all-female fish of the genus *Poeciliopsis. Biol. Bull.* **130,** 415–429.
Schultz, R. J. (1969). Hybridization, unisexuality and polyploidy in the teleost *Poeciliopsis* (Poeciliidae) and other vertebrates. *Am. Nat.* **103,** 605–619.
Shields, W. M. (1982). "Philopatry, Inbreeding and the Evolution of Sex." State Univ. of New York Press, Albany.
Simmon, V. F. (1979). *In vitro* assays for recombinogenic activity of chemical carcinogens and related compounds with *Saccharomyces cerevisiae* D3. *J. Natl. Cancer Inst.* **62,** 901–909.
Smith, K. C., and Meun, D. H. C. (1970). Repair of radiation-induced damage in *Escherichia coli.* I. Effect of rec mutations on post-replication repair of damage due to ultraviolet radiation. *J. Mol. Biol.* **51,** 459–472.

Smith-Sonneborn, J. (1979). DNA repair and longevity assurance in *Paramecium tetraurelia. Science* **203,** 1115–1117.

Stahl, F. W. (1979). "Genetic Recombination: Thinking about It in Phage and Fungi." Freeman, San Francisco.

Staub, R. W. (1984). The influence of B chromosomes on the susceptibility of maize to gamma irradiation induced DNA damage. Ph.D. thesis, University of Arizona.

Strobeck, C., Maynard Smith, J., and Charlesworth, B. (1976). The effects of hitchhiking on a gene for recombination. *Genetics* **82,** 547–558.

Sturtevant, A. H. (1915). No crossing over in the female of the silkworm moth. *Am. Nat.* **49,** 42–44.

Symonds, N., and Ritchie, D. A. (1961). Multiplicity reactivation after decay of incorporated radioactive phosphorus in phage T4. *J. Mol. Biol.* **3,** 61–70.

Szostak, J. W., Orr-Weaver, T. L., Rothstein, R. J., and Stahl, F. W. (1983). The double-strand break repair model for recombination. *Cell* **33,** 25–35.

Tazima, Y. (1964). "The Genetics of the Silkworm." Prentice Hall, Englewood Cliffs, N. J.

Teague, R. (1976). A result on the selection of recombination altering mechanisms. *J. Theor. Biol.* **59,** 25–32.

Tessman, I., and Ozaki, T. (1957). Multiplicity reactivation of bacteriophage T1. *Virology* **4,** 315–327.

Turner, J. R. G. (1967). On supergenes. I. The evolution of supergenes. *Am. Nat.* **101,** 195–221.

Uyenoyama, M. K. (1984). On the evolution of parthenogenesis: A genetics representation of the "cost of meiosis." *Evolution* **38,** 87–102.

Uyenoyama, M. K. (1985). On the evolution of parthenogenesis. II. Inbreeding and the cost of meiosis. *Evolution* **39,** 1194–1206.

Walker, G. C. (1984). Mutagenesis and inducible responses to deoxyribonucleic acid damage in *Escherichia coli. Microbiol. Rev.* **48,** 60–93.

Walker, I. (1978). The evolution of sexual reproduction as a repair mechanism. Part I. A model for self-repair and its biological implications. *Acta Biotheor.* **27,** 133–158.

Weisman, A. (1889). "Essays Upon Heredity and Kindred Biological Problems," pp. 268–277. Clarendon, Oxford.

West, S. C., Cassuto, E., and Howard-Flanders, P. (1982). Post-replication repair in *E. coli:* Strand exchange reactions of gapped DNA by *recA* protein. *Mol. Gen. Genet.* **187,** 209–217.

West, S. C., Countryman, J. K., and Howard-Flanders, P. (1983). Purification and properties of the *recA* proteins from *Proteus mirabilis. J. Biol. Chem.* **258,** 4648–4654.

Westerman, M. (1967). The effect of X-irradiation on male meiosis in *Schistocerca gregaria* (Forskal). I. Chiasma frequency response. *Chromosoma* **22,** 401–416.

White, M. J. D. (1945). "Animal Cytology and Evolution." Cambridge Univ. Press, London.

White, M. J. D. (1973). "Animal Cytology and Evolution," 3rd Ed. Cambridge Univ. Press, London.

Whitehouse, H. L. K. (1982). "Genetic Recombination." Wiley, New York.

Williams, G. C. (1975). "Sex and Evolution." Princeton Univ. Press, Princeton, N.J.

Williams, G. C. (1980). Kin selection and the paradox of sexuality. *In* "Sociobiology: Beyond Nature/Nurture? Reports, Definitions and Debate" (G. W. Barlow and J. Silverman, eds.), pp. 371–384. Westview Press, Boulder, Colorado.

Woese, C. R. (1983). The primary lines of descent and the universal ancestor. *In* "Evolution from Molecules to Men" (D. S. Bendall, ed.), pp. 209–233. Cambridge Univ. Press, London.

Yamamoto, H., and H. Shimojo. (1971). Multiplicity reactivation of human adenovirus type 12 and simian virus 40 irradiated by ultraviolet light. *Virology* **45**, 529–531.

Yarosh, D. B., Johns, V., Mufti, S., Bernstein, C., and Bernstein, H. (1980). Inhibition of UV and psoralen-plus-light mutagenesis in phage T4 by gene 43 antimutator polymerase alleles. *Photochem. Photobiol.* **31**, 341–350.

Zieve, F. J. (1973). Effects of the carcinogen *N*-acetoxy-2-fluorenylacetamide on the template properties of deoxyribonucleic acid. *Mol. Pharmacol.* **9**, 658–699.

GENE DOSAGE COMPENSATION IN *Drosophila melanogaster*

John C. Lucchesi* and Jerry E. Manning†

*Department of Biology, The University of North Carolina at Chapel Hill, Chapel Hill, North Carolina 27514

†Department of Molecular Biology and Biochemistry, University of California, Irvine, Irvine, California 92717

The comparative lack of sexual dimorphism of *apricot* and other [sex-linked] genes studied is due to "sex-limitation", i.e., to a compensatory influence of the dosage difference between the sexes in respect to other genes in the X-chromosome. The facts are of particular interest from an evolutionary standpoint [Muller *et al.*, 1931, *Anat. Rec. Suppl.* **51**, p. 110].

I. Introduction

A. The Seminal Observation

This article is concerned with the phenotypes produced by sex-linked genes in males and females. In many sexually reproducing organisms, these phenotypes are alike even though the heterogametic sex (with X and Y) has only one dose of X-linked genes and the homogametic sex (XX) has two doses of such genes. This was first noted by Herman J. Muller while studying several mutant genes of the fruit fly *Drosophila melanogaster*, and was reported by him in the abstract of a paper presented at a meeting of the American Society of Zoologists (Muller *et al.*, 1931). A year later, in the proceedings of the Sixth International Congress of Genetics, Muller discussed the phenotypic effects of *white apricot* (w^a), a mutant allele of an eye-pigment gene located on the X chromosome. The mutant in question allowed the deposition of less pigment than is normally present in wild type. Surprisingly, homozygous w^a females, with two doses of the allele, had the same level of pigmentation as males with a single dose. Yet, clearly, the mutant was able to show a phenotypic response to dosage changes: a female with three doses, i.e., carrying a small duplication with an allele of w^a in addition to the two w^a-bearing X chromosomes, had a darker eye phenotype than a female with two doses. A female with a single dose of w^a, i.e., with one X carrying the mutant allele and the other bearing a small deficiency encompassing the locus of the gene, had an eye color roughly half as dark as the two-dose female. Similarly, males with two doses of the mutant allele had significantly more pigment than males with a single dose (see Fig. 1). It appeared, therefore, that although the amount of pigment varied with dosage within each sex, the one-dose male was phenotypically identical to the two-dose female [the phenotypic equivalence of w^a males and females was eventually quantified by Smith and Lucchesi (1969), with spectrophotometric measurements of extracted eye pigments]. Muller concluded that there is a mechanism compensating for the difference in dosage of X-linked genes which normally occurs between the sexes.

Muller also suggested that, although dosage compensation was most easily observed with hypomorphic or *leaky* mutant alleles of X-linked genes, it must surely operate on the wild-type alleles of these genes.

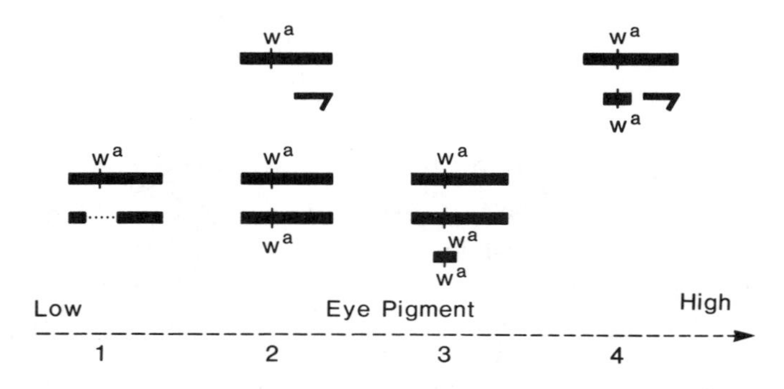

FIG. 1. Diagrammatic representation of H. J. Muller's observations on the level of pigmentation achieved by the X-linked eyecolor mutant allele w^a. Normal males and females have equivalent levels of pigmentation; a single dose of w^a in a female leads to half of the amount of pigment found in a normal male, while two doses of the allele in a male lead to twice the product found in a normal female. Note the gene dosage dependence of product within each sex. This phenotypic seriation would apply equally well to measurements of activity for enzymes, whose structural genes are present on the X chromosome, by replacing "pigment levels" with "specific activity" or "activity per unit of live weight."

This contention, beautifully developed in his Harvey Lecture (Muller, 1950), was later proved to be true.

B. Compensation Operates on Wild-Type Alleles

In *Drosophila,* the structural genes for the enzymes glucose-6-phosphate dehydrogenase (G6PD) and 6-phosphogluconate dehydrogenase (6PGD) are located on the X chromosome (Young, 1966; Young *et al.,* 1964). Komma (1966) and Seecof and Kaplan (1967) reported that males and females have the same activities for 6PGD and for G6PD and that their respective genes (Pgd^+ and Zw^+) are, therefore, dosage compensated. Within a sex, each gene exhibits the dosage response characteristic of w^a (Seecof *et al.,* 1969; Gvosdev *et al.,* 1970; Bowman and Simmons, 1973). Females carrying two normal X chromosomes and a small duplication (three-dose females) have significantly more activity than females with two X chromosomes (two-dose females). Females with a normal X and an X bearing a small deletion (one-dose females) have reduced levels of activity. Males bearing an extra dose of the gene exhibit a marked increase in enzyme activity. Excerpts of the data published by Seecof *et al.* (1969) are presented in Table 1.

Results similar to those just discussed have been obtained by Baillie and Chovnick (1971) and by Tobler *et al.* (1971) with respect to *vermillion* (v^+), the structural gene for tryptophan oxidase, and by Tobler and Grell (1978) with respect to Had^+, the structural gene for β-hydroxy acid dehydrogenase. Finally, Whitney and Lucchesi (1972) reported the equivalence in males and females of fumarase activity;

TABLE 1
Levels of 6PGD Activity for Various Doses of the Structural Gene[a]

Genotype[b]	Description	Number of *Pgd* genes	Enzyme activity[c]
$Pgd^+/Df(Pgd^+)$	Deficiency female	1	2.81 (0.04)
Pgd^+/Pgd^+	Normal female	2	4.66 (0.03)
Pgd^+/Y	Normal male	1	4.65 (0.01)
$Pgd^+/Dp(Pgd^+)/Y$	Duplication male	2	6.26 (0.26)

[a] After Seecof *et al.* (1969).

[b] $Df(Pgd^+)$ denotes an X chromosome with a small deficiency encompassing the locus of Pgd^+. $Dp(Pgd^+)$ is a small fragment of the X chromosome inserted into the centric heterochromatin region of chromosome 3.

[c] Expressed as moles $\times 10^2$ of NADP reduced/milliter/minute/milligram live weight; standard error is given in parentheses.

dosage response was documented for this enzyme's X-linked structural gene (Fuh^+) by Pipkin *et al.* (1977).

Another example of dosage compensation of wild-type alleles is provided by the X-linked gene encoding one of the polypeptides of the larval salivary gland secretion. There are from four to seven polypeptides in the secretion, depending on the technique employed to separate them (Korge, 1975; Beckendorf and Kafatos, 1976). Using naturally occurring electrophoretic variants, Korge was able to assign one of these polypeptides to an X-linked gene (*Sgs-4*), and to document that it is dosage compensated (Table 2).

All of the results just described consistently show equalization of X-linked gene products among males and females. Dosage responses (i.e., an increase in activity mediated by a duplication of the gene under study or a decrease in activity mediated by a corresponding deficiency), although qualitatively consistent, are not always of the expected magnitude. It is not always the case, for example, that a duplication in a male yields an increase in gene product of 100% and that a duplication in a female yields an increase of 50% (see Tables 1 and 2). This failure to achieve fully the expected values can be ascribed to position effects on the duplicated genes or to differential developmental effects caused by the presence of the duplications or deficiencies in the genome.

C. Certain Alleles of X-Linked Genes Do Not Exhibit Dosage Compensation

The wild-type alleles of the majority of X-linked genes studied to date are dosage compensated. Most hypomorphic or *leaky* alleles are also compensated. Yet, there exist specific mutant alleles of some of these genes which lead to a different phenotype in males and females. Furthermore, as we shall see below, there are on the X chromosome a few structural genes whose wild-type alleles are not compensated.

The first study of an X-linked allele exhibiting a different phenotype in males and females was performed by Muller (1932). Females homozygous for another allele of the *white* gene, called *white eosin* (w^e), have an eye color considerably darker than males. In fact, such females have twice as much pigment in their eyes as the hemizygous males (Smith and Lucchesi, 1969). Males with two doses of the mutant allele [$w^e/Dp(w^e)/Y$] are phenotypically similar to ordinary *eosin* females (w^e/w^e). Muller concluded that the sexual dimorphism of mutant males and females is due to the difference in dosage normally existing among the two sexes and not to a difference in the action of

TABLE 2
Dosage Compensation and Dosage Effects on the Synthesis of Sgs-4 Protein[a]

Genotype[b]	Description	Number of *Sgs-4* genes	Relative amount of protein 4[c]
Sgs-4$^+$/*Df*(*Sgs-4*$^+$)	Deficiency female	1	0.24 (0.02)
Sgs-4$^+$/*Sgs-4*$^+$	Normal female	2	0.44 (0.08)
Sgs-4$^+$/*Sgs-4*$^+$/*Dp*(*Sgs-4*$^+$)	Duplication female	3	0.60 (0.07)
Sgs-4$^+$/Y	Normal male	1	0.38 (0.04)
Sgs-4$^+$/*Dp*(*Sgs-4*$^+$)	Duplication male	2	0.62 (0.08)

[a] After Korge (1975).
[b] Symbols as in Table 1.
[c] Expressed as the ratio of polypeptide 4/polypeptide 3 (produced by the autosomal *Sgs-3* gene); standard error is given in parentheses.

the gene in males and females. An additional allele of *white* shown to lack compensation is *white cherry* (w^{ch}).

A temperature-sensitive allele of *vermillion* was recovered by Camfield (1974) and was shown to have the following characteristics: at the permissive temperature (22°C) males and females have slightly decreased but equivalent levels of tryptophan oxidase activity. At 29°C, enzyme activity is very substantially reduced and more so in males than in females. It would appear that at the restrictive temperature the regulation of this gene is affected so that it is no longer compensated.

Korge (1981) reported that two wild-type strains, Samarkand and Karsnäs, which have reduced levels of the X-linked salivary gland secretion polypeptide 4, also lack dosage compensation of the *Sgs-4* gene (see Section VI).

The first X-linked wild-type gene to show a lack of dosage compensation is the gene encoding the α-subunit of the larval serum protein 1, LSP1α (Roberts and Evans-Roberts, 1979; Brock and Roberts, 1982). A single dose of the gene, whether present in normal, hemizygous males or in females heterozygous for a deficiency encompassing the gene, leads to 50% of the quantity of α-chain polypeptide found in normal (two-dose) females.

Additional X-linked genes which might be expected to lack dosage compensation are those with sex-limited expression. For example, the three genes (*Yp1-3*) encoding the vitellogenins, or yolk protein precursors, are active only in the fat body and the ovaries of females. In order to determine if they would be dosage compensated, were they expressed in males, K. Oishi and his collaborators (Ota *et al.*, 1981) forced the activation of these genes in chromosomally male individuals. These workers used the mutant *double-sex* (*dsx*) to transform X/Y and XX individuals into morphologically indistinguishable intersexes. The amount of YP1-3 present in the hemolymph of X/X;*dsx*/*dsx* was approximately double the amount found in X/Y;*dsx*/*dsx* flies, confirming the occurrence, as expected, of dosage dependence or lack of compensation for a sex-limited gene.

D. Both X Chromosomes Are Active in the *Drosophila* Female Soma

It is well known that in mammals dosage compensation is achieved by the inactivation of one of the two X chromosomes in the somatic cells of female embryos, followed by the clonal inheritance of this event throughout tissue growth (for a review, see Gartler and Riggs, 1983). This is not the case in female *Drosophila,* for which three lines of evidence indicate that both X chromosomes are simultaneously active in somatic cells. The first line of evidence is the absence of phenotypic mosaicism in females heterozygous for X-linked mutant genes affecting cuticular structures. For example, females of the genotype *f*/+ (*f* is a recessive allele causing forked bristles) or *y*/+ (*y* is a recessive allele causing yellow hairs, bristles, and body color) are uniformly wild type. Yet these mutant phenotypes are highly autonomous, as seen by the occasional presence of single mutant bristles surrounded by wild-type tissue in mosaics generated by somatic recombination. If the X chromosome bearing the wild-type allele were inactivated in some cells, at some time during development, patches of mutant tissue would ensue, leading to a mosaic phenotype.

Further support for the absence of X-chromosome inactivation in *Drosophila* females was provided by Kazazian *et al.* (1965) using allozymes of 6PGD (Young, 1966). Most strains have either a fast (6PGD A) or a slow (6PGD B) electrophoretic variant of the enzyme. A cross between individuals of these two strains produces females with both parental allozymes and an additional enzyme of intermediate mobility. The simplest explanation is that the enzyme is a dimer and the parental slow and fast forms are the two homodimers, while the intermediate form is a heterodimer made up of one fast and one slow subunit. Kazazian and his co-workers were able to generate the intermediate form *in vitro* by mixing partially purified dissociated subunits of slow and fast enzymes. Mixtures of undissociated A and B enzymes yielded only those two forms, suggesting that in Pgd^{A}/Pgd^{B} heterozygotes, both alleles, each contributing a fast or slow subunit, are active in a given cell. These results are illustrated in Fig. 2.

Finally, evidence supporting the simultaneous function of the two

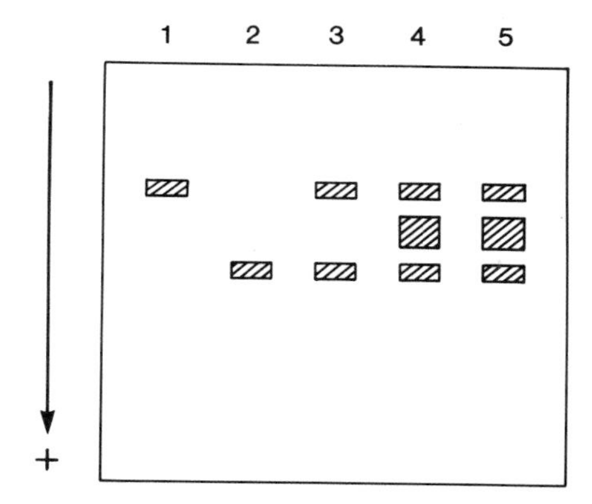

FIG. 2. Diagram of starch gel electrophoresis of 6PGD allozymes. Lanes 1 and 2 represent extracts of slow and fast variant-bearing strains, respectively. Lane 3 is a mixture of these two extracts. Lane 4 represents the enzyme activity zones in an extract from females heterozygous for the two variants. Lane 5 is a mixture of partially purified slow and fast variants treated with propanedithiol to cause the dissociation and reassociation of subunits. (After Kazazian *et al.*, 1965.)

X chromosomes in female somatic cells was provided by the cytological studies of Mukherjee and Beermann (1965) discussed in the following section.

II. Dosage Compensation Involves the Regulation of Transcription

A. EARLY CYTOLOGICAL OBSERVATIONS

The giant chromosomes found in several tissues of larvae and adults are produced by successive rounds of chromatid replication without the occurrence of mitosis (endoreplication), resulting in multistranded or polytenic units. The two members of a homologous pair are usually intimately associated throughout their length (this occurrence is known as somatic pairing). In the nuclei of males, the Y chromosome undergoes much less endoreplication than the rest of the genome and the single polytenic X chromosome remains unpaired. The width of this single X chromosome is comparable to the width of the paired X chromosomes in the nuclei of females or of paired autosomes in both sexes (Offermann, 1936).

In many interspecific hybrids, somatic pairing of the polytene chromosomes in the larval salivary gland nuclei does not take place. Making use of interspecific hybrids resulting from a cross between *Dro-*

sophila insularis × *Drosophila tropicalis*, Dobzhansky (1957) was able to compare an unpaired X chromosome in a female genome with the X chromosome in a male genome, confirming the sexual dimorphism reported by Offermann. Rudkin and his collaborators had already shown by Feulgen and ultraviolet microspectrophotometry that the X chromosome in the male had the same amount of DNA as each of the two X chromosomes in the female, in spite of the fact that its volume seemed to be equivalent to that of both (paired) X chromosomes (Aronson *et al.*, 1954). Dobzhansky was led to conclude that "the bloated appearance of the male X indicates accumulation in this chromosome of some products of gene activity. A male has only one X, but this single X works apparently twice as hard as does each of the two X's in the female." Rudkin later reported that the X chromosome in males contains approximately 10% more protein than either of the two X chromosomes in females (Rudkin, 1964).

B. Transcription Autoradiography Experiments

The validity of Dobzhansky's conclusion was established by Mukherjee and Beermann (1965) and Mukherjee (1966), using a technique developed by C. Pelling for the study of chromosomal RNA synthesis in chironomid midges. Salivary glands are briefly pulsed with [^{3}H]uridine by incubating them *in vitro* for a few minutes in a culture medium containing this nucleoside; chromosome squashes are prepared and the level of incorporation of the labeled uridine into RNA along the polytene chromosomes is monitored by means of autoradiography. Mukherjee and Beermann found that the distribution of silver grains over the two paired X chromosomes relative to a control region of paired autosomes in female cells was equal to the distribution over the single X chromosome relative to the control autosomal region in males. These investigators noted that the number of silver grains counted over regions of X chromosomes which were occasionally unpaired in female nuclei was substantially greater than half of the value obtained over the same regions when the X chromosomes were paired. Nevertheless, they emphasized that even if the grain counts over the X chromosome in males were corrected for its unpaired condition, these counts remained significantly greater than grain counts for either of the two female X chromosomes. Mukherjee and Beermann concluded that the transcriptional activity of a single X chromosome is significantly greater in males than in females.

Holmquist (1972) derived two separate methods to correct for the geometrical differences between a paired and an unpaired chromosome and calculated that the actual rate of RNA synthesis along the single male X chromosome was 43% greater than the rate along either of the two X chromosomes in a female. Holmquist noted that such a limited excess of RNA synthesis in the male is surprising in view of the equalization of phenotypic products among the sexes.

Maroni *et al.* (1974) analyzed the incompleteness of dosage compensation in their own data (Kaplan and Plaut, 1968; Kaplan, 1970; Maroni and Plaut, 1973a) and in those of others. Rather than expressing the activity of an X chromosome as the average of the ratio of silver grains on the X chromosome to silver grains on the autosomal control, they plotted the X-chromosome value against the autosomal value for each nucleus and used the slope of the best-fit straight line calculated by regression analysis as a measure of relative X-chromosome activity. When this calculation is performed on populations of nuclei obtained from several male or female glands, the regression coefficients are the same, indicating that the rate of chromosomal RNA synthesis is fully dosage compensated. This is true, as well, when the number of grains over the X chromosome and autosomal segments is averaged for all the nuclei in each gland and these averages are plotted and subjected to regression analysis. Surprisingly, if a regression line is calculated for the nuclei within each gland and the slopes of these individual regression lines are pooled to obtain an average slope, the value of this slope is substantially less in males than in females, suggesting dosage dependence rather than compensation. Although no biological explanation is available for these results, they indicate a difference among males and females in the variation of X-chromosome : autosomal grain counts among the nuclei of individual glands. This difference may be responsible for the incompleteness of compensation observed when the relative activity of the X chromosome is estimated by means of ratios.

Results similar to the ones just described for the whole X chromosome or for segments spanning several polytene bands have been obtained by various workers who compared the incorporation of [^{3}H]uridine over single bands or chromomeres or in single puffs in males and females. Korge (1970a,b) studied RNA synthesis along 11 active loci at the tip of the X chromosome. Although it was necessary to pool the grain counts over certain adjacent loci, the resulting five groups exhibited similar rates of RNA synthesis in males and females. Holmquist (1972) obtained equivalent results with seven X-chromosome intervals, each containing a single major band and interband region. Chatterjee and Mukherjee (1971) found that many but not all of the 17 puffing sites they studied along the X chromosome of *Dro-*

sophila hydei show the same level of [^{3}H]uridine incorporation in paired female X chromosomes as in males. Pierce and Lucchesi (1980) reported that two major puffs are induced by heat shock on the X chromosome of *Drosophila pseudoobscura*. They monitored RNA synthesis as a function of [^{3}H]uridine incorporation in these puffs and found it to be equivalent in males and females.

Although the incorporation of [^{3}H]uridine along polytene chromosomes is a valid measure of chromosomal RNA synthesis, i.e., of transcription, a more precise comparison of gene activity in the two sexes would require the direct assay of specific messenger RNAs. The results of these types of experiments are presented in the next section.

C. Steady-State Levels of Specific RNAs

Birchler *et al.* (1982) were the first to measure the level of a specific transcript in males and females. Unlike all of the genes discussed so far, the transcript in question was a transfer RNA and, therefore, an RNA polymerase III product. Transfer RNA was extracted from males and females and serine isoacceptors were labeled *in vitro* by aminoacylation with [^{14}C]serine. Following separation by reversed phase chromatography, the radioactivity of each isoacceptor was determined. Serine-4, encoded at a major X-linked site and at a major site on chromosome 2, was found to be equivalent in males and females and to respond to differences in dosage of the X-linked locus, in each sex. The magnitude of these responses was less than expected because of the contribution from the genes at the autosomal site to the total serine-4 level. Nevertheless, dosage effects were sufficient to support the conclusion that the X-linked serine-4 gene cluster is compensated in normal individuals.

With respect to RNA polymerase II transcripts, Ganguly *et al.* (1985) were first to publish that the level of G6PD messenger RNA is equivalent in adult males and females. Polyadenylated RNA was extracted, size fractionated by electrophoresis, blotted onto nitrocellulose filters [Northern transfer; Thomas (1983)], and allowed to hybridize with ^{32}P-labeled DNA fragments from the G6PD gene (Zw^+). The blots were also hybridized with labeled DNA from the alcohol dehydrogenase gene (Adh^+); because this gene is autosomal and is therefore represented in equal doses in male and females, its transcript serves as an internal check for variations in RNA extraction or quantitation. Ganguly *et al.* (1985) also performed an important control, made possible by gene-cloning techniques: they determined the relative Zw^+ gene copy number in adult males and females and found it to be 1 : 2, as expected. Their data on RNA measurements are reproduced in Table 3.

Analogous results were obtained by Breen and Lucchesi (1986) and by Kaiser *et al.* (1986) with respect to the salivary gland secretion polypeptide 4. Both sets of workers extracted total RNA from female or male third instar larvae, spotted the RNA onto nitrocellular filter membranes [dot blotting; Kafatos *et al.* (1979)], and allowed it to hybridize with labeled DNA fragments of the *Sgs-4* gene. As a control, the RNA encoded by the autosomal *Sgs-3* gene was quantitated in similar fashion. The Sgs-4 RNA levels were found to be compensated in normal one-dose males and in two-dose females. Furthermore, Breen and Lucchesi (1986) used a small duplication and an X chromosome bearing an extremely low-producing allele of *Sgs-4* [found by Korge (1977) in a strain called Hikone] to establish the customary dosage dependence within each sex.

III. The X/A Ratio Is the Primary Determinant of Dosage Compensation

A. X-Chromosome Activity Levels in Heteroploids

In this section, measurements of activity levels of X-linked genes in flies with numbers of X chromosomes or of sets of autosomes different from those found in normal diploid males and females will be reviewed (a haploid set of autosomes is designated as "A").

TABLE 3
Levels of G6PD and ADH mRNAs in Males and Females[a]

Probes	Poly(A)$^+$ RNA assayed[b] (μg)	Amount hybridized[c] (cpm)		G6PD/ADH	
		Males	Females	Males	Females
Zw$^+$ DNA	1.0	581	454	0.13	0.11
	2.0	1225	1146	0.27	0.29
Adh$^+$ DNA	0.5	4555	3976	—	—

[a] After Ganguly *et al.* (1985).

[b] Amount of poly(A)$^+$ RNA subjected to electrophoresis, Northern transfer, and hybridization.

[c] Areas of the Northern blots showing hybridization were cut out and their radioactivity determined.

1. *Triploid Females (3X;3A)*

All of the evidence reviewed so far supports the assertion that the activity of each X-linked gene dose in a two-X diploid individual is half of that of the single gene dose in a one-X diploid individual. Seecof *et al.* (1969) investigated whether this relationship extends to X-linked genes in a three-X triploid individual. They measured 6PGD and G6PD levels in triploid females and found them not to differ from those of control diploid females. Seecof and his collaborators interpreted their results to mean that a further reduction in activity per X-linked gene dose occurs in a triploid. To circumvent the potential imbalance between X-linked and autosomal gene products in these organisms, they invoked evidence that the addition of a whole chromosome set has little or no effect on most enzyme activities. Lucchesi and Rawls (1973a) tested the validity of these conclusions. X-Linked (6PGD and G6PD) and autosomal [α-glycerophosphate dehydrogenase (α-GPDH) and NADP-dependent isocitrate dehydrogenase (IDH–NADP)] enzyme activities were measured and correlated to DNA content in crude extracts, rather than to live weight as was done by Seecof and his co-workers. The results of these experiments showed that the contribution of each dose of a given gene to the level of enzyme activity is equal in diploid and triploid cells, i.e., triploids have 1.5 times the amounts of X-encoded products on a per-cell basis. The presence of an extra genome leads to a proportional increase in cell constituents, resulting in a proportional increase in cell size. The live weights of diploid and triploid females are comparable because triploid females have fewer but larger cells in their tissues.

Comparisons of X-linked gene activity in diploid and triploid females were also performed by Faizullin and Gvozdev (1973), Lucchesi and Rawls (1973b), and Maroni and Plaut (1973b), with results similar to those just discussed.

2. *Triploid Intersexes (2X;3A)*

Individuals with two X chromosomes and three sets of autosomes are referred to as triploid intersexes. They normally occur among the offspring of triploid females or they can be generated by crossing diploid females, with their two X chromosomes attached to the same centromere, to males whose major autosomes are in the form of four compound chromosomes, or isochromosomes. With respect to sex differentiation, triploid intersexes are mosaics of relatively large phenotypically male and female patches (Stern, 1966).

Lucchesi and Rawls (1973b) and Maroni and Plaut (1973b) com-

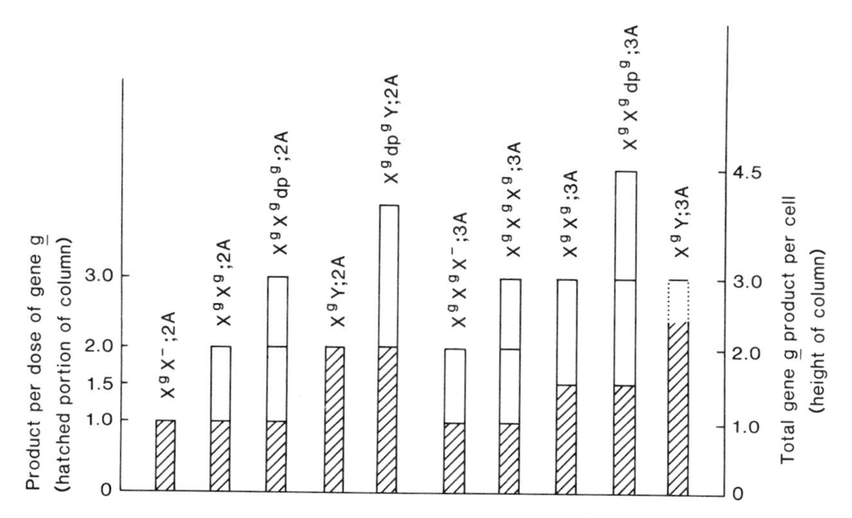

FIG. 3. Schematic representation of X-gene activity in individuals of various chromosomic constitutions. The height of the column represents total gene product per cell; a value of 2 was arbitrarily chosen for diploids. The shaded areas represent activity per gene dose; in diploid females this activity has a value of 1. The *g* denotes an X-linked gene whose product is being measured; X^g is a normal X chromosome bearing the gene under study, X^- is an X chromosome deficient for the locus of *g*, and dp^g is a small fragment of X containing the locus of *g*.

pared 6PGD and G6PD activities, respectively, in triploid intersexes and triploid females and found them to be equivalent. In addition, both groups measured gene activity as a function of gene dosage within each of these two types of flies. The conclusion was that gene activity expressed on a per-gene basis is greater in triploid intersexes than in triploid or diploid females. It is, nevertheless, less than the activity displayed by a diploid male (Fig. 3). Concordant results were published by Faizullin and Gvozdev (1973) with the following exception: intersexes selected for their more extreme malelike phenotype exhibited a 30% increase in X-chromosome activity over sexually intermediate intersexes.

Measurements of X-chromosome activity in triploid intersexes were also performed by transcription autoradiography of larval salivary gland polytene chromosomes with results similar to the ones just discussed: the relative number of silver grains over the bipartite X in intersex nuclei and over the tripartite X in triploid females is the same. Therefore, the rate of RNA synthesis per X chromosome in intersexes is intermediate between the rate in diploid males and the rate in diploid or triploid females (Maroni and Plaut, 1973a; Ananiev *et al.*, 1974).

Given the sexual mosaicism of intersexes, it is possible that the level of X-linked gene products that they exhibit is the result of a combination of male-level chromosome activity in those regions that differentiate into male tissue, and female-level activity in regions that differentiate into female tissue. Evidence for the absence of mosaicism at the level of the X-chromosome transcriptional rate was presented by Maroni and Plaut (1973a). These workers observed that the frequency distribution of the relative activity of the X chromosome in the salivary gland nuclei of triploid intersex larvae is unimodal. Mann *et al.* (1986) could find no evidence, in adult intersexes, of mosaicism for an X-linked gene product comparable to the sexual mosaicism exhibited by these organisms. In light of these results, the correlation between the sexual differentiation and X-chromosome activity levels reported by Faizullin and Gvozdev (1973) may be explained by the fact that they used measurements of specific activity of enzymes as a measure of the level of X-chromosome function. In diploids, the specific activity of many enzymes is greater in extracts from adult males than in extracts from adult females because of tissue-level differences between the sexes. It is possible that the higher specific-activity level reported for the sample of extreme malelike intersexes reflects a bias in the distribution of some tissues rather than a higher level of X-linked activity in all tissues.

Additional studies involving intersexes will be discussed in Sections IV,C,2 and V,A,2.

3. *Metamales (1X;3A)*

During the course of their study of triploid intersexes, Maroni and Plaut (1973b) recovered two metamales—individuals with a single X chromosome and three sets of autosomes—and observed that they exhibited G6PD activity levels only slightly lower than triploid females (3X;3A). Remembering the fact that triploid cells are proportionately larger than diploid cells (see above), the near equivalence of an X-linked gene product in metamales and triploid females indicates that the single X chromosome in metamales has a greater activity level than it does in diploid males (1X;2A).

In contrast to these results, Ananiev *et al.* (1974) measured the level of X-chromosome function in metamale larvae by transcription autoradiography and concluded that they synthesize chromosomal RNA at the same rate as diploid males. In an attempt to resolve these differences, Lucchesi *et al.* (1977) performed a more extensive set of measurements on metamales. A large number of 1X;3A larvae were generated with considerable difficulty since they are extremely poorly viable. Chromosomal RNA synthesis, determined once again by transcription autoradiography, was expressed in terms of the DNA content of the chromosomes (measured by chronic incorporation of [^{3}H]thymidine). This treatment corrects for the fact that a chromosome synapsed with a homologue emits fewer particles capable of producing silver grains and provides a measurement of X-chromosome activity on a per-gene basis. In addition, two X-linked (6PGD and G6PD) and two autosomal (α-GPDH and IDH–NADP) enzymes were monitored in crude extracts of metamale and control larvae of precisely comparable developmental stages (collected within 30 minutes of spiracle eversion). The results revealed that the rate of RNA synthesis by the single X chromosome in metamales is approximately 2.7 times that of a single X chromosome in a triploid or diploid female and is substantially higher than the level achieved in a diploid male (Fig. 3).

4. Metafemales (3X;2A)

These poorly viable individuals occur spontaneously as a consequence of X-chromosome nondisjunction. Their frequency can be greatly increased by the use of females with X chromosomes attached to a single centromere (compound X). Measurements of X-linked gene expression in metafemales have been performed by different investigators with very different results.

Stern (1960) reported that metafemales whose three X chromosomes are homozygous for w^a have an eye color closely resembling that of chromosomally normal mutant females. Lucchesi *et al.* (1974) performed a variety of enzyme activity determinations on whole larvae or adults, or on larval fat bodies. X-Linked enzyme activity measurements in metafemales, normalized in relation to autosomal enzyme activities, showed great variability, with values often in excess of those of control females. Nevertheless, activities per gene were predominantly smaller in metafemales than in diploid females, suggesting that the presence of a third X chromosome does not mediate a gene-dosage response. This conclusion was reinforced by the observation, through autoradiographic monitoring of X-chromosome activity, that the overall rate of RNA synthesis was equivalent in the two types of females. Just 1 year later, Stewart and Merriam (1975), in a study of the effects of segmental aneuploidy to be discussed below, generated metafemales and obtained data similar to those of Lucchesi *et al.* (1974) on enzyme activity measurements.

A different conclusion was reached by Faizullin and Gvozdev

(1973), who reported an increase of approximately 50% in 6PGD activity and 18–25% in G6PD activity in metafemales compared to normal females. On the basis of complementary transcription autoradiography experiments, Ananiev *et al.* (1974) stated that the transcription of each X chromosome in 3X;2A individuals proceeds at the same rate as in diploid females. Given the unresolved nature of the control of their X-linked gene activity, metafemales are not included in Fig. 3. They will be discussed further once the topic of autosomal dosage compensation (Section III,C) has been developed.

Examination of the experimental results summarized in Fig. 3 leads to the following considerations: (1) The products of X-linked genes are concordant with autosomal gene products. This is true whether autosomal products are generated at a level characteristic of diploids (2X;2A and 1X;2A) or triploids (3X;3A, 2X;3A, or 1X;3A). (2) In order to achieve a balance between X and autosomal gene products in diploids, the activity of X-linked genes is regulated by a mechanism for dosage compensation. Presumably, the same mechanism modulates the activity of X-linked genes in triploids. (3) X-Linked gene activity is influenced by the autosomes: if the number of X chromosomes is held constant, an extra set of autosomes produces an increase in activity (viz. 1X;2A versus 1X;3A and 2X;2A versus 2X;3A). (4) Conversely, given a constant number of sets of autosomes, the greater the number of X chromosomes present, the smaller the activity per gene dose (viz. 1X;3A versus 2X;3A versus 3X;3A and 1X;2A versus 2X;2A).

B. Aneuploidy for Large Segments of the X Chromosome

The results discussed in the previous sections establish that there are at least four different levels of activity possible for X-linked genes (left abscissa in Fig. 3). Do these levels represent selected points in an activity continuum? In other words, can any level of activity within the limits of the range represented occur? Experiments to address this question, carried out in different laboratories, were based on the following rationale. The addition of an X chromosome to a genome, without increasing the dosage of the gene under study (designated *g*), has an inverse effect on the activity of this gene (the activity per gene in X^gX^-;2A is half of what it is in X^g/Y;2A and the activity in $X^gX^gX^-$;3A is two-thirds of what it is in X^gX^g;3A). What, then, would be the effect of adding X-chromosome segments of different sizes?

Faizullin and Gvozdev (1973) tested the effect on 6PGD and G6PD activities of several duplications, some of which encompassed half an X chromosome or more. All measurements with the larger duplications were performed on adult females since males can only tolerate hyperploidy for very small regions of the X. Large duplications encompassing the 6PGD structural gene (Pgd^+) but not the G6PD gene (Zw^+) mediated the increase in 6PGD specific activity expected of small duplications (see Table 1 and Fig. 1, for examples) but had no decreasing effect on G6PD. Concordant results were obtained with large duplications covering Zw^+ but not Pgd^+. Stewart and Merriam (1975) generated 2.5X;2A females, hyperploid for either the distal half (where Pgd^+ is located) or the proximal half (where Zw^+ is located) of the X chromosome. Their results were generally similar to those just discussed. These authors proposed that, in order to reduce the activity of X-linked genes (see the rationale discussed above), the integrity of the X chromosome must be maintained: to achieve an effect, a whole X chromosome rather than segments, albeit large, must be added. This hypothesis, though, fails to take into account the observation by Faizullin and Gvosdev (1973) that the addition to a 2X;2A genome of a whole X chromosome deficient for Pgd^+ had no effect on the level of the 6PGD activity encoded by the two Pgd^+ resident genes. In contrast to the results just discussed, Williamson and Bentley (1983) found that the addition of large X-chromosome duplications to a basic diploid female genome decreased the activity of genes located on the nonduplicated region of the X chromosome. In addition to measuring the specific activity of X-linked enzymes, these authors determined the actual relative concentration of enzyme molecules by rocket immunoelectrophoresis.

Maroni and Lucchesi (1980) reexamined the effect of duplications for segments of the X chromosome on total X function. They measured the rate of [^{3}H]uridine incorporation into the chromosomal RNA of larval salivary gland chromosomes as an indicator of transcriptional activity. Using X–Y translocations synthesized by J. R. Merriam (Stewart and Merriam, 1974), they generated individuals with 1.25, 1.50, 1.62, and 1.85 X chromosomes, intermediate between diploid males and females. Their conclusions were as follows: (1) The total amount of X-chromosome transcription (relative to autosomal transcription) occurring in all of these segmental aneuploids as well as in normal males (1X;2A) and females (2X;2A) is constant [recall that the equivalence of chromosomal RNA synthesis in males and females had long been established by Mukherjee and Beermann (1965)]. (2) This constancy is achieved through a uniform reduction of the rate of RNA synthesis over all X-chromosome regions as the size of the duplication increases. A sample of the data on which this conclusion is based is reproduced in Fig. 4.

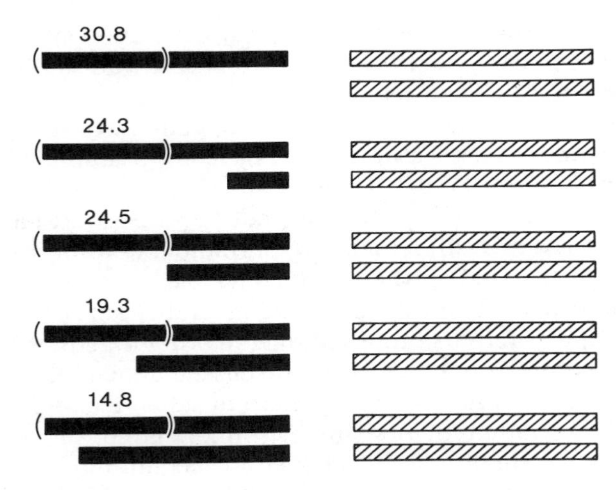

FIG. 4. Relative activity of a X-chromosome segment as a function of the total amount of X chromosome present in the genome (after Maroni and Lucchesi, 1980). Activity is expressed as a ratio of the number of autoradiographic silver grains on a particular X-chromosome segment to the total number of grains over the X-chromosome material plus the right arm of chromosome 2, times 100. Solid bars represent X chromosomes or fragments. Hatched bars represent a set of autosomes.

C. IS THERE AUTOSOMAL DOSAGE COMPENSATION?

The question of the existence of autosomal dosage compensation was not posed for several decades following Muller's discovery of the X-associated regulatory phenomenon. The major reason for this apparent lack of interest was the knowledge that autosomal aneuploidy simply does not lead to viable adults in natural populations of metazoans, obviating the circumstances necessary for the evolution of a regulatory mechanism to balance gene products in such organisms. This point of view was bolstered by the observation that the amount of product synthesized by autosomal structural genes appeared invariably to be directly proportional to gene dosage (see O'Brien and MacIntyre, 1978).

Devlin *et al.* (1982) decided to test for the presumed absence of autosomal compensation by measuring the level of gene expression in individuals trisomic for the left arm of chromosome 2 (such trisomics survive to late stages of pupal development). Two of the five enzymes monitored [alcohol dehydrogenase (ADH) or DOPA decarboxylase (DDC)] proved dosage dependent in trisomics. The other three enzymes [phosphoglycerate kinase (PGK), α-glycerophosphate dehydrogenase, and cytoplasmic malate dehydrogenase (cMDH)] showed no duplication effect if the extra gene dose was carried by a supernumerary whole chromosome 2 arm. Measurements of RNA synthesis by transcription autoradiography confirmed that approximately two-thirds of the loci in this chromosome arm compensate in trisomics (Devlin *et al.*, 1984). Similar results were obtained with trisomy for the left arm of chromosome 3, where four of the six genes examined showed levels of gene products equal to those in normal diploids (Devlin *et al.*, 1985a).

Soon after the publication of the first measurements of gene activity in autosomal trisomics, Lucchesi (1983) suggested that these individuals exhibit the same type of regulation displayed by X-chromosome trisomics, i.e., metafemales; this regulation is fundamentally different from the compensation of X-linked genes in normal males and females whereby the products of a single X chromosome are equalized with those of two X chromosomes. A similar view was reached by Baker and Belote (1983). Lucchesi pointed out that the proper test for "autosomal dosage compensation," admittedly very difficult to perform because of their inviability, would be to monitor gene activity in whole-arm monosomics (the single dose of autosomal genes present in such individuals would render them equivalent to males with their single dose of X-linked genes).

Devlin *et al.* (1985b) have provided evidence that compensation in metafemales (X-chromosome trisomics) occurs by a mechanism which is different from dosage compensation between normal males and females. They suggest a strong analogy between compensation in metafemales and in autosomal trisomics.

Using exogenous, heterologous RNA polymerase (from *Escherichia coli*), Bhadra and Chatterjee (1986) monitored autosomal template activity in trisomic individuals and found that the template activity in these aneuploids showed some level of dosage dependence (for a more complete description and evaluation of the technique, see Section VI,C,2).

IV. The Link between Dosage Compensation and Sex Determination

A. Lack of Effect of the Regulators of Sex Differentiation on Levels of X-Chromosome Activity

The discovery by Sturtevant (1945) of a mutant which he called *transformer* (*tra*) allowed the first test of the effect of sexual differentiation on the expression of X-linked genes. This mutant transforms chromosomally female individuals (X/X;*tra*/*tra*) into pseudomales or extreme malelike intersexes but has no effect on the development of chromosomally male (X/Y;2A) individuals. Sturtevant noted that pseudomales homozygous for the non-dosage-compensated mutant allele w^{e} had the same eye color as homozygous females. More significantly, using w^{a}, which is fully dosage compensated, Muller and Lieb (Muller, 1950) found that w^{a}/w^{a};*tra*/*tra* pseudomales had the same level of eye pigment as w^{a}/w^{a};+/+ females rather than the much darker level of $w^{a}/Dp(w^{a})$/Y;+/+ (two-dose males).

Some evidence favoring an influence of sex differentiation on X-linked gene expression was presented by Komma (1966). In addition to *tra*, this author used *dsx*, a mutant which transforms X/Y;2A as well as 2X;2A individuals into phenotypically identical intersexes (Hildreth and Lucchesi, 1963; Hildreth, 1965). Komma assumed that there is a physiological gradient from "femaleness" to "maleness" and that relative "femaleness" decreases X-linked gene activity while "maleness" increases it. Two-X pseudomales and intersexes should have higher gene activity than their sisters (since they are more malelike) while single-X intersexes should have less activity than their brothers (since they would be more femalelike). Measuring G6PD activity in normal and abnormal flies, Komma verified the expectation for two-X intersexes and pseudomales but not for single-X intersexes. Furthermore, he was unable to demonstrate any effect of sexuality on 6PGD activity.

Smith and Lucchesi (1969) reexamined more fully the possibility that sex physiology may play a role in establishing dosage compensation. These workers used *dsx* to alter sexual development, the spectrophotometric determination of red eye pigments as a measure of gene product, and a congenic genetic background to reduce the variability among the different types of flies compared. They concluded that sex differentiation differences between males and females cannot account for dosage compensation.

Lakhotia and Mukherjee (1969) investigated whether dosage compensation is dependent on the sex of an individual by generating gynandromorphs, individuals mosaic for male (X0;2A, where 0 indicates the absence of a homologue) and female (2X;2A) clones of cells. They examined the relative rate of chromosomal RNA synthesis in mosaic glands which were present in predominantly female larvae, and found that the single X chromosome of male nuclei was as active as the two paired X chromosomes of female nuclei, leading to the conclusion that dosage compensation is not dependent on sex differentiation but, rather, is a function of the number of X chromosomes present in a cell.

B. The Y Chromosome

The Y chromosome plays no role in the compensation of X-linked genes. This has been clearly and easily established by generating individuals with a single X chromosome and two sets of autosomes (X0;2A); such individuals develop as males whose only abnormality is a failure to complete spermiogenesis. Similarly, a Y chromosome can be added to two X chromosomes, yielding X/X/Y;2A zygotes which develop into perfectly normal and fertile females. The presence or absence of a Y chromosome in males or females does not alter the level of X-linked gene activity characteristic of each sex when activity is measured at the level of a terminal phenotypic product such as an eye pigment (Muller, 1950), at the level of an enzyme activity (Seecoff *et al.*, 1969; Lucchesi and Rawls, 1973a), or at the level of chromosomal RNA synthesis (Kaplan and Plaut, 1968).

C. Genes Controlling Dosage Compensation and Sex Differentiation

There are at least four genes which interact and contribute to proper sex differentiation as well as dosage compensation.

1. Daughterless (da)

Until very recently, the only *daughterless* mutant allele available was of spontaneous origin. This allele, when present in homozygous condition in females, causes the production of all-male progeny (Bell, 1954). At normal rearing temperatures (around 25°C), female zygotes die as embryos even if they carry a wild-type allele of the gene. This female-specific maternal-effect lethality can be overcome at lower rearing temperatures, although escaping females exhibit numerous abnormalities characteristic of cell death during development. At higher rearing temperatures, *da* is a recessive lethal: homozygous male and female individuals fail to emerge, even if their mothers were heterozygous (Cline, 1976). The *da* mutant must be hypomorphic (*leaky*) because it acts as a recessive lethal over a deficiency at normal rearing temperatures (Mange and Sandler, 1973). Recently, several ethyl methanesulfonate (EMS)-induced true null alleles of *da* have been isolated by C. Cronmiller in Cline's laboratory. Their analysis should clarify the relationship between the maternal and zygotic actions of this gene.

2. Sex-Lethal (Sxl)

In 1960, Muller and Zimmering described a sex-linked lethal mutation which acted in females but had no demonstrable effect in males; they named it *Female-lethal* (*Fl*). Phenotypic and genetic analyses of this mutant allele provided little information regarding the role of the gene's wild-type product (Zimmering and Muller, 1961; Marshall and Whittle, 1978). Thomas Cline had recovered from a *daughterless* stock an X-linked mutation which suppressed the daughterless phenotype by restoring full viability to females from *da*/*da* mothers. This mutation was lethal in males and was named *Sex-lethal, Male-specific* #1 ($Sxl^{M\#1}$). Cline (1978) found it to be so very closely linked to *Fl* that he renamed the latter *Sex-lethal, female-specific* #1 ($Sxl^{f\#1}$). The sex-specific phenotype of the two types of mutations made it worthwhile to determine if this gene plays some role in sex differentiation. Because the mutations kill early during development, a direct analysis was not possible and Cline (1979) resorted to the construction of mosaic individuals. These are mostly wild type with patches of mutant tissue resulting from the loss of an unstable Ring X chromosome bearing the wild-type allele of the gene or from somatic (mitotic) crossing-over, which produces homozygous mutant and homozygous wild-type daughter cells from a heterozygous parent cell. In mosaic individuals, mutant patches may occur, by chance, in areas where adults are normally sexually dimorphic, affording the opportunity to study the effect of the gene on sexual differentiation. The results of Cline's experiments clearly established that the female-specific sex-lethal mutation causes chromosomally female cells to differentiate male structures. A similar conclusion was reached by Sanchez and Nöthiger (1982, 1983). The opposite effect is true of the male-specific sex-lethal allele: it causes chromosomally male cells to differentiate into female structures.

Based on the observations that (1) *da* has a female-lethal maternal effect, (2) $Sxl^{f\#1}$ is female lethal, (3) the *da* maternal effect is suppressed by an apparent allele of *Sxl*, namely, $Sxl^{M\#1}$, and (4) this allele is lethal in males, Cline (1978) proposed a brilliant rationalization, represented in diagrammatic form in Fig. 5, which became the cornerstone for all subsequent dosage compensation and sex differentiation models. The *da* wild-type gene makes a factor which is present in the egg at fertilization and which interacts with a signal from the X/A ratio of the zygote. If this ratio is female, the factor activates the Sxl^+ locus

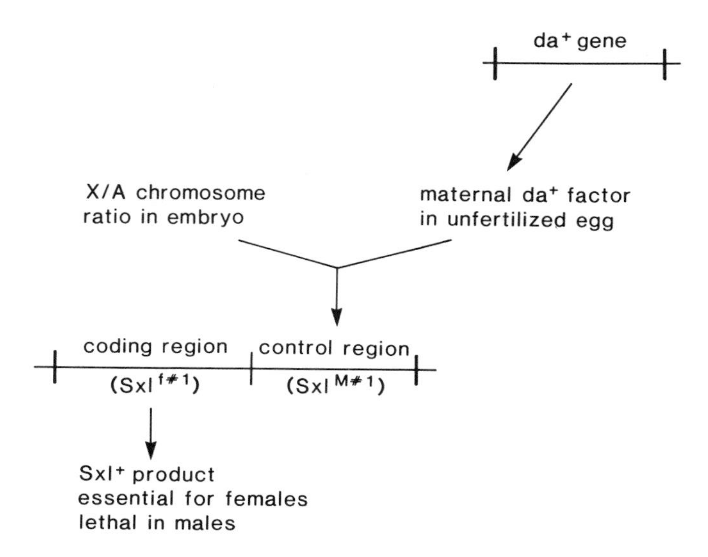

FIG. 5. Diagram of the explanation proposed by Cline (1978) to account for the interaction of the *da* and *Sxl* loci. The da^+ factor is present at fertilization as a maternal contribution to the egg cytoplasm. If the X/A ratio of the zygote is 1, the factor activates the Sxl^+ locus, whose product is necessary in females to achieve proper sexual differentiation and to maintain the proper level of X-chromosome activity. In males, in which the X/A ratio is 0.5, no Sxl^+ functional product is elaborated; sexual differentiation as well as X-chromosome activity proceed in the male mode. $Sxl^{M\#1}$ is a constitutive mutant allele; $Sxl^{f\#1}$ is a loss-of-function allele.

whose product is necessary in females but lethal in males. The $Sxl^{f\#1}$ mutant is a loss-of-function allele, lethal to females but viable in males, in which the Sxl^+ product is not present or active [this contention has recently been proved by the fact that males which are deleted for this gene are both viable and fertile (Maine *et al.*, 1985a)]. The $Sxl^{M\#1}$ mutant is a constitutive allele and can, therefore, circumvent the absence of the da^+ factor in females from *da/da* mothers; because it makes Sxl^+ product constitutively, it is lethal in males. Cline further proposed that the *Sxl* locus may be involved in the process of dosage compensation. In males which normally lack the Sxl^+ product, its presence may lead to a level of X-chromosome activity typical of each female X chromosome. In females, absence of this product may lead to a male level of transcription for each of their two X chromosomes.

Using transcription autoradiography to monitor X-chromosome activity, Lucchesi and Skripsky (1981) were able to involve Sxl^+ in dosage compensation, thereby establishing the link between compensation and sex differentiation. Partially complementing heteroallelic

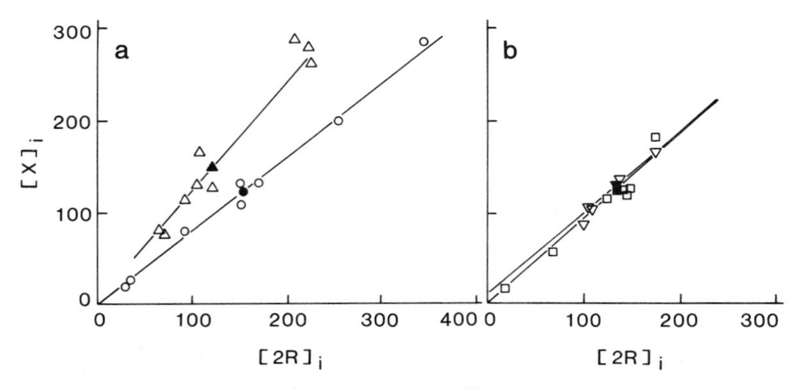

FIG. 6. Chromosomal RNA synthesis measured by transcription autoradiography as a function of [^{3}H]uridine incorporation and presented in the form of linear regressions of X-chromosome activity on autosomal activity (the right arm of chromosome 2 was used in these experiments). (a) Open triangles represent mutant *Sxl*^{f} females and open circles their male sibs. (b) Inverted open triangles and open squares represent control females and males, respectively. The solid symbols indicate the overall means. Note the similarity in the slopes of the male results in a and of the controls in b. In contrast, mutant females yield a line whose slope is significantly greater, indicating that the relative activity of their X chromosomes is greater than in the controls. (After Lucchesi and Skripsky, 1981.)

combinations were generated to circumvent the early lethality characteristic of most *Sxl* alleles. In such mutant female larvae, the relative level of X-chromosome transcription was clearly and significantly higher than in control females, leading to the conclusion that *Sxl*^{f} causes chromosomally female (2X;2A) cells not only to differentiate into adult male structures but to hyperactivate their X chromosomes as well (Fig. 6). Unfortunately, studies on the effect of *Sxl*$^{M\#1}$ were hampered by the failure to obtain third instar larvae with fully developed and cytologically favorable salivary gland chromosomes; in these larvae, the observed decrease in X-chromosome transcription could not be separated from an apparent differential under-replication of the chromosomes.

Cline (1983) used triploid intersexes to provide additional evidence that *Sxl*$^+$ controls both sex differentiation and dosage compensation. As previously described (Section III,A,2), all the cells of triploid intersexes have a 2X;3A chromosomal constitution, yet these individuals are phenotypic mosaics of relatively large sectors of male and female tissues. Cline noted that triploid intersexes homozygous for *Sxl*$^{f\#1}$ were completely masculinized while those carrying the *Sxl*$^{M\#1}$ allele were completely feminized. These observations are consistent with the effects of these mutations on sex differentiation in diploids. It is interesting to note that *da* also had the predicted effect: triploid intersexes produced by *da/da* mothers were completely masculinized. These intersexes presumably lack the ability to generate active Sxl^+ product during early development and are, therefore, equivalent to triploid intersexes homozygous for $Sxl^{f\#1}$ (see Fig. 5). To measure X-chromosome activity, Cline used two X-linked mutants, *Bar eyes* (*B*) and *Hairy wing* (*Hw*), which become more extreme with the addition of mutant alleles to the genome; he reasoned that the same effect should be achieved if the activity rather than the number of mutant alleles were increased, while a decrease in activity should lead to a more normal phenotype. The presence of $Sxl^{M\#1}$ in triploid intersexes led to a substantial decrease in the mutant effects of *B* and *Hw*, as might be expected of a mutation which would diminish the overall rate of X-linked gene expression. A similar effect, albeit of much smaller magnitude, was registered when three doses of Sxl^+ were present in intersexes instead of two. In contrast, neither the $Sxl^{f\#1}$ mutation nor a *da* maternal genotype had clear-cut effects on the expression of the "reporter" mutations.

Cline suggested that the dramatic results with respect to sex differentiation and the disappointing results with respect to dosage compensation could be explained if higher levels of Sxl^+ function were necessary for dosage compensation than for sex differentiation. Furthermore, triploid intersexes, with their intermediate X/A ratio of 0.67, appear to be remarkably buffered with respect to the deleterious effects of abnormal levels of X-chromosome activity: whereas $Sxl^{M\#1}$ is an early larval lethal for diploid males, it allows development of triploid intersexes to pharate adults; $Sxl^{f\#1}$ and the *da* maternal effect, which kill 2X;2A zygotes during embryonic development, have no apparent effect on the normal viability of intersexes. Although this lack of effect on viability could be interpreted as invalidating a role of Sxl^+ in dosage compensation, a similar observation made with an autosomal male-specific lethal mutant (see Section V,A,2) supports the contention that intersexes are more impervious or more refractory to alterations in the genetic control of this regulatory mechanism.

In an attempt to separate the effects of *Sxl* mutations on sex from their effect on dosage compensation, Cline (1984) subjected $Sxl^{M\#1}$ to γ-ray irradiation and recovered two new alleles ($Sxl^{M\#1,\,fm\#3}$ and $Sxl^{fm\#7,\,M\#1}$), which differ from $Sxl^{M\#1}$ in that they are viable in males and semilethal, in homozygous or heteroallelic combination, in females. Of significant interest is the observation that the 2X;2A mutant escapers are phenotypically male. Both alleles have retained the ability to rescue "daughters" from *da* mothers, i.e., to suppress the daughterless maternal effect, although with a much lower level of efficiency than the original *Sxl*$^{M\#1}$ mutation (quotation marks are used as a reminder that these escapers are 2X;2A phenotypic males). These facts can be interpreted as indicating that the new mutational events have changed the *Sxl*$^{M\#1}$ to a loss-of-function allele which is unable to mediate proper female sexual differentiation while retaining its ability to substantially suppress X-chromosome activity (a failure to do so is the cause of the female-lethal effect of common loss-of-function alleles such as *Sxl*$^{f\#1}$).

The two mutations just discussed have enabled Cline to gather some evidence of an important property of the *Sxl* locus: autoregulation. The ability of the mutant alleles to rescue "daughters" from *da* mothers, mentioned above, is greatly enhanced by the presence of a duplication bearing an *Sxl*$^+$ allele. The activation of this allele, supported by the fact that the "daughters" in question are no longer masculinized, could not have resulted from the presence of the da$^+$ maternal factor since the mothers were *da*. It must, therefore, have been caused by some residual activity of the new mutant alleles, acting in trans on the *Sxl*$^+$ of the duplication.

Two female-specific lethal mutations which fail to complement some *Sxl* mutant alleles while complementing others have been obtained by M. Torres and L. Sanchez (personal communication). Homozygous mutant clones for one of these mutations (*Sxl*^{fb}) show no alteration of sexual phenotype, suggesting that the dosage compensation function but not the sex differentiation function of *Sxl*$^+$ has been altered (L. Sanchez and D. Bachiller, personal communication).

The *Sxl*$^+$ gene has recently been cloned (Maine *et al.*, 1985a,b). It is a large gene, responsible for a complex transcriptional pattern with sex-specific characteristics. The discovery that it is active in both sexes, although, as stated above, males deficient for the gene are viable and fertile, need only impose, at this time, minor changes to all of the proposed models in which *Sxl*$^+$ plays a central role (see, for example, Baker and Belote, 1983, and Cline, 1985). Of course, a full elucidation of the function of the different gene products may, in time, mandate a substantial revision of the models.

3. *Sisterless-a (sis-a)*

Among the mutations recovered in a screen for EMS-induced X-linked maternal-effect mutations, Cline (1986) identified a female-specific lethal and named it *sisterless-a* (*sis-a*) to highlight both the similarities and an important difference between this gene and *da*. The new mutation is a female-specific zygotic lethal, i.e., it causes the death of homozygous individuals during embryonic or larval development. This lethality is completely suppressed by the presence of $Sxl^{M\#1}$ in the genotype of *sis-a* mutant females. $Sxl^{f\#1}$ and *sis-a* fail to complement in trans configuration (note that recombination and deficiency mapping places *sis-a* at region *10B4* on the cytological map while *Sxl* is at *6F*). Cline demonstrated that *sis-a* has the same masculinizing effect as $Sxl^{f\#1}$ on sex differentiation. Finally, by generating live $Sxl^{M\#1}$-suppressed *sis-a/sis-a* females he was able to show that *sis-a*, in contrast to *da*, lacks a lethal maternal effect.

4. *Daughterkiller (Dk)*

A new dominant maternal-effect mutation, *Daughterkiller* (*Dk*), has been discovered on the left arm of chromosome 3 (Steinmann-Zwicky *et al.*, 1987). As in the case of *da* and *sis-a*, the female-specific lethal effect is suppressed by $Sxl^{M\#1}$. The presence of a duplication for Dk^{+} in mutant-carrying females partially rescues the female-lethal maternal effect. The mutation transforms 2X;3A triploid intersexes into phenotypic males.

5. *Female Lethal(2)d [fl(2)d]*

An EMS-induced female-specific temperature-sensitive lethal mutation on chromosome 2 [*fl(2)d*] has been found by A. B. San Juan and L. Sanchez (personal communication). This mutation suppresses the lethal effect of $Sxl^{M\#1}$ in males.

6. *Hermaphrodite (her)*

Described in Baker and Belote (1983), *hermaphrodite* (*her*) is a temperature-sensitive mutant on chromosome 2 which affects sex differentiation: males are slightly feminized while females are transformed into intersexes. Because the viability of both sexes is greatly reduced at restrictive temperatures and because this aspect of the mutant phenotype as well as the sex transformations is influenced by a maternal effect, *her* is thought to participate in a process involved in sex differentiation as well as dosage compensation.

V. Trans-Acting Regulators of Compensation

A. Autosomal Male-Specific Lethal Genes

A significant step forward in the study of dosage compensation was the identification of autosomal genes whose wild-type product is necessary for the normal enhancement of X-chromosome function in males. These genes were identified by the male-specific lethality of their mutant alleles.

1. *Identification of Male-Specific Lethal Mutations*

The rationale underpinning these experiments was that mutations inactivating regulatory loci responsible for dosage compensation could well result in sex-specific lethality. For example, a mutation which prevents proper compensation could cause the death of single-X individuals due to a deficiency in X-linked gene products, while two-X individuals, although somewhat affected, could survive. A large screen for EMS-induced sex-specific lethals on the two major autosomes yielded four male-specific lethal mutations on chromosome 2 (Belote and Lucchesi, 1980a). One of these lethals was a temperature-sensitive allele of *maleless* (*mle*), previously described by Fukunaga *et al.* (1975) and Tanaka *et al.* (1976). Note that *mle* was earlier known as *male-killer* (*mak*), a mutation discovered by Golubovsky and Ivanov (1972) and shown to be allelic to *mle* by M. Ashburner (personal communication). Two lethals were allelic, mapped at *54* (*36F7–37B8*) on chromosome 2, and identified a new gene designated *male-specific lethal 1* (*msl-1*). The fourth lethal identified a third complementation group designated *male-specific lethal 2* (*msl-2*) located at 9 (*23E1,2–F6*) on chromosome 2. A subsequent screen for sex-specific lethals on chromosome 3 yielded two mutations with phenotypic characteristics similar to the chromosome 2 mutations (Lucchesi *et al.*, 1982). Both lethals are allelic to *maleless on the third-132* [*mle(3)132*] discovered by T. K. Watanabe (Uchida *et al.*, 1981).

Using the four male-specific lethal alleles they had recovered on chromosome 2, Belote and Lucchesi (1980a) demonstrated that lethality is not caused by the presence of the Y chromosome in the genome, since removing this chromosome from mutant males did not rescue them. It is not caused by sex differentiation, since the presence of sex-transforming genes such as *transformer* (*tra*) fails to alter the lethal phenotype whose specificity depends on the chromosomal sex of the mutant individual. Also, there does not appear to be a single particular region of the X chromosome that, when present in two doses as it is normally in females, would rescue mutant males.

2. *Developmental Effects*

All male-specific lethal mutant alleles exert their lethal effect during late larval or early pupal stages in males but have no detectable effect on the viability of mutant females. There is, however, some evidence for a maternal effect of the wild-type gene product: mutant males from heterozygous mothers develop further (die later) than sons of homozygous females (Belote and Lucchesi, 1980a; Tanaka *et al.*, 1976). It is, therefore, possible that these genes, although unnecessary to the well-being of the female soma, do need to be expressed in the female germ line. With respect to the activity of the male-specific lethal genes in males. Uenoyama *et al.* (1982b) and Belote (1983) report that gynandromorphs (XX/X0;2A mosaics) homozygous for mutant alleles of *mle*, *msl-2*, or *mle(3)132* are extremely inviable, suggesting that the lethality associated with these mutations is cell autonomous. In a complementary experiment, Belote (1983) induced, by somatic recombination, the loss of the wild-type allele of *msl-2* in heterozygous embryos or larvae of different ages. The paucity and small relative size of mutant patches indicated that $msl\text{-}2^{+}$ acts continuously and in all, or most, tissues of the male. Belote also noted that males simultaneously mutant at three male-specific lethal loci are no more severely affected than males homozygous for a single mutation.

Triploid intersexes, homozygous for one of the male-specific lethal mutations (*msl-2*), exhibit no observable reduction in viability (Mann *et al.*, 1986). This observation is surprising, given that these intersexes have been shown to compensate for the disparity in the number of X chromosomes relative to sets of autosomes in their genome. It is, nevertheless, concordant with the reported lack of effect of $Sxl^{M\#1}$ on the viability of intersexes which develop as far as the pharate adult stage.

3. *Germ-Line Effects*

In order to determine whether male-specific lethal mutations affect the germ line, Bachiller and Sanchez (1986) transplanted pole cells homozygous for *msl-1*, *msl-2*, or *mle* into wild-type host embryos to see if they were capable of forming functional sperm. The results of these experiments indicate that *msl-1* and *msl-2* are not needed in the germ line while the *mle* gene seems to be required for normal spermatogenesis.

These observations are particularly interesting in light of the knowledge that the mutations *tra*, *tra-2*, *dsx*, and *ix* have no effect on the sexual development of the germ line (Marsh and Wieschaus, 1978; Schüpbach, 1982). The only gene known to be required for normal female germ cell development is Sxl^{+} (Schüpbach, 1985).

4. *Role in Dosage Compensation*

Belote and Lucchesi (1980b) demonstrated that the basis of the sex-limited effect of the autosomal male-specific lethals is the result of a failure of the mechanism of dosage compensation, leading to a re-

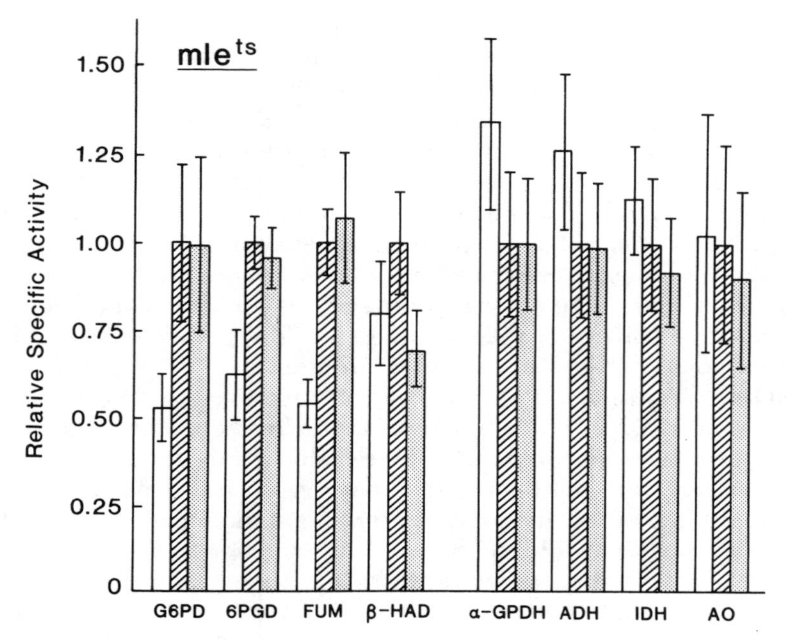

FIG. 7. Activities of X-linked and autosomal enzymes in *mle*^{ts} mutants and control larvae. The X-linked enzymes are G6PD (glucose-6-phosphate dehydrogenase), 6PGD (6-phosphogluconate dehydrogenase), β-HAD (β-hydroxyacid dehydrogenase), and FUM (fumarase); the autosomal enzymes are α-GPDH (α-glycerophosphate dehydrogenase), ADH (alcohol dehydrogenase), IDH (NADP-dependent isocitrate dehydrogenase), and AO (aldehyde oxidase). Open columns, Homozygous mutant males; hatched columns, control males; stippled columns, control females. Enzyme activities are presented as change in absorbance/minute/milligram of protein. The heterozygous males are given a relative specific-activity value of 1.0. (After Belote and Lucchesi, 1980b.)

duced level of X-chromosome gene activity in affected males. This conclusion was reached by determining that males homozygous for mutant alleles of *mle*, *msl-1*, or *msl-2* exhibit significantly reduced levels of X-linked enzymes [G6PD, 6PGD, fumarase (FUM), and, in one case, β-hydroxyacid dehydrogenase (β-HAD)] while autosomal enzymes [α-GPDH, ADH, IDH–NADP, and aldehyde oxidase (AO)] are not affected (Fig. 7). Furthermore, autoradiographic monitoring of RNA synthesis by larval salivary gland polytene chromosomes of males homozygous for one of these mutations (*mle*^{ts}) revealed that the relative rate of X-chromosome transcription is only 65% of that found in controls.

The effect of *mle* alleles on the level of a specific X-linked transcript was established by Breen and Lucchesi (1986) using the X-linked *Sgs-4* gene (see Section I,B) and a closely related autosomal gene (*Sgs-3*) as an internal control. Males heterozygous for a specific heteroallelic combination (mle^{ts}/mle^{185}) are fully viable when reared at 18°C, yet severely affected at 29°C. Sgs-4 RNA levels were substantially reduced in males reared at the restrictive temperature. These results show that the expression of a wild-type *mle* product is necessary for complete dosage compensation of the Sgs-4 transcript, lending support to the theory that *mle* exerts its lethal effect on males by acting to decrease the transcriptional activity of the X chromosome.

Although alleles of *mle(3)132* were found to significantly depress X-chromosome RNA synthesis in mutant males (Okuno *et al.*, 1984), the other two male-specific lethal genes have not yet been tested for effects on RNA synthesis. Yet, the fact that several of them exhibit similar effects on X-linked enzymes and on X-chromosome morphology (concordant with the early observations on this subject; see Section II,A), and the fact that all of them have the same effect on the viability of mutant males, suggest that all four of these loci specify products necessary for normal levels of X-chromosome transcription in males. This group of genes will, therefore, be referred to collectively as *msls*.

5. *Interactions with Mutants Controlling Both Dosage Compensation and Sex Differentiation*

As discussed above, the female-specific lethality of Sxl^{f} alleles is, in all probability, the result of the hyperactivation of the two X chromosomes in mutant females. Since the gene products of the *msls* are necessary for proper X-chromosome hyperactivation, the possible rescue of female-lethal embryos by male-specific lethal mutations was attempted by Skripsky and Lucchesi (1981, 1982). Although no clear-cut effects on the viability of mutant females were observed, an unexpected interaction between *Sxl* and *msl* mutant alleles was uncovered. Females heterozygous for a loss-of-function allele such as $Sxl^{f\#1}$ or carrying a loss-of-function allele and the sex-lethal homozygous-viable allele $Sxl^{fhv\#1}$ (this combination is semilethal), when made homozygous for mutant alleles of *mle, msl-1*, or *msl-2*, frequently developed as intersexes. The intersexuality is of a mosaic type, with clones of cells giving rise to patches of tissue differentiating into male sexual characteristics. These male patches are not caused by homozygosis of the loss-of-function *Sxl* allele via somatic crossing-over, nor by the loss of the X chromosome bearing the other allele (Baker and Belote, 1983). Skripsky and Lucchesi proposed that *Sxl*$^{f\#1}$/*Sxl*$^{fhv\#1}$ female embryos consist of clones of cells in which the level of Sxl product is sufficient for female differentiation and an appropriate level of X-chromosome function, and clones in which the level of Sxl product allows male differentiation and hyperactive X chromosomes. Cells of the latter type would be subjected to selection and replacement with cells in the normal range of X-chromosome activity. If hyperactivation of the X chromosomes were prevented by the presence of *msl* mutations, these cells could survive and may differentiate male structures. This explanation was endorsed by Cline (1984) and extended to the sex transformation of *Sxl*$^{f\#1}$/*Sxl*$^{+}$ females which, in the absence of *msls*, are fully viable and of normal appearance. He proposed that in such individuals, because of some error rate, a certain proportion of cells fail to activate their *Sxl*$^{+}$ gene. Because they occur in limited numbers, such cells would normally be eliminated, without any morphological consequences to the organism, unless *msl* mutations were present and allowed their survival. Because they lack Sx1^{+} product, these cells could then give rise to male structures. Sex-transforming interactions between *Sxl* and the *msls* were also reported by Uenoyama *et al.* (1982a) and Uenoyama (1984).

A similar interaction between *da* and the *msls* was briefly reported by Cline (1982) and was later more fully documented by this author (Cline, 1984). As previously mentioned (Section IV,C,1), *da* females reared at low temperatures produce a small proportion of live progeny females which exhibit numerous abnormalities characteristic of cell death during development (the death of these cells is presumably due to an insufficiency of da^{+} gene product, leading to a failure to activate *Sxl*$^{+}$ and, consequently, to the hyperactivation of the X chromosomes). The presence of homozygous *msl* mutations in daughters produce by *da* females under permissive conditions has no effect on their viability but transforms them into sexual mosaics. These intersexes offer no evidence of developmental damage, suggesting that lethal cells with hyperactive X chromosomes are rescued by the presence of the *msl* and are allowed to express their male sexual phenotype.

As one might expect, *sis-a* has been shown to interact with the *msls* in a manner concordant with its functional relatedness to *da* and *Sxl*$^{f\#1}$ (Cline, 1984). While no rescue of *sis-a* mutant females could be achieved, *sis-a*/+ females homozygous for *mle* exhibited some masculinization of their forelegs (presence of sex comb bristles), ascribable to the presence of the *sis-a* allele. Mention should be made that the mothers of these females were heterozygous for *da*, presumably in order to sensitize them to the effects of *sis-a*.

B. Other Trans-Acting Regulatory Genes

The screens designed to identify male-specific lethal mutations at new loci have yielded additional alleles of the four existing genes, leading to the conclusion that the number of these genes in the genome is not large and may, in fact, be limited to those already on hand (Belote and Lucchesi, 1980a; Lucchesi *et al.*, 1982; Uenoyama *et al.*, 1982a; T. W. Cline, personal communication). Arguments in support of the existence of additional genes contributing to the molecular mechanism responsible for X-chromosome hyperactivation have been advanced by Baker and Belote (1983) and by Cline (1984). Baker and Belote point out that X-chromosome transcription appears not to be set in a stepwise fashion at a male or female level; it is, instead, modulated in inverse proportion to the X/A, ratio which can be made to vary experimentally in less than unitary increments to the numerator. This would suggest that dosage compensation requires a level of control additional to that responsible for the simple activation of the *msls* or their products. Cline invokes the discrepancy in lethal periods between the male-specific lethal mutations (larval and pupal) and the lethal period of females from *da* mothers or homozygous for loss-of-function *Sxl* alleles (embryonic). Furthermore, he points out that the *msls* rescue somatic clones of cells in female-lethal embryos but never the organism as a whole. Cline proposes that the *msls* are responsible for dosage compensation late in development, acting in polytenic larval tissues and in imaginal disks and cells. These genes are distinct from another set of as yet undiscovered regulatory genes responsible for hyperactivation of the X chromosome in early embryos.

The rapid rate of discovery of new genetic elements involved in the regulation of sex differentiation and dosage compensation makes it virtually certain that all of the pieces of the puzzle are not yet available and justifies Baker and Belote's suggestion that an additional level of control may well exist for the modulation of the *msl* activity. The question of the existence of an additional gene or set of genes with a function comparable to that of the *msls*' but acting during early embryogenesis requires further discussion. Regarding the argument that their existence would explain the discrepancy in the time of lethality caused by *Sxl* and *da* mutations, on one hand, and *msl* mutations, on the other, a comparison of the lethal period of the *msls* with that of $Sxl^{M\#1}$ rather than $Sxl^{f\#1}$ would be more appropriate since this mutant allele is responsible for the same type of effect on X-linked gene activity, namely a failure to hyperactivate in males, as the *msl*s. In fact, $Sxl^{M\#1}$/Y males appear to develop further than $Sxl^{f\#1}$ females or females from *da* mothers.

Of practical significance to the question of the existence of early-acting regulators of compensation is the failure of the various mutant screens to identify mutations in these genes. The screens, after all, were designed to detect autosomal sex-specific lethal mutations, irrespective of the time of occurrence of the mediated lethality. It is possible, of course, that such mutations may be dominant lethals in males or that they may affect, in addition to the process of X hyperactivation in males, some fundamental step equally important to both sexes. If this were the case, very special screening procedures would have to be devised to identify these genes.

The possibility that dosage compensation may not operate at all in early embryogenesis or in the female germ line, for that matter, is raised by the observation that none of the homeo-box-containing genes (numbering well over a dozen, to date) are present on the X chromosome.

P. Gergen (personal communication) has attempted to obtain direct evidence that the *msl*s are late acting in regulating the level of X-chromosome activity. He made use of a leaky loss-of-function allele of the X-linked segmentation gene *runt*, which is known to be active in the blastoderm stage (Nüsslein-Volhard and Wieschaus, 1980). Gergen reports that female *runt* embryos carrying $Sxl^{f\#1}$ or produced by *da* mothers have a dramatically improved segmentation phenotype, consistent with a presumed hyperactivation of their X chromosomes. Simultaneous homozygosity for three *msl*s had no effect on male *runt* embryos, leading to the conclusion that these genes are not active during that early period of embryogenesis. Surprisingly, the $Sxl^{M\#1}$ allele had no effect either directly on the runt phenotype or on the improvement mediated by the *da* maternal effect.

C. A Model for the Action of Trans-Acting Genes

Dosage compensation could be achieved by a decrease of X-linked gene activity in the female such that the amount of product resulting from two gene doses is decreased to the level resulting from the single gene dose in the male. It could also be achieved by an increase of X-linked gene activity in the male such that the amount of product resulting from a single gene dose is increased to the level resulting from two doses in the female. Maroni and Plaut (1973a,b) and Schwartz (1973) proposed that equalization of gene products in males and females could be explained by postulating the existence of factors necessary for the transcription of all X-linked genes. If there were numerous sites on the X chromosome competing for these factors whose concentrations were limited, an increase in the number of X chromosomes per genome would effectively titrate the intranuclear concentration of the regulatory molecules. The probability of transcription of an X-linked gene would be directly proportional to the concentration of regulatory factors and inversely proportional to the number of X chromosomes in the genome.

The evidence derived from the study of the effects of the *msls* strongly suggests that dosage compensation is achieved by a mechanism which enhances gene activity in males. The lack of somatic effects of *msl* mutations in females may indicate that the wild-type products of these genes are not necessary in this sex. The regulatory mechanism would operate in genomes where the X/A ratio is less than 1. Such genomes include, in addition to normal diploid males, triploid intersexes, metamales, and basically diploid females or males with large deficiencies or duplications, respectively. In these individuals, numerous X-chromosome sites would compete for the regulatory factors whose synthesis would be modulated so that their concentration is inversely related to the magnitude of the X/A ratio.

VI. Evidence for Cis-Acting Regulatory Sequences Responsible for Dosage Compensation

Dosage compensation, i.e., the equalization of X-linked gene transcripts in males and females, is achieved by enhancing the relative rate of transcription in males. Is this due to an inherent regulatory feature of the genes themselves or is it a characteristic of the X chromosome, some environmental imperative imposed by the X chromosome on the structural genes which happen to be present on it? This question can be addressed by relocating X-linked genes to the autosomes or autosomal genes to the X chromosome and measuring the relative level of activity of these genes in males and females. Such an analysis can be performed at two different levels. In one type of experiment, the relocated genes are in a chromosome segment which, although cytologically small, nevertheless spans several polytene chromosome bands and several hundreds of kilobases of DNA. In another type of experiment, the relocated gene is included in a DNA segment

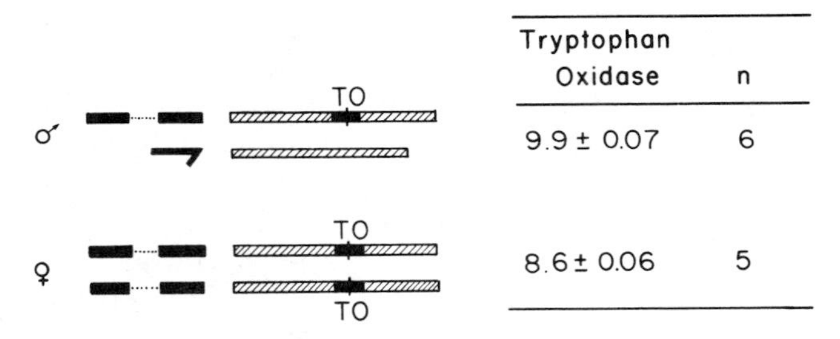

FIG. 8. Example of a test for dosage compensation of X-chromosome fragments translocated to an autosome. X-Chromosome material is represented by thick solid lines; the Y chromosome is a thin, solid J-shaped symbol; one pair of autosomes (hatching) is represented. The translocated fragment, which is cytologically small, contains, nevertheless, numerous genes; among these is v^+, the structural gene for the enzyme tryptophan oxidase (TO), whose activity is expressed in nanomoles of kynurenine/2 hours/milligram protein. Activity levels of the enzyme are the same in one-dose males and two-dose females, demonstrating that the gene has remained compensated. (After Tobler *et al.*, 1971.)

of less than a dozen kilobases. Each approach provides useful information.

A. Translocation of Small Chromosome Fragments

1. X-Chromosome Fragments to an Autosome

As mentioned in Section I,C, special stocks, in which small segments of the X chromosome are inserted into an autosome, can be used to produce females with three doses and males with two doses of a gene whose activity can be conveniently measured. The direct correlation between gene product levels and gene dosage in each sex (Seecof *et al.*, 1969; Gvosdev *et al.*, 1970; Baillie and Chovnick, 1971; Tobler *et al.*, 1971) indicates that genes on an X-chromosome interval exhibit dosage compensation regardless of their location in the genome. An example of this type of observation is presented in Fig. 8. The X-chromosome fragment translocated to chromosome 3 contains v^+, the structural gene of tryptophan oxidase. Activity levels of the enzyme were found to be the same in males with one dose and females with two doses of the relocated fragment, demonstrating that the v^+ gene has remained compensated.

Similar results were reported by Holmquist (1972), who compared, by means of transcription autoradiography of larval polytene chromosomes, the transcription rate of an X-chromosome interval translocated into the left arm of chromosome 3 with that of the same interval on a normal X chromosome.

2. *Autosomal Fragments to the X Chromosome*

Roehrdanz *et al.* (1977) used a special stock in which a small fragment of chromosome 3 containing the locus of *Aldox*$^+$, the structural gene for aldehyde oxidase, was inserted into the X chromosome. They found the same level of AO activity in males and females with a single dose of the relocated fragment. These results (illustrated in Fig. 9) offer evidence that the genetic activity of an autosomal segment is unaltered by its translocation to the X chromosome and is not modulated by the regulatory mechanism responsible for dosage compensation.

These results were concordant with the observation that in the salivary gland nuclei of male larvae, an autosomal segment inserted into the X chromosome retains the same appearance and degree of compactness as the homologous region in a normal autosome. It does not assume the enlarged and pale appearance characteristic of the X chromosome (Lakhotia, 1970).

Before leaving the discussion on translocated fragments, mention should be made of an older paper by Muller and Kaplan (1966) in which these authors reported that, in translocations between the X chromosome and chromosome 4, no influence of the rearrangement could be seen on the width or staining intensity (i.e., degree of compactness) of either of these chromosomes, in salivary gland nuclei of either sex. Muller and Kaplan concluded that the different chromosomal segments studied exhibited morphological and, therefore, probably functional autonomy.

B. Relocation of Cloned Genes

The observations discussed in the previous section lead to the conclusion that X-chromosome and autosomal fragments behave autonomously in translocations and that they do not respond to their new chromosomal environment. Recent experiments, in which genes contained in small cloned DNA segments are relocated by P-element-mediated germ-line transformation, suggest that the situation may not be quite so straightforward.

1. *Autosomal Sequences Transduced to the X Chromosome*

Spradling and Rubin (1983), using P-element-mediated transformation, constructed a series of strains that contain a single copy of the wild-type allele of *rosy* (*ry*$^+$), the structural gene for xanthine dehydrogenase (XDH), at different sites in the genome. XDH levels produced by autosomal insertions were measured against a null background, i.e., in lines homozygous for a null-activity indigenous *ry*

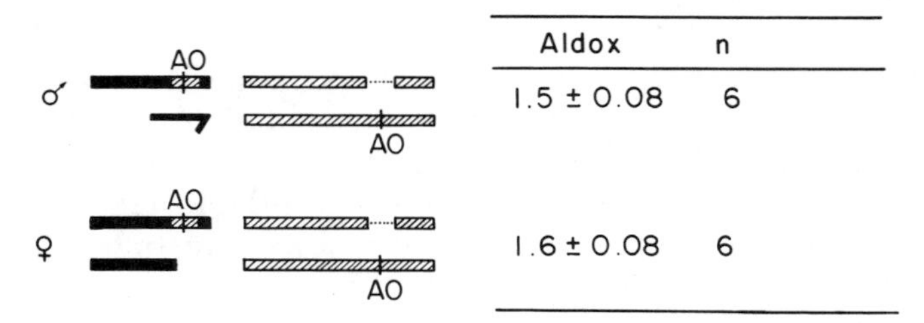

FIG. 9. Example of a test for dosage compensation of autosomal fragments translocated to the X chromosome. See Fig. 8 for an explanation of the symbols. The translocated fragment, which is cytologically small, contains, nevertheless, numerous genes; among these is *Aldox*$^+$, the structural gene for the enzyme aldehyde oxidase (AO), whose activity is expressed as change in absorbance/minute/milligram protein ($\times 10^2$). Enzyme activity levels are equivalent in males and females, demonstrating that the relocated gene has not become compensated. (After Roehrdanz *et al.*, 1977.)

allele. These lines yielded an average ratio of activity in males to females of 1.6 (the ratio was 1.7 in a wild-type control, reflecting the differential distribution of the enzyme in the two sexes). In contrast, the mean activity ratio in lines in which the transduced *ry*$^+$ gene is X linked was 2.6, indicating a substantial enhancement of activity in males and, therefore, a substantial level of compensation. Some of these data are reproduced in Table 4 and an example of a *ry*$^+$ gene transduced to an ectopic site are represented in Fig. 10.

Scholnick *et al.* (1983) reported that one copy of the structural gene (*Ddc*$^+$) for DOPA decarboxylase, relocated to the X chromosome from its normal position on chromosome 2 and measured against a null background, yielded an enzyme activity level which was 1.38-fold higher in males than in females at eclosion, and 1.64-fold higher in 30-hour-old flies. Marsh *et al.* (1985), in contrast, found little if any compensation for a *Ddc*$^+$ gene relocated to the X chromosome in prepupae or at eclosion, although, once again, 30-hour-old males had significantly higher activity levels than females.

Goldberg *et al.* (1983) found no significant difference in the activity levels of alcohol dehydrogenase in males and in females carrying one dose of *Adh*$^+$ transduced to the X chromosome at two different locations. A more extensive analysis of the activity of one of these two genes by Devlin *et al.* (1985b) revealed that it is, in fact, compensated. Additional cases of compensation of *Adh*$^+$ relocated to the X chromosome have been observed by C. Savakis and M. Ashburner (unpublished results).

It should be noted that, in all of the cases just discussed, the relocated genes often produce lower enzyme activity levels than do the

TABLE 4
Dosage Compensation of Transduced ry^+ Genes[a]

ry^+ insertion	Cytological position[b]	Male/female XDH activity[c]
Oregon-R	87D (3R)	1.7
R 301.1	42A (2R)	1.7
R 310.1	93B (3R)	1.3
R 703.1	22A (2L)	1.6
R 301.2	12D (X)	2.95
R 403.1	7D (X)	2.2
R 404.2	9A (X)	3.1
R 702.1	1F (X)	2.5

[a] After Spradling and Rubin (1983).
[b] Cytological position refers to polytene chromosome map; chromosome arms are given in parentheses.
[c] Ratio or mean ratios (two or three independent measurements) measured in crude extracts of 80 adult males or females containing a single copy of the ry^+ transposon.

wild type. In most instances, though, the enzyme activity is displayed with the correct pattern of tissue distribution.

A cautionary note, prompted by some unexpected sex differences in the expression of autosomal genes relocated to ectopic autosomal sites, was sounded by Bourouis and Richards (1985). In their experiments, the autosomal salivary gland secretion polypeptide gene *Sgs-3*, transduced to the X-chromosome, showed enhanced expression in males. However, similar results were obtained in some lines with an autosomal insertion of the gene. Bourouis and Richards stated that the nature of these sex differences in expression should be resolved before studying dosage compensation of the transduced *Sgs-3* gene.

2. *X-Chromosome Sequences Transduced to an Autosome*

These experiments have been performed with w^+, *Sgs-4*, and an X-linked heat-shock gene of *D. pseudoobscura*. Additional experiments utilizing Zw^+ and Pgd^+ are currently underway.

Hazelrigg *et al.* (1984) were the firsts to relocate a cloned X-linked gene to ectopic autosomal sites by P-element-mediated germ-line transformation. The gene in question, w^+, is responsible for pigment deposition in the eyes and other adult tissues, as well as larval tissues.

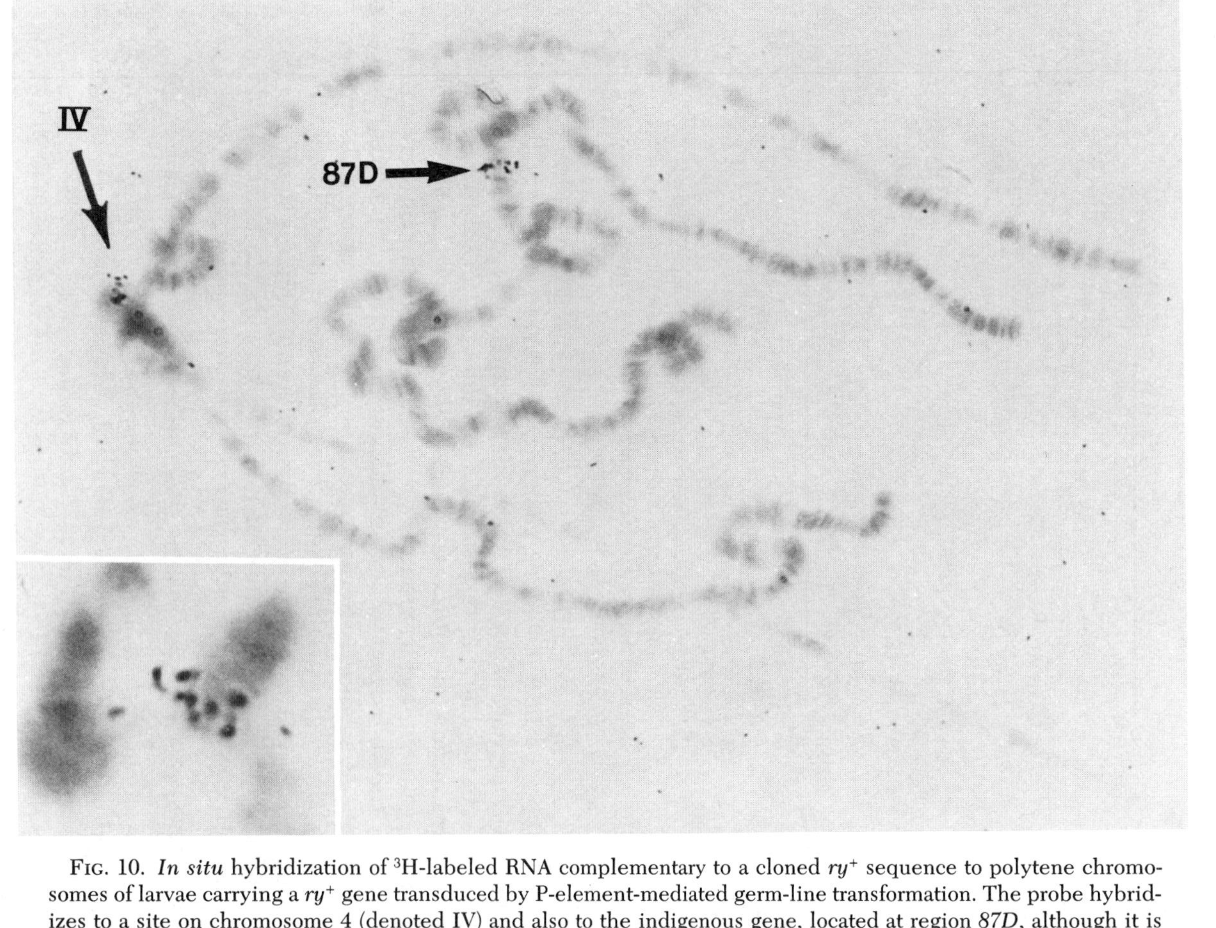

FIG. 10. *In situ* hybridization of ^{3}H-labeled RNA complementary to a cloned ry^{+} sequence to polytene chromosomes of larvae carrying a ry^{+} gene transduced by P-element-mediated germ-line transformation. The probe hybridizes to a site on chromosome 4 (denoted IV) and also to the indigenous gene, located at region *87D*, although it is represented by null alleles. [Courtesy of Allan Spradling, from a figure in Rubin and Spradling (1982).]

TABLE 5
Dosage Compensation of Transduced w^+ Genes[a]

w^+ insertion	Cytological position[b]	Male/female pigment ratio[c]
Canton S	3C (X)	1.46 (0.08)
A1-1	47A (2R)	1.54 (0.03)
A2-1	91C (3R)	1.92 (0.12)
A3-1	59B (2R)	1.93 (0.08)
B1-1	89A (3R)	1.53 (0.08)
B2-1	92BC (3R)	1.61 (0.02)
C1-1	87CD (3R)	1.60 (0.05)

[a] After Hazelrig *et al.* (1984).
[b] Cytological position as in Table 4.
[c] This ratio was determined from the average optical density of three independent pigment extractions from 10 heads of each genotype and sex; standard error is given in parentheses.

In all cases examined, a single dose of the relocated w^+ gene, in a null background, led to the deposition of more eye pigments in males than in females, indicating that the gene had retained the ability to compensate (Table 5). Since the smaller of the two transduced DNA segments was 11.7 k in length, Hazelrigg and her co-workers concluded that the cis-acting sequences reponsible for dosage compensation must be relatively small and tightly linked to the w^+ gene.

In a subsequent experiment, Levis *et al.* (1985) deleted varying lengths of the sequences normally flanking the w^+ gene and reintroduced the truncated DNA fragments into the germ line by transformation. They compared the eye pigmentation of males and females with a single dose of the transduced gene and determined that 400 bp of 5′-flanking DNA and 157 bp of 3′-flanking DNA are sufficient for dosage compensation. Somewhat surprisingly, the male value was as much as 2.6 times the female value, suggesting overcompensation. In a series of analogous experiments, published at precisely the same time, Pirrotta *et al.* (1985) were able to shorten the maximum length required for compensation to 210 bp upstream of the transcription start site. These workers also measured dosage effects of the transduced gene in males and females, with rather peculiar results. When one-dose males were compared to one-dose females, the same type of overcompensation found by Levis *et al.* (1985) was evident: one-dose males had up to a fourfold excess of pigment over one-dose females. Dosage effects also differed significantly from expectation, with two-dose males having as much as seven times the pigment level of one-dose males (rather than just twice as much) and two-dose females having up to nine times more pigment than one-dose females. Pirrotta *et al.* (1985) suggest that these effects may be due to a nonlinear response of pigmentation to gene dosage or to some influence of the genomic context on the transduced gene. We favor the second explanation, in light of the well-known effect of somatic pairing on the normal function of w^+, brought to light by the interaction of this gene with *zeste* mutations (Gans, 1953).

Krumm *et al.* (1985) relocated a 4.9-kb X-chromosome DNA fragment containing the *Sgs-4* coding region, with 2.6 kb of upstream and 1.3 kb of downstream sequences, to various new X-linked and autosomal sites in a strain which produces very low levels of Sgs-4 RNA or protein (Kochi-R strain). Males with one dose of the transduced gene at an X-linked ectopic site exhibited only 70–80% of the amount of Sgs-4 protein found in two-dose females. Males with two doses of the transduced gene on an autosome exhibited a 50% increase (rather than the expected 100% increase) in the protein over females. Furthermore, these values varied widely among the different ectopic sites. Nevertheless, taken *in toto*, these results allow the interpretation that sequences responsible for the hyperactivation of the *Sgs-4* gene, in males, are contained within the 4.9-kb fragment.

In the hope of identifying sequences wherein lesions could cause the loss of compensation of *Sgs-4*, Kaiser *et al.* (1986) sequenced a region of 600 bp, 5′ to the start of transcription, from two non-dosage-compensated strains (Korge, 1981; see Section I,C). A comparison of these sequences with those of three wild-type, compensated alleles (Muskavitch and Hogness, 1982; McGinnis *et al.*, 1983) revealed some single-base substitutions which, nevertheless, were not common to both noncompensated strains nor were found only in these strains. Kaiser *et al.* concluded that the cis-acting element responsible for dosage compensation of *Sgs-4* does not lie in the 600 bp immediately 5′ to the gene. McNabb and Beckendorf (1986) attempted to set a limit on how far upstream this element can reside. They measured the activity of *Sgs-4* genes, with 840 bp upstream of the coding region, transduced to autosomal sites. Here, again, the results were surprisingly variable with some autosomal sites and one X-linked site failing altogether to show compensation. Some sites appeared to show some level of transduced gene hyperactivity in males, leading McNabb and Beckendorf to suggest that the region controlling dosage compensation must lie 3′ to position −840. Kaiser *et al.* (1986) suggested that it must lie distal to −600, which would place this region in the 240-bp interval between these two coordinates.

An interspecific approach has been taken by Blackman and Meselson (1986) to identify the regulatory sequences responsible for the compensation of a heat-shock gene. In *D. melanogaster,* the *HSP82* gene is located on chromosome 3 while in *D. pseudoobscura* it is X linked and dosage compensated (Pierce and Lucchesi, 1980). Blackman and Meselson (1986) discovered that the *D. pseudoobscura* gene includes a region at −720 to −830 which represents a tandem repeat of the region at −20 to −130 and includes a second promoter. Since this repeat is absent from the autosomal *D. melanogaster* gene, they speculated that it may be responsible for dosage compensation. Jaffe and Laird (1986, personal communication) have identified a 7-bp sequence homology (GTCTCTT) between the 200-bp 5′ region of w^+ in *D. melanogaster* and the duplicated promoter region of *HSP82* of *D. pseudoobscura.* A consensus sequence of six of the above seven base pairs can be found upstream of the period gene (per^+) of *D. melanogaster.* It should be noted that the sequence is associated with at least one autosomal gene and is found at the end of P elements.

To enlarge the data base necessary for the identification of a consensus sequence responsible for compensation, we have begun the study of two X-linked genes—Zw^+ and Pgd^+—which differ from the genes just discussed in that they are "housekeeping" genes encoding enzymes responsible for sequential steps in the pentose shunt.

Pardue *et al.* (1987) have recently discovered that a synthetic polynucleotide (dC-dA)n·(dG-dT)n hybridizes to twice as many sites on the X chromosome than on the two major autosomes in salivary gland nuclei of *D. melanogaster* (no hybridization was seen on the dot chromosome). By analyzing other *Drosophila* species, these workers have established a correlation between the ability to compensate and the presence of high levels of the sequence. If it is, in fact, involved in this process, the sequence could achieve its function by acting as a binding site for a male-specific structural protein which would alter the conformation of chromatin, especially on the X chromosome.

C. Speculations on the Nature of Cis-Acting Regulatory Sequences

1. *Enhancement of Transcription*

The data just discussed lead to the general conclusion that a compensated X-chromosome gene relocated to an autosome remains compensated. This indicates the presence of cis-acting regulatory sequences closely linked to each gene. Yet, the relocation of autosomal genes to the X chromosome often results in their compensation, suggesting that these genes often come under the influence of the cis-acting elements of indigenous X-linked genes. A number of models, based on two types of organization of cis-acting elements (Fig. 11), can be derived to account for these observations. Three of these will be briefly discussed for illustrative purposes.

The first model proposes that there occurs along the X chromosome a number of enhancer-like sequences whose domains encompass multiple genes. These enhancers are activated in males, thereby achieving dosage compensation. Several examples exist in the literature of enhancers utilizing cell- or tissue-specific factors to mediate their effect on transcription (see Serfling *et al.*, 1985, for a review). It may not be unreasonable, therefore, to assume that sex-specific regulatory factors for control of enhancer function may be present in *Drosophila*. As expected, whenever an autosomal gene is relocated to the X chromosome, it comes under the influence of an enhancer unless it is surrounded by too long a portion of its original autosomal domain [as in the Roehrdanz *et al.* (1977) experiment with $Aldox^{+}$; Section VI,A,2]. Failure of transduced autosomal genes to hyperactivate in males could be caused by the inactivation of an enhancer sequence at the site of insertion; it could also result from insertion at the limit of the enhancer domain with vector sequences such as those of the P-element or of the marker gene (ry^{+}, Adh^{+}, etc.), placing the transduced gene of interest even further away. To account for the frequent retention of compensation by relocated X-chromosome genes, enhancer-like sequences should be densely distributed along the X chromosome so that the probability of including at least one of them in a DNA fragment containing a gene is high.

An alternate model postulates that each X-linked gene has its own regulatory sequence for compensation, equivalent to the sequences responsible for the temporal and spatial regulation of the gene. The fact that a significant number (perhaps a preponderance) of autosomal genes relocated to the X chromosome by transduction become compensated could be explained if P-element insertion were biased, i.e., if it showed a sequence preference which, more often than not, would place the construct under the influence of an X-linked gene regulatory site.

A variation of the model just described rationalizes the compensation of relocated autosomal genes by suggesting that enhanced transcription of the X chromosome in the male means a higher density of

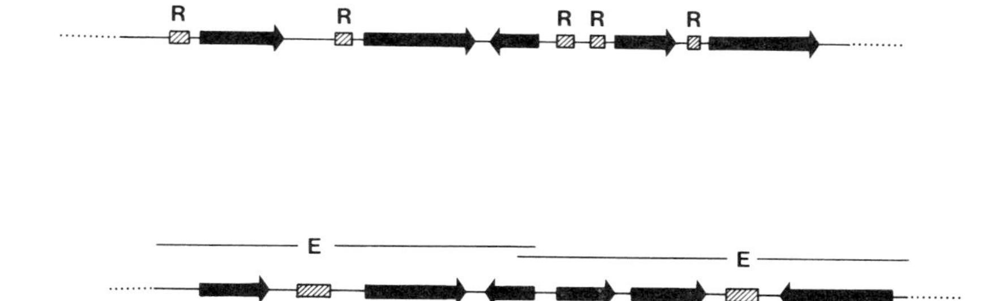

FIG. 11. Examples of the possible organization of cis-acting sequences responsible for dosage compensation. Transcriptional units on the X chromosome are indicated by the arrows. R, Regulatory sequences whose effect is specific to the gene with which they are associated. E, Enhancer-like sequences whose domain is indicated by straight lines.

RNA polymerase molecules and other transcriptional factors associated with (i.e., bound to or in the general vicinity of) X-linked transcriptional units. Any autosomal sequence transduced to this microenvironment would have an enhanced probability of being transcribed.

2. *Changes in Chromatin Conformation*

A current speculation is that enhancers function by providing a high density of RNA polymerase molecules over the upstream or downstream genes immediately adjacent to the enhancer locus. An alternative function for the hypothetical, X-linked enhancer-like sequences just discussed would be to alter the conformation of chromatin, thereby making X-linked genes generally more accessible to the transcriptional apparatus. Two types of experimental approaches have been taken to investigate this possibility.

Khesin and Leibovitch (1974) and Chatterjee and Mukherjee (1981) attempted to correlate the transcriptional characteristics of the X chromosome in males and females with chromatin organization by using a semi-*in vitro* experimental procedure. Devised by Umiel and Plaut (1973) and, independently, by Sederoff *et al.* (1973), the procedure consists of cytological preparations of fixed salivary gland polytene X chromosomes incubated with *E. coli* RNA polymerase in the presence of [^{3}H]UTP. Following a relatively short reaction time, the amount of chromosomal RNA synthesized by the heterologous polymerase with the polytene strands as template is measured by autoradiography. Under these conditions, the single X chromosome of males allowed the synthesis of as much RNA as was allowed by both X chromosomes in females. The rationale for interpretating these experimental results assumes that the binding of *E. coli* polymerase is directly proportional to the state of transcriptional activity of loci at the time of chromosome fixation. Chatterjee and Mukherjee (1981) used a high-salt treatment to extract various protein components from male and female polytene chromosome preparations prior to their use as *in vitro* templates. They found that, while the X chromosomes of females, as well as the autosomes of both sexes, showed a substantial increase in transcription, the male X chromosome was practically unaffected, suggesting that its organization is different from that of the other chromosomes.

Leibovitch *et al.* (1976) extended this type of measurement to triploid females (3X;3A), metafemales (3X;2A), and triploid intersexes (2X;3A), confirming the results which Ananiev *et al.* (1974) had obtained using *in vivo* transcription autoradiography (Section III,A). Chatterjee and Bhadra (1985) monitored template activity in segmental aneuploid larvae with findings generally similar to those of Maroni and Lucchesi (1980). Finally, Chatterjee and Bhadra (1985) measured the chromatin template activity of the X chromosome of larvae homozygous for mle^{ts}. They found a marked reduction of template activity for the X chromosome in males and no effect in females, in concordance with the results of Belote and Lucchesi (1980b).

A curious phenomenon was reported by Majumdar *et al.* (1978) and was recently studied in greater detail by Mukherjee and Ghosh (1986). These workers noted that, in a stock in which an X-chromosome inversion [$In(1)BM^2$] had reinverted to a cytologically normal sequence, approximately half of the nuclei of male larval salivary glands had an X chromosome 1.5–2 times as puffy as the X chromosome in a normal male; in an additional number of nuclei the X chromosome was intermittently puffy and nonpuffy. [^{3}H]Uridine incorporation *in vivo*, monitored by autoradiography, was found to be 40% higher in the puffy chromosomes than in the chromosomes of wild-type males or females homozygous for the reinverted X chromosome. This apparent increase in transcriptional activity was largely eliminated by the addition of fragments duplicating the proximal half of the X chromosome, for various lengths, and including the original breakpoints of $In(1)BM^2$. Surprisingly, a duplication for the distal tip of the X chromosome had the same effect. Mukherjee and Ghosh (1986) interpret these observations as evidence for the existence on the X chromosome of a locus responsible for the synthesis of an inhibitor of the autosomal gene products postulated to enhance X-chromosome activity. The reinverted chromosome would carry a residual *leaky* mutation at that locus, leading to an insufficiency of inhibitor and, therefore, to abnormally high X-chromosome hyperactivity. In a homozygous female, the two *leaky* alleles would produce enough inhibitor to titrate out the autosomal enhancers. It should be noted that no attempt was made to account for the suppressing effect of the distal tip duplication (whose presence eliminated the abnormally high levels of X-chromosome hyperactivation) into the general model. This duplication, extending from *1A* to *3E* on the cytological map, does not represent a region for which flies of either sex are senistive to variations in dosage (see Section VII,B).

All of the results just described have been interpreted to mean that the hyperactive X-chromosome has a DNA–protein organization different from that of a nonhyperactive X chromosome. Experiments performed by Eissenberg and Lucchesi (1982, 1983) failed to support this contention.

As previously mentioned (Section VI,B,2), two dosage-compensated heat-shock genes are present on the X chromosome of *D. pseudoobscura*. Eissenberg and Lucchesi tested whether a difference in the chromatin conformation of one of these genes, HSP83, could be detected among males and females by monitoring its sensitivity to nuclease digestion in the two sexes. This experimental approach stemmed from the original observation by Weintraub and Groudine (1976) that active genes retain a nuclease-sensitive conformation in relation to the surrounding chromatin (see Eissenberg *et al.*, 1985, for a review). Heat shock greatly increased the sensitivity of the *HSP83* gene to DNase I or micrococcal nuclease digestion, but to an equal extent in males and females, negating any difference in the configuration of the gene among the two sexes. Extraction of chromatin with 0.45 *M* salt prior to nuclease treatment removed the sensitivity to DNase I without revealing any covert differences among males and females. These data suggest that dosage compensation neither requires nor causes measurable changes in chromatin structure in order to effect hyperactivity of male X-linked genes.

What, then, of the correlation between the levels of X-chromosome function measured *in vivo* (transcription autoradiography and levels of steady-state mRNAs or of enzyme activity) and the levels of chromosomal RNA synthesis achieved *in vitro* with *E. coli* polymerase? Cytologically, a hyperactive polytenic X chromosome appears less compacted and more diffuse than an X chromosome with a lower activity level. This may be due to an unfolding of chromomeres on each individual chromatid of the polytenic bundle, which constitutes the chromosome, or it may be due to a looser association among the chromatids of the bundle. Therefore, it is possible that the greater level of transcription of the X chromosome in males by a heterologous polymerase reflects an easier access for polymerase molecules to the less densely associated chromatids of the male, rather than indicating a change in chromatin conformation of the active loci on individual chromatids.

VII. Overview of the Regulation of Dosage Compensation

A. The Regulatory Hierarchy

In the preceding sections, evidence has been presented that the primary determinant of dosage compensation is the ratio of the number of X chromosomes to the number of sets of autosomes present in the genome of a zygote or of an early embryo. A signal signifying this ratio is translated and transmitted by a group of regulatory genes, with maternal and/or zygotic functions, to another set of regulators whose products are specifically necessary for proper hyperactivation of the X chromosome in males. This regulatory hierarchy is diagrammed in Fig. 12.

A fundamental aspect of this hierarchy, currently under rather intensive investigation, concerns the identification of the genetic elements or molecular species which signal the number of X chromosomes and their ratio to the sets of autosomes in a developing embryo. The experimental progress achieved to date on this particular aspect will be reviewed in the next section.

B. How Does an Embryo Count Its X Chromosomes?

One would suspect that the simplest mechanism may require two components, one X linked and the other autosomal. Elements on the X chromosome and on the autosomes would elaborate their products in a dosage-dependent fashion. Female (2X;2A) zygotes could not tolerate a deficiency for the X-linked element since the X/A ratio of the components would be male and hyperactivation of their two X chromosomes would ensue. Males (1X;2A) could not tolerate a duplication since this would lead to a female level of X-linked component and repression of X-chromosome function. Both sexes would also be sensitive to dosage variation for the autosomal element. In this case, males with one dose should develop as females since their X/A ratio would be equal to 1. Females with three doses should develop as males (or intersexes).

There does not appear to exist, on the X chromosome, a single locus or segment that exhibits the characteristics just described, i.e., that is

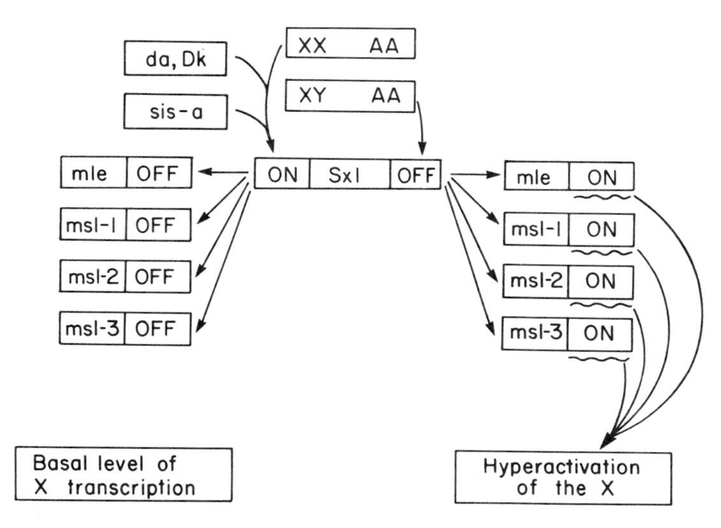

FIG. 12. Diagram illustrating the various levels of the regulatory hierarchy responsible for dosage compensation. The da and Dk gene products are maternal contributions to the egg cytoplasm. The action of the *sis-a* gene and of other signals emanating from the X/A ratio is zygotic. In combination, these elements activate the female-specific functions of *Sxl*, which, in turn, renders the *msl* genes or their products inactive. A male X/A ratio leads to a male-specific function for *Sxl* and the *msl* genes and to hyperactivation of the X chromosome. The symbolism of this diagram is similar to that used by Baker and Belote (1983; see Fig. 3) to illustrate the regulatory interactions leading to sex differentiation.

both lethal when present in single dose in females and lethal when duplicated in males. It may be reasonable to infer, therefore, that the X-linked component consists of at least two and most likely several elements. These elements may interact to produce a unique signal, may contribute additively to a unique signal, or may elaborate diverse signals. Reasonable candidates for these types of X-chromosome functional units have been identified at regions *11DE–12E* and *16E–17A*. The first of these small regions is apparently not tolerated in two doses in males (Stewart and Merriam, 1975; Belote and Lucchesi, 1980a), although females with a single dose are viable. Of interest is the recent discovery by Scott and Baker (1986) of a locus within this region involved in sex differentiation. The *16E–17A* region has an opposite effect: it is haplo-insufficient in females, while males with two doses do not appear to be affected. Curiously, this region does not seem to include the segment which suppresses the abnormally puffy appearance of the X chromosome in the In(1)BM[2] strain studied by Mukherjee and Ghosh (1986).

A third region, *3F3–4B1*, was recently reported by Steinmann-

Zwicky and Nöthiger (1985) to have the same dosage-related effects on viability as *11DE–12E*. Using a clever experimental design, these investigators suggested a correlation between the dosage of their region and the activity of *Sxl*$^+$. Mention should be made that, on the basis of his own experiments, Cline disputes this conclusion (personal communication). Additional observations suggest that other regions on the X chromosome contribute to the activation of this gene. The *sis-a*$^+$ gene, for example, is needed to help establish the activity level of *Sxl*$^+$ (Cline, 1986; see Section IV,C,3). Other repressor genes, analogous to *Sxl*$^+$ and activated by the appropriate dose of activator elements, could be present on the X chromosome. The nature of the activators themselves is still a matter of speculation. They could consist of inducers encoding molecules effecting a positive control on *Sxl*$^+$ and other analogous loci. Chandra (1985), on the other hand, proposed a model for sex determination in which the "activators" are noncoding stretches of DNA which perform their function by binding, with high affinity, a repressor of *Sxl*$^+$ elaborated by the autosomes. Waring and Pollack (1987) have isolated a 372-bp sequence repeated 300–400 times and localized to cytogenetic regions 4 to 14A on the X chromosome of *D. melanogaster* (it is also present at one minor autosomal site). They note that this sequence is analogous to the one proposed in Chandra's model and suggest that it may be involved in establishing the X/A ratio.

Irrespective of the nature of the activators or of the actual molecular reactions mediated by the product of the *Sxl*$^+$ gene, it would appear that X-chromosome transcription is regulated by a cascade of signals in females. Absence of these signals in males results in the expression of those genes whose products are necessary for X hyperactivation. Baker *et al.* (1986) suggest that this expression is "by default," in a manner analogous to that of autosomal genes, rather than by any sex-specific control. Although this view is well suited to the facts of sexual differentiation, it does not take into account that, in males, the rate of transcription of the X chromosome per unit of DNA (or cytological map division, etc.) is twice that of the X chromosome in females or of an autosome in either sex. In order to achieve this special rate of transcription by simply interacting with the general cellular transcriptional machinery, the X chromosome must have acquired, through evolution, special regulatory sequences that must respond in a sex-specific manner.

To date, the most promising candidate for the autosomal element of the X/A ratio is still the *triplo-lethal* (*Tpl*$^+$) locus, located at region *83DE* on chromosome 3 (Lindsley *et al.*, 1972). This locus appears to be unique in the genome in that diploid individuals of both sexes with a single dose or with three doses die early during development. The lethality caused by a duplication of the locus was found to be greatly diminished by the presence in females of a third X chromosome (X/X/X; $Tpl^+/Tpl^+/Dp(Tpl^+)$. A few adult escapers were obtained with an X-chromosome duplication encompassing the *7CD* region (Roehrdanz and Lucchesi, 1981). X-Linked enzyme activity measurements failed to provide evidence that Tpl^+ dosage is involved in some fundamental signal-regulating dosage compensation. Yet, because the experiments generally involved comparisons between females and metafemales, their significance is questionable, and a reexamination of Tpl^+ lethality and its interaction with an X-chromosome site may be in order.

VIII. Perspective

The unique features which distinguish dosage compensation in *Drosophila* from other model systems for the study of coordinate gene regulation are as follows. First, it is a regulatory mechanism which is superimposed on the controls responsible for the particular developmental programs followed by particular genes. Second, it involves the transcriptional modulation of a very sizable fraction of the genome. Last, it is achieved by the interaction of genes and signals which belong to a hierarchy of at least three levels of regulation. The most fundamental of these levels is the means by which an embryo assesses the number of its X chromosomes and their ratio to the other chromosomes of the genome.

It is possible that the molecular mechanism of dosage compensation per se represents a regulatory solution that evolved specifically to solve the problem in fruit flies and offers, therefore, little hope for generalization. The means by which an embryo counts its chromosomes may be, on the other hand, such a fundamental process as to have been conserved throughout evolution. The continued study of the phenomena discussed in this review should lead to the elucidation of this process in *Drosophila* long before it may be understood in higher forms.

Acknowledgments

We are very grateful to the following colleagues for generously sharing with us their unpublished data and/or thoughts and for allowing us to include them in this article: Drs. Bruce Baker, Thomas Cline, Peter Gergen, Elizabeth Jaffe, Rolf Nöthiger, Lucas Sanchez, and Monica Steinmann-Zwicky. Research on dosage compensation in our laboratories is supported by a research grant from the National Institutes of Health (GM 15691).

References

Ananiev, E. V., Faizullin, L. Z., and Gvozdev, V. A. (1974). The role of genetic balance in control of transcription rate in the X-chromosome of *Drosophila melanogaster. Chromosoma* **45,** 193–201.

Aronson, J. F., Rudkin, G. T., and Schultz, J. (1954). A comparison of giant X-chromosomes in male and female *Drosophila melanogaster* by cytophotometry in the ultraviolet. *J. Histochem. Cytochem.* **2,** 458–459.

Bachiller, D., and Sanchez, L. (1986). Mutations affecting dosage compensation in *Drosophila melanogaster:* Effects in the germline. *Dev. Biol.* **118,** 379–384.

Baillie, D. L., and Chovnick, A. (1971). Studies on the genetic control of tryptophan pyrrolase in *Drosophila melanogaster. Mol. Gen. Genet.* **112,** 341–353.

Baker, B. S., and Belote, J. M. (1983). Sex determination and dosage compensation in *Drosophila melanogaster. Annu. Rev. Genet.* **17,** 345–393.

Baker, B. S., Nagoshi, R. N., and Burtis, K. C. (1986). Molecular genetic aspects of sex determination in *Drosophila. Bioessays* **6,** 66–70.

Beckendorf, S. K., and Kafatos, F. C. (1976). Differentiation in the salivary glands of *Drosophila melanogaster:* Characterization of the glue proteins and their developmental appearance. *Cell* **9,** 365–373.

Bell, A. E. (1954). A gene in *Drosophila melanogaster* that produces all male progeny. *Genetics* **39,** 958–959.

Belote, J. M. (1983). Male-specific lethal mutations of *Drosophila melanogaster.* II. Parameters of gene action during male development. *Genetics* **105,** 881–896.

Belote, J. M., and Lucchesi, J. C. (1980a). Male-specific lethal mutations of *Drosophila melanogaster. Genetics* **96,** 165–186.

Belote, J. M., and Lucchesi, J. C. (1980b). Control of X chromosome transcription by the maleless gene in *Drosophila. Nature (London)* **285,** 573–575.

Bhadra, U., and Chatterjee, R. N. (1986). Dosage compensation and template organization in *Drosophila: In situ* transcriptional analysis of the chromatin template activity of the X and autosomes of *D. melanogaster* strains trisomic for the left arm of the second and third chromosomes. *Chromosoma* **94,** 285–292.

Birchler, J. A., Owenby, R. K., and Jacobson, K. B. (1982). Dosage compensation of serine-4 transfer RNA in *Drosophila melanogaster. Genetics* **102,** 525–537.

Blackman, R. K., and Meselson, M. (1986). Interspecific nucleotide sequence comparisons used to identify regulatory and structural features of the *Drosophila hsp 82* gene. *J. Mol. Biol.* **198,** 499–515.

Bourouis, M., and Richards, G. (1985). Remote regulatory sequences of the *Drosophila* glue gene *sgs3* as revealed by P-element transformation. *Cell* **41,** 349–357.

Bowman, J. T., and Simmons, J. R. (1973). Gene modulation in *Drosophila:* Dosage compensation of *Pgd* and *Zw* genes. *Biochem. Genet.* **10,** 319–331.

Breen, T. R., and Lucchesi, J. C. (1986). Analysis of the dosage compensation of a specific transcript in *Drosophila melanogaster. Genetics* **112,** 483–491.

Brock, H. W., and Roberts, D. A. (1982). The LSP1-gene is not dosage compensated in the *Drosophila melanogaster* species subgroup. *Biochem Genet.* **20,** 287–295.

Camfield, G. R. (1974). A genetic and biochemical study of a temperature-sensitive vermilion mutation in *Drosophila melanogaster.* Doctoral thesis, Department of Zoology, University of British Columbia, Vancouver, Canada.

Chandra, H. S. (1985). Sex determination: A hypothesis based on noncoding DNA. *Proc. Natl. Acad. Sci. U.S.A.* **82,** 1165–1169.

Chatterjee, R. N., and Bhadra, U. (1985). *In situ* transcription analysis of chromatin template activity of X chromosome of *Drosophila melanogaster* carrying maleless gene (*mle*[ts]). *Indian J. Exp. Biol.* **23**, 353–355.

Chatterjee, S. N., and Mukherjee, A. S. (1971). Chromosomal basis of dosage compensation in *Drosophila*. V. Puff-wise analysis of gene activity in the X-chromosome of male and female *Drosophila hydei*. *Chromosoma* **36**, 46–59.

Chatterjee, R. N., and Mukherjee, A. S. (1981). Chromosomal basis of dosage compensation in *Drosophila*. X. Assessment of hyperactivity of the male X in situ. *J. Cell Sci.* **47**, 295–309.

Chatterjee, R. N., Dube, D. K., and Mukherjee, A. S. (1981). In situ transcription analysis of chromatin template activity of the X-chromosome of *Drosophila* following high molar NaCl treatment. *Chromosoma* **82**, 515–523.

Cline, T. W. (1976). A sex-specific temperature-sensitive maternal effect of the daughterless mutation of *Drosophila melanogaster*. *Genetics* **84**, 723–742.

Cline, T. W. (1978). Two closely linked mutations in *Drosophila melanogaster* that are lethal to opposite sexes and interact with daughterless. *Genetics* **90**, 683–698.

Cline, T. W. (1979). A male-specific lethal mutation in *Drosophila* that transforms sex. *Dev. Biol.* **72**, 266–275.

Cline, T. W. (1983). The interaction between daughterless and Sex-lethal in triploids: A lethal sex-transforming maternal effect linking sex determination and dosage compensation in *Drosophila melanogaster*. *Dev. Biol.* **95**, 260–274.

Cline, T. W. (1984). Autoregulatory functioning of a *Drosophila* gene product that establishes and maintains the sexually determined state. *Genetics* **107**, 231–277.

Cline, T. W. (1985). Primary events in the determination of sex in *Drosophila melanogaster*. *In* "The Origin and Evolution of Sex" (H. Halvorson and A. Munroy, eds.), pp. 301–327. Liss, New York.

Cline, T. W. (1986). A female-specific lethal lesion in an X-linked positive regulator of the *Drosophila* sex determination gene, *Sex lethal*. *Genetics* **113**, 641–663.

Devlin, R. H., Holm, D. G., and Grigliatti, T. A. (1982). Autosomal dosage compensation in *Drosophila melanogaster* strains trisomic for the left arm of chromosome 2. *Proc. Natl. Acad. Sci. U.S.A.* **79**, 1200–1204.

Devlin, H. R., Grigliatti, T. A., and Holm, D. G. (1984). Dosage compensation is transcriptionally regulated in autosomal trisomies of *Drosophila melanogaster*. *Chromosoma* **91**, 65–73.

Devlin, H. R., Grigliatti, T. A., and Holm, D. G. (1985a). Gene dosage compensation in trisomies of *Drosophila melanogaster*. *Dev. Genet.* **6**, 39–58.

Devlin, R. H., Holm, D. G., and Grigliatti, T. A. (1985b). Regulation of dosage compensation in X-chromosomal trisomies of *Drosophila melanogaster*. *Mol. Gen. Genet.* **198**, 422–426.

Dobzhansky, T. (1957). The X-chromosome in the larval salivary glands of hybrids *Drosophila insularis* × *Drosophila tropicalis*. *Chromosoma* **8**, 691–698.

Eissenberg, J. C., and Lucchesi, J. C. (1982). Chromatin structure and dosage compensated heat-shock genes in "*Drosophila pseudoobscura*. *In* "Heat Shock from Bacteria to Man" (M. J. Schlesinger, M. Ashburner, and A. Tissieres, eds.), pp. 109–113. Cold Spring Harbor Laboratory, Cold Spring Harbor, New York.

Eissenberg, J. C., and Lucchesi, J. C. (1983). Chromatin structure and transcriptional activity of an X-linked heat-shock gene. *J. Biol. Chem.* **258**, 13986–13991.

Eissenberg, J. C., Cartwright, I. L., Thomas, G. H., and Elgin, S. R. C. (1985). Selected topics in chromatin structure. *Annu. Rev. Genet.* **19**, 485–536.

Faizullin, L. Z., and Gvozdev, V. A. (1973). Dosage compensation of sex-linked genes in *Drosophila melanogaster*. *Mol. Gen. Genet.* **126**, 233–245.

Fukunaga, A., Tanaka, A., and Oishi, K. (1975). Maleless, a recessive autosomal mutant of *Drosophila melanogaster* that specifically kills male zygotes. *Genetics* **81**, 135–141.

Ganguly, R., Ganguly, N., and Manning, J. E. (1985). Isolation and characterization of the glucose-6-phosphate dehydrogenase gene of *Drosophila melanogaster. Gene* **35**, 91–101.

Gans, M. (1953). Etude genetique et physiologique du mutant *z* de *Drosophila melanogaster. Bull. Biol. Fr. Belg. Suppl.* **38**, 1–90.

Gartler, S. M., and Riggs, A. D. (1983). Mammalian X-chromosome inactivation. *Annu. Rev. Genet.* **17**, 155–190.

Goldberg, D. A., Posakony, J. W., and Maniatis, T. (1983). Correct developmental expression of a cloned alcohol dehydrogenase gene transduced into the *Drosophila* germ line. *Cell* **34**, 59–73.

Golubovsky, M. D., and Ivanov, Y. N. (1972). Autosomal mutation in *Drosophila melanogaster* killing the males and connected with female sterility. *Drosophila Inf. Serv.* **49**, 117.

Gvosdev, V. A., Birnstein, V. I., and Faizullin, L, Z. (1970). Gene-dependent regulation of 6-phosphogluconate dehydrogenase activity of *Drosophila melanogaster. Dros. Inf. Serv.* **45**, 163.

Hazelrigg, T., Levis, R., and Rubin, G. M. (1984). Transformation of *white* locus DNA in *Drosophila:* Dosage compensation, *zeste* interaction, and position effects. *Cell* **36**, 469–481.

Hildreth, P. E. (1965). Doublesex, a recessive gene that transforms both males and females of Drosophila into intersexes. *Genetics* **51**, 659–678.

Hildreth, P. E., and Lucchesi, J. C. (1963). A gene which transforms males and females into intersexes. *Proc. Int. Congr. Genet., 11th* **1**, 171.

Holmquist, G. (1972). Transcription rates of individual chromosome bands: Effects of gene dose and sex in *Drosophila. Chromosoma* **36**, 413–452.

Jaffe, E., and Laird, C. (1986). Dosage compensation in *Drosophila. Trends Genet.* **2**, 316–321.

Kafatos, F. C., Jones, C. W., and Efstradiadis, A. (1979). Determination of nucleic acid sequence homologies and relative concentrations by a dot hybridization procedure. *Nucleic Acids Res.* **7**, 1541–1552.

Kaiser, K., Furia, M., and Glover, D. M. (1986). Dosage compensation at the *sgs*4 locus of *Drosophila melanogaster. J. Mol. Biol.* **187**, 529–536.

Kaplan, R. A. (1970). An autoradiographic study of dosage compensation in *Drosophila melanogaster.* Ph.D. thesis, University of Wisconsin, Madison.

Kaplan, R. A., and Plaut, W. (1968). A radioautographic study of dosage compensation in *Drosophila melanogaster. J. Cell Biol.* **39**, 71a.

Kazazian, H. H., Jr, Young, W. J., and Childs, B. (1965). X-linked 6-phosphogluconate dehydrogenase in *Drosophila:* Subunit associations. *Science* **150**, 1601–1602.

Khesin, R. B. (1973). Binding of thymus histone F1 and *E. coli* RNA polymerase to DNA of polytene chromosomes of *Drosophila. Chromosoma* **44**, 255–264.

Khesin, R. B., and Leibovitch, B. A. (1974). Synthesis of RNA by *Escherichia coli* RNA polymerase on the chromosomes of *Drosophila melanogaster. Chromosoma* **44**, 161–172.

Komma, D. J. (1966). Effect of sex transformation genes on glucose-6-phosphate dehydrogenase activity in *Drosophila* melanogaster. *Genetics* **54**, 497–503.

Korge, G. (1970a). Dosage compensation and effect for RNA synthesis in chromosome puffs of *Drosophila melanogaster. Nature (London)* **225**, 386–388.

Korge, G. (1970b). Dosiskompensation und dosiseffekt fur RSN-synthese in chromosomen-puffs von *Drosophila melanogaster. Chromosoma* **30**, 430–464.

Korge, G. (1975). Chromosome puff activity and protein synthesis in larval salivary glands of *Drosophila melanogaster. Proc. Natl. Acad. Sci. U.S.A.* **72**, 4550–4554.

Korge, G. (1977). Direct correlation between a chromosome puff and the synthesis of a larval saliva protein in *Drosophila melanogaster. Chromosoma* **62**, 155–174.

Korge, G. (1981). Genetic analysis of the larval secretion gene *Sgs-4* and its regulatory chromosome sites in *Drosophila melanogaster. Chromosoma* **84**, 373–390.

Krumm, A., Roth, G. E., and Korge, G. (1985). Transformation of salivary gland secretion protein gene *Sgs-4* in *Drosophila:* Stage- and tissue-specific regulation, dosage compensation, and position effect. *Proc. Natl. Acad. Sci. U.S.A.* **82**, 5055–5059.

Lakhotia, S. C. (1970). Chromosomal basis of dosage compensation in *Drosophila.* II. DNA replication patterns in an autosome-X insertion in *D. melanogaster. Genet. Res.* **15**, 301–307.

Lakhotia, S. C., and Mukherjee, A. S. (1969). Chromosomal basis of dosage compensation in *Drosophila.* I. Cellular autonomy of hyperactivity of the male X-chromosome in salivary glands and sex differentiation. *Genet. Res.* **14**, 137–150.

Leibovitch, B. A., Belyaeva, E. S., Zhimulev, I. F., and Khesin, R. B. (1976). Comparison of *in vivo* and *in vitro* RNA synthesis on polytene chromosomes of *Drosophila. Chromosoma* **54**, 349–362.

Levis, R., Hazelrigg, T., and Rubin, G. M. (1985). Separable *cis*-acting control elements for expression of the *white* gene of *Drosophila. EMBO J.* **4**, 3489–3499.

Lindsley, D. L., Sandler, L., Baker, B. S. Carpenter, A. T. C., Denell, R. E., Hall, J. C., Jacobs, P. A., Miklos, G. L. G., Davis, B. K., Gethman, R. C., Hardy, R. W., Hessler, A., Miller, S. M., Nozawa, H., Parry, D. M., and Gould-Somero, M. (1972). Segmental aneuploidy and the genetic gross structure of the *Drosophila* genome. *Genetics* **71**, 157–184.

Lucchesi, J. C. (1983). The relationship between gene dosage, gene expression, and sex in *Drosophila. Dev. Genet.* **3**, 275–282.

Lucchesi, J. C., and Rawls, J. M., Jr. (1973a). Regulation of gene function: A comparison of enzyme activity levels in relation to gene dosage in diploids and triploids of *Drosophila melanogaster. Biochem. Genet.* **9**, 41–51.

Lucchesi, J. C., and Rawls, J. M., Jr. (1973b). Regulation of gene function: A comparison of enzyme activity levels in normal and intersexual triploids of *Drosophila melanogaster. Genetics* **73**, 459–464.

Lucchesi, J. C., and Skripsky, T. (1981). The link between dosage compensation and sex differentiation in *Drosophila melanogaster. Chromosoma* **82**, 217–227.

Lucchesi, J. C., Rawls, J. M., Jr., and Maroni, G. (1974). Gene dosage compensation in metafemales (3X;2A) of *Drosophila. Nature (London)* **248**, 564–567.

Lucchesi, J. C., Belote, J. M., and Maroni, G. (1977). X-linked gene activity in metamales (XY;3A) of *Drosophila. Chromosoma* **65**, 1–7.

Lucchesi, J. C., Skripsky, T., and Tax, F. E. (1982). A new male-specific lethal mutation in *Drosophila melanogaster. Genetics* **100**, s42.

Maine, E. M., Salz, H. K., Cline, T. W., and Schedl, P. (1985a). The *Sex-lethal* gene of *Drosophila:* DNA alterations associated with sex-specific lethal mutations. *Cell* **43**, 521–529.

Maine, E. M., Salz, H. K., Schedl, P., and Cline, T. W. (1985b). Sex-lethal, a link between sex determination and sexual differentiation in *Drosophila melanogaster. Cold Spring Harbor Symp. Quant. Biol.* **50**, 595–604.

Majumdar, D., Ghosh, M., Das, M., and Mukherjee, A. S. (1978). Extra hyperactivity of the X chromosome in spontaneous occurring mosaic salivary glands of *Drosophila. Cell Chromosome News Lett.* **1**, 8–11.

Mange, A. P., and Sandler, L. (1973). A note on the maternal effect mutants daughterless and abnormal oocyte in *Drosophila melanogaster. Genetics* **73**, 73–86.

Mann, R., Hake, L., and Lucchesi, J. C. (1986). Phenogenetics of triploid intersexes in *Drosophila melanogaster. Dev. Genet.* **6**, 247–255.

Maroni, G., and Lucchesi, J. C. (1980). X-chromosome transcription in *Drosophila. Chromosoma* **77**, 253–261.

Maroni, G., and Plaut, W. (1973a). Dosage compensation in *Drosophila melanogaster* triploids. I. Autoradiographic study. *Chromosoma* **40**, 361–377.

Maroni, G., and Plaut, W. (1973b). Dosage compensation in *Drosophila melanogaster* triploids. II. Glucose-6-phosphate dehydrogenase activity. *Genetics* **74**, 331–342.

Maroni, G., Kaplan, R., and Plaut, W. (1974). RNA synthesis in *Drosophila melanogaster* polytene chromosomes. *Chromosoma* **47**, 203–212.

McGinnis, W., Shermoen, A. W., Heemskerk, J., and Beckendorf, S. K. (1983). DNA sequence changes in an upstream DNase I-hypersensitive region are correlated with reduced gene expression. *Proc. Natl. Acad. Sci. U.S.A.* **80**, 1063–1067.

McNabb, S. L., and Beckendorf, S. K. (1986). *Cis*-acting sequences which regulate expression of the *Sgs-4* glue protein gene of *Drosophila. EMBO J.* **5**, 2331–2340.

Marsh, J. L., and Wieschaus, E. (1978). Is sex determination in germ-line and soma controlled by separate genetic mechanisms? *Nature (London)* 272, 249–251.

Marsh, J. L., Gibbs, P. D. L., and Timmons, P. M. (1985). Developmental control of transduced dopa decarboxylase genes in *D. melanogaster. Mol. Gen. Genet.* **198**, 393–403.

Marshall, T., and Whittle, J. R. S. (1978). Genetic analysis of the mutation female-lethal in *Drosophila melanogaster. Genet. Res.* **32**, 103–111.

Mukherjee, A. S. (1966). Dosage compensation in *Drosophila:* An autoradiographic study. *Nucleus* **9**, 83–96.

Mukherjee, A. S., and Beermann, W. (1965). Synthesis of ribonucleic acid by the X-chromosomes of *Drosophila melanogaster* and the problem of dosage compensation. *Nature (London)* **207**, 785–786.

Mukherjee, A. S., and Ghosh, M. (1986). A different level of X-chromosomal transcription in an In(1)BM^2 (reinverted) strain and in its hyperploid derivatives resolves an X-coded regulatory activity for dosage compensation in *Drosophila. Genet. Res.* **48**, 65–75.

Muller, H. J. (1932). Further studies on the nature and causes of gene mutations. *Proc. Int. Congr. Genet., 6th* **1**, 213–255.

Muller, H. J. (1950). Evidence of the precision of genetic adaptation. *Harvey Lect. Ser.* **XLIII**, 165–229.

Muller, H. J., and Kaplan, W. D. (1966). The dosage compensation of *Drosophila* and mammals as showing the accuracy of the normal type. *Genet. Res.* **8**, 41–59.

Muller, H. J., League, B. B., and Offermann, C. A. (1931). Effects of dosage changes of sex-linked genes, and the compensatory effects of other gene differences between male and female. *Anat. Rec. Suppl.* **51**, 110 (Abstr.).

Muskavitch, M. A. T., and Hogness, D. S. (1982). An expandable gene that encodes a *Drosophila* glue protein is not expressed in variants lacking remote upstream sequences. *Cell* **29**, 1041–1051.

Nüsslein-Volhard, C., and Wieschauss, E. (1980). Mutations affecting segment number and polarity in *Drosophila. Nature (London)* **287**, 795–801.

O'Brien, S. J., and MacIntyre, R. J. (1978). Genetics and biochemistry of enzymes and specific proteins of *Drosophila*. *In* "The Genetics and Biology of *Drosophila*" (M. Ashburner and T. R. F. Wright, eds.), Vol. 2A, pp. 396–551. Academic Press, New York.

Offermann, C. A. (1936). Branched chromosomes as symmetrical duplications. *J. Genet.* **32**, 103–116.

Okuno, T., Satou, T., and Oishi, K. (1984). Studies on the sex-specific lethals of *Drosophila melanogaster*. VII. Sex-specific lethals that do not affect dosage compensation. *Jpn. J. Genet.* **59**, 237–247.

Ota, T., Fukunaga, A., Kawabe, M., and Oishi, K. (1981). Interactions between sex-transformation mutants of *Drosophila melanogaster*. I. Hemolymph vitellogenins and gonad morphology. *Genetics* **81**, 429–441.

Pardue, M. L., Lowenhaupt, K., Rich, A., and Nordheim, A. (1987). (dC-dA)n·(dG-dT)n sequences have evolutionarily conserved chromosomal locations in *Drosophila* with implications for roles in chromosome structure and function. *EMBO J.* **6**, 1781–1789.

Pierce, D. A., and Lucchesi, J. C. (1980). Dosage compensation of X-linked heat-shock puffs in *Drosophila melanogaster*. *Chromosoma* **76**, 245–254.

Pipkin, S. B., Chakrabartty, P. K., and Bremner, T. A. (1977). Location and regulation of *Drosophila* fumarase. *J. Hered.* **68**, 245–252.

Pirrotta, V., Steller, H., and Bozzetti, M. P. (1985). Multiple upstream regulatory elements control the expression of the *Drosophila white* gene. *EMBO J.* **4**, 3501–3508.

Roberts, D. B., and Evans-Roberts, S. (1979). The X-linked-chain gene of *Drosophila* LSP-1 does not show dosage compensation. *Nature* (*London*) **280**, 691–692.

Roehrdanz, R. L., and Lucchesi, J. C. (1981). An X chromosome locus in *Drosophila melanogaster* that enhances survival of the triplo-lethal genotype, *Dp-(Tpl)*. *Dev. Genet.* **2**, 147–158.

Roehrdanz, R. L., Kitchens, J. M., and Lucchesi, J. C. (1977). Lack of dosage compensation for an autosomal gene relocated to the X chromosome in *Drosophila melanogaster*. *Genetics* **85**, 489–496.

Rubin, G. M., and Spradling, A. C. (1982). Genetic transformation of *Drosophila* with transposable element vectors. *Science* **218**, 348–353.

Rudkin, G. T. (1964). The proteins of polytene chromosomes. *In* "The Nucleohistones" (J. Bonner and P. Ts'o, eds.), pp. 184–192. Holden Day, San Francisco.

Sanchez, L., and Nöthiger, R. (1982). Clonal analysis of Sex-lethal, a gene needed for sexual development in *Drosophila melanogaster*. *Wilhelm Roux Arch. Entwicklungsmech. Org.* **191**, 211–214.

Sanchez, L., and Nöthiger, R. (1983). Sex determination and dosage compensation in *Drosophila melanogaster*: Production of male clones in XX females. *EMBO J.* **2**, 485–491.

Scholnick, S. B., Morgan, B. A., and Hirsh, J. (1983). The cloned dopa decarboxylase gene is developmentally regulated when reintegrated into the *Drosophila* genome. *Cell* **34**, 37–45.

Schüpbach, T. (1982). Autosomal mutations that interfere with sex determination in somatic cells of *Drosophila* have no direct effect on the germline. *Dev. Biol.* **89**, 117–127.

Schüpbach, T. (1985). Normal female germ cell differentiation requires the female X chromosome to autosome ratio and expression of sex-lethal in *Drosophila melanogaster*. *Genetics* **109**, 529–548.

Schwartz, D. (1973). The application of the maize-derived gene competition model to the problem of dosage compensation in *Drosophila*. *Genetics* **75**, 639–641.

Scott, T. N., and Baker, B. S. (1986). Genetic identification of a new locus involved in sex determination in *Drosophila melanogaster*. *Genetics* **113**, s35.

Sederoff, R., Clynes, R., Poncz, M., and Hachtel, S. (1973). RNA synthesis by exogenous RNA polymerase on cytological preparations of chromosomes. *J. Cell Biol.* **57**, 538–550.

Seecof, R. L., and Kaplan, W. D. (1967). Effect of gene dosage on G6PD and 6PGD activities in *D. melanogaster*. *Genetics* **56**, 587.

Seecof, R. L., Kaplan, W. D., and Futch, D. G. (1969). Dosage compensation for enzyme activities in *Drosophila melanogaster*. *Proc. Natl. Acad. Sci. U.S.A.* **62**, 528–535.

Serfling, E., Jasin, M., and Shaffner, W. (1985). Enhancers and eukaryotic gene transcription. *Trends Genet.* **1**, 224–230.

Skripsky, T., and Lucchesi, J. C. (1980). Females with sex combs. *Genetics* **94**, s98–s99.

Skripsky, T., and Lucchesi, J. C. (1982). Intersexuality resulting from the interaction of sex-specific lethal mutations in *Drosophila melanogaster*. *Dev. Biol.* **94**, 153–162.

Smith, P. D., and Lucchesi, J. C. (1969). The role of sexuality in dosage compensation in *Drosophila*. *Genetics* **61**, 607–618.

Spradling, A. C., and Rubin, G. M. (1983). The effect of chromosomal position on the expression of the *Drosophila* xanthine dehydrogenase gene. *Cell* **34**, 47–57.

Steinmann-Zwicky, M., and Nöthiger, R. (1985). A small region on the X chromosome of *Drosophila* regulates a key gene that controls sex determination and dosage compensation. *Cell* **42**, 877–887.

Steinmann-Zwicky, M., Bernhardsgrutter, E., and Nöthiger, R. (1987). In preparation.

Stern, C. (1960). Dosage compensation—Development of a concept and new facts. *Can. J. Genet. Cytol.* **2**, 105–118.

Stern, C. (1966). Pigmentation mosaicism in intersexes of *Drosophila*. *Rev. Suisse Zool.* **73**, 339–355.

Stewart, B. R., and Merriam, J. R. (1974). Segmental aneuploidy and enzyme activity as a method for cytogenetic localization in *Drosophila melanogaster*. *Genetics* **76**, 301–309.

Stewart, B. R., and Merriam, J. R. (1975). Regulation of gene activity by dosage compensation at the chromosomal level in *Drosophila*. *Genetics* **79**, 635–647.

Sturtevant, A. H. (1945). A gene in *Drosophila melanogaster* that transforms females into males. *Genetics* **30**, 297–299.

Tanaka, A., Fukunaga, A., and Oishi, K. (1976). Studies on the sex-specific lethals of *Drosophila melanogaster*. II. Further studies on a male-specific lethal gene, maleless. *Genetics* **84**, 257–266.

Thomas, P. S. (1983). Hybridization of denatured RNA transferred or dotted to nitrocellulose paper. *In* "Methods in Enzymology" (R. Wu, L. Grossman, and K. Moldave, eds.), Vol. 100, pp. 255–266. Academic Press, New York.

Tobler, J., and Grell, E. H. (1978). Genetic and physiological expression of β-hydroxy acid dehydrogenase in *Drosophila*. *Biochem. Genet.* **16**, 333–342.

Tobler, J., Bowman, J. T., and Simmons, J. R. (1971). Gene modification in *Drosophila:* Dosage compensation and relocated v^+ genes. *Biochem. Genet.* **5**, 111–117.

Uchida, S., Uenoyama, T., and Oishi, K. (1981). Studies on the sex-specific lethals of *Drosophila melanogaster*. III. A third chromosome male-specific lethal mutant. *Jpn. J. Genet.* **56**, 523–527.

Uenoyama, T. (1984). Studies on the sex-specific lethals of *Drosophila melanogaster*. VIII. Enhancement and suppression of *Sxl*. *Jpn. J. Genet.* **59**, 335–348.

Uenoyama, A., Fukunaga, A., and Oishi, K. (1982a). Studies on the sex-specific lethals of *Drosophila melanogaster*. V. Sex transformation caused by interactions between a female-specific lethal, *Sxl,* and the male-specific lethals *mle(3)132, msl-2,* and *mle. Genetics* **102**, 233–243.

Uenoyama, T., Uchida, S., Fukunaga, A., and Oishi, K. (1982b). Studies on the sex-specific lethals of *Drosophila melanogaster.* IV. Gynandromorph analysis of three male-specific lethals, *mle, msl-2,* and *mle(3)132. Genetics* **102**, 223–231.

Umiel, N., and Plaut, W. (1973). Interaction of poly-L-lysine with chromatin. *J. Cell Biol.* **56**, 139–144.

Weintraub, H., and Groudine, M. (1976). Chromosomal subunits in active genes have an altered conformation. *Science* **193**, 848–856.

Waring, G. L., and Pollack, J. C. (1987). Cloning and characterization of a dispersed, multicopy, X chromosome sequence in *Drosophila melanogaster. Proc. Natl. Acad. Sci. U.S.A.* **84**, 2843–2847.

Whitney, J. B., III, and Lucchesi, J. C. (1972). Ontogenetic expression of fumarase activity in *Drosophila melanogaster. Insect Biochem.* **2**, 367–370.

Williamson, J. H., and Bentley, M. M. (1983). Dosage compensation in Drosophila: NADP-enzyme activities and cross reacting material. *Genetics* **103**, 649–658.

Young, W. J. (1966). X-Linked electrophoretic variation in 6-phosphogluconate dehydrogenase. *J. Hered.* **57**, 58–60.

Young, W. J., Porter, J. E., and Childs, B. (1964). Glucose-6-phosphate dehydrogenase in *Drosophila:* X-linked electrophoretic variants. *Science* **143**, 140–141.

Zimmering, S., and Muller, H. J. (1961). Studies on the action of the dominant female-lethal *Fl* and a seemingly less extreme allele Fl^{S}. *Dros. Inf. Serv.* **35**, 103–104.

DEVELOPMENTAL MORPHOGENESIS AND GENETIC MANIPULATION IN TISSUE AND CELL CULTURES OF THE GRAMINEAE

Fionnuala Morrish, Vimla Vasil, and Indra K. Vasil

Department of Botany, University of Florida, Gainesville, Florida 32611

I. Introduction

Members of the family Gramineae are an important source of food and animal fodder for mankind. Since the initial cultivation of cereals and grasses, plant breeders have continually developed new breeding techniques and breeding lines. These have been used successfully to develop plants with increased yield, protein quality, disease resistance, and other features, resulting in the more widespread cultivation and increased productivity of many crop plants. Notwithstanding the major improvements made with traditional breeding methods, certain limitations have prevented or slowed down the development of superior lines that would be more disease resistant and could be grown in a greater range of environments. These have been discussed in detail by Simmonds (1983, 1984), Sybenga (1983), Duvick (1984), and, more recently, Larkin (1985) and include (1) the limited gene pool available due to sexual incompatibility, (2) the low level of variability present in many breeding lines, and (3) the time scale of present breeding programs, which may extend over many years due to the need to break up undesirable genetic linkages and conversely to create and identify desired assemblages of useful traits.

A possible solution to these problems has, however, appeared with the development of procedures for regenerating plants *in vitro* from tissues or cells of the Gramineae (Bright and Jones, 1985; I. K. Vasil and Vasil, 1986; I. K. Vasil, 1987). These procedures, when combined with techniques which have been used in dicotyledonous plants to integrate selected genetic information into cells (Fraley *et al.*, 1986) and isolate mutants (Chaleff, 1981; Maliga, 1984; Meredith, 1984; Negrutiu *et al.*, 1984) or somaclonal variants (Larkin and Scowcroft, 1981, 1983b; Orton, 1984), give us potential access to a limitless pool of genetic variation. Furthermore, the rapid rate of propagation possible with tissue cultures of the Gramineae (Chandler and Vasil, 1984a; Dalton and Dale, 1985) and the proposed stability of some pathways of regeneration (Dalton and Dale, 1981; Lawrence, 1981; I. K. Vasil, 1983a, 1985) suggest that we should be able to tailor plant improvement and reduce the time scale involved in developing and propagating new varieties.

It must be emphasized that the procedures and potentials outlined above, and in other reviews on the application of tissue culture to plant breeding (e.g., Skirvin, 1978; Larkin and Scowcroft, 1981, 1983b; Sybenga, 1983; Orton, 1984; I. K. Vasil, 1984, 1985; Larkin, 1985), are primarily based on results with model plant species such as tobacco, and have yet to be successfully used in the development of valuable new crop plants. Extensive research conducted on the tissue culture of the Gramineae in the last few years has brought into focus the many problems which must be overcome before we can utilize tissue culture as a practical adjuvant to plant breeding.

The production of improved crop plants *in vitro* requires not only the successful alteration of genetic information or its transfer and integration into plant cells but also the regeneration of these cells to form plants which then reproduce and pass on the new trait(s) to their progeny. Presently there are bottlenecks encountered at each of these stages (King *et al.*, 1978; Thomas *et al.*, 1978; Jones, 1985; Larkin, 1985; Maddock, 1985; I. K. Vasil and Vasil, 1986). Fundamental to our ability to solve these problems is a greater understanding of the genetic and epigenetic changes occurring at the cellular level, both *in vivo* and *in vitro*, which may play a major role in controlling morphogenic competence, genome interactions, and genome stability. Recent reviews have given an overview of the tissue culture of the Gramineae and discussed some of the problems involved (Bright and Jones, 1985; I. K. Vasil and Vasil, 1986; I. K. Vasil, 1987). In this article we wish to bring into focus the importance of relating with a greater awareness the phenomena encountered in tissue culture to the events occurring *in vivo* at the whole plant level, particularly with reference to developmental morphogenesis and genetic stability. We also wish to emphasize the importance of understanding and recognizing how environmental, genetic, and epigenetic factors influence the phenotype of plant cells maintained *in vitro* and place limitations on our present ability to genetically manipulate and regenerate plants from cell cultures of the Gramineae.

II. Developmental Morphogenesis

Plants consist of a heterogeneity of cell phenotypes which arise as a consequence of differentiation and development. This heterogeneity is evident from observations on tissue histology, DNA (qualitative and quantitative), and studies on gene expression as manifested by changes in protein and isozyme production during differentiation and development (Scandalios, 1974; Nagl, 1978, 1979; D'Amato, 1977, 1985; Scandalios and Baum, 1982; Raghavan, 1983, 1986). With continuous developments in molecular biology we have gradually increased our understanding of the factors that may control the changes which occur during plant development and lead to heterogeneity. The suggestions put forward for mechanisms that may control gene repression and derepression during development include transposable elements (McClintock, 1951; Nevers *et al.*, 1986), DNA amplification (Wardell and Skoog, 1973), and DNA methylation (Razin and Riggs, 1983), among others (see Nagl, 1978, 1979). Changes in gene expression, and hence cell phenotype, may have a major influence on a cell's capacity to produce a callus or regenerate plants *in vitro*. It is for this reason that a knowledge of molecular, ultrastructural, and biochemical changes occurring during plant development is important in our attempts to culture and regenerate plants from tissues and cells maintained *in vitro*.

Environmental factors also play a role in controlling gene expression and cellular differentiation *in vivo*. Changes in photoperiod stimulate or inhibit flowering (e.g., Jacobs, 1979) and increases in temperature, reduction in soil water, and nutrient deficiencies can cause premature senescence (Thimann, 1980). Hence these factors too must be borne in mind when culturing plant tissues, particularly from donor plants grown under field conditions where environmental controls are not as stringent as those possible in the greenhouse. Environmental and developmental parameters do not, however, cease to control a cell's responses once it has been established *in vitro*. The evidence available indicates that morphogenic competence *in vitro* is dependent on the culture environment and the physiological and organizational state of the cells (i.e., calluses, cells, and protoplasts) (e.g., Cornejo-Martin *et al.*, 1979; V. Vasil *et al.*, 1983; I. K. Vasil, 1985).

The influence of environmental and developmental constraints on the responses of tissues or cells in culture has not been traditionally regarded as a major controlling factor in the success of plant regeneration *in vitro*. For this reason the following discussion will cover some of the many examples which demonstrate the influence of differential gene expression, of alterations in endogenous factors resulting from environmental and developmental changes, and of physiological gradients on cellular responses *in vitro*.

Because tissue culture is primarily aimed at the regeneration of plants, we first provide an overview of the regeneration pathways possible in cereals and grasses both *in vivo* and *in vitro*.

A. Plant Regeneration Pathways *in Vivo* and *in Vitro*

At the whole plant level, depending on the genotype, members of the Gramineae are propagated either vegetatively or from seed, which may be produced with or without fertilization (apomixis). While the majority are propagated from seed, some grasses and cereals must be vegetatively propagated. Others are apomictic, and many perennial species, as well as being propagated from seed, also produce rhizomes from which new tillers emerge every season. Seeds produced through sexual reproduction arise from a single cell (the zygote) and develop into a new plant through embryogenesis. Embryos may also arise through apomixis, in which they are produced from an unfertilized egg, from any other cell in the embryo sac, or from an individual somatic cell of the nucellus or integuments. In asexual or vegetative propagation, spontaneously or artifically excised plant parts produce buds, form roots, and grow into new plants. The young shoot is connected via procambial strands to the vascular system of the mother plant. In contrast to a bud the embryo does not establish vascular connections with the mother plant. Therefore, the most distinctive characteristic of any embryo is its anatomically discrete and closed vascular system. In all cases, plant regeneration through any of these pathways is triggered and controlled by both environmental and developmental influences (Wareing and Phillips, 1978; Evenari, 1984). Because of their influence on plant regeneration, adverse environmental conditions or genetic mutations can have a detrimental effect on the initiation and development of zygotic embryos, shoot primordia, or somatic embryos. For example, environmental factors have been shown to induce vivipary in grasses (Barnard, 1966) and developmental mutants are known for each stage in the development of the maize embryo (Sheridan and Neuffer, 1982).

Thus, as outlined above, the pathway of and competence for plant regeneration *in vivo* is controlled by genotype and by physiological and developmental constraints. A recognition of the role played by these factors on morphogenic capacity in plant cells is important in our attempts to understand why certain cell types and species appear recalcitrant *in vitro*, and must be borne in mind when assessing the propensity for, and the pathway of, morphogenesis in cell and tissue cultures of the Gramineae.

Plant regeneration from tissues of the Gramineae cultured *in vitro* has not been an easy task, but the use of young, immature tissues at specific stages of development and potent auxins such as 2,4-dichlorophenoxyacetic acid (2,4-D) has allowed plant regeneration from cells of all the major cereals and grasses (I. K. Vasil and Vasil, 1986). In the tissue culture of totipotent cells it is generally agreed that it is primarily the interaction between media constituents and endogenous factors in cultured tissues or cells that determines whether cells will divide and differentiate (Cassells *et al.*, 1982; Rajasekaran *et al.*, 1987a,b). Hence, provided cells are morphogenically competent and contain adequate levels of growth regulators and nutrients, subtle changes in exogenous hormones will direct cells along a pathway of proliferation or differentiation. Details of explants, media, and environmental conditions required for cell division and differentiation in many cereals and grasses are available in publications cited in I. K. Vasil and Vasil (1986).

Histological and morphological studies on tissue cultures of a number of different cereal and grass species indicate that like plants *in vivo*, tissues or cells *in vitro* may produce plants by a number of different pathways. These pathways, illustrated in Fig. 1, demonstrate that plants can be regenerated either directly from the explant or indirectly after an intervening callus stage. Direct regeneration (Fig. 1a) has been reported in *Dactylis glomerata* (Hanning and Conger, 1982), *Oryza sativa* (Wernicke *et al.*, 1981), and *Sorghum* (Dunstan *et al.*, 1978). Also, a number of grasses (Dalton and Dale, 1981), *Zea mays* (Raman *et al.*, 1980), and *Triticum aestivum* (Tanzarella and Greco, 1985) have been shown to undergo axillary bud proliferation when shoot meristems are cultured *in vitro*. This mode of regeneration is referred to as "microtillering" (Dalton and Dale, 1981). In *Lolium* this procedure allows the rapid and large-scale propagation of plants (Dalton and Dale, 1985), and due to the absence of a callus phase, which is believed to increase the risk of genome destabilization (D'Amato, 1985), this pathway of regeneration would appear suitable for clonal propagation.

EXPLANT RESPONSE	CALLUS TYPE	PLANT REGENERATION	STABILITY
a	 none	yes	stable
b	 rooty	no	
c	 shooty	yes	unstable
d	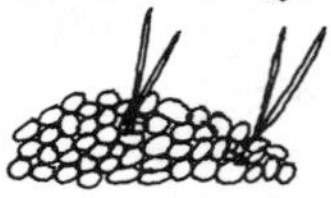 organogenic	yes	unstable
e	 embryogenic	yes	stable
f	 nonmorphogenic	no	

FIG. 1. Explant response, plant regeneration capacity, and genetic stability in tissue cultures of the Gramineae. (a) Direct embryogenesis from single epidermal cells of an explant, (b) callus with preformed root primordia, (c) callus with preformed shoot primordia, (d) callus generating shoot meristems *de novo*, (e) embryogenic callus forming embryos, (f) friable nonmorphogenic callus.

Indirect regeneration after the production of a callus from tissues or cells *in vitro* is the most common pathway of regeneration reported in the Gramineae (I. K. Vasil and Vasil, 1986). Detailed morphological and histological studies allow us to identify distinct callus types (see Fig. 1). These calluses have also been shown to differ in their cellular ultrastructure, organization, source of origin, and ability to regenerate plants from both primary and subcultured calluses. In the past a variety of terms have been used to describe these different calluses (King *et al.*, 1978; I. K. Vasil and Vasil, 1986). However, due to lack of agreement on a classification scheme there has been some confusion regarding the terminology ascribed to different callus types. A prime example of this is the erroneous description of an embryogenic callus as a callus which has the potential to regenerate plants either via organogenesis or embryogenesis (Tomes, 1985).

Because the nature of a callus can determine both its ability to regenerate and the stability of the regenerants (Maddock, 1985; I. K. Vasil and Vasil, 1986), it is important to be able to identify different callus types and selectively culture those which suit the required purpose of a given experiment, be it clonal propagation or the production of somaclonal variants. For this reason we have herein described and characterized the morphogenic capacity and stability of regenerants of five of the most commonly reported callus types in cell cultures of the Gramineae. The morphology of these calluses and their totipotency and stability are illustrated in Fig. 1.

Culture of mature and differentiated tissues of a number of cereals and grasses produces a callus which has a tendency to form roots when the 2,4-D concentration of the medium is lowered (called rooty callus, Fig. 1b) (e.g., Cure and Mott, 1978; O'Hara and Street, 1978; Ozias-Akins and Vasil, 1983). This has led some authors to suggest that these outgrowths do not represent true callus tissues but may in fact be proliferations of preformed root primordia (King *et al.*, 1978). Histological evidence exists for the presence of proliferating root primordia in cultures of wheat derived from roots (Cure and Mott, 1978). However, in calluses derived from wheat embryos, meristematic zones present do not show any organized structures during the early stage of growth (Ozias-Akins and Vasil, 1983) and hence it would not be correct to label these as proliferating root primordia. Calluses which contain presumptive shoot primordia may be termed shooty calluses (Fig. 1c). These have been reported, for example, in cultures derived from immature tissue of *Sorghum* (Dunstan *et al.*, 1979). In both rooty and shooty calluses it is suggested (King *et al.*, 1978) that the exogenous supply of hormones merely induces proliferation of preformed meristems; the dilution of shoot meristerms on subculture possibly explains the rapid reduction in regenerative capacity noted in shooty calluses over time. The tendency of these calluses to form roots or clumps of cells when transferred to liquid culture prevents the establishment of "true" suspension cultures and the isolation of viable protoplasts, a capability deemed essential for genetic manipulation *in vitro* (Section III). Furthermore, as plants regenerated from such calluses are derived primarily from cells of the original explant and not *de novo* from callus cells, these calluses are not suitable for use in studies on the factors controlling *de novo* morphogenesis or for mutagenic studies. Such studies are best conducted on calluses for which the *de novo* origin of shoots or embryos has been proved by histological evidence (Rangan, 1973; Nakano and Maeda, 1979; Shimada and Yamada, 1979; Springer *et al.*, 1979; V. Vasil and Vasil, 1981a, 1982; Botti and Vasil, 1984; Ho and Vasil, 1983b; Lu and Vasil, 1985). These morphogenic calluses may be divided into two distinct types, those undergoing organogenesis (OR) to produce adventitious shoots (Fig. 1d) and those producing somatic embryos (SE) (Fig. 1e). The morphology of these calluses, when established, is quite distinct but is difficult to distinguish during the early stages of development. These calluses may occur alone or together in mixed cultures. Organogenic callus has been identified in many different species and is produced by a number of different organs (Table 1 in I. K. Vasil and Vasil, 1986). This type of callus consists of parenchymatous and vacuolated cells, forming compact centers surrounded by soft and friable tissues, and has a rapid rate of growth. These calluses have been used to establish suspension cultures in a number of cereal and grass species. Such suspensions generally consist of a mixture of meristematic clumps, callus pieces, and nondividing single cells. Where plants have been regenerated from these suspensions they have been produced from callus clumps (e.g., Zimny and Lorz, 1986) and not from single cells.

To induce plant regeneration, calluses are generally transferred to media with lower auxin levels. Regeneration may be further improved by the addition of cytokinins. Plants are produced by the organization of shoot meristems which develop into adventitious shoots, form roots, and grow into new plants. During subculture the potential for shoot differentiation from these calluses generally decreases. But there are some reports of maintenance of regenerative capacity after prolonged periods of callus maintenance *in vitro* (e.g., Cummings *et al.*, 1976; Rines and McCoy, 1981; McCoy *et al.*, 1982; Sears and Deckard, 1982; Ahloowalia, 1982). These regenerants, however, are not all true to type and those with variant phenotypes have been classified as somaclonal variants (Larkin and Scowcroft, 1981).

Because of the *de novo* origin of shoots, these organogenic calluses have been used in many studies for the selection of mutants (Bright, 1985; see also Section III,C). However, histological studies (Springer *et al.*, 1979) and the production of chimeral and aneusomatic regenerants (e.g., Heinz and Mee, 1971; Bennici and D'Amato, 1978; Mix *et al.*, 1978; Lupi *et al.*, 1981) indicated that plants regenerated from these calluses may be multicellular in origin. Furthermore, all reported cases of variability in plant populations obtained from tissue cultures of various species of the Gramineae involve organogenic or mixed (organogenic + embryogenic) calluses. For these reasons organogenic calluses are unsuitable for routine clonal propagation or the production of solid mutants but evidently provide a source of somaclonal variants, the stability and utility of which must be evaluated by continuous studies on sexually or vegetatively propagated progeny.

Production of embryogenic callus has been reported in many cereal and grass species (Table 1 in I. K. Vasil and Vasil, 1986) and it has been suggested that this may be the most common pathway of plant regeneration *in vitro* for this group of plants (Vasil, 1982). Embryogenic callus is predominantly produced by immature and young tissues of infloresences, embryos, and leaves (e.g., Dale, 1980; V. Vasil and Vasil, 1980, 1981a; Lu and Vasil, 1981b; Wernicke *et al.*, 1981; Hanning and Conger, 1982; Lu *et al.*, 1982; Thomas and Scott, 1985), and reports of production from mature tissues are rare (Abe and Futsuhara, 1985). Embryogenic competence is attained or expressed by only a few selected cells. Cell division prior to embryogenic callus production in complex tissue explants appears to be initiated only in certain cells at specific loci. These have been identified as cells adjoining the procambial strand in the scutellum of immature embryos (V. Vasil and Vasil, 1982b; Lu and Vasil, 1985), parenchymatous cells near the vascular tissue or meristematic cells in the floral primordia of infloresence explants (Botti and Vasil, 1984), and mesophyll cells situated between the lower epidermis and vascular tissue in the basal portion of leaves (Haydu and Vasil, 1981; Lu and Vasil, 1981b; Ho and Vasil, 1983b). The embryogenic callus cultures derived from these cells are typically white to pale white in color, compact and organized in nature, and contain large numbers of small, richly cytoplasmic, starch-containing meristematic cells. In *Z. mays* two distinctly different types of embryogenic calluses have been described. These calluses differ from each other in their morphology and maintenance requirements on subculture. Callus type I has been assigned to compact, nodular, and organized embryogenic callus. This type of callus has been described in a wide variety of hybrid and inbred maize lines (e.g., Lu *et al.*, 1982, 1983; Tomes and Smith, 1985) and is similar to embryogenic callus described in other gramineous species (I. K. Vasil, 1985). While type I calluses of other species can be maintained in long-term culture (e.g., Chandler and Vasil, 1984a), type I calluses of maize are difficult to maintain beyond a few subcultures (Lu *et al.*, 1982). In contrast, maize type II calluses are soft and friable, grow rapidly, and yet maintain their embryogenic capacity in long-term culture (Green, 1982; Tomes and Smith, 1985; V. Vasil and Vasil, 1986).

Due to the relatively slow growth rate of embryogenic calluses, as compared to other callus types, it is important to identify and selectively culture them at an early stage, before they are outgrown by other callus types, such as nonmorphogenic callus (Fig. 1f). This is most important when establishing embryogenic suspension cultures, where monitoring of subculture intervals and dilution ratios is essential if embryogenic cells are to predominate in culture (V. Vasil and Vasil, 1984). While type I calluses have been used successfully in the establishment of embryogenic cell suspensions in a number of different species (I. K. Vasil, 1985), type I calluses of maize have proved unsuitable (V. Vasil and I. K. Vasil, unpublished results). However, type II maize calluses have been used with success (Green *et al.*, 1983; V. Vasil and Vasil, 1986). Because of their meristematic nature, embryogenic calluses produce rapidly dividing suspension cultures. These cultures are an ideal source of protoplasts which can be successfully cultured to produce plants (V. Vasil and Vasil, 1980; Srinivasan and Vasil, 1986) and are thus of major importance in the establishment of systems for genetic manipulation *in vitro* (Section III).

In general, embryo development and germination from embryogenic calluses or suspensions require sequential transfer of embryogenic cultures to lower concentrations of 2,4-D. Other additives, such as abscisic acid, gibberellic acid, and casein hydrolysate, among others (I. K. Vasil, 1983a; Gray and Conger, 1985), have also been included to enhance development and germination. Quantitative studies on the number of embryos produced and the percentage germination from embryogenic cultures are rare, but results from studies on *Pennisetum purpureum* indicate that embryogenic callus from leaf tissue readily develops into embryos and a higher percentage of these germinate and give rise to plants. Moreover, the higher regenerative capacity of the callus can be maintained over subculture and studies by Chandler and Vasil (1984a) indicate that, given optimal conditions for germination and development, a single leaf explant has the potential to produce 25,000 plants within 6 months.

Unlike plants regenerated from organogenic calluses it is generally believed that somatic embryos, like their zygotic counterparts, arise from single cells either directly or after the formation of a mass of proembryogenic cells. This belief is based on detailed histological studies (Steward *et al.*, 1964; McWilliam *et al.*, 1974; Haccius, 1978; Tisserat *et al.*, 1979; Conger *et al.*, 1983; V. Vasil and Vasil, 1982b; Ho and Vasil, 1983b; Magnusson and Bornman, 1985; Lu and Vasil, 1985). In addition to histological evidence the absence of chimeras in regenerants from embryogenic callus of *Panicum maximum* (Hanna *et al.*, 1984), *Pennisetum americanum* (Swedlund and Vasil, 1985), *P. purpureum* (Rajasekaren *et al.*, 1986) and *Z. mays* (Kamo *et al.*, 1985; Armstrong and Green, 1985), at least in the first generation, does give evidence which suggests a single-cell origin.

As yet we do not know why in one case nonzygotic embryos develop directly from segmenting single cells, and in another, indirectly from a proembryonal cell complex. A number of authors have ascribed to the proposal that some cells in the donor plant are more likely to undergo direct regeneration because of their phenotype while other cells only acquire suitable phenotypes after a period of cell division (Sharp *et al.*, 1980; Williams and Maheswaran, 1986).

It has been suggested that the embryogenic pathway of regeneration is less likely to result in the regeneration of somaclonal variants (I. K. Vasil, 1983a, 1984, 1985). This view is based first on the fact that the level of genetic variability present in embryogenic cultures *in vitro*, as judged from chromosome counts (Swedlund and Vasil, 1985; Karlsson and Vasil, 1986b) and DNA content (Karlsson and Vasil, 1986b), is less than that found in other cultures (e.g., Heinz *et al.*, 1969; Orton, 1980b; Ahloowalia, 1982) and second, because of the lack of appreciable variability in regenerants as reported by Hanna *et al.* (1984), Swedlund and Vasil (1985), and Armstrong and Green (1985). This lack of variability may be due to a selection for normal cells during the process of embryogenesis (I. K. Vasil, 1983a; Swedlund and Vasil, 1985) and this would appear to be verified by a recent report on the detrimental influence of genetic variability on embryo development in carrot (Toncelli *et al.*, 1985). However, it must be recognized that the aforementioned studies on regenerants from embryogenic callus were conducted on small numbers of plants (20–100) and only the first generations of plants were assessed. Recent detailed studies of both the biomass and phenotype of plants produced from embryogenic callus cultures of *P. americanum* × *P. purpureum*, compared with controls, indicate that a high level of phenotypic uniformity (96%) was present in the tissue-culture regenerants. In all, 23 of the 524 plants regenerated were found to show variation in phenotype (height, leaf width, etc.). This variability, furthermore, proved to be stable after a vegetative propagation cycle (Rajasekaran *et al.*, 1986). The variability produced in this study could be grouped into nine distinct phenotypes which may have been present in the original explant and were merely perpetuated and magnified *in vitro*. It is also possible that the variants produced may be transient (Larkin and

Scowcroft, 1983a; Irvine, 1984) and were caused by epigenetic rather than genetic change. This cannot be confirmed, however, due to the sexual sterility of hybrid Napier grass.

If regeneration via somatic embryogenesis is to be used as a method for producing stable, solid mutants or cloning genetically engineered plants, future studies are needed to determine the level of stability attained using this pathway of regeneration. Such a study will require detailed analysis of the somatic variation present in the original explant, of that induced *in vitro,* and of the heritability of any variation produced in regenerants. These studies must involve analysis of both quantitative and qualitative genetic traits, as chromosome studies alone will not be sufficient proof of stability in view of the potential for chromosome rearrangement in culture in the absence of numerical change (Gould, 1982; D'Amato, 1985). The pathway of regeneration will also have to be strictly monitored to ensure that plants are only produced from purely embryogenic cultures and not mixtures of embryogenic + organogenic cultures as found in wheat (Karp and Maddock, 1984), for which the variability reported may reside in regenerants from organogenic callus (Maddock, 1985).

A final callus type found in combination with morphogenic calluses is nonmorphogenic callus (Fig. 1f). This callus has also been referred to as nonembryogenic. As the term suggests, this callus does not undergo morphogenesis and despite numerous attempts (V. Vasil and I. K. Vasil, unpublished results) may not be maintained as a pure line. Nonmorphogenic callus is typically friable, with no distinct growth pattern, and is generally translucent and watery and consists of elongated, vacuolated cells. The rapid growth rate of this callus in many cases prevents the successful isolation of morphogenic calluses if these are not identified and isolated at an early stage. The amount of nonmorphogenic callus produced is dependent upon the age of the initial explant. Such callus is the predominant type recovered from cultures of mature tissues (V. Vasil and Vasil, 1981a). It is not clear what type of callus was originally used in the establishment of the nonmorphogenic suspensions, which have been isolated in *Sorghum* (Chourey and Sharpe, 1985) and *Zea mays* (Chourey and Zurawski, 1981; Ludwig *et al.,* 1985) (Table 1). It is possible that the callus used arose as a "sport" which was nonmorphogenic but could be maintained by subculture.

While it has been suggested that nonmorphogenic callus can become embryogenic (Nabors *et al.,* 1983), this claim remains unsubstantiated. The formation of scattered sectors of embryogenic callus in nonmorphogenic callus is more likely to be the result of divisions in a

TABLE 1
Culture of Protoplasts in the Gramineae

Species	Source of protoplasts	Result	Reference
Bromus inermis	Embryogenic cell suspension	Callus, embryoids, plantlets	Kao *et al.* (1973)
Hordeum vulgare	Nonmorphogenic callus	Nonmorphogenic callus	Koblitz (1976)
Lolium multiflorum	Cell suspension	Nonmorphogenic callus	Jones and Dale (1982)
Oryza sativa	Leaf sheath, callus	Callus, roots	Deka and Sen (1976)
Oryza sativa	Cell suspension	Nonmorphogenic callus	Cai *et al.* (1978)
Oryza sativa	Callus	Nonmorphogenic callus	Wakasa *et al.* (1984)
Oryza sativa	Cell suspension	Nonmorphogenic callus	Toriyama and Hinata (1985)
Oryza sativa	Callus	Callus, plantlets	Coulibaly and Demarly (1986)
Oryza sativa	Cell suspension	Callus, plants	Fujimura *et al.* (1985)
Oryza sativa	Cell suspension	Callus, plants	Yamada *et al.* (1986)
Panicum maximum	Embryogenic cell suspension	Callus, embryoids, plantlets	Lu *et al.* (1981)
Panicum miliceum	Embryogenic cell suspension	Callus, embryoids, albino plantlets	Heyser (1984)
Pennisetum americanum	Nonmorphogenic suspension	Nonmorphogenic callus	V. Vasil and Vasil (1979)
Pennisetum americanum	Embryogenic cell suspension	Callus, embryoids, plantlets	V. Vasil and Vasil (1980)
Pennisetum purpureum	Embryogenic cell suspension	Callus, embryoids, plantlets	V. Vasil *et al.* (1983)
Saccharum officinarum	Nonmorphogenic suspension	Nonmorphogenic callus	Maretzki and Nickell (1973)
Saccharum officinarum	Nonmorphogenic suspension	Nonmorphogenic callus	Yan and Li (1984)
Saccharum officinarum	Cell suspension	Callus, embryoids	Yan *et al.* (1985)
Saccharum officinarum	Shoot tips	Nonmorphogenic callus	Evans *et al.* (1980)
Saccharum officinarum	Embryogenic cell suspension	Callus, embryoids, plants	Srinivasan and Vasil (1986)
Sorghum bicolor	Nonmorphogenic suspension	Nonmorphogenic callus	Brar *et al.* (1980); Chourey and Sharpe (1985)
Triticum monococcum	Nonmorphogenic suspension	Nonmorphogenic callus	Nemet and Dudits (1977)
Zea mays	Shoot tips	Nonmorphogenic callus	Potrykus *et al.* (1977)
Zea mays	Nonmorphogenic suspension	Nonmorphogenic callus	Potrykus *et al.* (1979); Chourey and Zurawski (1981); Ludwig *et al.* (1985)
Zea mays	Embryogenic cell suspension	Embryogenic callus, embryoids	V. Vasil and Vasil (1987)

few embryogenic cells which were maintained in a suppressed state.

As indicated in the above discussion a range of different calluses may be produced from explants cultured *in vitro*. The factors which control the production of callus, its nature, and the subsequent regeneration of plants is the topic of Section II,B.

B. Factors Controlling Tissue and Cell Responses *in Vitro*

1. *Introduction*

The growth and differentiation of tissues or cells *in vitro* are dependent on the receptivity of individual cells and the interaction of endogenous and exogenous factors. Hence the response of a given cell to an external stimulus will depend on its biochemical and physiological status, features which are controlled by genotype, ontogeny, and prevalent environmental conditions. In initial studies on the Gramineae in tissue culture the influence of these factors on responses *in vitro* was not fully acknowledged. Many attempts have been made to induce responses from recalcitrant tissues by altering exogenous hormones, nutrients, or the *in vitro* environment. More recently, however, there has been a greater awareness of the role played by genotype, ontogeny, and genetic and epigenetic effects on tissue and cell responses *in vitro* (Maddock, 1985; Jones, 1985; I. K. Vasil and Vasil, 1986). This awareness has prompted detailed studies on the genetic basis of genotype responses (Nesticky *et al.*, 1983; Beckert and Qing, 1984; Tomes and Smith, 1985; Mathias *et al.*, 1986; Mathias and Fukui, 1986) and on the factors which may influence cell phenotype and genotype in donor plants (Rajasekaran *et al.*, 1987a,b; Taylor and Vasil, 1987) and callus tissue *in vitro* (Rajasekaran *et al.*, 1987b). These studies and examples which demonstrate how environmental, genetic, epigenetic, and developmental constraints influence a cell's morphogenic competence *in vitro* are discussed below.

2. *Developmental Influences*

The role played by a cell's developmental status in tissue responses *in vitro* is most clearly evident in the selective differentiation of morphogenic callus from immature tissues (e.g., V. Vasil and Vasil, 1982b) and the sequential loss of morphogenic competence with leaf age and distance from the apical meristem (e.g., Wernicke and Brettell, 1980; Ho and Vasil, 1983b; Wernicke and Milkovits, 1984).

Immature organs that have been observed to produce embryogenic callus in culture include embryos, infloresences, and leaves (I.K. Vasil

EMBRYO SIZE

1mm< 1 TO 1.5mm >2mm

no response ---------- embryogenic callus --------- germination

< competence window >

FIG. 2. Embryo responses *in vitro* relative to developmental state.

and Vasil, 1986). For each of these organs tissue-culture studies demonstrate that there is a particular stage of development to which the production of embryogenic callus is confined. This suggests that there is a selective phase or "competence window" wherein cells are capable of producing embryogenic callus. This phase is particularly well defined in the embryo and leaf. By studying the responses of embryos at difference stages of development, when cultured on a single, defined medium, Lu *et al.* (1983) found a distinct sequential pattern in the capacity to produce embryogenic callus (Fig. 2). Very young embryos (<1 mm) showed no response when cultured, embryos with a differentiated scutellum (1–1.5 mm) produced embryogenic callus, whereas mature embryos (>2 mm) germinated. Similar differences in the production and regenerative capacity of organogenic + embryogenic and embryogenic calluses have been recorded in embryos from *T. aestivum* (Maddock *et al.*, 1982), *Z. mays* (Beckert, 1982), *P. americanum* (V. Vasil and Vasil, 1982b), and *Triticale* (Stolarz and Lorz, 1986; C. J. Boyes, F. Morrish, and I. K. Vasil, unpublished results). The underlying causes of these differential responses from embryos at different stages of development remain to be elucidated. Due to the constancy of exogenous levels of hormones applied it is evident that the controlling factors lie in the endogenous levels of hormones and/or nutrients or the differentiated state of the cells of the explant placed *in vitro*. Because osmotic values change during the growth of the embryo (Ryczkowski, 1970) and explant size has been shown to influence responses *in vitro* (Negrutiu and Jacobs, 1978; Chandler and Vasil, 1984a; Ram and Nabors, 1984), it is possible that these factors too may influence embryo response in culture. Globular embryos consist of a small population of relatively undifferentiated cells. *In vivo* these are bathed in liquid endosperm. The sudden transfer to the dehydrating conditions of the culture dish may cause osmotic shock which could irreversibly damage cells and thereby render them incapable of producing any callus. In the case of older embryos such an explanation does not apply and here it would appear that cells may have become irreversibly committed to germination, possibly by gene repression, or they may be unable to respond to exogenous stimuli because endogenous levels of some hormones override those applied exogenously.

The development and differentiation of embryos involves histochemical, ultrastructural, molecular, and hormonal changes (Raghavan, 1986) and these changes, either singly or in combination, may be responsible for the tissue-culture responses found in developing embryos. The distinct pattern of accumulation and depletion of growth regulators observed in developing embryos would suggest that they may be responsible for the fluctuating morphogenic competence found in embryos at different developmental stages. Studies of wheat indicate that cytokinin levels rise sharply from zero to a maximal value just past anthesis and then fall sharply. Auxin concentrations increase a little during anthesis, decrease, and then increase markedly to reach a maximum value at 30–35 days postanthesis. By maturity, however, all auxin activity has disappeared (Wheeler, 1972). In wheat embryos abscisic acid (ABA) has also been shown to increase rapidly during the middle and later stages of terminal development, reaching a maximum value 25–40 days postanthesis. This is followed by a rapid drop in concentration (King, 1976). Similar results have been reported for barley (Goldbach and Michael, 1976) and *Triticale* (King *et al.*, 1979).

Fluctuations also occur at the molecular level. Studies on storage protein of cotton cotyledons indicate that mRNA sequences for storage proteins decrease and suffer almost total obliteration as embryos progress from the cell division phase into phases of cell expansion and maturation (Dure and Galau, 1981). Research has also been conducted on changes in gene expression in soybean embryos, in which Goldberg *et al.* (1981) found accumulation and decay in the expression of four genes for certain subunits of glycine and conglycine. Reports on somatic embryogenesis *in vitro* also demonstrate that a range of changes occur during embryo maturation (Nuti Ronchi and Giuliano, 1984; Raghavan, 1983, 1986). These changes include alterations in enzyme and protein synthesis (Sung and Okimoto, 1981, 1983; Ashihara *et al.*, 1981; Pitto *et al.*, 1983; Caligo *et al.*, 1985) and histone composition (Fujimura *et al.*, 1981). As gene activation during the transformation of a somatic cell into an embryoid is quite different from the information processing that occurs after genetic recombination at fertilization, changes occurring *in vitro* do not necessarily parallel those occurring *in vivo*. However, *in vitro* techniques have been developed to synchronize somatic embryo development (Fukuda and Komamine, 1985) and thereby allow the assessment of large samples of embryos at similar developmental stages, a feat not possible *in vivo*.

Another feature of the response of embryos *in vitro* is the selective production of callus by individual tissues of the differentiated embryo. Embryo development is accompanied by the differentiation of different tissue types, such as the scutellum, coleoptile, epiblast, and coleorhiza. Individually, the cells of these tissues within the embryo appear to show different levels of competence. This has been demonstrated by the selective production of embryogenic callus by the scutellum (V. Vasil and Vasil, 1982b; Lu *et al.*, 1982). In these cases orientation of the scutellum is important and callus proliferation is only achieved when the scutellum is orientated away from the medium. Cells of the embryo axis in contact with the medium tend to produce nonembryogenic callus. Such differential cell competence may be a result of the selective distribution of receptive loci due to endogenous gradients in hormones, osmotic potential, and nutrients, or may be related to the meristematic state of the cells in these tissues. It is also possible that changes in nuclear DNA, which occur as a natural consequence of embryo development (see D'Amato, 1977; Nagl, 1979), may also influence tissue responses, as genetic change may cause loss of competence (Smith and Street, 1974).

Developmental changes are also deemed responsible for the gradual decline in morphogenic competence observed in leaves of a number of cereals and grasses, including *O. sativa* (Wernicke *et al.*, 1981), *Sorghum bicolor* (Wernicke and Brettell, 1980), *D. glomerata* (Hanning and Conger, 1982), *P. purpureum* (Haydu and Vasil, 1981; Rajasekaran *et al.*, 1987a), *T. aestivum* (Alfinetta *et al.*, 1983; Wernicke and Milkovits, 1984), *P. maximum* (Lu and Vasil, 1981), *Secale cereale* (Linacero and Vazquez, 1986), *Lolium multiflorum* (Joarder *et al.*, 1986), and *Z. mays* (Wenzler and Meins, 1986). In these plants, as in all Gramineae, increasing temporal and spatial separation from the apical meristem is a natural consequence of growth due to the absence of secondary meristem formation during ontogenesis. This separation is believed to account for a decrease in cell competence which is manifest as a gradual decline in the percentage of callus produced with increasing leaf age and distance from the apical meristem. This lack of competence may result from qualitative or quantitative changes in DNA, the cell cycle status of cells, or alterations in the endogenous levels of hormones or nutrients due to leaf senescence and aging.

Qualitative and quantitative changes in DNA are a natural consequence of development and aging in many plants (D'Amato and Hoffman-Ostenhof, 1959; D'Amato, 1977; Nagl, 1978, 1979; Osborne *et al.*, 1984) and may also arise due to environmental stress (McClintock, 1978; Cullis, 1981, 1983; Durrant, 1981; Walbot and Cullis, 1985). Such changes result in plants which have a complex genetic architecture consisting of a population of genetically heterogeneous cells. These genetic changes in plant cells may be invoked by a number of different mechanisms, including genome endoreduplication, transposable elements, and somatic mutations (Dyer, 1976; Walbot and Cullis, 1985; Nevers *et al.*, 1986). Much of the evidence for changes in nuclear DNA during differentiation comes from polysomatic species, which account for 90% of the total angiosperm population (cf. D'Amato, 1985). In cereals and grasses there are relatively few studies on the changes which may occur in leaf cell DNA. As with other plant families (D'Amato, 1977), the nature and degree of variability present may be genotype dependent. Studies on *S. cereale* by Hesemann and Schroder (1982) suggest a decrease in nuclear DNA with increasing leaf age. However, this study is not conclusive as no comprehensive statistical analysis of the data was presented. In a report on *T. aestivum* Beaulieu *et al.* (1985) found that the DNA obtained from older leaves appeared to be fragmented when compared to the DNA obtained from younger leaves. Again the reliability of these data is in doubt because of the possibility that increased nuclease activity during sample isolation may have been responsible for the degradation of DNA. These studies need to be repeated and comprehensively evaluated in order to show conclusively that changes occur in DNA during leaf maturation, particularly when recent studies by Taylor and Vasil (1987) indicate no change in DNA content or in fragment size of mature leaves and the basal and distal leaf zones of young leaves of *P. purpureum*.

While it is possible that qualitative changes in DNA may be responsible for the recalcitrance found in mature leaf tissues, this recalcitrance may equally be a result of changes in endogenous hormone levels, cell senescence, or the cell cycle status of the constituent cells.

Recent studies by Rajasekaren *et al.* (1987a) indicate that distinct spatial and temporal gradients exist for the hormones ABA and indole acetic acid (IAA) in young and mature leaf tissue of *P. purpureum* which were found to correspond to similar gradients in embryogenic competence. Studies conducted by Taylor and Vasil (1987) did not find any distinct relationship between decreasing embryogenic competence and the cell cycle. They concluded also that this loss of competence was not necessarily related to the loss of mitotic activity. Similar results have been reported in *L. multiflorum* (Joarder *et al.*, 1987) and hence it would appear that morphogenic competence, while requiring actively dividing cells, is not directly controlled by mitotic activity.

The role of endogenous ABA levels in the production of embryogenic callus in leaves of *P. purpureum* has been further confirmed by the correlated reduction in embryogenic capacity of explants after pretreatment with fluridone, which reduced endogenous levels of ABA (Rajasekaran *et al.*, 1987b).

It is not clear how differences in the relative percentage of cells in G_1 and G_2 in *P. purpureum* leaves may influence morphogenic capacity. The role of the cell cycle in the differentiation of plant cells is a topic of much controversy and has been discussed in a number of reviews on differentiation (Roberts, 1976; Fukuda and Komamine, 1985) and cell culture (Gould, 1983). In order for cells to differentiate, a quantal cell cycle must occur, where the cells generated are different from the mother cell. This is in contrast to a proliferation cell cycle, which generates copies of the mother cell (Holtzer and Rubenstein, 1977). According to Dodds (1981) the lag phase prior to a cell's programming for differentiation will depend on the stage of the cell cycle in which a cell is stationed prior to receiving a hormonal stimulus. Studies with *Zinnia* cells, however, suggest that the process of cytodifferentiation may be independent of the progression of the cell cycle (Fukuda and Komamine, 1985). Hence the question of whether the initiation of cytodifferentiation is limited to a particular phase of the cell cycle remains unanswered.

3. *Environmental Influences*

In addition to ontogeny, the environment also plays a major role in determining the phenotype of cells introduced into and maintained in culture. Plants under stress, such as increased temperature or nutrient deficiencies, show distinctive differences in the level and distribution of endogenous hormones. Alterations in the photoperiod or intensity of light also have major consequences for the biochemical status of individual cells (Letham *et al.*, 1978). While environmental effects have been noted in tissue-culture studies on the Gramineae (e.g., Lu *et al.*, 1983; Armstrong and Green, 1985; Santos and Torne, 1986), the consequences resulting from directed changes in the environment on tissue-culture responses in the Gramineae have received little attention. In ornamental plants seasonal effects are believed to be linked to photoperiod, light intensity, and temperature, as plants grown under environments where these factors were defined and altered similarly showed alterations in regenerative responses *in vitro* (Hilding and Welander, 1976; Appelgren, 1985). Studies on tobacco by Cassells *et al.* (1982) suggest that endogenous hormone levels may be responsible for the altered responses produced by different photoperiods, as pretreatments of donor plants in the dark for 48 hours decreased the number of shoots produced from petiole explants. This was found to be correlated with an increase in endogenous IAA levels.

It is accepted dogma that plant health and survival are dependent on adequate nutrients and water. Excesses too can influence plant health and the physiological condition of a plant. The influence of these factors on cell viability and response *in vitro* have only recently received attention. Studies by Cassells and Tamma (1986) indicate that the stress caused by waterlogging and drought influences the viability of tobacco protoplasts *in vitro* and is related to an increase in the production of ethane and ethylene. Similar detailed studies on the influence of the donor plant environment on responses of explants or cells *in vitro* have not been conducted on members of the Gramineae. However, there are numerous reports demonstrating the effect of environment on the biochemical state of cells in cereals and grasses (e.g., Preger and Gepstein, 1985; Kasperbauer and Karlen, 1986). As these differences may have major consequences for the successful initiation and regeneration of plants from cultures of the Gramineae, further studies on the factors which induce stress in cells and the consequential changes in the biochemical and osmotic status and responses of cells *in vitro* are warranted.

As found *in vivo*, environmental conditions *in vitro* and the organizational status of cells have a distinct influence on cell proliferation and differentiation. In cell cultures of the Gramineae, regenerative competence decreases during subculture (e.g., Heyser *et al.*, 1983) and more drastically with a change in organizational state, morphogenic competence being reduced stepwise from callus, to cell cultures, to isolated protoplasts (V. Vasil *et al.*, 1983). This loss of competence is one of the major problems faced in the development of systems for the genetic manipulation of the Gramineae, for which regeneration from protoplasts and long-term culture are essential (Section III). Presently the underlying cause for these changes in cell competence is believed to be either genetic or physiological (Smith and Street, 1974). The changes may equally be the result of epigenetic effects (Orton, 1984; D'Amato, 1985). According to the physiological hypothesis, cells retain their totipotency, but this is masked by the development of a new and relatively stable metabolism in response to culture conditions. The dedifferentiation associated with culture initiation gives rise to cells programmed for continuous growth in culture and hence not competent to express their totipotency. However, this loss of competency is reversible and appropriate changes in media and environmental conditions will restore cell competence (e.g., Reinert and Backs, 1968; Inoue and Maeda, 1981).

The genetic hypothesis implies not only cytological instability but the production of deviant cells which are lacking in totipotency and at a selective advantage in culture. Cells proliferating *in vitro* are in an alien environment and their normal rate of division is greatly increased by the hormones provided in the medium. Both these factors may be responsible for the genetic drifts that occur in cell and callus cultures during subculture and which may have a detrimental influence on cell competence. The range of genetic variability found in cell cultures has been extensively reviewed by Sunderland (1977), Bayliss (1980), and D'Amato (1985). From these reviews we know that cells in culture may undergo numerous genetic changes, including chromosome loss, polyploidization, chromosome rearrangement, and gene mutation. These changes may also arise in somatic cells during differentiation and aging and have already been discussed in relation to their influence on recalcitrance (Section II,B). In many calluses, due to the detrimental influence of genetic changes occurring *in vitro*, there is an attentuation of variation on regeneration from callus *in vitro* (Orton, 1984). However, plants regenerated from tissue cultures derived from polyploids and hybrids have been found to bear a wide range of genetic changes (Larkin and Scowcroft, 1981; Orton, 1984; D'Amato, 1985). The absence of detrimental influence on cell competence in these calluses is believed to be due to the greater buffering capacity of these plants (Lorz, 1984; Orton, 1984). In diploids, loss, rearrangement, or gain of chromosomes may render cells permanently noncompetent (e.g., Murashige and Nakano, 1967; Gould, 1978, 1982); detailed studies on the influence of chromosomal rearrangement and chromosome loss on embryo regeneration in carrot demonstrate how such changes may directly influence morphogenesis (Toncelli *et al.*, 1985). Gene mutations too may halt cellular differentiation as found *in vivo* in developmental mutants of maize (Sheridan and Neuffer, 1982).

Very little is known about the physiological state of cells *in vitro* and much of the information available is indirect and is based on responses to exogenous hormones, nutrients, temperature changes, gaseous elements, or osmotic agents (Inoue and Maeda, 1981; Siriwardana and Nabors, 1983; Chandler and Vasil, 1984a; Gray and Conger, 1985). In a callus there are distinct gradients of nutrients and hormones which arise as a consequence of zonation in cell division centers and the subsequent channeling of metabolites. These gradients vary with callus size and growth pattern and it is perhaps for this reason that there may be a critical callus size which is most appropriate for regeneration (Negrutiu and Jacobs, 1978). During subculture it is possible that the relative levels of hormones and other metabolites present in cells may alter as a consequence of changes induced in gene expression due to the *in vitro* environment. This is most clearly evident in the habituation of tobacco callus, wherein callus cells after subculture or environmental treatments can be maintained on a cytokinin-free medium because they have gained the ability to produce adequate supplies of cytokinin (Meins and Binns, 1977, 1978). Because this capacity is not transmitted through a sexual cycle this change is regarded as being epigenetic. Such changes, while having the appearance of mutations, are not permanent, are not sexually transmitted, and merely involve changes in gene expression (Chaleff, 1981; Meins, 1983). Similar epigenetic changes may be responsible for reduced competence in callus and suspension cultures. Changes in endogenous factors will become manifest as recalcitrance if they significantly alter the original endogenous/exogenous balance of hormones or nutrients required for regeneration. Such changes occur in embryogenic callus cultures of *P. purpureum* after extended periods of subculture in which reduction in ABA and IAA levels was correlated with reduction in embryogenic competence (Rajasekaran *et al.*, 1987a). Subtle changes in exogenous factors or a return of cells to the phenotypic state of the original culture may be required to return regenerative capacity. While the former approach has been successful in some cases (Chandler and Vasil, 1984a), many calluses and cell suspensions still remain nonmorphogenic despite extensive media manipulations.

The ability to avert or limit the genetic and epigenetic drifts occurring in culture will depend on our development of a greater understanding of how *in vitro* conditions influence genetic and epigenetic changes in plant cells. An attempt to return to the phenotype of the initial competent culture will require characterization of both genotypic and phenotypic differences which may exist among freshly cultured and subcultured callus cells and awaits further study.

4. *Genotypic Influences*

Traditionally, plant genotypes are characterized by unique traits or subtle alterations in a given trait, such as yield, which identify their individuality. Differences in genotypic traits may extend to response in tissue culture. Results from *in vitro* cultures of different genotypes of many crop plants have caused many investigators to suggest that differences exist among genotypes in the ability to produce callus, in callus proliferation rate, in requirements for callus maintenance on subculture, and in regenerative ability.

The relative influence of genotype on tissue-culture responses has been a source of much controversy. Because some genotypes have proved recalcitrant or have a low regenerative ability on selected media they have been judged as unsuitable for use in tissue culture. However, as discussed by I. K. Vasil and Vasil (1986), it is unlikely that any genotype would *not* possess the ability to regenerate and in many cases the apparent genotypic effect may in fact have been a combination of both genotypic and epigenotypic influences. It is not our intention here to dispute the view that genotypic influences exist but we do wish to emphasize that in a number of cases the extent of genotypic influence may have been misjudged as a result of epigenotypic effects.

The epigenotype (i.e., environment and ontogeny), as already discussed (Sections II,B,2 and 3), has a major influence on the response of tissues *in vitro*. Hence it is possible that in some instances genotype recalcitrance could have been overcome if suitable explants had been taken from donor plants grown under environmental conditions conducive to optimal response *in vitro*, or if alterations had been made in the exogenous supply of hormones, sucrose, or nutrients provided *in vitro*. Results from a number of studies indicate the relevance of this argument. For example, in a study of the response of both embryo and inflorescence explants of 25 genotypes of wheat by Maddock *et al.* (1982), individual genotypes were shown to vary in their responses to tissue culture. However, differences also existed in the responses of the two organs cultured and there were genotypes in which one organ gave a low response while the second organ was highly responsive. Environment is also known to be a factor in the response of maize genotypes. The normally highly responsive line A188, when grown under low light and at high population densities, was shown to have a reduced response in tissue culture (Tomes and Smith, 1985).

Developmental influences on genotype responses in maize have been demonstrated by studies on a range of maize genotypes (Lu *et al.*, 1982, 1983). By selection of the appropriate developmental stage the response of a given genotype could be increased by nearly 80%. For example, only 3.3% of embryos from the maize line Dekalb XL 80 produced embryogenic callus in the absence of selection, while this was increased to 83% when embryos of a defined stage of development were used.

The true influence of the genotype is only discernible when both the developmental state of the organ used and the conditions under which the different genotypes were grown are similar. Results from such studies on a number of different species indicate that genotype does influence responses *in vitro* (Rao and Nitzsche, 1984; Hanzel *et al.*, 1985). In rice, recent studies by Abe and Futsuhara (1986) on callus production and regenerative responses of mature seeds of 60 varieties belonging to three subspecies (*japonica, indica,* and *javanica*) indicate both intra- and intervarietal differences. These authors suggest that the differences manifest may be attributed to differences in the components and concentrations of endogenous phytohormones and differences in the susceptibility of different rice varieties to 2,4-D. Indirect evidence for the role of endogenous factors as the causal agents in genotype responses has been demonstrated in maize, where increasing the sucrose or 2,4-D concentration in the medium overcame the apparent differences observed in embryogenic callus production when genotypes were compared on a single, defined medium (Lu *et al.*, 1983). Changes in media nutrients as well as the level of hormones have also been shown to influence the response of barley genotypes, where such changes have resulted in a 50-fold increase in response in some apparently recalcitrant genotypes (Hanzel *et al.*, 1985).

The reports outlined above indicate that it may be possible to induce callus and regenerate plants from genotypes deemed recalcitrant simply by determining the optimum media, environment, and developmental status of the explant required for culture initiation. Should such manipulations fail and if the genotype in question is important as a breeding line, then regenerative capacity may possibly be improved by breeding.

To establish a breeding program to develop lines which are especially suited to a particular tissue-culture system an understanding of the genetic basis of culture behavior is essential. Stimulated by the success of Bingham *et al.* (1975), who increased regenerative ability in alfalfa from 12 to 67% by backcrossing over three generations, there has been a good deal of research on the genetic determinants of callus growth and development in the Gramineae. Genetic analysis by

Tomes and Smith (1985) on maize inbred lines B75 and G39 shows a strong negative maternal effect for embryogenic callus production. In other lines of maize similar maternal effects have been noted for callus growth from mature embryos (Nesticky *et al.*, 1983), for endosperm callus growth (Tabata and Motoyoshi, 1965), and for other culture characteristics (Beckert and Qing, 1984). It has been suggested that the underlying genetic basis of genotype differences in *in vitro* responses may be cytoplasmic (e.g., mtDNA) (Nesticky *et al.*, 1983; Beckert and Qing, 1984) or due to segregation of nuclear factors (Tomes and Smith, 1985). Recent studies on the effect of cytoplasmic substitutions in wheat on *in vitro* responses indicate no major influence on regenerative ability (Mathias *et al.*, 1986). However, the substitution of a single chromosome (Cap48B) in the same wheat line was found to have a significant influence on regenerative capacity (Mathias and Fukui, 1986). This is the first report of the involvement of a specific chromosome in the control of shoot regeneration in tissue culture and these authors propose to establish which arm of the chromosome carries the gene(s) responsible for increased regenerative competence and whether the effects result from the addition of a promotor or the removal of an inhibitor. These studies should give an answer to the question of the source of control of genotype responses in this wheat line and will perhaps allow speculation on the potential for isolating regeneration genes, which could be cloned and transferred to other wheat lines.

5. *Conclusions*

The preceding discussion illustrates how the calluses produced from individual members of the Gramineae and different tissue explants can show considerable variation in morphology, regenerative competence, and genetic stability. This knowledge coupled with an increased awareness of the factors influencing the responses of tissues and cells *in vitro* should help us in determining suitable genotypes and *in vivo/in vitro* environments for the successful growth and regeneration of plants from cells of the Gramineae. Such strategies are necessary if we are to improve our ability to regenerate plants and genetically manipulate cells *in vitro*.

III. Genetic Manipulation and Plant Improvement

The establishment of procedures for the culture and regeneration of plants from the Gramineae (Section II) has laid the foundation required to begin to experiment with tissue culture as an adjuvant to traditional breeding methods. Tissue-culture procedures, which have been shown to have potential application to genetic manipulation and plant improvement using model dicot species, are now being tested for their application to the Gramineae. Such procedures include somatic hybridization, transformation, somaclonal variation, and mutant selection. The following discussion will outline the development and potential application of these techniques in the Gramineae.

A. Protoplast Culture, Somatic Hybridization, and Genetic Transformation

1. *Introduction*

The availability of potent commercial cell wall-degrading enzymes in the late 1960s made it possible for the first time to obtain large numbers of viable plant protoplasts (Cocking, 1960; Takebe *et al.*, 1968). The ability of these naked plant cells to fuse, take up macromolecules, regenerate a new cell wall, and then undergo divisions to form a cell colony from which plants could be recovered (I. K. Vasil, 1976) paved the way for what is now popularly called *in vitro* genetic modification of plant cells.

The fusion of protoplasts of different species or varieties to produce somatic hybrids and/or cybrids can overcome the natural barriers of sexual incompatibility which prevent crossing between unrelated varieties and species. Similarly, the introduction and integration of alien genes into plant cells by *in vitro* transformation can be used to transfer defined genetic material directly into cells. These methods are thus of much interest in the genetic manipulation and improvement of plants. However, the success of these strategies depends largely on the ability to recover plants from cultured protoplasts. Plants have been regenerated from cultured protoplasts in many species since Takebe *et al.* (1971) first succeeded in obtaining plants from mesophyll protoplasts of tobacco (see I.K. Vasil and Vasil, 1980; Gleba and Sytnik, 1984). Success has been limited, however, largely to herbaceous dicotyledonous species (Binding, 1986).

The difficulties encountered in the culture of protoplasts were further compounded by the severe problems faced initially in the regeneration of plants from cell and tissue cultures of gramineous species. As described in Section II,A, reliable and efficient methods for the regeneration of plants from tissue cultures of a wide variety of gramineous species were not developed until the early 1980s. Successful plant regeneration in the Gramineae has been accompanied by progress in the culture of protoplasts (Table 1), somatic hybridization (Table 2), and *in vitro* genetic transformation (Table 3).

2. *Systems Used for Protoplast Culture*

a. Intact Plant Tissues. Protoplasts can be isolated from a variety of explant sources in the Gramineae, such as leaf, leaf sheath, shoot tip, stem, root, endosperm, immature embryo, etc. (I. K. Vasil, 1983b). High yields are obtained only from leaf tissues, following either the removal of the abaxial epidermis or the abrasion of the leaf surface with carborundum powder. Protocols used for the isolation and culture of gramineous protoplasts are no different from those commonly used for dicotyledonous species. The most effective enzyme mixtures are those containing varying concentrations of cellulase R10 or RS, pectolyase, and pectinase (V. Vasil and Vasil, 1984). In spite of very extensive efforts involving tens of thousands of variations of media and other parameters of isolation and culture, no sustained divisions have ever been obtained from cultured mesophyll protoplasts in any species of the Gramineae (Potrykus *et al.*, 1976; Potrykus, 1980; Dale, 1983; I. K. Vasil, 1983b). It has been suggested that the mesophyll protoplasts of grasses have a mitotic block or are otherwise incapable of division (Potrykus, 1980; Flores *et al.*, 1981). However, these unfounded speculations have been rendered moot by the numerous recent demonstrations of the totipotency of immature grass leaf tissues (Section II). Clearly, there are other unknown factors which prevent cell wall regeneration and division in isolated mesophyll protoplasts of the Gramineae.

There are only three reports of sustained divisions in protoplasts isolated directly from intact plant tissues or organs. About 30% of the protoplasts isolated from leaf sheaths of rice seedlings was said to give rise to a root-forming callus (Deka and Sen, 1976). Sustained divisions in protoplasts isolated from young stem tissues of *Z. mays* resulted in the formation of a nonmorphogenic callus (Potrykus *et al.*, 1977). In *Saccharum* spp. about 10% of the protoplasts isolated from young shoot tips formed cell colonies and callus (Evans *et al.*, 1980). None of these findings, however, is reproducible and none has been substantiated by later workers. It is possible that these were either rare events or that the protoplast preparations were contaminated by undigested but viable meristematic cells.

b. Callus Cultures. The yield of protoplasts from callus cultures is generally very low, particularly from organized morphogenic cultures which consist of starch-containing cells (see Section II,A for a description of different types of calluses). Nevertheless, protoplasts isolated

TABLE 2

Somatic Hybridization in the Gramineae

Species	Source of protoplasts	Result/evidence	Reference
Glycine max + *Hordeum vulgare*	Cell suspension/leaf	Cell colonies/cytological	Kao and Michayluk (1974)
Glycine max + *Zea mays*	Cell suspension/leaf	Cell colonies/cytological	Kao *et al.* (1974)
Daucus carota + *Oryza sativa*	Cell suspension/cell suspension	Callus/DNA analysis	Sala *et al.* (1985)
Glycine max + *Oryza sativa*	Cell suspension/cell suspension	Callus/isozymes, fraction 1 protein	Niizeki *et al.* (1985)
Sorghum bicolor + *Zea mays*	Cell suspension/leaf	Heterokaryons/cytological	Brar *et al.* (1980)
Daucus carota + *Hordeum vulgare*	Cell suspension/leaf	Heterokaryon/cytological	Dudits *et al.* (1976)
Panicum maximum + *Pennisetum americanum*	Cell suspension/cell suspension	Callus/isozyme, DNA hybridization	Ozias-Akins *et al.* (1986)
Pennisetum americanum + *Saccharum officinarum*	Cell suspension/cell suspension	Callus, embryoids/isozyme, DNA hybridization	Tabaeizadeh *et al.* (1986)
Pennisetum americanum + *Triticum monococcum*	Cell suspension/cell suspension	Callus, isozyme, DNA hybridization	V. Vasil *et al.* (1987)

TABLE 3
Genetic Transformation in the Gramineae

Species	Method	Transient or stable/selection	Reference
Triticum monococcum	Direct DNA uptake	Stable/kanamycin	Lorz *et al.* (1985)
Lolium multiflorum	Direct DNA uptake	Stable/G418	Potrykus *et al.* (1985)
Zea mays	Electroporation	Transient	Fromm *et al.* (1985)
Zea mays	Electroporation	Stable/kanamycin	Fromm *et al.* (1986)
Oryza sativa	Fusion with *Agrobacetrium* spheroplasts	Stable (?)/growth on hormone-free media	Baba *et al.* (1986)
Triticum monococcum; Panicum maximum; Pennisetum purpureum; Pennisetum americanum × *Pennisetum purpureum* × *Pennisetum squamulatum* hybrid; *Saccharum officinarum*	Electroporation	Transient	Hauptmann *et al.* (1987a)
Panicum maximum; Triticum monococcum	Electroporation	Stable/hygromycin, methotrexate	Hauptmann *et al.* (1987b)

from nonmorphogenic calluses of *Hordeum vulgare* (Koblitz, 1976) and *O. sativa* (Deka and Sen, 1976; Cai *et al.*, 1978; Wakasa *et al.*, 1984) have been cultured successfully to obtain calluses which were also found to be nonmorphogenic. More recently, Coulibaly and Demarly (1986) have reported recovering plantlets from protoplasts isolated from the primary callus formed at the base of the coleoptile of germinating seeds of *O. sativa.* In this study the age of the donor plants as well as the callus was found to be the most important factor in obtaining totipotent protoplasts. Unfortunately, this report does not provide sufficient details of the protocols used and the results obtained, nor photographic documentation of the early stages of protoplast culture and division.

c. Nonmorphogenic Cell Suspension Cultures. Rapidly growing and finely dispersed cell suspension cultures provide an efficient and practically unlimited source of viable protoplasts. Such suspensions have been used with considerable success to obtain sustained cell divisions in protoplasts and the formation of calluses in several species of the Gramineae (Table 1). Although such cultures tend to be genetically unstable and consist largely of aneuploid and polyploid cells, they do offer many advantages. The cell lines grow rapidly and yield large numbers of protoplasts following incubation in enzyme mixtures for a relatively brief period of time. Rapid subculture protocols avoid the formation of large senescent cells, which do not form protoplasts owing to their thickened secondary walls. Such undigested cells, when present, contaminate otherwise excellent protoplast preparations. Protoplasts isolated from nonmorphogenic cell suspension cultures of several species have been cultured to obtain calluses: *Saccharum officinarum* (Maretzki and Nickell, 1973; Yan and Li, 1984; Yan *et al.*, 1985), *Triticum monococcum* (Nemet and Dudits, 1977), *P. americanum* (V. Vasil and Vasil, 1979), *Z. mays* (Potrykus *et al.*, 1979; Chourey and Zurawski, 1981; Ludwig *et al.*, 1985), *S. bicolor* (Brar *et al.*, 1980; Chourey and Sharpe, 1985), and *O. sativa* (Wakasa *et al.*, 1984; Toriyama and Hinata, 1985). The reported division frequency of protoplasts in these studies ranges from 10 to 25%. The protoplast-derived calluses, like the suspension cultures from which the protoplasts were isolated, were also nonmorphogenic.

d. Morphogenic Cell Suspension Cultures. The failure of the mesophyll protoplasts of the Gramineae to divide and the generally satisfactory division frequencies obtained in protoplasts isolated from nonmorphogenic suspension cultures have prompted attempts to establish morphogenic suspension cultures as a source of totipotent protoplasts. Such suspensions have been available since the late

1960s in some dicotyledonous species. However, serious difficulties have been faced in the establishment of similar suspensions in the Gramineae due to the compact and organized nature of their morphogenic tissue cultures. Embryogenic and/or morphogenic cell suspension cultures have been obtained only in a few species of cereals and grasses: *Bromus inermis* (Gamborg *et al.*, 1970), *P. americanum* (V. Vasil and Vasil, 1981b), *P. maximum* (Lu and Vasil, 1981a), *P. purpureum* (V. Vasil *et al.*, 1983), *Panicum miliaceum* (Heyser, 1984), *O. sativa* (Fujimura *et al.*, 1983; Yamada *et al.*, 1986), *S. officinarum* (Ho and Vasil, 1983a; Yan *et al.*, 1985; Srinivasan and Vasil, 1986), *L. multiflorum* (Jones and Dale, 1982), and *Z. mays* (Green *et al.*, 1983; V. Vasil and Vasil, 1986).

The histology, cytology, and morphology of some embryogenic suspension cultures have been characterized in detail (e.g., V. Vasil and Vasil, 1982a; Karlsson and Vasil, 1986a,b). These suspensions consist of small aggregates of thin-walled, richly cytoplasmic cells. An important feature of these suspensions is the absence of any organized tissues, for example, meristems and roots. These suspensions have proved very amenable to the isolation of protoplasts. To release protoplasts, suspensions are incubated for 4–6 hours in appropriate enzyme mixtures (V. Vasil and Vasil, 1984). During this time almost all the cells in the suspension are converted into protoplasts. The yield and quality of protoplasts obtained as well as their division efficiency are closely related to the quality of cells present in the suspension culture at the time of protoplast isolation. By changing the dilution ratio and the duration of each subculture, the suspension can be manipulated to improve protoplast isolation and division frequencies. The highly cytoplasmic and nonvacuolated nature of the embryogenic cells and their presence in tight groups result in a considerable amount of spontaneous fusion during the process of protoplast isolation. This phenomenon can be largely avoided, however, by carefully monitoring the suspension culture.

Plating efficiencies of about 25% can be obtained by using cells from 4- to 5-day-old suspension cultures. Both the plating density (1–5 $\times$ 10^6 protoplasts/ml) and the method of protoplast culture (1–2 ml of protoplast suspension in a shallow layer in a 5 $\times$ 35-mm petri dish) significantly affect plating efficiency.

Within 2–3 days after plating, cell wall regeneration is completed and the newly regenerated cells undergo their first divisions during the first 3–5 days of culture. Within 3 weeks, protocolonies, resembling cell aggregates found in suspension cultures, and somatic embryos can be seen in culture. Further development of the somatic embryos and plant regeneration takes place only upon transfer of the protocolonies and the embryos to agar-solidified media. Although plantlets have been obtained from protoplasts of several species, they generally fail to grow into mature plants (Table 1). Therefore, the recent success in recovering mature plants from protoplasts of *S. officinarum* (Srinivasan and Vasil, 1986) and *O. sativa* (Fujimura *et al.*, 1985; Yamada *et al.*, 1986) is of considerable importance. Nevertheless, it should be pointed out that at the present time the efficiency of plant regeneration is sporadic and low.

3. *Somatic Hybridization*

The development of efficient procedures for the chemical and electrical fusion of protoplasts has led to the recovery of both inter- and intraspecific somatic hybrids and cybrids in many dicotyledonous species (Schieder and Vasil, 1980; Gleba and Sytnik, 1984). One of the essential requirements for the success of somatic hybridization is the identification and preferential selection of heterokaryons/hybrids. This has generally been achieved by using morphological markers, wherein green mesophyll protoplasts are fused with colorless suspension-culture protoplasts, or by the use of antimetabolites and drugs that preferentially allow the growth of only the hybrid cells. These strategies have been unavailable with gramineous protoplasts because of the difficulties encountered in the culture of mesophyll protoplasts and the unavailability of resistant/variant lines.

Early attempts at somatic hybridization in the Gramineae involved the fusion of nondividing mesophyll protoplasts of grasses with dividing protoplasts from suspension cultures of other grasses or dicotyledonous species (Table 2). The heterokaryons were identified by the presence of green plastids as well as colorless cytoplasm. Although cell divisions were observed and putative somatic hybrid cell colonies were formed, no convincing evidence of somatic hybridization was present in any of these early studies (Kao and Michayluk, 1974; Kao *et al.*, 1974; Dudits *et al.*, 1976; Brar *et al.*, 1980). Somatic hybrid colonies were also obtained by fusion of *O. sativa* protoplasts with those of *Daucus carota* (Sala *et al.*, 1985) or *Glycine max* (Niizeki *et al.*, 1985). DNA hybridization analysis of rice–carrot protoplasts showed that only a fraction of the rice nuclear genome was present. In the rice–soybean somatic hybrids, the only evidence of hybridization was the presence of both rice and soybean chloroplasts, as shown by fraction 1 protein analysis.

Compelling evidence of somatic hybridization is provided in three recent reports. In each instance protoplasts isolated from a cell line of

P. americanum resistant to *S*-(2-aminoethyl)-L-cysteine (AEC) and treated with iodoacetate were fused with protoplasts isolated from suspension cultures of *P. maximum* (Ozias-Akins *et al.*, 1986), *S. officinarum* (Tabaeizadeh *et al.*, 1986), and *T. monococcum* (Vasil *et al.*, 1987). The fusion procedure was a modification of the method used by Menczel and Wolfe (1984) which included dimethyl sulfoxide in the polyethylene glycol (PEG) solution as originally proposed by Haydu *et al.* (1977). Selection of putative somatic hybrid colonies was made on AEC-supplemented media. The somatic hybrid nature of several of the recovered calluses was confirmed by analysis of dimeric isozymes which formed heterodimer bands in the hybrids (alcohol dehydrogenase for *P. americanum* + *P. maximum* and 6-phosphogluconate dehydrogenase for *P. americanum* + *S. officinarum* and *P. americanum* + *T. monococcum*). Additional evidence for somatic hybridization was obtained from the analysis of several monomeric enzymes and hybridization with a maize ribosomal DNA probe. Morphogenesis was only observed in the *P. americanum* + *S. officinarum* hybrid lines, where many somatic embryos were formed. Isozyme analysis of these embryos confirmed their hybrid nature. Embryos did not, however, develop to produce mature plants.

Restriction analysis of the mitochondrial DNA of the *P. americanum* + *P. maximum* (Ozias-Akins *et al.*, 1987a,b) and *P. americanum* + *S. officinarum* (Tabaeizadeh *et al.*, 1987) somatic hybrids showed the presence of novel bands which may have arisen either through rearrangement or recombination of the parental mitochondrial genomes.

4. *Genetic Transformation*

The soil pathogen *Agrobacterium tumefaciens* has been used as a vector system for the transformation of dicotyledonous species (Fraley *et al.*, 1986). Upon infection, *A. tumefaciens* transfers a defined piece of DNA (T-DNA) from its tumor-inducing (Ti) plasmid into a chromosome of the plant cell (Bevan and Chilton, 1982). The native T-DNA contains genes which code for phytohormone production, causing an autonomous proliferation of cell growth (Barry *et al.*, 1981; Schroeder *et al.*, 1984), and production of opines such as octopine and nopaline. These genes are not necessary for transformation (Garfinkel *et al.*, 1981; Leemans *et al.*, 1982) and can be removed, allowing the insertion of engineered genes to obtain normal, transformed plants with new traits. These engineered genes can be mapped to specific chromosomes and, in most cases, are inherited in a simple Mendelian manner (Wallroth *et al.*, 1986).

Agrobacterium tumefaciens has provided an effective DNA delivery system for the transformation of several dicot species. Unfortunately, infection in monocots is greatly reduced and apparently lacking in members of the Gramineae. There have been only two reports of the successful use of *A. tumefaciens* as a vector system in monocots. *Narcissus* stem segments were infected and developed small swellings containing octopine and nopaline at the wound sites (Hooykaas-Slogteren *et al.*, 1984). In *Asparagus officinalis,* tumorous proliferations developed on stem segments cultured on hormone-free medium. These tumors demonstrated hormone autonomous growth and produced nopaline (Hernaalsteens *et al.*, 1984).

Because of the inability of *Agrobacterium* to infect gramineous monocots, free-DNA delivery systems have provided an important method for plant transformation. Recently direct DNA uptake procedures using PEG and calcium phosphate (Krens *et al.*, 1982) have been successfully applied for the recovery of stably transformed calluses from protoplasts of *T. monococcum* (Lorz *et al.*, 1985) and *L. multiflorum* (Potrykus *et al.*, 1985). Putative transformants were selected on the basis of resistance to kanamycin or G418 and the demonstration of neomycin phosphotransferase (NPTII) activity. Because nonmorphogenic cell lines were used as a source of protoplasts in both of these studies, there is no likelihood of recovering transformed plants.

Another method of DNA delivery, electroporation, employs the use of an electrical pulse to create transient pores in protoplast membranes (Zimmerman and Vienken, 1982). Macromolecules can be taken up through the pores into the protoplast and, in the case of DNA, can be expressed. Because the introduced DNA is not immediately integrated into the chromatin, its expression peaks 36 hours after delivery and decays as the DNA is degraded in the cell (Fromm *et al.*, 1985; Hauptmann *et al.*, 1987a).

Fromm *et al.* (1985) used this procedure for the delivery of DNA into protoplasts of a nonmorphogenic Black Mexican Sweet line of maize, and demonstrated the transient expression of the chloramphenicol acetyltransferase (CAT) gene, which was driven by the CaMV 35 S promoter with the nopaline synthase 3′-end polyadenylation signals. Subsequently they used the same system to obtain stable transformants of maize using kanamycin (100 μg/ml) as the selection agent (Fromm *et al.*, 1986). The highest frequency was 161 kanamycin-resistant colonies per 2×10^6 initial protoplasts. Hauptmann *et al.* (1987a,b) have also used electroporation for the delivery of DNA and the transient expression of CAT activity in protoplasts of *T. monococcum, P. purpureum, P. maximum, S. officinarum,* and a double cross, trispecific hybrid between *P. purpureum*, *P. americanum*, and *Pennisetum squamulatum*. All the species tested, including *Z. mays*, were found to be highly resistant to growth inhibition by kanamycin, which indicated that kanamycin might not be useful as a universal selection agent in the Gramineae.

Other selectable markers which are available for the direct selection of stable transformants are methotrexate or hygromycin resistance. Methotrexate resistance is obtained from a mutant dihydrofolate reductase (DHFR) gene created by site-specific mutagenesis of a wild-type mouse DHFR gene (Eichholtz *et al.*, 1986). DHFR converts dihydrofolate to tetrahydrofolate, which is a key step in the biosynthesis of purines and the pyrimidine thymine. Enzymatic activity is inhibited by methotrexate, a folic acid analog. The engineered mouse dehydrofolate reductase encodes an altered protein that does not bind methotrexate and therefore confers resistance to methotrexate. Another marker available is the hygromycin phosphotransferase (HPT) gene, which codes for resistance to the antibiotic hygromycin (Waldron *et al.*, 1985).

The species which were found to be resistant to kanamycin were sensitive to both methotrexate and hygromycin. Recently, these markers have been used for the direct selection of stable transformants, confirmed by Southern blot analysis, in *P. maximum* and *T. monococcum* (Hauptmann *et al.*, 1987b).

5. *Conclusions*

The foregoing account provides convincing record of substantial progress in the genetic manipulation of cereal and grass species. Nevertheless, many difficulties remain. Foremost among these are the problems being encountered in the efficient regeneration of mature plants from protoplasts. Recent recovery of somatic hybrid cell lines/somatic embryos and genetically transformed cells should further stimulate efforts toward the development of reliable protocols for the regeneration of plants from gramineous protoplasts.

B. Expression of Preexisting Somatic Variation and Generation of *de Novo* Variation *in Vitro*

1. *Introduction*

Tissue culture was initially conceived as a method of cloning plants but in many cases plants regenerated from tissue culture were found to be phenotypically different from the donor plant. This is particularly true where plants have been derived from callus culture, but variants may equally arise directly from the explant. At first this variation was regarded purely as an impediment to clonal propagation. It was only as a result of detailed studies on agronomic crops that the potential application of this variation to the production of new varieties was first brought into focus. Preliminary studies on potato somaclones (Shepard *et al.*, 1980) and regenerants from sugarcane tissue cultures (Heinz and Mee, 1971) suggested that tissue culture may provide a source of novel variants for complex traits such as yield and disease resistance. As these characters are governed by groups of genes (pleiotropic) they are difficult and time consuming to produce using traditional breeding methods. This fact was drawn to the attention of a wider audience by a review written by Larkin and Scowcroft (1981) in which they outlined the widespread occurrence of variability in regenerants from tissue culture. These authors coined the term "somaclonal variation" to describe tissue-culture-derived variation which arises in the absence of any direct selection pressure. The prospects of generating novel and potentially useful variants and their possible use in plant breeding have kindled enthusiasm, leading to the numerous studies which exist today on the characterization of regenerants from tissue culture (D'Amato, 1985).

The extensive and diverse studies conducted on somaclonal variation both in dicotyledonous and monocotyledonous plants in recent years have concentrated on a number of objectives. The determination of heritability and utility of somaclonal variation has been of primary interest. It has also been deemed important to establish the source of this variation and the factors which control it, as this information will be essential to those wishing to avoid the production of somaclonal variation when regenerating and cloning genetically engineered plant cells. Here we briefly outline our understanding of the phenomenon of somaclonal variation and discuss its potential application to plant improvement in the Gramineae. While many of the examples cited here will relate to studies on the Gramineae, examples from dicot cultures will also be included where appropriate.

2. *Perpetuation and Generation of Somatic Variation*

By broadly defining somaclonal variation as tissue-culture-derived variation, Larkin and Scowcroft (1981) have led many to the misconception that tissue culture is directly responsible for the variation produced in regenerants. However, from the studies conducted to date on the origin of somaclonal variation it is generally agreed that this variation may be the result of inherent variation present in the explant used, which is selectively perpetuated *in vitro,* and/or the generation of new variation as a result of genome instability induced by culture media or environmental conditions *in vitro* (D'Amato, 1985). The latter type of variation results from disruptions in the regulation of the mitotic cell cycle leading to the production of polyploids, aneuploids, or hypo-/hyperdiploids (Dyer, 1976). Similarly, chromosomal rearrangements resulting from breakage–fusion–bridge cycles (Sunderland, 1977; Toncelli *et al.,* 1985), Robertsonian fusion (Murata and Orton, 1983), or somatic crossovers (Lapitan *et al.,* 1984) induced *in vitro* may result in pseudodiploids (e.g., Gould, 1982) or other karyotypic changes (e.g., McCoy *et al.,* 1982; Armstrong *et al.,* 1983).

An important feature of somaclonal variation revealed in these studies is the fact that the degree and nature of variation produced from a given plant are dependent on the plant genotype (McCoy *et al.,* 1982) and the type of explant introduced into culture (D'Amato, 1985), as well as media constituents (Dale and Dembrogio, 1979), subculture duration (McCoy *et al.,* 1982; Ahloowalia, 1982), callus type (I. K. Vasil and Vasil, 1986), and the prevailing environmental and physiological conditions imposed on cells during culture (Orton, 1980b). If a given plant genotype undergoes qualitative or quantitative changes in DNA during differentiation and development (D'Amato, 1977; Nagl, 1978, 1979), possesses transposable elements (Nevers *et al.,* 1986), or has unstable genetic loci (Durrant, 1981; Cullis, 1983; Walbot and Cullis, 1985), it will evidently be more susceptible to the production of variants from tissue culture than is a plant genotype which does not provide this inherent source of somatic variation. This is evident from the fact that polysomatic plants (Van Harten *et al.,* 1981), polyploids (McCoy *et al.,* 1982; Ahloowalia, 1982; Karp and Maddock, 1984), chimeras (Skirvin and Janick, 1976; Cassells *et al.,* 1986), and plants with unstable genetic loci (Cullis, 1981; Cullis and Cleary, 1985) or putative transposable elements (Groose and Bingham, 1986a,b) produce a greater range and number of variants than do diploids, nonpolysomatic plants, or plants with a high level of genome stability (Sheridan, 1977; Brossard, 1978; McCoy and Phillips, 1982; Cassells and Morrish, 1985).

The expression of the somatic variation which may exist in cells within the explant introduced into culture is dependent on their participation in the production of callus and subsequent involvement in plant regeneration. This is only possible if culture media and environmental conditions can induce the division and differentiation of variant cells. In some polysomatic plants the division of endopolyploid cells is dependent on the presence of kinetin in the medium (Torrey,

1961). In culture these cells may form the basis for a genetically heterogeneous population. For example, endopolyploid cells may undergo amitosis (nuclear fragmentation) to produce aneuploid, polyploid, or haploid cells (D'Amato, 1985). Cells with an unstable genetic locus or transposable element may, depending on the level of stress imposed by culture conditions, continually produce variant cells (McClintock, 1978; Walbot and Cullis, 1985). For this reason it is possible that different callus cultures may produce different levels of variation at different stages of subculture due to the fluctuating population of variant cells, and this may explain the differences found in regenerants derived from different subcultures (McCoy *et al.*, 1982; Ahloowalia, 1982) and the range of genetically different variants which can arise from a single callus (Edallo *et al.*, 1981; Fukui, 1983).

The influence of callus type on the level of variation produced in regenerants has already been discussed (Section II,A). From the studies conducted to date there appears to be a distinct difference in the level of genetic stability maintained in different callus types. However, it must be emphasized that none of these studies compared the level of variation produced in regenerants derived from different callus types of a single genotype. Due to the variable levels of tissue-culture stability which may be encountered in plants with different genetic architectures, it is not possible at this time to conclusively state that regeneration from an individual callus type will preferentially result in true-to-type regenerants. This will only be possible when a comparison has been made between regenerants of a selected genotype using different callus types (e.g., embryogenic versus organogenic). Presently the evidence for differences in genetic stability in individual callus types in the Gramineae is limited to the cell level. Different callus types produced in *Hordeum* cultures show characteristic cytological differences which appear to influence their rate of growth and ability to regenerate (Orton, 1980b). Genetic drift in these callus lines was found to increase when they were cultured in suspension, demonstrating how changing the culture environment may also alter the rate of somatic variation induced in cell cultures.

To date the nature of the variation produced from tissue cultures of the Gramineae is extremely diverse (Table 4) and there is evidence for the production of homozygous (Cooper *et al.*, 1986) as well as heterozygous (Oono, 1983; Cooper *et al.*, 1986) mutants. Meiotically identified rearrangements have been documented in barley (Orton, 1980b), wheat (Karp and Maddock, 1984), oats (McCoy *et al.*, 1982), and *Triticale* (Armstrong *et al.*, 1983). These include reciprocal and nonreciprocal interchanges, inversions, partial chromosome loss, and duplications. Polyploids (Orton, 1980b; Oono, 1985), aneuploids (Edallo *et al.*, 1981; Karp and Maddock, 1984), and other chromosome number variants have also been recorded (Heinz and Mee, 1971; Liu and Chen, 1979). Variation has also been found in the cytoplasmic genome in corn, of which variants for mtDNA have been regenerated (Pring *et al.*, 1981; see Section III,C). More recent studies have closely examined the molecular basis of somaclonal variation and demonstrate the presence of single-base changes (Brettell *et al.*, 1986), genome amplification (De Paepe *et al.*, 1983), and modification in the copy number of repeat sequences (Cullis and Cleary, 1985; Landsmann and Uhrig, 1985; Brettell *et al.*, 1986). These results have led advocates of somaclonal variation to cite as its major advantages (1) the potential for somatic recombination which would be useful in the introgression of alien genes in wide crop hybrids (Orton, 1980a,b; Larkin and Scowcroft, 1981, 1983b), (2) an increased incidence of mutation (Orton, 1984), (3) the introduction of variability into crops by allowing the expression of preexisting variation and hence eliminating the introduction of deleterious genes from other plants, which would have to be removed by backcrossing in traditional breeding programs (Larkin, 1985), and (4) recovery of aneuploid stocks, ditelosomics, and monosomic lines (McCoy *et al.*, 1982). On closer examination, however, it has been shown that tissue culture, like traditional breeding, also has limitations and furthermore exhibits its own unique anomalies.

Because of the potential for somatic crossovers and alloploidization, which would restore fertility, the application of tissue culture to the introgression of alien genes has been studied in cultures of wide-cross crop hybrids in a number of cereals and grasses (Table 5). A number of these studies indicate that chromosome interchanges between genomes occur *in vitro* (Orton, 1980b; Lapitan *et al.*, 1984). Fertility has also been restored in regenerants from *L. multiflorum* × *Festuca arundinacea* after spontaneous polyploidization of the genome (Kasperbauer *et al.*, 1979). However, these changes do not occur in all hybrids studied (Table 5). Studies on the chromosome number present in callus as compared with regenerants indicate that in some hybrids the regeneration process selects against polyploid cells and to a lesser extent against cells exhibiting aneuploidy and structural changes (Orton, 1980b). Furthermore, attempts to induce amphiploids by *in vitro* colchicine treatment of hybrid calluses have not proved successful generally and result in the production of chimeras (Nakamura *et al.*, 1981) or aneuploids (Orton and Steidl, 1980). These results indicate that the problems of fertility and the introgression of

TABLE 4
Somaclonal Variation in the Gramineae

Species	Variant Character	Heritability[a]	Reference
Avena sativa	Maturation, sterility, variegation	NT	Cummings *et al.* (1976)
Avena sativa	Aneuploids, chromosome breaks, interchanges	H	McCoy *et al.* (1982)
Zea mays	Polyploids, aneuploids, endosperm, seedling mutants	H	Edallo *et al.* (1981)
Zea mays	*Helminthosporium maydis* race-T toxin resistance	H	Brettell *et al.* (1980)
Zea mays	mtDNA sequence rearrangement, fertility	H	Kemble *et al.* (1982)
Zea mays	Resistance to *Phyllostica maydis*	NT	Gengenbach *et al.* (1981)
Zea mays	Defective kernel mutants	H	McCoy and Phillips (1982)
Zea mays	Plant height, albinism, flowering date, yield	H	Beckert *et al.* (1983)
Zea mays	Alcohol dehydrogenase mutant	H	Brettell *et al.* (1986a)
Zea mays (monosomic)	Altered karyotype, translocations, chromosome loss, genomic doubling, coenocytic microsporocytes	H	Rhodes *et al.* (1986)
Saccharum spp.	Plant morphology, chromosome number, isozymes	NT	Heinz and Mee (1971)
Saccharum spp.	Auricle length, sugar content, chromosome number	NT	Liu and Chen (1979)
Saccharum spp.	Eyespot toxin resistance, plant height, chlorophyll mutants	VT	Larkin and Scowcroft (1983a)

Saccharum spp.	Marker remission	VT	Irvine (1984)
Oryza sativa	Tiller number, panicle size, seed fertility, plant height, grain weight, chlorophyll mutations, polyploids	H	Oono (1983, 1985)
Oryza sativa	Tiller number, chlorophyll mutants, plant height	H	Zong-xin *et al.* (1983)
Oryza sativa	Early heading, albinism, sterility	H	Fukui (1983)
Triticum aestivum	Plant height, stem thickness, pollen fertility, seed set, chromosome inversions and deletions	NT	Ahloowalia (1982)
Triticum aestivum	Fertility, chlorophyll pigmentation, spike length, flowering	NT	Sears and Deckard (1982)
Triticum aestivum	Plant height, spike shape, leaf wax, grain color, α-amylase	H	Larkin *et al.* (1984)
Triticum aestivum	Interchanges, aneuploids	NT	Karp and Maddock (1984)
Triticum aestivum	Grain yield, haploids, aneuploids, mixoploids	H	Ahloowalia and Sherington (1985)
Triticum aestivum	Gliadins, alcohol dehydrogenase mutant	H	Cooper *et al.* (1986); Davies *et al.* (1986)
Triticum durum	Aneusomaty, spike length, fertility	DS	Lupi *et al.* (1981)
Festuca arundinacea	Aneuploids, trisomics	H	Reed and Conger (1985)
Lolium spp.	Chromosome number, leaf shape and size, floral development, growth vigor	NT	Ahloowalia (1983)

[a] Abbreviations: H, heritable through sexual cycle; NT, not tested; DS, diplontic selection; VT, vegetative transmission.

TABLE 5
Tissue Culture of Wide-Cross Hybrids

Hybrid	Result[a]	Reference
Hordeum distichum × *Secale cereale*	No improvement	Cooper *et al.* (1978)
Lolium multiflorum × *Festuca arundinacea*	Doubling (fertile)	Kasperbauer *et al.* (1979)
Hordeum vulgare × *Hordeum jubatum*	Doubling (C); chromosomal rearrangements; microhybridity, haploids; aneuploids (C)	Orton (1979); Orton (1980a); Orton (1980b); Orton and Steidl (1980)
Hordeum vulgare × *Secale cereale*	No improvement	Shumnyi and Pershina (1979)
Hordeum vulgare × *Triticum aestivum*	Mixoploids (PR)	Chu *et al.* (1984)
Triticum crassum × *Hordeum vulgare*	Chimeras	Nakamura *et al.* (1981)
Saccharum × *Zea*	Mosaics, dicentrics	Sreenivasan and Jalaja (1982)
Saccharum × *Sclerostachya fusca*	Translocation	Sreenivasan and Sreenivasan (1984)
Lolium multiflorum × *Festuca arundinacea*	Doubling (fertility)	Rybczynski *et al.* (1983)
Lolium rigidum × *Festuca arundinacea*	Doubling (fertility)	
Secale cereale × *Triticum crassum*	Translocations; amphidiploids, translocations, deletions	Armstrong *et al.* (1983); Lapitan *et al.* (1984)
Lolium multiflorum × *Festuca pratensis*	Mixoploids, tetraploids	Zwierzykowski *et al.* (1985)

[a] Abbreviations: C, colchicine treatment; PR, preliminary report.

alien genes and elimination of deleterious genes are as great in tissue culture as with conventional attempts to utilize alien germ plasm.

While it is true that tissue culture has resulted in the recovery of a number of mutants, most of them are not novel and none of the mutants produced to date is of major agronomic importance. Many are recessive and some are deleterious, such as the endosperm mutants found in maize (Edallo *et al.*, 1981). Furthermore, the postulated unique advantages of using tissue culture for the production of variants with altered pleiotropic traits may have been overestimated. This view is based on the absence of somaclonal variants which bear characters uniquely different from those which may be generated as sports in normal propagation cycles. An example is the absence of significant morphological or isozyme differences between "bolter" populations (sports) and protoclones of Russet Burbank (Sanford *et al.*, 1984).

The study of the heritability of somaclonal variation has revealed some of the anomalies of tissue-culture-derived variation. We now know that variation induced *in vitro* can result from either genetic, epigenetic, or nongenetic change (Meins, 1983). Nongenetic variation includes chimeral breakdown, physiological effects, and the elimination of virus or viruslike agents (Cassells, 1985). There are a number of reviews which outline the implications and consequences of this form of variation (e.g., Larkin and Scowcroft, 1981; Meins, 1983; Cassells, 1985).

Reviews by Orton (1984) and more recently by D'Amato (1985) indicate that there is evidence for the production of genetic variation as judged from studies on Mendelian inheritance of variant traits. However, the reversion of many variants after a propagation cycle (Larkin and Scowcroft, 1983a, Irvine, 1984) has heightened our awareness of the epigenetic changes which may occur *in vitro*. Epigenetic change results from a change in gene expression and not gene mutation (Chaleff, 1981; Meins, 1983). These variants, therefore, will revert when sexually propagated.

Because of their multicellular origin, many of the variants induced *in vitro* are chimeral and aneusomatic (Heinz and Mee, 1971; Liu and Chen, 1979; Bennici and D'Amato, 1978; Lupi *et al.*, 1981; McCoy and Phillips, 1982). In some cases this fact is not immediately obvious and the production of variants in progeny from plants which were both phenotypically and cytologically normal in the first generation demonstrates the fact that plants regenerated from tissue culture may have a complex genetic architecture as a result of their derivation from multicellular initials in a genetically heterogeneous callus (Sree Ramulu *et al.*, 1984). Hence many so-called variants or normal plants are likely to break down or segregate when propagated.

3. *Conclusions*

For the reasons enumerated above and because there is as yet no report of the production of an agronomically useful somaclonal variant, there are many who dispute the relative merits of using somaclonal variation as a means of producing new crop plants. Those in favor of somaclonal variation, however, emphasize its relatively recent recognition as a means of crop improvement and the overall lack of large-scale studies on a selected crop to study the nature and heritability of such variation (Larkin, 1985). It is certainly true that when compared with the field trials used in traditional breeding, the studies so far undertaken in the assessment of somaclonal variation have not been extensive. Hence it is difficult at the present time to conclusively state that somaclonal variation cannot be regarded as a potential source of useful genetic variation. In view of the influence of genotype, explant source, regeneration procedure, and culture conditions in controlling the level and nature of variation produced, this issue will not be easily resolved.

C. Positive Selection for Somatic Variation *in Vitro*

1. *Introduction*

Natural plant populations undergo continuous selection and adaptation. The selective pressures of a given environment will only allow the survival of unique genotypes which have the capacity to grow in such environments. This principle may also be applied to tissues and cells *in vitro,* where inherent somatic variation (Selection III,B) is more likely to be expressed in regenerants from tissue cultures subject to selection pressure. The use of *in vitro* selection systems has been widely applied in dicots, of which numerous mutants have been isolated using calluses, single cells, and protoplasts (Maliga, 1980, 1984; Negrutiu *et al.*, 1984). Because procedures for the regeneration of plants from cell cultures and protoplasts have only recently been devised (Section III,A), mutant selection systems for the Gramineae are limited to callus cultures (Bright, 1985).

Selection for mutants at the tissue level has the obvious advantage of saving space and time. However, the selective growth of certain cell lines *in vitro* may not always be due to genetic change, as enhanced growth may equally be due to physiological or epigenetic adaptation

(Chaleff, 1981; Meins, 1983; Meredith, 1984). Furthermore, the protocols used for regenerating plants from variant cell lines may result in the production of chimeras if plants are of multicellular origin. For this reason assessment for the production of true mutants from cell lines selected *in vitro* must involve progeny testing through a number of generations to test the heritability of this variation.

In vitro selection systems are limited to phenotypes which are expressed in cultured cells and for which suitable selection strategies can be designed. Hence, complex traits which are governed by groups of genes or traits which are developmentally regulated are not amenable to *in vitro* selection. Selection systems tested in the Gramineae include the use of toxins for disease resistance, amino acids or their analogs for amino acid overproduction, and sodium chloride for salt tolerance (Table 6).

2. *Characters Tested for Selection in Vitro*

a. Protein Quality. The improvement of protein quality in the cereal grains requires increases in the concentration of the limiting amino acids, which include lysine, threonine, and tryptophan. *In vitro* selection systems used to select cell lines and plants which produce increased concentrations of these amino acids involve the use of amino acids or their analogs (Bright and Shewry, 1983). Generally, biosynthesis of amino acids is controlled by feedback regulation. This may be circumvented by either a decrease in the concentration of an inhibitory end product or mutational alteration of the enzyme(s) to a condition less sensitive to feedback effects, i.e., mutation to feedback resistance. *In vitro*, three selection agents have been used to identify mutants overproducing aspartate-derived amino acids: (1) the lysine analog AEC, (2) a combination of lysine plus threonine (LT), and (3) the tryptophan analog 5-methyltryptophan (5-MT).

AEC is an analog which will compete with lysine for uptake and amino acetylation of tRNA and incorporation into protein (Bright *et al.*, 1979). Because inhibition due to AEC can be relieved by externally supplied lysine, mutants resistant to AEC can accumulate lysine. AEC-resistant cell lines have been isolated in *O. sativa* (Schaeffer, 1981) and *P. americanum* (Boyes and Vasil, 1987). Both the cell lines and their regenerants contained higher levels of lysine than controls. The cytoplasmic male sterility of the Gahi 3 *P. americanum* cultivar used in the latter study has prevented the determination of the heritability of lysine overproduction; however, in rice, heritability of the increased lysine concentration was confirmed.

Lysine and threonine, when supplied singly to cells, tissues, or

TABLE 6
Selection Pressures and Cell, Plant, and Progeny Mutants Isolated from Tissue Cultures of the Gramineae

		Mutant[a]			
Selection pressure	Plant	Cells	Regenerants	Progeny	Reference
High osmoticum					
NaCl	*Oryza sativa*[b]	+	#	#	Wong *et al.* (1983)
	Oryza sativa[b]	+	+	C	Woo *et al.* (1985)
	Oryza sativa	+	+	?	Wong and Woo (1986)
	Oryza sativa	+	+	#	Reddy and Vaidyanath (1986)
	Oryza sativa	+	+	?	Nabors and Dykes (1985)
	Sorghum	+	+	#	Bhaskaran *et al.* (1983)
	Sorghum	+	+	?	Bhaskaran *et al.* (1986)
	Pennisetum americanum	+	#	#	Rangan and Vasil (1983)
	Pennisetum purpureum	+	N	#	Chandler and Vasil (1984a)
	Avena sativa	+	+	?	Nabors and Dykes (1985)
Analogs					
S-(2-aminoethyl)-L-cysteine (AEC)	*Oryza sativa*	+	+	+	Schaeffer (1981)
AEC	*Pennisetum americanum*	+	+	*	Boyes and Vasil (1987)
5-Methyltryptophan	*Zea mays*	+	+	+	Hibberd *et al.* (1986)
Feedback inhibitors					
Threonine + lysine	*Zea mays* (D33)	+	+	*	Hibberd *et al.* (1980)
	Zea mays (Lt 19)	+	+	+	Hibberd and Green (1982)
	Hordeum vulgare	+	+	#	Miflin *et al.* (1983)
Antibiotics					
Streptomycin	*Zea mays*	+	+	C	Umbeck and Gengenbach (1983)
Pathotoxins					
Helminthosporium T toxin	*Zea mays*	+	+	+	Gengenbach *et al.* (1977, 1981)
	Zea mays	+	+	+	Brettell *et al.* (1980)
Helminthosporium sacchari	*Saccharum*	+	+	V	Larkin and Scowcroft (1983)
Phyllostica maydis	*Zea mays*	#	#	+	Gengenbach *et al.* (1981)
Victorin toxin	*Avena sativa*	+	+	+	Rines and Luke (1985)

[a] Notation: +, mutants recovered; #, not tested; *, sterile; ?, preliminary report; C, some chimeral progeny; N, nontolerant plants recovered; V, vegetatively transmitted.

plants, can inhibit growth by interfering with arginine or nitrate metabolism (Filner, 1969). When supplied together they interact synergistically to close down the operation of the aspartate pathway and starve the plant of methionine (Green and Donovan, 1980). Mutants with decreased feedback regulation of the pathway are expected to accumulate end-product amino acids. In maize, regenerating callus cultures have been used to obtain two LT-resistant lines, one of which gave rise to fertile plants (Hibberd *et al.*, 1980; Hibberd and Green, 1982). When evaluated, resistance to lysine plus threonine was found to be due to an altered form of aspartate kinase with decreased sensitivity to feedback regulation by lysine (Hibberd *et al.*, 1980). A feature of LT resistance is the associated accumulation of soluble threonine in vegetative tissue (plant or callus). Threonine accumulation of two- to sixfold greater than methionine accumulation has been recorded. In mature seed, threonine accumulation in the maize mutant LT-19 (Hibberd and Green, 1982) and some barley mutants (Bright, 1985) led to an increase of 33–59% in the total seed content of threonine, thus improving nutritional quality.

Metabolite flow in the tryptophan pathway of higher plants is apparently regulated by tryptophan through the feedback inhibition of the enzyme anthranilate synthase (AS). Hence changes in this enzyme may alter the level of tryptophan synthesized. Alterations in AS allow cells to grow in inhibitory concentrations of 5-methyltryptophan or tryptophan. Procedures for the isolation of maize mutants for increased tryptophan synthesis have been patented by Hibberd *et al.* (1986). In this report maize cell lines selected on 5-MT were found to produce an altered anthranilate synthase enzyme which conferred resistance to inhibition by certain tryptophan analogs and permitted growth of callus cultures, plant tissue, plants, and seeds in the presence of the same analogs at concentrations which normally inhibit growth. Mutant plant seeds, tissue cultures, and plants had an endogenous free-tryptophan content of at least 10 times that of controls.

The procedures used in the studies outlined above may also be applied to other crops. Hence, such selection systems may be used for isolating mutants that overproduce tryptophan, threonine, and lysine.

b. Salt Tolerance. Saline soils prevent the growth of crop plants in many parts of the world. In conventional breeding programs existing cultivars are screened for salt tolerance and are used as a basis for further breeding. Often the range of tolerance within commercial varieties is relatively restricted, as salt tolerance has not been a factor in the selection of modern crop plants. An alternative approach is to exploit the possible natural salt tolerance which may be present in the wild relatives (Wyn Jones and Gorham, 1986). These procedures are time consuming as they involve extensive backcrossing programs, hence the potential for isolating salt-tolerant lines using *in vitro* selection has received much attention. Direct selection for salt tolerance *in vitro* has been attempted in a number of species of the Gramineae and some salt-resistant cell lines have been isolated (Table 6). Plants have been regenerated from NaCl-tolerant cells of oats (Nabors *et al.*, 1982), rice (e.g., Yano *et al.*, 1982; Woo *et al.*, 1985; Reddy and Vaidyanath, 1986), *P. purpureum* (Chandler and Vasil, 1984b), and *Sorghum* (Bhaskaran *et al.*, 1983, 1986). In some of these reports tolerance to NaCl was expressed in a few of the regenerated plants but as yet there is no conclusive genetic evidence to demonstrate the production of salt-tolerant plants from tissue cultures of the Gramineae. These disappointing results may possibly be a consequence of a number of factors, including (1) the polygenic nature of salt tolerance and (2) the use of NaCl alone as a stress rather than in combination with other ions, as may occur in natural environments, or *in vitro* anomalies, including (3) the physiological adaptation of cultured cells, (4) the nonregeneration of selected cell types, (5) the phenotypic changes in regenerated plants, and (6) the developmental regulation of cellular expression of salt tolerance.

Organisms resistant to saline conditions have been found to accumulate high levels of proline and glycinebetaine (e.g., Le Rudulier *et al.*, 1984). For this reason the isolation of cell lines with increased levels of proline or glycinebetaine has been suggested as an indirect method of selecting salt-tolerant cell lines. High-proline mutants of barley have been isolated from callus cultures but these show little or no improvement in stress resistance (Kueh and Bright, 1982). Barley plantlets regenerated from embryo cultures accumulate exogenously applied proline and glycinebetaine and improve the salt tolerance of axenic plants (Pervez *et al.*, 1986). The accumulation of these metabolites, however, may simply be a response to stress, and the relative utility of selecting for overproducers of these metabolites in a search for salt-tolerant plants remains to be elucidated.

c. Disease Tolerance. The most obvious selection strategy for isolating disease-resistant cells is to challange a cell population with the pathogen. This approach, however, is not practical *in vitro* due to the fact that a pathogen introduced into culture will often overgrow plant cells. Also, resistance to disease may be associated with certain organized structures *in vivo* which would not be expected to operate in *in vitro* systems. The absence of cuticle and wax in rice callus, for example, has been associated with increased susceptibility to infection by several fungal pathogens (Uchiyama and Ogasawara, 1977). The expression of disease resistance which is associated with secondary compounds may also be reduced. Reduced resistance of rice callus to infection by fungal pathogens has been attributed to the absence of lignin in this tissue (Uchiyama *et al.*, 1983). To avoid the difficulties of introducing a pathogen into culture, toxins have been used as selective agents. Host-selective and nonselective toxins are important in symptom formation in a number of cereal diseases (Strobel, 1982) and such pathotoxins may be used as *in vitro* selective agents by incorporating lethal or sublethal doses into culture media.

The incorporation of toxin into the medium has been used successfully to obtain maize (*Z. mays* L.) cultures (Gengenbach and Green, 1975; Brettell *et al.*, 1979) and regenerated plants and progeny (Gengenbach *et al.*, 1977, 1981) that were resistant to the fungal pathotoxin produced by *Helminthosporium maydis* race T. Resistance to *H. maydis* toxin was transmitted to sexual progeny as an extranuclear trait. In one study, male fertility and resistance to the *Phyllosticta maydis* pathoxin were cotransferred with *H. maydis* resistance to progeny of male-fertile resistant plants (Gengenbach *et al.*, 1981).

The maize lines used for tissue-culture selection in these studies had Texas male-sterile cytoplasm (T cytoplasm), which conditions a high-susceptibility interaction with *H. maydis* and its toxin (Liu and Hooker, 1972). In contrast, male-fertile or normal (N) cytoplasm maize plants and tissue cultures are highly resistant to *H. maydis* toxin (Gengenbach and Green, 1975; Brettell *et al.*, 1979). Proof that mutants produced from tissue culture did not originate from selection of N-mitochondrial genomes coexisting with T genomes in T cytoplasm has been provided by the distinct mitochondrial DNA organization found in mutants after restriction endonuclease analysis (Gengenbach *et al.*, 1981; Kemble *et al.*, 1982).

Maize plants resistant to *H. maydis* have also been selected by incorporation of streptomycin in the medium (Umbeck and Gengenbach, 1983). In this study, however, a lower number of variants was isolated, only 5% as compared with 50% in the study of Brettell *et al.* (1980). The change to male fertility and/or resistance to *H. maydis* pathotoxin found by Umbeck and Gengenbach (1983) was unexpected, as the cultures had not been selected for pathotoxin resistance. This report also provides an example of the problem posed by chimerism in regenerants. Some of the regenerated plants were not solid mutants for cytoplasmic alterations. These plants were characterized by (1) sectoring for fertility, (2) fertile tassels but toxin-susceptible leaves, and (3) maternally derived R1 progeny that segregated for fertility, resistance, and sterility with toxin susceptibility. Brettell *et al.* (1980) also report the production of a putative chimeral male-fertile toxin-susceptible R0 plant. These plants could be the result of (1) a multicellular origin in culture and/or (2) a heterogeneous cytoplasmic condition that assorts somatically during plant regeneration and differentiation (cf. Umbeck and Gengenbach, 1983).

Selection for pathotoxin resistance has also proved successful in isolating cell lines, plants, and progeny of oats (*Avena sativa*) resistant to *Helminthosporium victoriae* (Rines and Luke, 1985) and cell lines and plants of *Saccharum* spp. resistant to the toxin produced by *Helminthosporium sacchari* (Heinz *et al.*, 1977; Nickell, 1977; Larkin and Scowcroft, 1983a). In an analysis of the stability of this trait in vegative propagation cycles Larkin and Scowcroft (1983a) found that 19% of plants originally classified as variants segregated, suggesting chimerism. The 8% which reverted was deemed to have shown physiological adaptation to the toxin rather than true mutation to tolerance.

3. *Conclusions*

It is clear from the preceding discussion that the production of novel genotypes with heritable mutations may be achieved with *in vitro* screening. Furthermore, the variants isolated using this approach can make some contribution to crop improvement. There are still, however, many limitations to the application of *in vitro* selection for crop improvement in the Gramineae. Available evidence suggests that *in vitro* selection is limited to traits governed by single genes, hence many agriculturally desirable phenotypes governed by multiple genes cannot be selected for *in vitro*. Also, the use of callus cultures creates an inherent problem due to the possibility of ranges of tolerance and the presence of intolerant residual cells, which may enhance the frequency of regeneration of nontolerant or chimeral plants. Further advances in the application of *in vitro* selection will therefore require an advance in cell-culture technology to allow the routine regeneration of plants from protoplasts or single cells as well as the development of selection strategies which prevent growth of all but the desired variant cells.

IV. General Conclusions

By the selection of appropriate explants, media, and culture conditions it is now possible to regenerate plants from all the major cereal and grass species. The seasonal limitations imposed by the use of immature tissues and the possible detrimental influences of environmental conditions on *in vitro* responses (e.g., Santos and Torne, 1986), however, have imposed limitations on the frequency and reproducibility with which cultures may be established and plants regenerated. Ideally, the maintenance of genome stability and regenerative competence in callus or suspension culture would provide the best method of providing cells for genetic manipulation or propagation. At present we know very little about the causes underlying the loss of competence found in long-term cultures. It is evident that some change occurs which alters the capacity of cells to differentiate. Such changes may involve DNA methylation or amplification, or the translocation of transposable elements. Recently Jones (1985) addressed the question of the possible role played by DNA methylation in tissue and cell recalcitrance and results from studies in this area should give some guidelines toward resolving the role played by this variable on the competence of gramineous cells *in vitro*. The possible role played by other quantitative variables also deserves attention. By developing an understanding of the underlying causes of recalcitrance it may be possible in the future to overcome this problem.

The recent interest in establishing possible correlations between changes in endogenous hormones and quantitative and/or qualitative changes in DNA with recalcitrance of tissues *in vitro* represents a starting point for increasing our understanding of the biochemical and genetic status of cells *in vivo* in relation to their competence *in vitro*. Concentrated efforts should be made in the future to continue such studies and establish if correlative changes occur in other members of the Gramineae. Our ability to detect such alteration would be greatly enhanced by the development of methods which allow efficient detection of changes in a wider range of hormones. Developments in molecular biology continue to increase our ability to detect changes at the DNA level. Future application of these techniques to the study of changes occurring in DNA during development, differentiation, and exposure to stress should give some promising pointers to the genetic factors which may influence cell competence.

The use of genetic manipulation in the Gramineae is in its infancy, but the successful isolation of somatic hybrid and transformed cell lines indicates that these procedures work in the Gramineae. The success of this approach to crop improvement, however, will depend on the development of procedures for the routine regeneration of plants from protoplasts as well as more effective methods for the induction and selection of transformed cell lines.

The production of somaclonal variants by nonselective regeneration of plants from tissue cultures has not provided any useful variants. In contrast, *in vitro* selection has resulted in the regeneration of novel and potentially useful mutants (e.g., Hibberd *et al.*, 1986). This suggests that the somatic variation present in tissue cultures is best exploited by creating selective pressures which encourage the growth of variant cells and increase the likelihood of regenerating useful mutants.

The field of molecular biology has undergone spectacular developments in the last decade. The developmental biology of plants, however, has not received the attention warranted to provide the systems presently needed for the genetic manipulation of plants. A dearth of information on the basic phenomenon of plant regeneration needs to be accounted for before we can conceivably begin to produce useful, novel crops by genetic manipulation.

Acknowledgments

The authors wish to thank Drs. P. Ozias-Akins, K. Rajasekaran, R. Hauptmann, Z. Tabaeizadeh, and Mark Taylor and Charlene Boyes for making available preprints of their publications and for useful suggestions and discussion. Special thanks are due to Dr. Randy Hauptmann for his suggestions for Section III,4. The research reported from this laboratory was supported by a joint project between the Gas Research Institute (Chicago, Illinois) and the Institute of Food and Agricultural Sciences (University of Florida), and by the Monsanto Company (St. Louis, Missouri).

References

Abe, T., and Futsuhara, Y. (1985). Efficient plant regeneration by somatic embryogenesis from root callus tissues of rice. *J. Plant Physiol.* **121,** 111–118.

Abe, T., and Futsuhara, Y. (1986). Genotypic variability for callus formation and plant regeneration in rice (*Oryza sativa* L.). *Theor. Appl. Genet.* **72,** 5–10.

Ahloowalia, B. S. (1982). Plant regeneration from callus cultures of wheat. *Crop Sci.* **22,** 405–410.

Ahloowalia, B. S. (1983). Spectrum of variation in somaclones of triploid ryegrass. *Crop. Sci.* **23,** 1141–1147.

Ahloowalia, B. S., and Sherington, J. (1985). Transmission of somaclonal variation in wheat. *Euphytica* **34,** 525–537.

Alfinetta, B., Zamora, Z. A., and Scott, K. J. (1983). Callus formation and plant regeneration from wheat leaves. *Plant Sci. Lett.* **29,** 183–189.

Appelgren, M. (1985). Effect of supplementary light of mother plants on adventitious shoot formation in flower peduncle segments of *Begonia* × *hiemalis* Fotsch *in vitro*. *Sci. Hortic.* **25,** 77–83.

Armstrong, C. L., and Green, C. E. (1985). Establishment and maintenance of friable embryogenic maize callus and the involvement of L-proline. *Planta* **164,** 207–214.

Armstrong, K. C., Nakamura, C., and Keller, W. A. (1983). Karyotype instability in tissue culture of regenerates of *Triticale* (× *Triticosecale* Wittmark) cv. Welsh from 6-month-old callus cultures. *Z. Pflanzenzuecht.* **91**, 233–245.

Ashihara, H., Fujimura, T., and Komamine, A. (1981). Pyrimidine nucleotide changes during somatic embryogenesis in a carrot cell-suspension culture. *Z. Pflanzenphysiol.* **104**, 129–137.

Baba, A., Hasezawa, S., and Syono, K. (1986). Cultivation of rice protoplasts and their transformation mediated by *Agrobacterium* spheroplasts. *Plant Cell. Physiol.* **27**, 463–471.

Barnard, C. (1966). "Grasses and Grasslands." Macmillan, London.

Barry, G. F., Rogers, S. G., Fraley, R. T., and Brand, L. (1981). Identification of a cloned cytokinin biosynthetic gene. *Proc. Natl. Acad. Sci. U.S.A.* **81**, 477–4789.

Bayliss, M. W. (1980). Chromosomal variation in plant tissues in culture. *Int. Rev. Cytol.* **11A**, 113–144.

Beaulieu, G. C., Rogers, S. O., and Bendich, H. J. (1985). DNA extracted from wheat leaves is highly degraded: A possible basis for the difficulty in establishing leaf cell cultures in the Gramineae. *Int. Congr. Plant Mol. Biol., 1st, Savannah, Georgia* Abstr. OR-03-11, p. 11.

Beckert, M. (1982). Role du scutellum dans l'obtention de plantes neoformees *in vitro* chez le mais. *Agronomie* **2**, 611–615.

Beckert, M., and Qing, C. M. (1984). Results of a dialle trial and a breeding experiment for *in vitro* aptitude in maize. *Theor. Appl. Genet.* **68**, 247–251.

Beckert, M., Pollacsek, M., and Cainen, M. (1983). Etude de la variabilité genetique obtenue chez le mais après callogenese et regeneration de plantes *in vitro*. *Agronomie* **3**, 9–18.

Bennici, A., and D'Amato, F. (1978). *In vitro* regeneration of Durum wheat plants. I. Chromosome number of regenerated plantlets. *Z. Pflanzenzuecht.* **81**, 305–311.

Bevan, M. W., and Chilton, M. D. (1982). T-DNA of the *Agrobacterium* Ti and Ri plasmids. *Annu. Rev. Genet.* **16**, 357–384.

Bhaskaran, S., Smith, R. H., and Schertz, K. (1983). Sodium chloride tolerant callus of *Sorghum bicolor* (L.) Moench. *Z. Pflanzenphysiol.* **112**, 459–463.

Bhaskaran, S., Smith, R. H., and Schertz, K. F. (1986). Progeny screening of *Sorghum* plants regenerated from sodium tolerant selected callus for salt tolerance. *J. Plant Physiol.* **122**, 205–210.

Binding, H. (1986). Regeneration from protoplasts. *In* "Cell Culture and Somatic Cell Genetics of Plants" (I. K. Vasil, ed.), Vol. 3, pp. 259–274. Academic Press, Orlando, Florida.

Bingham, E. T., Hurley, L. V., Kaatz, D. M., and Saunders, J. (1975). Breeding alfalfa which regenerates from callus tissue in culture. *Crop Sci.* **15**, 719–721.

Botti, C., and Vasil, I. K. (1984). The ontogeny of somatic embryos of *Pennisetum americanum* (L.). II. In immature inflorescences. *Can. J. Bot.* **62**, 1629–1635.

Boyes, C. J., and Vasil, I. K. (1987). *In vitro* selection for tolerance to S-(2-aminoethyl)-L-cysteine and overproduction of lysine by embryogenic calli and regenerated plants of *Pennisetum americanum* (L.). *Plant Sci.* **50**, 197–205.

Brar, D. S., Rambold, S., Gamborg, O., and Constabel, F. (1979). Tissue culture of corn and *Sorghum*. *Z. Pflanzenphysiol.* **95**, 377–388.

Brar, D. S., Rambold, D., Constabel, F., and Gamborg, O. L. (1980). Isolation, fusion and culture of *Sorghum* and corn protoplasts. *Z. Pflanzenphysiol.* **96**, 269–275.

Brettell, R. I. S., Goddard, B. V. D., and Ingram, D. S. (1979). Selection of Tms cytoplasm maize tissue cultures resistant to *Drechslera maydis* T-toxin. *Maydica* **24**, 203–213.

Brettell, R. I. S., Thomas, E., and Ingram, D. S. (1980). Reversion of Texas male-sterile cytoplasmic maize in culture to give fertile T-toxin resistant plants. *Theor. Appl. Genet.* **58**, 55–58.

Brettell, R. I. S., Dennis, E. S., Scowcroft, W. R., and Peacock, W. Y. (1986a). Molecular analysis of a somaclonal mutant of maize alcohol dehydrogenase. *Mol. Gen. Genet.* **202**, 235–239.

Brettell, R. I. S., Pallota, M. A., Gustafson, J. P., and Appels, R. (1986b). Variation at the Nor loci in triticale derived from tissue culture. *Theor. Appl. Genet.* **71**, 637–643.

Bright, S. W. J. (1985). Selection *in vitro*. *In* "Cereal Tissue and Cell Culture" (S. W. J. Bright and M. G. K. Jones, eds.), pp. 231–261. Nijhoff/W. Junk, Dordrecht.

Bright, S. W. J., and Jones, M. G. K., eds. (1985). "Cereal Tissue and Cell Culture." Nijhoff/W. Junk, Dordrecht.

Bright, S. W. J., and Shewry, P. R. (1983). Improvement of protein quality in cereals. *Crit. Rev. Plant Sci.* **1**, 49–93.

Bright, S. W. J., Featherstone, L. C., and Miflin, B. J. (1979). Lysine metabolism in a barley mutant resistant to *S*-(2-aminoethyl)-L-cysteine. *Planta* **146**, 629–633.

Brossard, D. (1978). Microspectrophotometric and ultrastructural analysis of a case of cell differentiation without endopolyploidy: The pith of *Crepis capillaris*. *Protoplasma* **93**, 369–380.

Cai, Q., Quain, Y., Zhou, Y., and Wu, S. (1978). A further study on the isolation and culture of rice (*Oryza sativa* L.) protoplasts. *Acta Bot. Sin.* **20**, 97–102.

Caligo, M. A., Nuti Ronchi, V., and Nozzolini, M. (1985). Proline and serine affect polarity and development of carrot somatic embryos. *Cell Differ.* **17**, 193–198.

Cassells, A. C. (1985). Genetic, epigenetic and non-genetic variation in tissue culture derived plants. *In "In Vitro* Techniques. Propagation and Long-term Storage" (A. Schafer-Menher, ed.), pp. 111–120. Nijhoff/W. Junk, Dordrecht.

Cassells, A. C., and Morrish, F. M. (1985). Growth measurements of *Begonia rex* Putz. plants regenerated from leaf cuttings and *in vitro* from leaf petioles, axenic leaves, re-cycled axenic leaves and callus. *Sci. Hortic.* **27**, 113–121.

Cassells, A. C., and Tamma, L. (1986). Ethylene and ethane release during tobacco protoplast isolation and protoplast survival potential *in vitro*. *Physiol. Plant.* **66**, 303–308.

Cassells, A. C., Long, R. D., and Mousdale, D. A. (1982). Endogenous IAA and morphogenesis in tobacco petiole cultures. *Physiol. Plant.* **56**, 507–512.

Cassells, A. C., Farrell, G., and Goetz, E. M. (1986). Variation on the tissue culture progeny of the chimeral potato (*Solanum tuberosum* L.) variety Golden Wonder. *In* "Somaclonal Variation and Plant Improvement" (J. Semal, ed.), pp. 201–212. Nijhoff/W. Junk, Dordrecht.

Chaleff, R. S. (1981). "Genetics of Higher Plants." Cambridge Univ. Press, London.

Chandler, S. F., and Vasil, I. K. (1984a). Optimization of plant regeneration from long term embryogenic callus cultures of *Pennisetum purpureum* Schum. (Napier grass). *J. Plant Physiol.* **117**, 147–156.

Chandler, S. F., and Vasil, I. K. (1984b). Selection and characterization of NaCl tolerant cells from embryogenic cultures of *Pennisetum purpureum* Schum. (Napier grass). *Plant Sci. Lett.* **37**, 157–164.

Chourey, P. S., and Sharpe, D. Z. (1985). Callus formation from protoplasts of *Sorghum* cell suspension cultures. *Plant Sci.* **39**, 171–175.

Chourey, P. S., and Zurawski, D. B. (1981). Callus formation from protoplasts of a maize cell culture. *Theor. Appl. Genet.* **59**, 341–344.

Chu, C. C., Sun, C. S., Chen, X., Zhang, W. X., and Du, Z. H. (1984). Somatic embryogenesis and plant regeneration in callus from infloresences of *Hordeum vulgare* × *Triticum aestivum* hybrids. *Theor. Appl. Genet.* **68**, 375–379.

Cocking, E. C. (1960). A method for the isolation of plant protoplasts and vacuoles. *Nature (London)* **187**, 827–929.

Conger, B. V., Hanning, G. E., Gray, D. J., and McDaniel, J. K. (1983). Direct embryogenesis from mesophyll cells of orchard grass. *Science* **221**, 850–851.

Cooper, D. B., Sears, R. G., Lookhart, G. L., and Jones, P. C. (1986). Heritable somaclonal variation in gliadin proteins of wheat plants derived from immature embryo callus culture. *Theor. Appl. Genet.* **71**, 784–790.

Cooper, K. V., Dale, J. E., Dyer, A. F., Lyne, R. C., and Walker, J. T. (1978). Hybrid plants from the barley × rye cross. *Plant Sci. Lett.* **12**, 293–298.

Cornejo-Martin, M. J., Mingo-Castel, A. M., and Primo-Millo, E. (1979). Organ redifferentiation in rice callus: Effects of C_2H_4, CO_2 and cytokinins. *Z. Pflanzenphysiol.* **94**, 117–123.

Coulibaly, M. Y., and Demarly, Y. (1986). Regeneration of plantlets from protoplasts of rice (*Oryza sativa* L.). *Z. Pflanzenzuecht.* **96**, 78–81.

Cullis, C. A. (1981). Environmental induction of heritable changes in flax: Defined environments inducing changes in rDNA and peroxidase isozyme band pattern. *Heredity* **47**, 87–94.

Cullis, C. A. (1983). Environmentally induced DNA changes in plants. *Crit. Rev. Plant Sci.* **1**, 117–131.

Cullis, C. A., and Cleary, W. (1985). DNA variation in flax tissue culture. *Can. J. Genet. Cytol.* **28**, 247–251.

Cummings, D. P., Green, C. E., and Stuthmann, D. D. (1976). Callus induction and plant regeneration in oats. *Crop Sci.* **16**, 465–470.

Cure, W. W., and Mott, R. L. (1978). A comparative anatomical study of organogenesis in cultured tissues of maize, wheat and oats. *Physiol. Plant.* **42**, 91–96.

Dale, P. J. (1980). Embryoids from cultured immature embryos of *Lolium multiflorum*. *Z. Pflanzenphysiol.* **100**, 73–77.

Dale, P. J. (1983). Protoplast culture and plant regeneration of cereals and other recalcitrant species. *Experientia Suppl.* **46**, 31–41.

Dale, P. J., and Deambrogio, E. (1979). A comparison of callus induction and plant regeneration from different explants of *Hordeum vulgare*. *Z. Pflanzenphysiol.* **94**, 67–77.

Dalton, S. J., and Dale, P. J. (1981). Induced tillering of *Lolium multiflorum in vitro*. *Plant Cell Tissue Org. Cult.* **1**, 57–64.

Dalton, S. J., and Dale, P. J. (1985). The application of *in vitro* tiller induction in *Lolium multiflorum*. *Euphytica* **34**, 897–904.

D'Amato, F. (1977). "Nuclear Cytology in Relation to Development." Cambridge Univ. Press, London.

D'Amato, F. (1985). Cytogenetics of plant cell and tissue cultures and their regenerates. *Crit. Rev. Plant Sci.* **3**, 73–112.

D'Amato, F. and Hoffman-Ostenhof (1959). Metabolism and spontaneous mutations in plants. *Adv. Genet.* **8**, 1–28.

Davies, P. A., Pallotta, M. A., Ryan, S. A., Scowcroft, W. R., and Larkin, P. J. (1986). Somaclonal variation in wheat: Gentic and cytogenetic characterisation of alcohol dehydrogenase 1 mutants. *Theor. Appl. Genet.* **72**, 644–653.

Deka, P. C., and Sen, S. K. (1976). Differentiation in calli originating from isolated protoplasts of rice (*Oryza sativa* L.) through plating technique. *Mol. Gen. Genet.* **145**, 239–243.

De Paepe, R., Prat, D., and Hugnet, T. (1983). Heritable nuclear DNA changes in doubled haploid plants obtained by pollen culture of *Nicotiana sylvestris*. *Plant Sci. Lett.* **28,** 11–28.

Dodds, J. H. (1981). Relationship of the cell cycle to xylem cell differentiation: A new model. *Plant Cell Environ.* **4,** 145–146.

Dudits, D., Kao, K. N., Constabel, F., and Gamborg, O. L. (1976). Fusion of carrot and barley protoplasts and division of heterokaryocytes. *Can. J. Genet. Cytol.* **18,** 263–269.

Dunstan, D. I., Short, K. C., and Thomas, E. (1978). The anatomy of secondary morphogenesis in cultured scutellum tissues of *Sorghum bicolor*. *Protoplasma* **97,** 251–260.

Dunstan, D. I., Short, K. C., Dhaliwal, H., and Thomas, E. (1979). Further studies on plantlet production from cultured tissues of *Sorghum bicolor*. *Protoplasma* **101,** 355–351.

Dure, L., and Galau, G. A. (1981). Developmental biochemistry of cottonseed embryogenesis and germination. XIII. Regulation of biosynthesis of principal storage proteins. *Plant Physiol.* **68,** 187–194.

Durrant, A. (1981). Unstable genotypes. *Philos. Trans. R. Soc. London* **292,** 467–474.

Duvick, D. N. (1984). Progress in conventional plant breeding. *In* "Gene Manipulation in Plant Improvement" (J. P. Gustafson, ed.), pp. 17–33. Plenum, New York.

Dyer, A. F. (1976). Modification and errors of mitotic cell divisions in relation to differentiation. *In* "Cell Division in Higher Plants" (M. M. Yeoman, ed.), pp. 199–253. Academic Press, London.

Earle, E. D., and Demarly, Y., eds. (1982). "Variability in Plants Regenerated from Tissue Culture." Praeger, New York.

Edallo, S., Zucchinali, C., Perenzi, M., and Salamini, F. (1981). Chromosomal variation and frequency of spontaneous mutations associated with *in vitro* culture and plant regeneration in maize. *Maydica* **26,** 39–56.

Eichholtz, D. A., Rogers, S. G., Horsch, R. B., Klee, H., Hayford, H. J., Hoffmann, N. L., Goldberg, S. B., Fink, C., O'Connell, K., and Fraley, R. T. (1986). Expression of a mouse dihydrofolate reductase gene confers methotrexate resistance in transgenic petunia plants. *Somat. Cell Genet.* **13,** 67–76.

Evans, D. A., Crocomo, O. J., and de Carvalho, M. T. V. (1980). Protoplast isolation and subsequent callus regeneration in sugarcane. *Z. Pflanzenphysiol.* **98,** 355–358.

Evenari, M. (1984). Seed physiology: From ovule to maturing seed. *Bot. Rev.* **50,** 143–170.

Filner, P. (1969). Control of nutrient assimilation, a growth-regulating mechanism in cultured plant cells. *Dev. Biol. Suppl.* **3,** 206–226.

Flores, H. E., Kaur-Sawhney, R., and Galston, A. W. (1981). Protoplasts as vehicles for plant propagation and improvement. *Adv. Cell Cult.* **1,** 241–279.

Fraley, R. T., Rogers, S. G., and Horsch, R. B. (1986). Genetic transformation in higher plants. *Crit. Rev. Plant Sci.* **4,** 1–46.

Fromm, M., Taylor, L. P., and Walbot, V. (1985). Expression of genes transferred into monocot and dicot plant cells by electroporation. *Proc. Natl. Acad. Sci. U.S.A.* **82,** 5824–5828.

Fromm, M., Taylor, L. P., and Walbot, V. (1986). Stable transformation of maize after gene transfer by electroporation. *Nature (London)* **319,** 791–793.

Fujimura, T., Komamine, A., and Matsumoto, H. (1981). Changes in chromosomal proteins during early stages of synchronized carrot embryogenesis. *Z. Pflanzenphysiol.* **102,** 293–298.

Fujimura, T., Sakurai, M., and Akagi, H. (1985). Regeneration of rice plants from protoplasts. *Plant Tissue Cult. Lett.* **2,** 74–75.

Fukuda, H., and Komamine, A. (1985). Cytodifferentiation. *In* "Cell Culture and Somatic Cell Genetics of Plants" (I. K. Vasil, ed.), Vol. 2, pp. 151–212. Academic Press, Orlando, Florida.

Fukui, K. (1983). Sequential occurrence of mutations in a growing rice callus. *Theor. Appl. Genet.* **65,** 225–230.

Gamborg, O. L., Constabel, F., and Miller, R. A. (1970). Embryogenesis and the production of albino plants from cell cultures of *Bromus inermis*. *Planta* **95,** 355–358.

Garfinkel, D. J., Simpson, R. B., Ream, L. W., White, F. F., Gordon, M. P., and Nester, E. W. (1981). Genetic analysis of crown gall: Fine structure map of the T-DNA by site-directed mutagenesis. *Cell* **27,** 143–153.

Gengenbach, B. G., and Green, C. E. (1975). Selection of T-cytoplasm maize callus cultures resistant to *Helminthosporium maydis* race T pathotoxin. *Crop Sci.* **15,** 645–649.

Gengenbach, B. G., Green, C. E., and Donovan, C. M. (1977). Inheritance of selected pathotoxin resistance in maize plants regenerated from cell cultures. *Proc. Natl. Acad. Sci. U.S.A.* **74,** 5113–5117.

Gengenbach, B. G., Connelly, J. A., Pring, D. R., and Conde, M. F. (1981). Mitochondrial DNA variation in maize plants regenerated during tissue culture selection. *Theor. Appl. Genet.* **59,** 161–167.

Gleba, Y. Y., and Sytnik, K. M. (1984). "Protoplast Fusion. Genetic Engineering in Higher Plants." Springer-Verlag, Heidelberg.

Goldbach, H., and Micheal, G. (1976). Abscisic acid content of barley grains during ripening as affected by temperature and variety. *Crop Sci.* **16,** 797–799.

Goldberg, R. B., Horschek, G., Tam, S. H., Ditta, G. S., and Breidenbach, R. W. (1981). Abundance, diversity, and regulation of mRNA sequence sets in soybean embryogenesis. *Dev. Biol.* **83,** 201–217.

Gould, A. R. (1978). Diverse pathways of morphogenesis in tissue cultures of the composite *Brachycome lineriloba* ($2n = 4$). *Protoplasma* **97,** 125–135.

Gould, A. R. (1982). Chromosome instability in plant tissue cultures studied with the banding technique. *In* "Plant Tissue Culture" (A. Fujiwara, ed.), p. 431. Maruzen, Tokyo.

Gould, A. R. (1983). Control of the cell cycle in cultured plant cells. *Crit. Rev. Plant Sci.* **1,** 315–344.

Gray, D. J., and Conger, B. V. (1985). Influence of dicamba and casein hydrolysate on somatic embryo number and culture quality in suspension culture of *Dactylis glomerata*. *Plant Cell Tissue Org. Cult.* **4,** 123–133.

Green, C. E. (1982). Somatic embryogenesis and plant regeneration from the friable callus of *Zea mays*. *In* "Plant Tissue Culture" (F. Fujiwara, ed.), pp. 107–108. Maruzen, Tokyo.

Green, C. E., Armstrong, C. L., and Anderson, P. C. (1983). Somatic cell genetic systems in corn. *In* "Advances in Gene Technology: Molecular Genetics of Plants and Animals" (K. Downey, J. Voellmy, and F. Ahmad, eds.), pp. 147–157. Academic Press, New York.

Groose, R. W., and Bingham, E. T. (1986a). An unstable anthocyanin mutation recovered from tissue culture of alfalfa (*Medicago sativa*). 1. High frequency of reversion upon reculture. *Plant Cell Rep.* **5,** 104–107.

Groose, R. W., and Bingham, E. T. (1986b). An unstable anthocyanin mutation recovered from tissue culture of alfalfa (*Medicago sativa*). 2. Stable non-revertant derived from reculture. *Plant Cell Rep.* **5,** 108–110.

Haccius, B. (1978). Question of unicellular origin of nonzygotic embryos of callus cultures. *Phytomorphology* **24**, 74–81.

Hanna, W. W., Lu, C., and Vasil, I. K. (1984). Uniformity of plants regenerated from somatic embryos of *Panicum maximum* Jacq. (Guinea grass). *Theor. Appl. Genet.* **6**, 155–159.

Hanning, G. E., and Conger, B. V. (1982). Embryoid and plantlet formation from leaf segments of *Dactylis glomerata* L. *Theor. Appl. Genet.* **63**, 155–159.

Hanzel, J. J., Miller, J. P., Brinkman, M. A., and Fendos, E. (1985). Genotype and media effects on callus formation and regeneration in barley. *Crop Sci.* **25**, 27–31.

Hauptmann, R., Ozias-Akins, P., Vasil, V., Tabeizadeh, Z., Rogers, R. B., Horsch, R. B., Vasil, I. K., and Fraley, R. T. (1987a). Transient expression of electroporated DNA in monocotyledonous and dicotyledenous species. *Plant Cell Rep.* (in press).

Hauptmann, R. M., Vasil, V., Ozias-Akins, P., Tabaeizadeh, Z., Rogers, S. G., Fraley, R. T., Horsch, R. B., and Vasil, I. K. (1987b). Evaluation of selectable markers for obtaining stable transformants in the Gramineae. Submitted.

Haydu, Z., and Vasil, I. K. (1981). Somatic embryogenesis and plant regeneration from leaf tissues and anthers of *Pennisetum americanum*. *Theor. Appl. Genet.* **59**, 269–273.

Haydu, Z., Lazar, G., and Dudits, D. (1977). Increased frequency of polyethylene glycol induced protoplast fusion by dimethyl-sulfoxide. *Plant Sci. Lett.* **10**, 357–360.

Heinz, D. J., and Mee, G. W. P. (1971). Morphological, cytogenetic and enzymatic variation in *Saccharum* species hybrids and their cell suspension cultures. *Am. J. Bot.* **58**, 257–262.

Heinz, D. J., Krishnamurthi, M., Nickell, L. G., and Maretzki, A. (1977). Cell, tissue and organ culture in sugarcane improvement. *In* "Applied and Fundamental Aspects of Plant, Cell, Tissue and Organ Culture" (J. Reinert and Y. P. S. Bajaj, eds.), pp. 3–17. Springer-Verlag, Berlin.

Hernaalsteens, J. P., Thia-toong, L., Schell, J., and Van Montagu, M. (1984). An *Agrobacterium*-transformed cell culture from the monocot *Asparagus officinalis*. *EMBO J.* **3**, 3039–3041.

Hesemann, C. U., and Schroder, G. (1982). Loss of nuclear DNA from leaves of rye. *Theor. Appl. Genet.* **62**, 325–328.

Heyser, J. W. (1984). Callus and shoot regeneration from protoplasts of Proso millet (*Panicum maximum* L.). *Z. Pflanzenphysiol.* **113**, 292–299.

Heyser, J. W., Dykes, T. A., DeMott, K. J., and Nabors, M. W. (1983). High frequency long term regeneration of rice from callus culture. *Plant Sci. Lett.* **29**, 175–182.

Hibberd, K. A., and Green, C. (1982). Inheritance and expression of lysine plus threonine resistance. *Proc. Natl. Acad. Sci. U.S.A.* **79**, 559–563.

Hibberd, K. A., Walter, C. E., Green, C. E., and Gengenbach, B. G. (1980). Selection and characterization of feedback-insensitive tissue culture of maize. *Planta* **148**, 183–187.

Hibberd, K. A., Anderson, P. C., and Barker, M. (1986). Tryptophan overproducer mutants of cereal crops. United States Patent No. 4,581,847.

Hilding, A., and Welander, T. (1976). Effect of some factors on propagation of *Begonia* × *himealis in vitro*. *Swed. J. Agric. Res.* **6**, 191–199.

Ho, W. J., and Vasil, I. K. (1983a). Somatic embryogenesis in sugarcane (*Saccharum officinarum* L.). II. The growth of and plant regeneration from embryogenic cell suspension cultures. *Ann. Bot.* **51**, 719–726.

Ho, W. J., and Vasil, I. K. (1983b). Somatic embryogenesis in sugarcane (*Saccharum officinarum* L.). I. The morphology and physiology of callus formation and the ontogeny of somatic embryos. *Protoplasma* **118**, 169–180.

Holtzer, H., and Rubenstein, N. (1977). Binary decisions, quantal cell cycles, and cell diversification. *In* "Cell Differentiation in Microorganisms, Plants and Animals" (L. Nover and K. Mothes, eds.), pp. 424–437. Fisher, Jena/Elsevier, Amsterdam.

Hooykaas-Van Slogteren, G., Hooykaas, P. J. J., and Schilperoort, R. A. (1984). Expression of Ti plasmid genes in monocotyledonous plants infected with *Agrobacterium tumefaciens*. *Nature (London)* **311,** 763–764.

Inoue, M., and Maeda, E. (1981). Stimulation of shoot bud and plantlet formation in rice callus cultures by two-step culture method using abscisic acid and kinetin. *Jpn. J. Crop Sci.* **50,** 318–327.

Irvine, J. E. (1984). The frequency of marker changes in sugarcane plants regenerated from callus culture. *Plant Cell Tissue Org. Cult.* **3,** 201–201.

Jacobs, W. P. (1979). Inhibition of flowering in short-day plants. *In* "Plant Growth Substances" (F. Skoog, ed.), pp. 301–309. Springer-Verlag, New York.

Joarder, O. I., Joarder, N. H., and Dale, P. J. (1986). *In vitro* response of leaf tissues from *Lolium multiflorum*—A comparison with leaf segment position, leaf age and *in vivo* mitotic activity. *Theor. Appl. Genet.* **73,** 286–291.

Jones, M. G. K. (1985). Cereal protoplasts. *In* "Cereal Tissue and Cell Culture" (S. W. J. Bright and M. G. K. Jones, eds.), pp. 204–230. Nijhoff/W. Junk, Dordecht.

Jones, M. G. K., and Dale, P. J. (1982). Reproducible regeneration of callus from suspension culture protoplasts of the grass *Lolium multiflorum*. *Z. Pflanzenphysiol.* **105,** 267–274.

Kamo, K. K., Becwar, M. R., and Hodges, T. K. (1985). Regeneration of *Zea mays* L. from embryogenic callus. *Bot. Gaz.* **146,** 327–334.

Kao, K. N., and Michayluk, M. R. (1974). A method for high frequency intergeneric fusion of plant protoplasts. *Planta* **115,** 355–367.

Kao, K. N., Gamborg, O. L., Michayluk, M. R., Keller, W. A., and Miller, R. A. (1973). The effect of sugars and inorganic salts on cell regeneration and sustained division in plant protoplasts. *Colloq. Int. C. N. R. S.* **212,** 207–213.

Kao, K. N., Constabel, F., Michayluk, M. R., and Gamborg, O. L. (1974). Plant protoplast fusion and growth of intergeneric hybrids. *Planta* **120,** 215–227.

Karlsson, S. B., and Vasil, I. K. (1986a). Morphology and ultrastructure of embryogenic cell suspension cultures of *Panicum maximum* Jacq. (Guinea grass) and *Pennisetum purpureum* Schum. (Napier grass). *Am. J. Bot.* **73,** 894–901.

Karlsson, S. B., and Vasil, I. K. (1986b). Growth, cytology and flow cytometry of embryogenic cell suspension cultures of *Panicum maximum* Jacq. (Guinea grass) and *Pennisetum purpureum* Schum. (Napier grass). *J. Plant Physiol.* **123,** 211–227.

Karp, A., and Maddock, S. E. (1984). Chromosome variation in wheat plants regenerated from cultured immature embryos. *Theor. Appl. Genet.* **67,** 249–256.

Kasperbauer, M. J., and Karlen, D. L. (1986). Light-mediated bioregulation of tillering and photosynthate partitioning in wheat. *Physiol. Plant.* **66,** 159–163.

Kasperbauer, M. J., Buckner, R. C., and Bush, L. P. (1979). Tissue culture of annual ryegrass × tall fescue F1 hybrids: Callus establishment and plant regeneration. *Crop Sci.* **19,** 457–460.

Kemble, R. J., Flavell, R. B., and Brettell, R. I. S. (1982). Mitochondrial DNA analysis of fertile and sterile maize plants derived from tissue culture with the texas maize sterile cytoplasm. *Theor. Appl. Genet.* **62,** 213–217.

King, P. J., Potrykus, I., and Thomas, E. (1978). *In vitro* genetics of cereals: Problems and perspectives. *Physiol. Veg.* **16,** 381–399.

King, R. W. (1976). Abscisic acid in developing wheat grains and its relationship to grain growth and maturation. *Planta* **132,** 43–51.

King, R. W., Salminen, S. O., Hill, R. D., and Higgins, T. J. V. (1979). Abscisic acid and gibberellin action in developing kernels of triticale (cv. 6A 190). *Planta* **146,** 249–255.

Koblitz, H. (1976). Isolation and cultivation of protoplasts from callus cultures of barley. *Biochem. Physiol. Pflanz.* **170,** 287–293.

Krens, F. A., Molendijk, L., Wullems, G. L., and Schilperoort, R. A. (1982). *In vitro* transformation of plant protoplasts with Ti plasmid DNA. *Nature (London)* **296,** 72–74.

Kueh, J. S. H., and Bright, S. W. J. (1982). Biochemical and genetical analysis of three proline accumulating barley mutants. *Plant Sci. Lett.* **27,** 233–236.

Landsmann, J., and Uhrig, H. (1985). Somaclonal variation in *Solanum tuberosum* detected at the molecular level. *Theor. Appl. Genet.* **71,** 500–505.

Lapitan, N. L. V., Sears, R. G., and Gill, B. S. (1984). Translocation and other karyotypic structural changes in wheat × rye hybrids regenerated from tissue culture. *Theor. Appl. Genet.* **68,** 547–554.

Larkin, P. J. (1985). *In vitro* culture and cereal breeding. *In* "Cereal Tissue and Cell Culture" (S. W. J. Bright and M. G. K. Jones, eds.), pp. 273–296. Nijhoff/W. Junk, Dordrecht.

Larkin, P. J., and Scowcroft, W. R. (1981). Somaclonal variation—A novel source of variability for plant improvement. *Theor. Appl. Genet.* **60,** 197–214.

Larkin, P. J., and Scowcroft, W. R. (1983a). Somaclonal variation and eyespot toxin tolerance in sugarcane. *Plant Cell Tissue Org. Cult.* **2,** 111–121.

Larkin, P. J., and Scowcroft, W. R. (1983b). Somaclonal variation and crop improvement. *In* "Genetic Engineering of Plants" (T. Kosuge, C. P. Meredith, and A. Hollaender, eds.), pp. 289–314. Plenum, New York.

Larkin, P. J., Ryan, S. A., Brettell, R. I. S., and Scowcroft, W. R. (1984). Heritable somaclonal variation in wheat. *Theor. Appl. Genet.* **67,** 443–455.

Lawrence, R. H. (1981). *In vitro* plant cloning systems. *Environ. Exp. Bot.* **21,** 209–300.

Leemans, J., Deblaere, R., Willmitzer, L., DeGreve, H., Hernalsteens, J. P., VanMontagu, M., and Schell, J. (1982). Genetic identification of functions of Ti-DNA transcripts in octopine crown galls. *EMBO J.* **1,** 147–152.

Le Rudulier, D., Strom, A. R., Dandekar, A. M., and Smith, C. T. (1984). Molecular biology of osmoregulation. *Science* **224,** 1064–1074.

Letham, D. S., Goodwin, P. B., and Higgins, T. J. V. (1978). "Phytohormones and Related Compounds: A Comprehensive Treatise," Vols. I and II. Elsevier, Amsterdam.

Linacero, R., and Vazquez, A. M. (1986). Somatic embryogenesis and plant regeneration from leaf tissues of rye (*Secale cerale* L.). *Plant Sci.* **44,** 219–222.

Liu, M. C., and Chen, W. C. (1979). Tissue and cell culture as aids to sugarcane breeding. 1. Creation of genetic variation through callus culture. *Euphytica* **25,** 393–403.

Liu, S. U., and Hooker, A. L. (1972). Disease determinant of *Helminthosporium maydis* race T. *Phytopathology* **62,** 968–971.

Lorz, H. (1984). Variability in tissue culture derived plants. *In* "Genetic Manipulation. Impact on Man and Society" (W. Arber, K. Illmensee, W. J. Peacock, and P. Starlinger, eds.), pp. 103–114. Garden City Press, Hertfordshire.

Lorz, H., Baker, B., and Schell, J. (1985). Gene transfer to cereal cells mediated by protoplast transformation. *Mol. Gen. Genet.* **199,** 178–182.

Lu, C., and Vasil, I. K. (1981a). Somatic embryogenesis and plant regeneration from freely suspended cells and cell groups of *Panicum maximum in vitro*. *Ann. Bot.* **47,** 543–548.

Lu, C., and Vasil, I. K. (1981b). Somatic embryogenesis and plant regeneration from leaf tissue of *Panicum maximum* Jacq. *Theor. Appl. Genet.* **59**, 275–280.

Lu, C., and Vasil, I. K. (1985). Histology of somatic embryogenesis in *Panicum maximum* (Guinea grass). *Am. J. Bot.* **72**, 1908–1913.

Lu, C., Vasil, V., and Vasil, I. K. (1981). Isolation and culture of protoplasts of *Panicum maximum* Jacq. (Guinea grass): Somatic embryogenesis and plantlet formation. *Z. Pflanzenphysiol.* **104**, 311–318.

Lu, C., Vasil, I. K., and Ozias-Akins, P. (1982). Somatic embryogenesis in *Zea mays* L. *Theor. Appl. Genet.* **62**, 109–112.

Lu, C., Vasil, V., and Vasil, I. K. (1983). Improved efficiency of somatic embryogenesis and plant regeneration in tissue cultures of maize (*Zea mays* L.). *Theor. Appl. Genet.* **66**, 285–289.

Ludwig, S. R., Somers, D. A., Peterson, W. L., Pohlman, B. F., Zarovitz, B. G., Gengenbach, B. G., and Messing, J. (1985). High frequency callus formation from maize protoplasts. *Theor. Appl. Genet.* **71**, 344–350.

Lupi, M. C., Bennici, A., Baroncelli, S., Gennari, D., and D'Amato, F. (1981). *In vitro* regeneration of durum wheat plants. II. Diplontic selection of aneusomatic plants. *Z. Pflanzenzuecht.* **87**, 167–171.

McClintock, B. (1951). Chromosome organization and genic expression. *Cold Spring Harbor Quant. Biol.* **16**, 13–47.

McClintock, B. (1978). Mechanisms that rapidly reorganize the genome. *Stadler Genet. Symp.* **10**, 25–47.

McCoy, T. J., and Phillips, R. L. (1982). Chromosome stability in Maize (*Zea mays*) tissue cultures and sectoring in some regenerated plants. *Can. J. Genet. Cytol.* **24**, 559–565.

McCoy, T. J., Phillips, R. L., and Rines, H. W. (1982). Cytogenetic analysis of plants regenerated from oat (*Avena sativa*) tissue cultures; high frequency of partial chromosome loss. *Can. J. Genet. Cytol.* **24**, 37–50.

McWilliam, A. A., Smith, S. M., and Street, H. E. (1974). The origin and development of embryoids in suspension cultures of carrot (*Daucus carota*). *Ann. Bot.* **38**, 243–250.

Maddock, S. E. (1985). Cell culture, somatic embryogenesis and plant regeneration in wheat, barley, oats, rye and triticale. *In* "Cereal Tissue and Cell Culture" (S. W. J. Bright and M. G. K. Jones, eds.), pp. 131–175. Nijhoff/W. Junk, Dordrecht.

Maddock, S. E., Lancaster, V. A., Risiott, R., and Franklin, J. (1982). Plant regeneration from cultured immature embryos and infloresences of 25 cultivars of wheat (*Triticum aestivum*). *J. Exp. Bot.* **34**, 915–926.

Magnusson, I., and Bornman, C. H. (1985). Anatomical observations on somatic embryogenesis from scutellar tissues of immature embryos of *Triticum aestivum*. *Physiol. Plant.* **63**, 137–145.

Maliga, P. (1980). Isolation, characterization and utilization of mutant cell lines in higher plants. *Int. Rev. Cytol.* **11A**, 225–250.

Maliga, P. (1984). Isolation and characterization of mutants in plant cell culture. *Annu. Rev. Plant Physiol.* **35**, 519–546.

Maretzki, A., and Nickell, L. G. (1973). Formation of protoplasts from sugarcane cell suspensions and the regeneration of cell cultures from protoplasts. *Colloq. Int. CNRS* **212**, 51–63.

Mathias, R. J., and Fukui, K. (1986). The effect of specific chromosome and cytoplasmic substitutions on the tissue culture response of wheat (*Triticum aestivum*) callus. *Theor. Appl. Genet.* **71**, 797–800.

Mathias, R. J., Fukui, K., and Law, C. N. (1986). Cytoplasmic effects on the tissue culture response of wheat (*Triticum aestivum*). *Theor. Appl. Genet.* **72,** 70–75.
Meins, F. (1983). Heritable variation in plant cell cultures. *Annu. Rev. Plant Physiol.* **34,** 327–348.
Meins, F., and Binns, A. (1977). Epigenetic variation in cultured somatic cells: Evidence for gradual changes in the requirement for factors promoting cell division. *Proc. Natl. Acad. Sci. U.S.A.* **74,** 2928–2932.
Meins, F., and Binns, A. (1978). Epigenetic clonal variation in the requirement of plant cells for cytokinins. *In* "The Clonal Basis of Development" (S. Subtelny and I. M. Sussex, eds.), pp. 185–201. Academic Press, New York.
Menczel, L., and Wolfe, K. (1984). High frequency of fusion induced in freely suspended protoplast mixtures by polyethylene glycol and dimethylsulfoxide at high pH. *Plant Cell Rep.* **3,** 196–198.
Meredith, C. P. (1984). Selecting better crops from cultured cells. *In* "Gene Manipulation in Plant Improvement" (J. P. Gustafson, ed.), pp. 503–529. Plenum, New York.
Miflin, B. J., Bright, S. W. J., Rognes, S. E., and Kueh, J. H. S. (1983). Amino acids, nutrition and stress: The role of biochemical mutants in solving problems of crop quality. *In* "Genetic Engineering of Plants—An Agricultural Perspective" (T. Kosuge, A. Hollaender, and C. P. Meredith, eds.), pp. 391–414. Plenum, New York.
Mix, G., Wilson, H. W., and Foroughi-Wehr, B. (1978). The cytological status of plants of *Hordeum vulgare* L. regenerated from microspore callus. *Z. Pflanzenzuecht.* **80,** 89–99.
Murashige, T., and Nakano, R. (1967). Chromosome complement as a determinant of the morphogenic potential of tobacco cells. *Am. J. Bot.* **54,** 963–970.
Murata, M., and Orton, T. J. (1983). Chromosomal structural changes in cultured callus cells. *In vitro* **19,** 83–89.
Nabors, M. W., and Dykes, T. A. (1985). Tissue culture of cereal cultivars with increased salt, drought and acid tolerance. *In* "Biotechnology in International Agricultural Research" (F. J. Zapata, ed.), pp. 121–138. IRRI, Manila.
Nabors, M. W., Kroskey, C. S., and McHugh, D. M. (1982). Green spots are predictors of high callus growth rates and shoot formation in normal and in salt-stressed tissue cultures of oats (*Avena sativa* L.). *Z. Pflanzenphysiol.* **157,** 341–349.
Nabors, M. W., Heyser, J. W., Dykes, T. A., and de Mott, K. J. (1983). Long-duration, high-frequency plant regeneration from cereal tissue cultures. *Planta* **157,** 385–391.
Nagl, W. (1978). "Endopolyploidy and Polyteny in Differentiation and Evolution: Toward an Understanding of Quantitative and Qualitative Variation in Nuclear DNA in Ontogeny and Phylogeny." North-Holland Publ., Amsterdam.
Nagl, W. (1979). Differential replication in plants: A critical review. *Z. Pflanzenphysiol.* **95,** 283–314.
Nakamura, C., Keller, W. A., and Fedak, G. (1981). *In vitro* propagation and chromosome doubling of a *Triticum* × *Hordeum vulgare* intergeneric hybrid. *Theor. Appl. Genet.* **60,** 89–96.
Nakano, H. M., and Maeda, E. (1979). Shoot differentiation in callus of *Oryza sativa* L. *Z. Pflanzenphysiol.* **93,** 449–458.
Negrutiu, I., and Jacobs, M. (1978). Factors which enhance *in vitro* morphogenesis of *Arabidopsis thaliana*. *Z. Pflanzenphysiol.* **90,** 423–430.
Negrutiu, I., Jacobs, M., and Caboche, M. (1984). Advances in somatic cell genetics of higher plants—The protoplast approach in basic studies on mutagenesis and isolation of biochemical mutants. *Theor. Appl. Genet.* **67,** 289–304.
Nemet, G., and Dudits, D. (1977). Potential of protoplast, cell and tissue culture in

cereal research. *In* "Use of Tissue Cultures in Plant Breeding" (F. J. Novak, ed.), pp. 145–163. Czechoslovak Academy of Sciences, Prague.

Nesticky, M., Novak, F. J., Piovarci, A. and Dolezelova, M. (1983). Genetic analysis of callus growth of maize (*Zea mays* L.) *in vitro*. *Z. Pflanzenzuecht.* **91,** 322–328.

Nevers, P., Sheperd, N. S., and Saedler, H. (1986). Plant transposable elements. *Adv. Bot. Res.* **12,** 104–194.

Nickell, L. G. (1977). Crop improvement in sugarcane: Studies using *in vitro* methods. *Crop. Sci.* **17,** 717.

Niizeki, M., Tanaka, M., Akada, S., Hirai, A., and Saito, K. (1985). Callus formation of somatic hybrid of rice and soybean and characteristics of the hybrid callus. *Jpn. J. Genet.* **60,** 81–92.

Nuti Ronchi, V., and Giuliano, G. (1984). Temporal dissection of carrot somatic embryogenesis. *In* "Plant Tissue and Cell Culture Application to Crop Improvement" (F. J. Novak, L. Havel, and J. Dolezel, eds.), pp. 77–87. Czechoslovak Academy of Sciences, Prague.

O'Hara, J. F., and Street, H. E. (1978). Wheat callus culture: The initiation, growth and organogenesis of callus derived from various explant sources. *Ann. Bot.* **42,** 1029–1038.

Oono, K. (1983). Genetic variability in rice plants regenerated from cell culture. *In* "Cell and Tissue Culture Techniques for Cereal Crop Improvement," pp. 95–104. Gordon & Breach, New York.

Oono, K. (1985). Putative homozygous mutations in regenerated plants of rice. *Mol. Gen. Genet.* **198,** 377–384.

Orton, T. J. (1979). A quantitative analysis of growth and regeneration from tissue cultures of *Hordeum vulgare, H. jubatum* and their interspecific hybrid. *Environ. Exp. Bot.* **19,** 319–335.

Orton, T. J. (1980a). Haploid barley regenerated from callus cultures of *Hordeum* × *H. jubatum*. *J. Hered.* **71,** 280–282.

Orton, T. J. (1980b). Chromosomal variability in tissue cultures and regenerated plants of *Hordeum*. *Theor Appl. Genet.* **56,** 101–112.

Orton, T. J. (1984). Somaclonal variation: Theoretical and practical considerations. *In* "Gene Manipulation in Plant Improvement" (J. P. Gustafson, ed.), pp. 427–469. Plenum, New York.

Orton, T. J., and Steidl, R. P. (1980). Cytogenetic analysis of plants regenerated from colchicine-treated callus cultures of an interspecific *Hordeum* hybrid. *Theor. Appl. Genet.* **57,** 89–95.

Osborne, D. J., Del-Aquila, A., and Elder, R. H. (1984). DNA repair in plant cells. An essential event of early embryo germination in seeds. *Folia Biol.* **31,** 153–169.

Ozias-Akins, P., and Vasil, I. K. (1983). Callus induction and growth from the mature embryo of *Triticum aestivum* (wheat). *Protoplasma* **115,** 104–113.

Ozias-Akins, P., Ferl, R. J., and Vasil, I. K. (1986). Somatic hybridization in the Gramineae: *Pennisetum americanum* (L.) K. Schum. (Pearl millet) + *Panicum maximum* Jacq. (Guinea grass). *Mol. Gen. Genet.* **203,** 365–370.

Ozias-Akins, P., Pring, D. R., and Vasil, I. K. (1987a). Rearrangements in the mitochondrial genome of somatic cell lines of *Pennisetum americanum* (L.) K. Schum. + *Panicum maximum* Jacq. *Theor. Appl. Genet.* (in press).

Ozias-Akins, P., Tabaeizadeh, Z., Pring, D. R., and Vasil, I. K. (1987b). Mitochondrial DNA amplification in intergeneric somatic hybrids of the Gramineae. Submitted.

Pervez, I., Kueh, J. S. H., Wyn Jones, R. G., and Bright, S. W. J. (1986). Cited in Wyn-Jones and Gorham (1986).

Pitto, L., Lo Schiavo, F., Giuliano, G., and Terzi, M. (1983). Analysis of the heat-shock pattern during somatic embryogenesis of carrot. *Plant Mol. Biol.* **2,** 231–237.

Potrykus, I. (1980). The old problem of protoplast culture: Cereals. *In* "Advances in Protoplast Research" (L. Ferenczy, G. L. Farkas, and G. Lazar, eds.), pp. 243–254. Akademiai Kiado, Budapest.

Potrykus, I., Harms, C. T., and Lorz, H. (1976). Problems in culturing cereal protoplasts. *In* "Cell Genetics of Higher Plants" (D. Dudits, G. L. Farkas, and P. Maliga, eds.), pp. 129–140. Akademiai Kiado, Budapest.

Potrykus, I., Harms, C. T., Lorz, H., and Thomas, E. (1977). Callus formation from stem protoplasts of corn (*Zea mays* L.). *Mol. Gen. Genet.* **156,** 347–350.

Potrykus, I., Harms, C. T., and Lorz, H. (1979). Callus formation from cell culture protoplasts of corn (*Zea mays* L.). *Theor Appl. Genet.* **54,** 209–214.

Potrykus, I., Saul, M. W., Petruska, J., Paszkowski, J., and Shillito, R. D. (1985). Direct gene transfer to cells of gramineous monocots. *Mol. Gen. Genet.* **199,** 183–188.

Preger, R., and Gepstein, S. (1985). Regulation of ethylene biosynthesis during senescence of oat leaf segments. *Physiol. Plant.* **65,** 163–166.

Pring, D. R., Conde, M. F., and Gengenbach, B. G. (1981). Cytoplasmic genome variability in tissue culture derived plants. *Environ. Exp. Bot.* **21,** 369–377.

Raghavan, V. (1983). Biochemistry of somatic embryogenesis. *In* "Handbook of Plant Cell Culture" (D. A. Evans, W. R. Sharp, P. V. Ammirato, and Y. Yamada, eds.), Vol. 1, pp. 654–671. Macmillan, New York.

Raghavan, V. (1986). "Embryogenesis in Angiosperms." Cambridge Univ. Press, London.

Rajasekaran, K., Schank, S. C., and Vasil, I. K. (1986). Characterization of biomass production and phenotypes of plants regenerated from embryogenic callus cultures of *Pennisetum americanum* × *P. purpureum* (hybrid Napiergrass). *Theor. Appl. Genet.* **73,** 4–10.

Rajasekaran, K., Hein, M. B., Davis, G. C., Carnes, M. G., and Vasil, I. K. (1987a). Endogenous growth regulators in leaves and tissue cultures of *Pennisetum purpureum* Schum. *J. Plant Physiol.* (in press).

Rajasekaran, K., Hein, M. B., and Vasil, I. K. (1987b). Endogenous abscisic acid and indole-3-acetic acid and somatic embryogenesis in cultured leaf explants of *Pennisetum purpureum* Schum.: Effects *in vivo* and *in vitro* of Glyphosate, Fluridone and Paclobutrazol. *Plant Physiol.* **84,** 47–52.

Ram, N. V. R., and Nabors, M. W. (1984). Cytokinin mediated long-term high frequency plant regeneration in rice tissue cultures. *Z. Pflanzenphysiol.* **113,** 315–323.

Raman, K., Watten, D. B., and Greyson, R. T. (1980). Propagation of *Zea mays* by shoot tip culture. A feasibility study. *Ann. Bot.* **45,** 183–189.

Rangan, T. S. (1973). Morphogenic investigations on tissue cultures of *Panicum miliaceum*. *Z. Pflanzenphysiol.* **72,** 456–459.

Rangan, T. S., and Vasil, I. K. (1983). Sodium chloride tolerant embryogenic cell lines of *Pennisetum americanum* (L.) K. Schum. *Ann. Bot.* **52,** 59–64.

Rao, M. V., and Nitzsche, W. (1984). Genotypic differences in callus growth and organogenesis of eight pearl millet lines. *Euphytica* **33,** 923–928.

Razin, A., and Riggs, A. (1983). DNA methylation and gene function. *Science* **210,** 604–610.

Reddy, P. J., and Vaidyanath, K. (1986). *In vitro* characterization of salt stress effects and the selection of salt tolerant plants in rice (*Oryza sativa* L.). *Theor. Appl. Genet.* **71,** 757–760.

Reed, J. N., and Conger, C. (1985). Meiotic analysis of tall fescue plants regenerated from callus cultures. *Environ. Exp. Bot.* **25,** 277–284.

Reinert, J., and Backs, D. (1968). Control of totipotency in plant cells growing *in vitro*. *Nature (London)* **220**, 1340–1341.

Rhodes, C. A., Phillips, R. L., and Green, C. E. (1986). Cytogenetic stability of aneuoploid maize tissue cultures. *Can. J. Genet. Cytol.* **28**, 374–384.

Rines, H. W., and Luke, H. H. (1985). Selection and regeneration of toxin-insensitive plants from tissue cultures of oats (*Avena sativa*) susceptible to *Helminthosporium victoriae*. *Theor. Appl. Genet.* **71**, 16–21.

Rines, H. W., and McCoy, T. J. (1981). Tissue culture initiation and plant regeneration in hexaploid species of oats. *Crop Sci.* **21**, 837–842.

Roberts, L. W. (1976). "Cytodifferentiation in Plants. Xylogenesis as a Model System." Cambridge Univ. Press, London.

Rybczynski, J. J., Zwierzykowski, Z., and Slusarkiewicz-Jarzina, A. (1983). Plant regeneration with doubled chromosome numbers in tissue culture of F1 *Lolium* × *Festuca* hybrids. *Genet. Pol.* **24**, 1–8.

Ryczkowski, M. (1970). Changes in osmotic values in the central vacuole and endosperm sap during the growth of the embryo and ovule. *Z. Pflanzenphysiol.* **61**, 422–429.

Sala, C., Biasini, M. G., Morandi, C., Nielsen, E., Parisi, B., and Sala, F. (1985). Selection and nuclear DNA analysis of cell hybrids between *Daucus carota* and *Oryza sativa*. *J. Plant Physiol.* **118**, 409–419.

Sanford, J. C., Weeden, N. F., and Chyi, Y. S. (1984). Regarding the novelty and breeding value of protoplast-derived variants of Russet burbank (*Solanum tuberosum* L.). *Euphytica* **33**, 709–715.

Santos, M. A., and Torne, J. M. (1986). A comparative analysis between totipotency and growth environment conditions of the donor plants in tissue culture of *Zea mays* L. *J. Plant Physiol.* **123**, 299–305.

Scandalios, J. G. (1974). Isozymes in development and differentiation. *Annu. Rev. Plant Physiol.* **25**, 225–258.

Scandalios, J. G., and Baum, J. A. (1982). Regulatory gene variation in higher plants. *Adv. Genet.* **21**, 347–370.

Schaeffer, G. W. (1981). Mutations and cell selections: Increased protein from regenerated rice tissue cultures. *Environ. Exp. Bot.* **21**, 333–345.

Schieder, O., and Vasil, I. K. (1980). Protoplast fusion and somatic hybridization. *Int. Rev. Cytol. Suppl.* **11B**, 21–46.

Schroder, G., Waffenschmidt, S., Weiler, E. W., and Schroder, J. (1984). The T region of Ti plasmid codes for an enzyme synthesizing IAA. *Eur. J. Biochem.* **132**, 387–392.

Sears, R. G., and Deckard, E. L. (1982). Tissue culture variability in wheat: Callus induction and plant regeneration. *Crop Sci.* **22**, 546–550.

Sharp, W. R., Sondhal, M. R., Caldas, L. S., and Maraffa, S. B. (1980). The physiology of *in vitro* asexual embryogenesis. *Hortic. Rev.* **2**, 268–310.

Shepard, J. F., Bidney, D., and Shahin, E. (1980). Potato protoplast in crop improvement. *Science* **208**, 17–24.

Sheridan, W. F. (1977). Plant regeneration and chromosome stability in tissue cultures. *In* "Genetic Manipulation with Plant Materials" (L. Ledoux, ed.), pp. 263–295. Plenum, New York.

Sheridan, W. F., and Neuffer, M. G. (1982). Maize developmental mutants. *J. Hered.* **73**, 318–329.

Shimada, T., and Yamada, Y. (1979). Wheat plants regenerated from embryo cell cultures. *Jpn. J. Genet.* **54**, 379–385.

Shumnyi, V. K., and Pershina, L. A. (1979). Production of barley-rye hybrids and their clonal propagation by the method of isolated tissue cultivation. *Dokl. Akad. Nauk. SSSR* **249**, 218–220.

Simmonds, N. W. (1983). Plant breeding: The state of the art. *In* "Genetic Engineering of Plants. An Agricultural Perspective" (T. Kosuge, C. P. Meredith, and A. Hollander, eds.), pp. 5–27. Plenum, New York.

Simmonds, N. W. (1984). Gene manipulation and plant breeding. *In* "Gene Manipulation and Plant Improvement" (J. P. Gustafson, ed.), pp. 637–655. Plenum, New York.

Siriwardana, S., and Nabors, M. W. (1983). Tryptophan enhancement of somatic embryogenesis in rice. *Plant Physiol.* **73,** 142–146.

Skirvin, R. M. (1978). Natural and induced variation in tissue culture. *Euphytica* **27,** 241–266.

Skirvin, R. M., and Janick, J. (1976). Tissue culture induced variation in scented *Pelargonium* spp. *J. Am. Soc. Hortic. Sci.* **101,** 281–290.

Smith, S. M., and Street, H. (1974). The decline of embryogenic potential as callus and suspension callus of carrot (*Daucus carota* L.) are serially subcultured. *Ann. Bot.* **38,** 223–241.

Springer, W. D., Green, C. E., and Kohn, K. A. (1979). A histological examination of tissue culture initiation from immature embryos of maize. *Protoplasma* **101,** 269–281.

Sreenivasan, T. V., and Jalaja, N. C. (1982). Production of subclones from the callus culture of *Saccharum-Zea* hybrid. *Plant Sci. Lett.* **21,** 255–259.

Sreenivasan, J., and Sreenivasan, T. V. (1984). *In vitro* propagation of a *Saccharum officinarum* (L.) and *Sclerostachya fusca* (Roxb.) A. Camus hybrid. *Theor. Appl. Genet.* **67,** 171–174.

Sree Ramalu, K., Dijkhuis, P., and Roest, S. (1984). Genetic instability in protoclones of potato (*Solanum tuberosum* L. cv. 'Bintje'): New types of variation after vegetative propagation. *Theor. Appl. Genet.* **68,** 515–519.

Srinivasan, C., and Vasil, I. K. (1986). Plant regeneration from protoplasts of sugarcane (*Saccharum officinarum* L.). *J. Plant Physiol.* **126,** 41–48.

Steward, F. C., Mapes, M. O., Kent, A. E., and Holsten, R. D. (1964). Growth and development of cultured plant cells. *Science* **143,** 20–27.

Stolarz, A., and Lorz, H. (1986). Somatic embryogenesis, *in vitro* multiplication and plant regeneration from immature embryo explants of hexaploid *Triticale* (× *Triticosecale* Wittmark). *Z. Pflanzenzuecht.* **96,** 353–362.

Stroble, G. A. (1982). Phytotoxins. *Annu. Rev. Biochem.* **51,** 309–333.

Sunderland, N. (1977). Nuclear cytology. *In* "Plant Cell and Tissue Culture" (H. E. Street, ed.), pp. 177–206. Univ. of California Press, Berkeley.

Sung, Z. R., and Okimoto, R. (1981). Embryonic proteins in somatic embryos of carrot. *Proc. Natl. Acad. Sci. U.S.A.* **78,** 3683–3687.

Sung, Z. R., and Okimoto, R. (1983). Coordinate gene expression during somatic embryogenesis in carrot. *Proc. Natl. Acad. Sci. U.S.A.* **80,** 2661–2665.

Swedlund, B., and Vasil, I. K. (1985). Cytogenetic characterization of embryogenic callus and regenerated plants of *Pennisetum americanum* (L.) K. Schum. *Theor. Appl. Genet.* **69,** 575–581.

Sybenga, J. (1983). Genetic manipulation in plant breeding: Somatic versus generative. *Theor. Appl. Genet.* **66,** 179–201.

Tabaeizadeh, Z., Ferl, R. J., and Vasil, I. K. (1986). Somatic hybridization in the Gramineae: *Saccharum officinarum* L. (sugarcane) + *Pennisetum americanum* (L.) K. Schum. (pearl millet). *Proc. Natl. Acad. Sci. U.S.A.* **83,** 5616–5619.

Tabaeizadeh, Z., Pring, D. R., and Vasil, I. K. (1987). Analysis of mitochondrial DNA from somatic hybrids of *Saccharum officinarum* (sugarcane) and *Pennisetum americanum* (pearl millet). *Plant Mol. Biol.* **8**, 509–515.

Tabata, M., and Motoyoshi, F. (1965). Heredity control of callus formation in maize endosperm cultured *in vitro*. *Jpn. J. Genet.* **40**, 343–355.

Takebe, I., Otsuki, Y., and Aoki, S. (1968). Isolation of tobacco mesophyll cells in intact and active state. *Plant Cell Physiol.* **9**, 115–124.

Takebe, I., Labib, G., and Melchers, G. (1971). Regeneration of whole plants from isolated mesophyll protoplasts of tobacco. *Naturwissenschaften* **58**, 318–320.

Tanzarella, O. A., and Greco, B. (1985). Clonal propagation of *Triticum durum* Desf. from immature embryos and shoot base explants. *Euphytica* **34**, 273–277.

Taylor, M. G., and Vasil, I. K. (1987). DNA size, content and cell cycle analysis in leaves of Napiergrass (*Pennisetum purpureum*) Schum. *Theor. Appl. Genet.* (in press).

Thiman, K. V. (1980). "Senescence in Plants." CRC Press, Boca Raton, Florida.

Thomas, E., King, P. J., and Potrykus, I. (1978). Improvement of crop plants via single cells *in vitro*—An assessment. *Z. Pflanzenzuecht.* **82**, 1–30.

Thomas, M. R., and Scott, K. J. (1985). Plant regeneration by somatic embryogenesis from callus initiated from immature embryos and immature infloresences of *Hordeum vulgare*. *J. Plant Physiol.* **121**, 159–169.

Tisserat, B., Esau, E. B., and Murashige, T. (1979). Somatic embryogenesis in angiosperms. *Hortic. Rev.* **1**, 1–78.

Tomes, D. T. (1985). Cell culture, somatic embryogenesis and plant regeneration in maize, rice, sorghum and millet. *In* "Cereal Tissue and Cell Culture" (S. W. J. Bright and M. G. K. Jones, eds.), pp. 175–203. Nijhoff/W. Junk, Dordrecht.

Tomes, D. T., and Smith, D. S. (1985). The effect of parental genotype on initiation of embryogenic callus from elite maize (*Zea mays* L.) germplasm. *Theor. Appl. Genet.* **70**, 505–509.

Toncelli, F., Martini, G., Giovianazzo, G., and Nuti Ronchi, V. (1985). Role of permanent dicentric systems in carrot somatic embryogenesis. *Theor. Appl. Genet.* **70**, 345–348.

Toriyama, K., and Hinata, K. (1985). Cell suspension and protoplast culture in rice. *Plant Sci.* **41**, 179–183.

Torrey, J. G. (1961). Kinetin as a trigger for mitosis in mature endomitotic plant cells. *Exp. Cell Res.* **23**, 281–299.

Uchiyama, T., and Ogasawara, N. (1977). Disappearance of the cuticle and wax in outermost layers of callus cultures and decrease of protective ability against microorganisms. *Agric. Biol. Chem.* **41**, 1401–1405.

Uchiyama, T., Sata, J., and Ogasawara, N. (1983). Lignification and qualitative changes in phenolic compounds in rice callus tissues inoculated with plant pathogenic fungi. *Agric. Biol. Chem.* **47**, 1–10.

Umbeck, P. F., and Gengenbach, B. G. (1983). Reversion of male sterile T-cytoplasm maize to male fertile in tissue culture. *Crop Sci.* **23**, 584–588.

Van Harten, A. M., Bouter, H., and Broertjes, C. (1981). *In vitro* adventitious bud technique for vegetative propagation and mutation breeding of potato (*Solanum tuberosum* L.). II. Significance of mutation breeding. *Euphytica* **30**, 1–8.

Vasil, I. K. (1976). The progress, problems and prospects of plant protoplast research. *Adv. Genet.* **28**, 119–160.

Vasil, I. K. (1982). Somatic embryogenesis and plant regeneration in cereals and grasses. *In* "Plant Tissue Culture" (A. Fujiwara, ed.), pp. 101–104. Maruzen, Tokyo.

Vasil, I. K. (1983a). Regeneration of plants from single cells of cereals and grasses. *In*

"Genetic Engineering in Eukaryotes" (P. F. Lurquin and A. Kleinhofs, eds.), pp. 233–252. Plenum, New York.
Vasil, I. K. (1983b). Isolation and culture of protoplasts of grasses. *Int. Rev. Cytol. Suppl.* **16,** 79–88.
Vasil, I. K. (1984). Developing biotechnology for the improvement of cereal and grass crops—The consequences of somatic embryogenesis. *In* "Plant Tissue and Cell Culture Application to Crop Improvement" (F. J. Novak and L. Havel, eds.), pp. 67–75. Czechoslovak Academy of Sciences, Praha.
Vasil, I. K. (1985). Somatic embryogenesis and its consequences in the Gramineae. *In* "Tissue Culture in Forestry and Agriculture" (R. R. Henke, K. W. Hughes, M. P. Constantin, and K. W. Hollaender, eds.), pp. 31–47. Plenum, New York.
Vasil, I. K. (1987). Developing cell and tissue culture systems for the improvement of cereal and grass crops. *J. Plant Physiol.* **128,** 193–218.
Vasil, I. K. and Vasil, V. (1980). Isolation and culture of protoplasts. *Int. Rev. Cytol. Suppl.* **11B,** 1–19.
Vasil, I. K., and Vasil, V. (1986). Regeneration in cereals and other grass species. *In* "Cell Culture and Somatic Cell Genetics of Plants" (I. K. Vasil, ed.), Vol. 3, pp. 121–150. Academic Press, Orlando, Florida.
Vasil, V., and Vasil, I. K. (1979). Isolation and culture of cereal protoplasts. I. Callus formation from pearl millet (*Pennisetum americanum*) protoplasts. *Z. Pflanzenphysiol.* **92,** 379–383.
Vasil, V. and Vasil, I. K. (1980). Isolation and culture of cereal protoplasts. II. Embryogenesis and plantlet formation from protoplasts of *Pennisetum americanum. Theor. Appl. Genet.* **56,** 97–99.
Vasil, V., and Vasil, I. K. (1981a). Somatic embryogenesis and plant regeneration from tissue cultures of *Pennisetum americanum* and *P. americanum* × *P. purpureum* hybrid. *Am. J. Bot.* **68,** 864–872.
Vasil, V., and Vasil, I. K. (1981b). Somatic embryogenesis and plant regeneration from suspension cultures of pearl millet (*Pennisetum americanum*). *Ann. Bot.* **47,** 669–678.
Vasil, V., and Vasil, I. K. (1982a). Characterization of an embryogenic cell suspension culture derived from inflorescences of *Pennisetum americanum* (Pearl millet; Gramineae) *Am. J. Bot.* **69,** 1441–1449.
Vasil, V., and Vasil, I. K. (1982b). The ontogeny of somatic embryos of *Pennisetum americanum* (L.) K. Schum.: In cultured immature embryos. *Bot. Gaz.* **143,** 454–465.
Vasil, V., and Vasil, I. K. (1984). Isolation and culture of embryogenic protoplasts of cereals and grasses. *In* "Cell Culture and Somatic Cell Genetics of Plants" (I. K. Vasil, ed.), Vol. 1, pp. 398–403. Academic Press, Orlando, Florida.
Vasil, V., and Vasil, I. K. (1986). Plant regeneration from friable embryogenic callus and cell suspension cultures of *Zea mays. J. Plant Physiol.* **124,** 399–408.
Vasil, V., and Vasil, I. K. (1987). Formation of callus and somatic embryos from protoplasts of a commercial hybrid of maize (*Zea mays* L.). *Theor. Appl. Genet.* **73,** 793–798.
Vasil, V., Wang, D., and Vasil, I. K. (1983). Plant regeneration from protoplasts of *Pennisetum purpureum* Schum. (Napier grass). *Z. Pflanzenphysiol.* **111,** 319–325.
Vasil, V., Ferl, R. J., and Vasil, I. K. (1987). Somatic Hybridization in the Gramineae: *Pennesitum americanum + Triticum monococcum. J. Plant Physiol.* (in press).
Wakasa, K., Kobayashi, M., and Kamada, H. (1984). Colony formation from protoplasts of nitrate reductase-deficient rice cell lines. *J. Plant Physiol.* **117,** 223–231.
Walbot, V., and Cullis, C. (1985). Rapid genomic change in higher plants. *Annu. Rev. Plant Physiol.* **36,** 367–441.

Waldron, C., Murphy, E. B., Roberts, J. L., Gustafson, G. D., Armour, S. L., and Malcolm, S. K. (1985). Resistance to hygromycin B. *Plant Mol. Biol.* **5**, 103–392.

Wallroth, M., Gerats, A. G. M., Rogers, S. G., Fraley, R. T., and Horsch, R. B. (1986). Chromosomal localization of foreign genes in *Petunia hybrida. Mol. Gen. Genet.* **202**, 6–15.

Wardell, W. L., and Skoog, F. (1973). Flower formation in excised tobacco stem segments. III. DNA content in stem tissues of vegetative and flowering tobacco plants. *Plant Physiol.* **52**, 215–220.

Wareing, P. F., and Phillips, I. D. (1978). "The Control of Growth and Differentiation in Plants." Pergamon, Oxford.

Wenzler, H., and Meins, F. (1986). Mapping regions of the maize leaf capable of proliferation in culture. *Protoplasma* **131**, 103–105.

Wernicke, W., and Brettell, R. (1980). Somatic embryogenesis from *Sorghum bicolor* leaves. *Nature (London)* **287**, 138–139.

Wernicke, W., and Brettell, R. I. S. (1982). Morhogenesis from cultured leaf tissue of *Sorghum bicolor*—Culture initiation. *Protoplasma* **111**, 19–27.

Wernicke, W., and Milkovits, L. (1984). Developmental gradients in wheat leaves—Responses of leaf segments in different genotypes cultured *in vitro*. *J. Plant Physiol.* **115**, 49–58.

Wernicke, W., Brettell, R., Wakizuka, T., and Potrykus, I. (1981). Adventitious embryoid and root formation from rice leaves. *Z. Pflanzenphysiol.* **102**, 361–365.

Wheeler, D. C. (1972). Changes in growth substance contents during growth of wheat grains. *Ann. Appl. Biol.* **72**, 327–334.

Williams, E. G., and Maheswaran, G. (1986). Somatic embryogenesis: Factors influencing coordinated behavior of cells as an embryogenic group. *Ann. Bot.* **57**, 443–463.

Wong, C. K., and Woo, S. L. (1986). Production of rice plantlets on NaCl-stressed medium and evaluation of their progenies. *Bot. Bull. Acad. Sin.* **27**, 11–23.

Wong, C. K., Ko, S. W., and Woo, S. C. (1983). Regeneration of rice plantlets on NaCl-stressed medium by anther culture. *Bot. Bull. Acad. Sin.* **24**, 59–64.

Woo, S. C., Ko, S. W., and Wong, C. K. (1985). *In vitro* improvement of salt tolerance in a rice cultivar. *Bot. Bull. Acad. Sin.* **26**, 97–104.

Wyn Jones, R. B., and Gorham, J. (1986). The potential for enhancing the salt tolerance of wheat and other important crop plants. *Outlook Agric.* **15**, 33–39.

Yamada, Y., Yang, Z., and Tang, D. (1986). Plant regeneration from protoplast-derived callus of rice (*Oryza sativa* L.). *Plant Cell Rep.* **5**, 85–88.

Yan, Q., and Li, X. (1984). Isolation and culture of sugarcane protoplasts and callus formation. *Kexue Tongbao* **29**, 381–385.

Yan, Q., Zang, X., and Chen, Z. (1985). Organogenesis from sugarcane protoplasts. *Kexue Tongbao* **30**, 1392–1395.

Yano, S., Ogawa, M., and Yamada, Y. (1982). Plant formation from selected rice cells resistant to salts. *In* "Plant Tissue Culture" (A. Fujiwara, ed.), pp. 495–496. Maruzen, Tokyo.

Zimmermann, U., and Vienken, J. (1982). Electric field-induced cell-to-cell fusion. *J. Membr. Biol.* **67**, 165–182.

Zimny, J., and Lorz, H. (1986). Plant regeneration and initiation of cell suspensions from root tip derived callus of *Oryza sativa* L. (rice). *Plant Cell Rep.* **5**, 89–92.

Zong-xiu, S., Cheng-zhang, Z., Kang-le, Z., Xiu-fang, Q., and Ya-ping, F. (1983). Somaclonal genetics of rice, *Oryza sativa* L. *Theor. Appl. Genet.* **67**, 67–73.

Zwierzykowski, Z., Slusarkiewicz-Jarzina, A., and Rybczynski, J. J. (1985). Regeneration of plants and chromosome doubling in callus cultures of *Lolium multiflorum* Lam. ($2n = 14$) × *Festuca pratensis* Huds. ($2n = 14$) hybrid. *Genet. Pol.* **26**, 187–197.

INDEX

B

C

D

E

F

G

Y

Z

CURRICULUM VITAE: ERNST W. CASPARI

Born:	Berlin, Federal Republic of Germany (October 24, 1909) Son of Wilhelm Caspari, M.D., Sc.D. (honorary) and his wife Gertrud Gerschel Caspari
Married:	Hermine B. Abraham (August 16, 1938–May 31, 1979)
Immigration:	United States (October 29, 1938)
Naturalization:	Easton, Pennsylvania (May 22, 1944)
Positions held	
	Studied at the University of Freiburg, the University of Frankfurt, and the University of Göttingen, Federal Republic of Germany
1933	Ph.D., University of Göttingen, Federal Republic of Germany
1933–1935	Assistant in Zoology, University of Göttingen, Göttingen, Federal Republic of Germany
1935–1938	Assistant in Microbiology, University of Istanbul, Medical Faculty, Turkey
1938–1941	Fellow in Biology, Lafayette College, Easton, Pennsylvania
1941–1944	Assistant Professor of Biology, Lafayette College, Easton, Pennsylvania
1944–1945	Assistant Professor of Zoology, University of Rochester, Rochester, New York
1945–1946	Research Associate (Manhattan Project), University of Rochester, Rochester, New York
1946–1947	Associate Professor of Biology, Wesleyan University, Middletown, Connecticut
1947–1949	Staff Member, Department of Genetics, Carnegie Institution of Washington, Cold Spring Harbor, New York

1949–1960	Professor of Biology, Wesleyan University, Middletown, Connecticut
1956–1960	Daniel Ayres Professor of Biology, Wesleyan University, Middletown, Connecticut
1960–1966	Chairman, Department of Biology, University of Rochester, Rochester, New York
1960–1975	Professor of Biology, University of Rochester, Rochester, New York
1975–1976	Visiting Professor, Department of Genetics, Justus-Liebig-University, Giessen, Federal Republic of Germany
1975–Present	Professor Emeritus of Biology, University of Rochester, Rochester, New York
1981–1982	Research, Department of Genetics, Justus-Liebig-University, Giessen, Federal Republic of Germany

Membership in societies and offices

Genetics Society of America: Treasurer (1951–1953), Vice President (1965), and President (1966)

American Society of Naturalists: Vice President (1960) and Honorary Member (1978)

Behavior Genetics Association

American Society of Zoologists

Honorary memberships and appointments

1942	Fellow, American Association for the Advancement of Sciences
1947–Present	Sigma Xi
1950	M.A. ad eundem gradum, Wesleyan University, Middletown, Connecticut
1950–Present	Phi Beta Kappa
1956–1957	Fellow, Center for Advanced Studies in the Behavioral Sciences, Palo Alto, California
1958	Chairman, Program Committee and Member, General Organizing Committee, Tenth International Congress of Genetics, Montreal, Canada
1959–1963	Member, Scientific Advisory Committee and Board of Directors, Long Island Biological Association
1959–Present	American Academy of Arts and Sciences
1962	Vice President and Chairman, Section F (Zoological Sciences), American Association for the Advancement of Sciences

1962–1967	Member, Board of Trustees, Associated Universities Inc.
1965–1966	Fellow, Center for Advanced Studies in the Behavioral Sciences, Palo Alto, California
1978	Honorary Member, American Society of Naturalists
1979	Th. Dobzhansky Award for Research in Behavior Genetics
1981	Award for Senior United States Scientists, Alexander von Humboldt–Stiftung
1983	Dr. rer. nat. h.c. (honorary), Justus-Liebig-University, Giessen, Federal Republic of Germany
	Golden Ph.D. Anniversary, University of Göttingen, Göttingen, Federal Republic of Germany

Editorial activities

1960–Present	Editor, *Advances in Genetics*
1961–1963	Member, Editorial Board, *American Naturalist*
1963–Present	Member, Editorial Board, *Behavioral Science*
1968–1972	Editor, *Genetics* (Volumes 58–72)
1969	Caspari, E., and Ravin, A. W., eds. "Genetic Organization I." Academic Press, New York and London.
1972	Ehrman, L., Omenn, G. S., Caspari, E., eds. "Genetics, Environment and Behavior." Academic Press, New York and London.
1972–Present	Member, Editorial Board, *Behavior Genetics*

LIST OF PUBLICATIONS: ERNST W. CASPARI

I. Articles and Books

Caspari, E. W. (1933). Ueber die Wirkung eines pleiotropen Gens bei der Mehlmotte *Ephestia kühniella* Zeller. *Wilhelm Roux' Arch. Entwicklungsmech. Org.* **130**, 353–381.

Caspari, E. W. (1934). Die Wirkungsweise der Erbfaktoren. Die Umschau.

Caspari, E. W., and Plagge, E. (1935). Versuche zur Physiologie der Verpuppung von Schmetterlingsraupen D. *Naturwissenschaften* **23**, 751–752.

Kühn, A., Caspari, E. W., and Plagge, E. (1935). Ueber hormonale Genwirkungen bei *Ephestia kühniella* Z. *Nachr. Ges. Wiss. Göttingen, Math.–Phys. Kl. Fachgruppe 1* **2**, 1–29.

Caspari, E. W. (1936). Zur Analyse der Matroklinie der Vererbung in der a-Serie der Augenfarbenmutationen bei der Mehlmotte *Ephestia kühniella. Z. Induk. Abstamm. Vererbungsl. 1.* **71**, 546–555.

Braun, H., and Caspari, E. W. (1937). Ueber Paratyphusinfektion der Kanarienvögel. *Istanbul Seririyati* **6**, 1–6.

Braun, H., and Caspari, E. W. (1938). Können durch Culices und Wanzen bakterielle Infektionskrankheiten verbreitet werden? *Istanbul Seririyati* **3**, 1–7. (In Ger. and Turk.)

Braun, H., and Caspari, E. W. (1938). Sur la propagation de maladies d'origine bacterienne par des *Culex* et des punaises. *Ann. Parasitol. Hum. Comp.* **16**, 543–547.

Caspari, E. W. (1939). Kann die Bettwanze bakterielle Infektionserreger verbreiten? *Proc. Fac. Med. Istanbul* **1**, 572–583.

Braun, H., and Caspari, E. W. (1939). Kann *Culex pipiens* bakterielle Infektionserreger verbreiten? *Schweiz. Arch. Allg. Pathol. Bakteriol.* **2**, 175–193.

Caspari, E. W., and David, P. R. (1940). The inheritance of a tail abnormality in the house mouse. *J. Hered.* **31**, 427–431.

Caspari, E. W. (1941). The influence of low temperature on the pupation of *Ephestia kühniella* Zeller. *J. Exp. Zool.* **86**, 321–331.

Caspari, E. W. (1941). The morphology and development of the wing pattern of Lepidoptera. *Q. Rev. Biol.* **16**, 249–273.

Dunn, L. C., and Caspari, E. W. (1942). Close linkage between mutations with similar effects. *Proc. Natl. Acad. Sci. U.S.A.* **28**, 205–210.

Caspari, E. W. (1943). Genes affecting testis color in *Ephestia kühniella* Z. *Genetics* **28,** 286–294.

Caspari, E. W. (1943). The influence of hatching order on the intensity of testis pigmentation in *Ephestia kühniella* Z. *J. Exp. Zool.* **94,** 241–260.

Caspari, E. W. (1943). The tryptophane content of a^+a^+ and aa *Ephestia kühniella* Z. *Science* **26,** 478–479.

Caspari, E. W. (1944). A color abnormality in the slate-colored Junco. *Auk* **61,** 576–580.

Dunn, L. C., and Caspari, E. W. (1945). A case of neighboring loci with similar effects. *Genetics* **30,** 543–568.

Caspari, E. W. (1946). On the effects of the gene *a* on the chemical composition of *Ephestia kühniella* Zeller. *Genetics* **31,** 454–474.

Caspari, E. W. (1946). Oxidation of tryptophan by homogenized a^+a^+ and aa *Ephestia* tissue. *Nature (London)* **158,** 555.

Caspari, E. W., and Stern, C. (1948). The influence of chronic irradiation with gamma rays at low dosages on the mutation rate in *Drosophila melanogaster. Genetics* **33,** 75–95.

Caspari, E. W. (1948). Cytoplasmic inheritance. *Adv. Genet.* **2,** 1–66.

Caspari, E. W., and Richards, J. (1948). On the proteins of a^+a^+ and *aa Ephestia. Proc. Natl. Acad. Sci. U.S.A.* **34,** 587–594.

Caspari, E. W. (1948). Production of mutations by ionizing radiations. Brookhaven Conf. Rep. BNL-C-4. *Biol. Appl. Nucl. Phys.*, pp. 21–26.

Caspari, E. W., and Richards, J. (1948). Genic action. *Year Book Carnegie Inst. Washington* **47,** 183–189.

Caspari, E. W. (1949). On the action of genes in development. *Port. Acta Biol., Ser. A.*, pp. 147–160. (Volume in honor of R. B. Goldschmidt.)

Caspari, E. W., and Dalton, H. C. (1949). Genic action. *Year Book Carnegie Inst. Washington* **48,** 188–201.

Caspari, E. W. (1949). Physiological action of eye color mutants in the moths *Ephestia kühniella* and *Ptychopoda seriata. Q. Rev. Biol.* **24,** 185–199.

Fano, U., Caspari, E. W., and Demerec, M. (1950). Genetics. *In* "Medical Physics" (O. Glasser, ed.), Vol. 2, pp. 365–385. Yearbook Publ., Chicago, Illinois.

Caspari, E. W. (1950). On the selective value of the alleles *Rt* and *rt* in *Ephestia kühniella. Am. Nat.* **84,** 367–380.

Caspari, E. W. (1951). The lectures of Professor Robert Pohl in Göttingen. *Am. J. Phys.* **19,** 61–63.

Caspari, E. W. (1951). Quantitative Biology. *Science* **114** (2958). (Editorial page.)

Caspari, E. W. (1951). On the biological basis of adaptedness. *Am. Sci.* **39,** 441–451.

Bernheimer, A. W., Caspari, E. W., and Kaiser, A. D. (1952). Studies on antibody formation in caterpillars. *J. Exp. Zool.* **119,** 23–35.

Caspari, E. W. (1952). Pleiotropic gene action. *Evolution* **6,** 1–18.

Caspari, E. W., and Santway, R. (1954). Differences in composition between mitochondria from two mouse strains. *Exp. Cell Res.* **7,** 351–367.

Caspari, E. W. (1954). Genetic constitution and cytoplasmic particles in animals. *Proc. Int. Congr. Genet., 9th,* pp. 167–181. [*Caryologia* (Suppl.) **6.**]

Caspari, E. W. (1955). The role of genes and cytoplasmic particles in differentiation. *Ann. N.Y. Acad. Sci.* **60,** 1026–1037.

Caspari, E. W. (1955). On the pigment formation in the testis sheath of *Rt* and *rt Ephestia kühniella* Z. *Biol. Zentralbl.* **74,** 585–602.

Caspari, E. W. (1956). The inheritance of the difference in the composition of the liver mitochondria between two mouse strains. *Genetics* **41,** 107–117.

Caspari, E. W., and Blomstrand, I. (1956). The effects of nuclear genes on the structure and differentiation of cytoplasmic particles. *Cold Spring Harbor Symp. Quant. Biol.* **21**, 291–301.
Caspari, E. W. (1958). Genetic basis of behavior. *In* "Behavior and Evolution" (A. Roe and G. G. Simpson, eds.), pp. 103–127. Yale Univ. Press, New Haven, Connecticut. German translation: (1969). Genetische Grundlagen des Verhaltens. *In* "Evolution und Verhalten" (A. Roe and G. G. Simpson, eds.), pp. 36–69. Frankfurt/Main, Suhrkamp.
Caspari, E. W., and Blomstrand, I. (1958). A yellow pigment in the testis of *Ephestia*: Its development and its control by genes. *Genetics* **43**, 679–694.
Caspari, E. W., and Gottlieb, F. J. (1959). On a modifier of the gene a in *Ephestia kühniella. Z. Vererbungsl.* **90**, 263–272.
Caspari, E. W., and Watson, G. S. (1959). On the evolutionary importance of cytoplasmic sterility in mosquitoes. *Evolution* **13**, 568–570.
Caspari, E. W. (1959). The making of man. "Wesleyan Univ. Alumnus," pp. 10–12. November 1959.
Watson, G. S., and Caspari, E. W. (1960). The behavior of cytoplasmic pollen sterility in populations. *Evolution* **14**, 56–63.
Caspari, E. W. (1960). Cytoplasmic inheritance. *In* "Encyclopedia of Science and Technology," pp. 658–661. McGraw-Hill, New York.
Caspari, E. W. (1960). Maternal influence. *In* "Encyclopedia of Science and Technology," p. 171. McGraw-Hill, New York.
Caspari, E. W. (1960). Genetic control of development. *Perspect. Biol. Med.* **4**, 26–39.
Egelhaaf, A., and Caspari, E. W. (1960). Ueber die Wirkungsweise and genetische Kontrolle des Tryptophanoxydasesystems bei *Ephestia kühniella. Z. Vererbungsl.* **91**, 373–379.
Caspari, E. W. (1961). Some genetic implications of human evolution. *In* "Social Life of Early Man" (S. Washburn, ed.), pp. 267–277. Wenner Gren Foundation, New York.
Caspari, E. W., and Thoday, J. M., eds. (1961). *Adv. Genet.* **10**, 1–429.
Caspari, E. W. (1962). On the conceptual basis of the biological sciences. *In* "Frontiers of Science and Philosophy" (R. G. Colodny, ed.), pp. 131–145. Univ. of Pittsburgh Press, Pittsburgh, Pennsylvania. (Also abridged in *Lehigh Univ. Alumni Bull.* **49**, 4–8.)
Caspari, E. W., and Thoday, J. M., eds. (1962). *Adv. Genet.* **11**, 1–394.
Caspari, E. W. (1963). Selective forces in the evolution of man. *Am. Nat.* **97**, 5–14.
Caspari, E. W. (1963). Genes and the study of behavior. *Am. Zool.* **3**, 97–100.
Caspari, E. W. (1963). Symposium on principles and methods of phylogeny: Introductory remarks. *Am. Nat.* **97**, 261–263.
Caspari, E. W. (1964). Symposium on behavior genetics: Synthesis. *Genet. Today, Proc. Int. Congr., 11th* **2**, 21–29.
Caspari, E. W. (1964). Refresher course on behavior genetics: Introductory remarks and synthesis and outlook. *Am. Zool.* **4** (2), 97–99 and 169–173.
Caspari, E. W., and Thoday, J. M., eds. (1964). *Adv. Genet.* **12**, 1–388.
Caspari, E. W. (1964). The problem of development. *Brookhaven Lect. Ser.* **35**, 1–12.
Caspari, E. W., Muth, W., and Pohley, H.-J. (1965). Effects of DNA base analogues on the scales of the wing of *Ephestia. Genetics* **51**, 771–794.
Caspari, E. W., and Nawa, S. (1965). A method to demonstrate transformation in *Ephestia. Z. Naturforsch. B* **20**, 281–284.
Caspari, E. W., and Marshak, R. E. (1965). The rise and fall of Lysenko. *Science* **149**, 275–278.

Caspari, E. W. (1965). The evolutionary importance of sexual processes and of sexual behavior. *In* "Sex and Behavior" (F. Beach, ed.), pp. 34–62. Wiley, New York.

Caspari, E. W., Watson, S. G., and Smith, W. (1965). The influence of cytoplasmic pollen sterility on gene exchange between populations. *Genetics* **53**, 741–746.

Caspari, E. W., and Thoday, J. M., eds. (1965). *Adv. Genet.* **13**, 1–378.

Caspari, E. W. (1967). Behavioral consequences of genetic differences in man: A summary. *In* "Genetic Diversity and Human Behavior" (J. N. Spuhler, ed.), pp. 269–278. Aldine, Chicago, Illinois.

Caspari, E. W. (1967). Remarks on evolutionary aspects of behavior. *In* "Behavior–Genetic Analysis" (J. Hirsch, ed.), pp. 3–9. McGraw-Hill, New York.

Caspari, E. W. (1967). Gene action as applied to behavior. *In* "Behavior–Genetic Analysis" (J. Hirsch, ed.), pp. 112–134. McGraw-Hill, New York.

Caspari, E. W. (1968). Genetic endowment and environment in the determination of human behavior: Biological viewpoint. *Am. Educ. Res. J.* **5**, 43–55.

Caspari, E. W., ed. (1968). *Adv. Genet.* **14**, 1–418.

Caspari, E. W. (1969). Genetic changes in the evolution of man. *Can. J. Genet. Cytol.* **11**, 468–476.

Caspari, E. W., and Ravin, A. W., eds. (1969). "Genetic Organization I," 525 pp. Academic Press, New York.

Caspari, E. W., ed. (1970). *Adv. Genet.* **15**, 1–393.

Caspari, E. W., and Eicher, E. M. (1970). Mutations to sex-linked lethals in the mouse. Report NY00-2902-14 on research carried out under contract No. AT (30-2902), U.S. Atomic Energy Commission.

Caspari, E. W. (1971). Differentiation and pattern formation in the development of behavior. *In* "The Biopsychology of Development" (E. Tobach, L. R. Aronson, and E. Shaw, eds.), pp. 3–15. Academic Press, New York.

Caspari, E. W., ed. (1971). *Adv. Genet.* **16**, 1–393.

Miller, H. C., and Caspari, E. W. (1972). Ureteral reflux as a genetic trait. *J. Am. Med. Assoc.* **220**, 842–843.

Caspari, E. W. (1972). Sexual selection in human evolution. *In* "Sexual Selection and the Descent of Man 1871–1971" (B. Campbell, ed.), pp. 332–356. Aldine, Chicago, Illinois.

Ehrman, L., Omenn, G. S., and Caspari, E. W., eds. (1972). "Genetics, Environment, and Behavior. Implications for Educational Policy," pp. 1–324. Academic Press, New York.

Caspari, E. W., ed. (1973). *Adv. Genet.* **17**, 1–516.

Caspari, E. W. (1975). Problems in modern behavioral genetics. *Verh. Dtsch. Zool. Ges. 1975,* 13–17. (In Ger.; Eng. sum.)

Caspari, E. W., and Gottlieb, F. J. (1975). The Mediterranean meal moth, *Ephestia kühniella. In* "Handbook of Genetics" (R. C. King, ed.), Vol. 3, pp. 125–147. Plenum, New York.

Caspari, E. W., ed. (1976). *Adv. Genet.* **18**, 1–437.

Caspari, E. W., and Eicher, E. (1977). Induction of somatic mutations by 5-bromodeoxyuridine in the wing scale system of *Ephestia kühniella. In* "Problems of Experimental Biology" (D. K. Belyaev and O. G. Stroeva, ed.), pp. 162–174. Nauka, Moscow. (In Russ.; Eng. sum.)

Caspari, E. W. (1977). Genetic mechanisms of behavior. *In* "Genetics, Environment and Intelligence" (A. Oliverio, ed.), pp. 3–22. Elsevier, Amsterdam.

Caspari, E. W. (1977). Thoughts on recombinant DNA. *Proc. Int. Conf. Unity Sci., 6th,* San Francisco, pp. 675–681.

Caspari, E. W., ed. (1977). *Adv. Genet.* **19**, 1–585.
Caspari, E. W. (1978). The biological basis of female hierarchies. *In* "Female Hierarchies" (L. Tiger, ed.), pp. 87–122. Aldine, Chicago, Illinois.
Teicher, L. S., and Caspari, E. W. (1978). The genetics of blind—a lethal factor in the mouse. *J. Hered.* **69**, 86–90.
Caspari, E. W. (1978). Saltatory events in human evolution. *Proc. Int. Conf. Unity Sci., 7th,* Boston, pp. 743–752.
Vankin, G. L., and Caspari, E. W. (1979). Developmental studies of the lethal gene *Bld* in the mouse. I. Postimplantation development of the lethal homozygote. *J. Embryol. Exp. Morphol.* **49**, 1–12.
Caspari, E. W. (1979). Evolutionary theory and the evolution of the human brain. *In* "Development and Evolution of Brain Size" (M. E. Hahn, C. Jensen, and B. C. Dudek, eds.), pp. 9–26. Academic Press, New York.
Caspari, E. W. (1979). The goals and future of behavior genetics. *In* "Theoretical Advances in Behavior Genetics" (J. R. Royce and L. P. Mos, eds.), pp. 661–679. Sythoff and Noordhoff, Alpen/Rign.
Caspari, E. W., ed. (1979). *Adv. Genet.* **20**, 1–464.
Caspari, E. W. (1980). Richard Goldschmidt, an evaluation of his work after twenty years. *In* "Richard Goldschmidt, Controversial Geneticist and Creative Biologist" (L. K. Piternick, ed.). Birkhaeuser, Basel.
Caspari, E. W. (1981). Istanbul Tip Fakültesi'nden (1935–1938) bir Zoologun anilari. [Memories of a zoologist on the Medical Faculty of Istanbul (1935/38).] *In* "Atatürk 6; Anilardan tasşan." Work and Credit Bank, Istanbul.
Caspari, E. W. (1982). Maternal influence. "Encyclopedia of Science and Technology," 5th Ed., pp. 247–248. McGraw-Hill, New York.
Caspari, E. W. (1983). Extinction. *In* "Environment and Population. Problems of Adaptation" (J. B. Calhoun, ed.), pp. 16–18. Praeger, New York.
Caspari, E. W. (1987). Curt Stern, Ernst Hadorn, and Karl Belar. *In* "Dictionary of Scientific Biographics, Supplement II" (S. Sayre, ed.). Scribner's Sons, New York, in press.

II. Abstracts, Book Reviews, and Obituaries

Caspari, E. W. (1940). The inheritance of kinky tail and choreic behavior in a strain of the house mouse. *Collect. Net* **15** (Abstr.).
Caspari, E. W., and Richards, J. (1948). Differences in protein constitution of a^+a^+ and *aa Ephestia. Genetics* **33**, 605–606 (Abstr.).
Caspari, E. W. (1948). Genetic and environmental conditions affecting a behavior trait in *Ephestia kühniella. Anat. Rec.* **101**, 690 (Abstr.).
Caspari, E. W. (1949). A synopsis of contemporary evolutionary thinking. *Evolution* **3**, 377–378. [Review of Genetics, Paleontology and Evolution (G. L. Jepsen, G. G. Simpson, and E. Mayr, eds.).]
Caspari, E. W. (1950). Serological differences between a^+a^+ and *aa Ephestia. Genetics* **35**, 100–101 (Abstr.).
Caspari, E. W. (1950). Pigment formation in the eye of *Ephestia* and its genic determination. *Zoologica Pigm. Cell Conf., 2nd,* p. 17 (Abstr.).
Caspari, E. W. (1950). Principles of Genetics, by E. W. Sinnott, L. C. Dunn, and Th. Dobzhansky. 4th Ed. *Science* **112**, 725 (Rev.).

Caspari, E. W. (1951). Visible plasmagenes. *Evolution* **5**, 362–363. (Rev. of Problems of morphogenesis of ciliates. The kinetosomes in development, reproduction and evolution, by A. Lwoff.)

Caspari, E. W. (1952). Differences in mitochondria between different strains of the mouse. *Genetics* **37**, 572–573 (Abstr.).

Caspari, E. W. (1953). General Genetics, by A. M. Srb and R. D. Owen. *Science* **117**, 45–46 (Rev.).

Caspari, E. W. (1954). New directions in genetics. *Quant. Rev. Biol.* **29**, 245–247. [Rev. of "Genes and Mutations" (M. Demerec, ed.). *Cold Spring Harbor Symp. Quant. Biol.* **16**.]

Caspari, E. W. (1954). Advances in Genetics, Vol. 5. *Science* **119**, 646–647 (Rev.).

Caspari, E. W. (1954). Textbook of Genetics, by W. Hovanitz. *Science* **119**, 560–561 (Rev.).

Caspari, E. W. (1955). Grundriss der Allgemeinen Zoologie, by A. Kühn. *Sci.* Mon. **81** (4) (Rev.).

Caspari, E. W. (1955). Studies in motivation, by D. C. McClelland. Wesleyan Univ. Alumnus, p. 36. August 1955 (Rev.).

Caspari, E. W., and Arndt, W. F. (1955). Mutation patterns in the alga *Scenedesmus*. *Genetics* **40**, 566 (Abstr.).

Caspari, E. W. (1957). K. Henke, Developmental Biologist. *Science* **125**, 1076 (Obituary).

Caspari, E. W. (1957). Theoretical Genetics, by R. B. Goldschmidt. *Science* **125**, 816–817 (Rev.).

Caspari, E. W. (1958). The genic control of riboflavin in the testis of *Ephestia*. *Proc. Int. Congr. Genet. II, 10th*, pp. 45–46 (Abstr.).

Caspari, E. W. (1958). Advances in Genetics, Vol. 9, edited by M. Demerec. *Arch. Biochem. Biophys.* **78**, 272 (Rev.).

Caspari, E. W. (1960). Richard Benedict Goldschmidt. *Genetics* **45**, 1–5 (Obituary).

Caspari, E. W. (1961). Implications of genetics for psychology. *Contemp. Psychol.* **6**, 337–340. (Rev. of Behavior Genetics, by J. L. Fuller and W. R. Thompson.)

Caspari, E. W. (1963). Phylogeny: Principles and methods. *Science* **139**, 773–774 (Symp. Rep.).

Caspari, E. W. (1963). New Patterns in Genetics and Development, by C. H. Waddington. *Science* **139**, 580–581 (Rev.).

Caspari, E. W. (1963). Genetics, by R. C. King. Genetics, by E. G. White. *Q. Rev. Biol.* **38**, 385–386.

Caspari, E. W. (1964). Hubert Baker Goodrich. *Genetics* **50**, s12–s13 (Obituary).

Eicher, E. M., and Caspari, E. W. (1968). Genic control of the disappearance of a protein in *Ephestia*. *Genetics* **60**, 175 (Abstr.).

Caspari, E. W., Friedman, H. S., and Young, E. M. (1975). Genetic changes induced by DNA in *Ephestia*. *Genetics* **80**, s19–s20 (Abstr.).

Caspari, E. W. (1983). Genetics of behavior updated. "Behavior, Genetics and Evolution," by L. Ehrman and P. A. Parsons. *BioScience* **33**, 9 and 599.

III. Articles by Collaborators

Vankin, G. L. (1956). The embryonic effects of "*Blind*," a new early lethal mutation in mice. *Anat. Rec.* **125**, 648 (Abstr.).

Watson, G. S. (1960). The cytoplasmic "sex-ratio" condition in *Drosophila*. *Evolution (Lawrence, Kans.)* **14,** 256–265.
Muth, W. (1961). Effect of 5-fluorouracil on pupal development in *Ephestia*. *Am. Zool.* **2,** 543 (Abstr.).
Pohley, H. J. (1961). Interactions between the endocrine system and the developing tissue in *Ephestia kühniella*. *Wilhelm Roux' Arch. Entwicklungsmech. Org.* **153,** 443–458.
Muth, W. (1965). The effect of 5-fluorouracil on the eye pigmentary system in *Ephestia kühniella*. *Exp. Cell Res.* **37,** 54–64.
Pierce, B. L. S. (1966). The effect of a bacterial mutator gene upon mutation rates in bacteriophage T4. *Genetics* **54,** 657–662.
Eicher, E. M. (1967). Genetic extent of the insertion involved in the flecked translocation in the mouse. *Genetics* **55,** 203–212.
Imberski, R. B. (1967). The effect of 5-fluorouracil on the development of the adult eye in *Ephestia kühniella*. *J. Exp. Zool.* **166,** 151–161.
Stumpf, H. (1967). Ueber die Lagebestimmung der Kutikularzonen innerhalb eines Segments von *Galleria melonella*. *Dev. Biol.* **16,** 144–167.
Eicher, E. M. (1970). X-autosome translocations in the mouse: Total inactivation versus partial inactivation of the X-chromosome. *Adv. Genet.* **15,** 175–259.
Eicher, E. M. (1970). The position of *ru-2* and *qv* with respect to the flecked translocation in the mouse. *Genetics* **64,** 495–510.
Champlin, A. K. (1971). Suppression of oestrus in grouped mice: The effects of various densities and the possible nature of the stimulus. *J. Reprod. Fertil.* **27,** 233–241.
Eicher, E. M. (1971). The identification of the chromosome bearing linkage group XII in the mouse. *Genetics* **69,** 267–271.